THE MAJOR SURGERY

OF GUY DE CHAULIAC

SURGEON AND MASTER IN MEDICINE OF THE
UNIVERSITY OF MONTPELLIER

WRITTEN IN 1363

HERE RE-EDITED AND COLLATED FROM LATIN AND FRENCH EDITIONS
AND COMPLEMENTED WITH ILLUSTRATIONS

SUPPLEMENTED WITH NOTES AND AN HISTORICAL INTRODUCTION
ABOUT THE MIDDLE AGES AND THE LIFE AND THE WORKS OF
GUY DE CHAULIAC

BY

E. NICAISE
PROFESSOR IN THE FACULTY OF MEDICINE OF PARIS
SURGEON OF THE LAENNEC HOSPITAL
SENIOR MEMBER OF THE SUPERVISORIAL COUNCIL OF PUBLIC ASSISTANCE

"Science grows by additions made possible by beginnings and further developments.
We are as children perched on the shoulders of a giant. Therefore we are at an
advantage because we can see farther than the giant." Guy de Chauliac, Prologue

PARIS
ANCIENNE LIBRAIRIE GERMER, BAILLI´ERE ET CIE.
FELIX ALCAN, EDITOR
108 BOULEVARD SAINT-GERMAIN

1890

AN ENGLISH TRANSLATION
BY
LEONARD D. ROSENMAN, M.D.
2005

To order additional copies of this book, contact:
Xlibris Corporation
1-888-795-4274
www.Xlibris.com
Orders@Xlibris.com
38809

THE MAJOR SURGERY OF GUY DE CHAULIAC
FRONTISPIECE

A LECTURE-ROOM IN THE 14TH AND 15TH Cs

A Reproduction of a miniature placed at the beginning of a Manuscript of
Guy' de Chauliac's Surgery
Latin Ms. 6966. Bibliotheque National, Paris

a

FOR

LILY, JESSE, CUYLER, AMELIA, AND SHAY

Who Stand High On Our Shoulders And See Far Beyond

and

TO MY ACCUMULATING TREASURE

I would write the names of all of the Persons of Four Generations of my Extended Family, far and near, if only I had a Magic Pen that Could use the finest Honey as ink.

ENGLISH TRANSLATOR'S PREFACE

E dward Nicaise, the indefatigable surgeon-historian-editor, provided wonderful editions in French of three giants in the history of French Surgery: Guy de Chauliac in 1890, Henri de Mondeville in 1893, and Pierre Franco in 1895. In that remarkable series he completed what Malgaigne did for Ambroise Paré in 1840-42, and together they opened the eyes of curious readers during the latter half of the 19th C, to see the accomplishments of the Medieval and Renaissance surgeons who were the torch-bearers of our Art until the 18th C. Yet, as Ulysses spoke for Shakespeare,

"Time hath, my lord, a wallet at his back wherein he puts alms for oblivion."[1]. Even the brightest lights grow dim if they are not attended and the great torch-bearers and their works need refreshment. After Nicaise all the Europeans, except the English speaking people, had new editions and translations of the three, and recent English editions of Henri and Pierre are available. Yet, when we seek modern English editions of Guy de Chauliac we find only one, a translation of a small part of Guy's *Major Surgery* by W. A. Brennan, published in 1923. And here now is a complete new English Edition.[2]

Let me explain the title. Guy's Latin Manuscript was named the *Chirurgia Magna,* and a small probably spurious French work was titled the *Chirurgie Petite.* My terms refer to the sizes of the books: the *Major Surgery* and the *Minor Surgery,* and in no way indicate the extent, or the difficulties, or the perils of certain manual procedures performed by surgeons, as we today call them major or minor operations. Certainly, in Guy's era there were few operations, perhaps excepting amputations, repairs of hernias, and removal of bladder stones that we would call major.

Little need be said to supplement the ample and appropriate comments provided by Nicaise and Joubert in their long discursive Introductions and footnotes, and I have inserted only a few footnotes and explanatory interjections. Indeed, as Nicaise wrote in the historical Introduction to his edition of Pierre Franco, if we were to combine the introductions in his editions of Henri de Mondeville, Pierre Franco, and Guy de Chauliac with Malgaigne's Introduction to his Ambroise Paré, we will have a complete history of European Surgery before the modern era.

---

[1]   Troilus and Cressida, Act III, Scene 3 (LDR).

[2]   This Edition abbreviates some of Nicaise's repetitions, and eliminates most of his section of Supporting Documents (ie Pieces Justificatives), and all of his Glossary of Old French Terms. They are redefined in English as they appear in the text (LDR).

Nicaise took pains to explain his difficulties with 14ᵗʰ C Latin and 16ᵗʰ C French. I will add a note of apology for my own misconceptions of what Nicaise called obscurities. The Reader should know that this book is a translation of a work that was transformed from what had been written in 1353 in the strange Latin of southern France (see Fn. 135 and page 127). It was submitted to scribes and copyists through several centuries, and no two manuscripts were alike in all details, in the Latin, old French or other languages. Many printed editions appeared after the 15ᵗʰ C, and the many variances were retained, based on the various manuscripts, which were their sources. And they were printed in many languages. Joubert reviewed several of them and published several editions. In 1659 he edited the book used by Nicaise, who based his own selection after a careful research among all the mss and printed editions available to him. He chose Joubert's printed edition for his own of 1890. Joubert's edition used the French language that was spoken during his own epoch, and Nicaise repeated what he found there with few emendations. Joubert had encountered many passages and terms which he translated with apologies, and Nicaise did the same; both of them submitted many footnotes that provided the reader with variants from which to choose. I accept and follow their paths.

Many passages seem obscure, clumsy, and difficult to translate, even to transliterate. The Reader must blame. The Author must explain.

Guy hoped to produce a treatise that summarized the best of what he could find in the surgical lore of the ancient and more recent treatises to which he had access, and he would indicate what he thought were the better methods, as they were recommended by the Masters whom he revered, or which he himself had tested. He is prolix and repetitive. His conservative, almost reluctant acceptance of manual surgical procedures, and his explicit details of treatments with diets and medications, repeated often ad nauseam, reflects his situation among the church-educated physicians who dominated Academic Medicines in his epoch. That position was strong in the Schools, and it explains why Guy's treatise was the bible for surgeons during the ensuing two centuries, and why it was referred to with respect until the late 18ᵗʰ C.

Guy's era was that of the Great Plague, and his book was written afterwards. Its respectful attitude toward his colleagues and his suave gentility secured him as an officer of the Church, and as a Surgeon for The Popes at Avignon. However, he lacked the vigor, the wit, and the critical outlook of Henri de Mondeville whose bitter reflections on the churchmen, the surgeons who were less critical than himself, and all the other qualities of a burdened surgeon who wrote—but did not live to complete—his remarkable *Surgery*. If only one great medieval surgical text was available to a curious Reader, I would name Henri de Mondeville's.

FORMAT

Nicaise's text is in a standard type face.
All titles of Books and Manuscripts are in Italics.

Manuscripts in general are in lower case (ms and mss). A particular Manuscript
when used as a title or with other identification is Ms

All notes in the text proper, by Nicaise or by me, are in Italics.

Small insertions by me to offer a date, et al. are in brackets with (ie)

Footnotes by Nicaise are identified by (EN). They represent comments written in 1891

Footnotes by Joubert are identified by (J). They represent comments written in 1563

Footnotes by me are identified by (LDR). They represent comments written in 2007

BIBLIOGRAPHY OF AUTHORS CITED HERE

Allbutt, TC: The Historical Relations Of Medicine And Surgery To The End Of The Sixteenth Century. 1905. London, Macmillan and Company.

Baas, JH: Outlines Of The History Of Medicine. 1889. Translated and enlarged by HE Handerson, 1891. Reprinted by Robert Krieger Publishing Co., Huntington, NY. 1971.

Bruno da Longoburgo (1252): An Italian Surgeon of the 13thC; (an annotated transl. of his text); M Tabanelli, 1970, Florence, Leo S. Olschki. English Edition by LD Rosenman, Pittsburgh, Dorrance Publishing Co. 2003.

Celsus, Cornelius Aurelius: De Medicina (8 books, ca 20 AD); Translated by WG Spencer, Loeb Classical Library, Cambridge, Harvard Univ. Press.1935.

Daremberg, Charles: Essay on The Glosses of the Four Salernitan Masters. See Roland, below.

Pierre Franco (1556): The Surgery; French Edition by Nicaise, 1895. English Translation by LD Rosenman, Philadelphia, Xlibris Publishing Co. 2006.

Galen: The Usefulness of The Parts of The Body; English Translation by MT May. Itaca. Cornell Univ. Press. 1968

Garrison, Fielding: An Introduction To The History Of Medicine, 3rd Edition, Revised. Philadelphia, W.B. Saunders Co., 1921

Guy de Chauliac (1563): On Wounds and Fractures; Translation by W.A. Brennan, Published by the Translator. Chicago, 1923.

The Glosses of the Four Salernitan Masters—an Essay by Charles Daremberg in De Renzi's Collectio Salernitana, 1855. Here it is published with the English Edition of Roland, see below.

Henri de Mondeville (1315): The Surgery of H.de M.: French Edition by E Nicaise; 1893, Paris. Germer-Balliere. English Edition in two volumes, by LD Rosenman. Philadelhia, Xlibris Corp. 2003.

Hippocrates: The Genuine Works; Translated from the Greek by F. Adams. Baltimore, Williams and Wilkins Co.,1939

Lanfranchi of Milan (1295): Chirurgia Magna; Middle English Edition, as 'Lanfrank's Science of Cirurgie' in 1380. Edited. for The Early English Text Society. by R von Fleischhacker.1894. English Edition. by LD Rosenman. Philadelphia, Xlibris Co. 2003

Malgaigne, J.F, and Hamby, W: Surgery and Ambroise Paré. Hamby's English Translation of Malgaigne's Introduction to his edition of Complete Works of Ambroise Paré (1840-42), Norman, Univ. of Oklahoma Press, 1965.

Paul of Aegina: The Seven Books, (ca 640 AD). Transl. by F. Adams, 1844-47. London, The Sydenham Society. Vol. III.

Pliny, Gaius Secundus: Naturalis Historia. Book I (ca 70 AD); Translated by H. Rackham. Loeb Classical Library, Cambridge, Harvard Univ. Press. 1938.

Roger Frugard (1170), Chirurgia. Italian Translation e by L Stroppiana and D Spallone. 1957. Roma, Istituto di Storia della Medicina della Universita di Roma. English Edition by LD Rosenman. Philadelphia, XLibris Co. 2002

Roland of Parma (1250); Chirurgia; a part of La Chirurgia Italiana Nel'Alto Medievo, by M. Tabanelli. Florence, Leo S. Olschki, 1965. English Edition by LD Rosenman. Philadelphia, XLibris Co. 2002. This Edition includes Daremberg's Essay on the Glosses of the Four Salernitan Masters.

Sarton, G., An Introduction to the History of Science, Three Vols.; Baltimore, Williams and Wilkins, 1926-1947.

Theodoric (1265): The Surgery of Theodoric, Two Vol. English Edition by E Campbell and J Colton; New York, Appleton-Century-Crofts, Inc. 1955 and 1960

William of Saliceto (1275): Chirurgie, French Edition by P Pifteau;Toulouse, Imprim. Saint Cyprien.1898. English Edition.by LD Rosenman. Philadelphia, Xlibris Co.2003

Jehan Yperman (1320): The Surgery. French Edition by Doctor A. DeMets; Paris, Editions Hippocrat.1936, and an Italian Edition by M.Tabanelli, Jehan Yperman, Padre della Chirurgia Fiamminga. English.Edition by LDRosenman is taken from both. Philadelphia. Xlibris Co. 2003

ACKNOWLEDGMENTS

This Edition owes much to Dr. Haskell Norman, who gave me my copy of Nicaise's *Guy* almost fifty years ago and with it encouraged my interests in medieval Surgery. I hope that it will serve as a memorial to him. As always, I must acknowledge the assistance and patience of the wonderful Librarians at the Library at the Mount Zion Hospital Campus of the University of California in San Francisco. And again I thank Professor Orlo Clark, Chairman of the Department of Surgery for his support during my years of exploration in those medieval centuries of European Surgery.

I offer a special note of appreciation to Jeffrey Pearl, M.D. He taught me to use a word-processing computer, which enabled me to work at translating, and he has encouraged and supported me at the work for many years. He has been the son who is the father of the man.

NICAISE'S NOTE ON THE RUBRICS:

G uy placed the Rubrics for all the seven treatises at this site as a general summary, and as did Guy, I place a Total Table of Contents here at the beginning of the entire Work. However, at the end of this edition I have added a general alphabetized Index, and I have placed special Rubrics to precede each treatise. And for sake of completeness, I will insert what Guy wrote as his prelude: (EN)

> *"In order to simplify a Reader's search for a particular item in the text, I will place here the rubrics (ie table of contents) for the entire book. If one forgets the first letter of a title he still can find the desired page. That arrangement is important in every scientific work, as Averroes insisted (G de Ch).*

The phrase "the first letter is lost . . . ." requires an explanation. The printed edition of 1559 says, "Even if the leading letter is rubbed away, the rest of the page will remain." Joubert interpreted that to mean the first word of the title for the entire chapter, and he repeats what is said in an Hebrew Ms of Guy. Thus Guy meant that if the first letter of the title of each chapter was rubbed out or lost, or if the rubricator had not used red paint, one could be unsure of finding what he wanted, he could still find his way using the general rubrics at the beginning of the main text.

The mss prepared for the Royalty or for dignitaries of the Church were written on vellum stained purple, with writing in gold ink. Later on, only a few pages were stained—the margins or the frontispiece—and red ink was used for the titles of chapters and for words with special emphasis. The rubricators were specialist-scribes who painted the red letters in spaces left for him by the general copyists. The early printers also left spaces for the red letters.

Guy denied the need of the rubriques for the speculative sciences, as Averroes had insisted. According to Joubert, "Avicenna had said that he had divided his book into many chapters, and had devoted an entire chapter to every topic. The sages of Andalusia in Spain (ie Arabs) and most of the sages in the past. (a jibe at Avicenna) have vainly honored him, and they honored the Father of Philosophy in the same fashion. In doing so, they have demonstrated their own weaknesses as scholars, failing to recognize the differences between what is first in importance and what is second, at the time that they read their books."

Guy divided every Treatise into two sections; a doctrine is a system of instruction and treatment. The first doctrines deal with general matters. The second doctrines deal with particulars, as to regions of the body, organs, etc, (EN).

TRANSLATOR'S NOTE ABOUT RUBRICS

T he Aphabetical Index at the end of Nicaise's Edition, lacks many entries, and
that deficiency defeats his purpose. This translator no longer can generate the
energy, the skill, and the patience necessary to provide an extensive word-index. In lieu
of such, I have provided here a long and detailed section of Rubrics, with clearly labeled
sections for Treatises, Doctrines, Sections, Special Notes, and Illustrations. Every item
is paginated.

Every Treatise in the text is preceded by a duplicate set of Rubrics. The Reader may
be led to the words that he seeks when he uses the Rubrics as the guide to the pages that
display them. (LDR).

GENERAL TABLE OF CONTENTS

HERE BEGINS NICAISE'S LONG INTRODUCTION

GUY DE CHAULIAC'S TEXT

TREATISE I: THE ANATOMY

TREATISE II THE APOSTHEMS

TREATISE III: ON WOUNDS

TREATISE IV ON ULCERS

TREATISE V: ON FRACTURES AND DISLOCATIONS

TREATISE VI:
MISCELANEOUS MALADIES

TREATISE VII: THE ANTIDOTARY

ENGLISH TRANSLATOR'S NOTE

Guy's Antidotary is more than a pharmacopoeia. He discusses a few manual treatments in addition to medications. Although he opened Treatise VII with a chapter on Bleedings and he devoted Chapter 3 to Cauteries, most of the Antidotary gives full attention to the medicaments used by surgeons, always with the approval or oversight of the Academic Physicians, He agreed with most of the earlier authors that a manual surgical procedure is called for only when the medications fail. That insistence induced the priestly physicians to favor him for his convervatism during ensuing centuries, while the lay surgeons and the barbers undertook to use their hands and instruments, and they became the Master Surgeons!

RUBRICS

TWO APPENDICES

NICAISE'S DEDICATION

TO

THE ILLUSTRIOUS FACULTY OF MEDICINE

OF MONTPELLIER

Guy de Chauliac is the most famous Surgeon to be claimed by The University of Montpellier. Indeed, he is the Founding Father of Academic Surgery, and until the 18th C his *Major Surgery* was the official basis for instruction in that field of Medicine. And even today, his Work continues to interest Physicians and Surgeons. And that is why I here publish a new edition, and rightfully dedicate it to The Faculty Of Medicine, as did he in 1363, as a proud product of its School.

I offer it as part of the National Festival that celebrates the six centuries of existence of the University of Montpellier, and I renew the Homage paid by Guy de Chauliac more than five hundred years ago to his beloved School at Montpellier, which has always been one of the glories of our French Patrimony.

E. Nicaise
Professor of the Faculty of Medicine at Paris

PREFACE[3]

T he history of Surgery is closely linked to that of Medicine. During many centuries the two parts of the healing arts were as one; in ancient times and in the early middle ages the surgeon had to be educated as a physician and know the current concepts of the nature of disease and its pathology. At times, Surgery contributed to advances in Medicine by demonstrating the lack of merit in some of the medical dogmas. In that respect, what happened in the medieval epoch is happening today, and the parallels should be of interest to us now.

Surgery did not always bring about brilliant results, nor were the surgeons usually as venturesome as were professional artisans in other fields during the same eras. However, the success of conservatism in surgery really was the basis for its benefits for mankind. It led to great advances. For example, antisepsis in surgery led to the introduction of the science of microbiology into Medicine. In our epoch the role of surgery is analogous to what happened in medieval times, when Aristotelian scholasticism came to dominate the schools and everything was reduced to syllogisms. The medical physician had no real direct sources of anatomic or physiologic information, and apriori dogmas were the rule. On the other hand, the surgeons who daily faced reality in those matters, turned away from the syllogisms and scholastic dogmas that were different from their own experiences. Their treatises were the first to resist and to oppose the dogmatic denial of direct observation. Among them was the premier surgeon of the middle ages, Guy de Chauliac, who published his *Chirurgia* in 1363.[4]

I shall try to explore the Surgery of the Middle Ages by focusing on the principal author. I will edit his work after a complete study of it, as it relates to the medical doctrines of its epoch, as I find the materials in the books available to me.

Guy de Chauliac was a scholar as well as a practicing surgeon and he scorned nothing that belonged to surgery. His was a free and independent spirit, and his judgment was

---

3   Nicaise here reprinted the quotation taken from Guy's Prologue, printed on the Title Page (see). He failed to add that Chauliac had copied Lucian, who probably took it from Aristotle. The wonderful metaphor has been used by Robert Burton, Herbert, Coleridge, Montaigne, and others since (LDR).

4   Nicaise's enthusiasm in 1890 lessened when he encountered Henri de Mondeville's great but incomplete treatise which he edited and published in 1893. He then recognized the importance of Henri as the forerunner and model for Guy (LDR).

straightforward. He ascribed very little to astrology and he completely resisted other superstitions that were then current. He was a true scholastic when he adhered to Aristotle, but he took him as the disciple of Plato, who also was his true friend. Guy's book in places is a treasure trove of original matters, but he denied his own priorities. The surgeon of the 14th C strongly favored healing by first intention.[5] He used dry dressings and applied desiccative medications such as wine and balsams. But his surgery was timid[6], that is to say prudent, in an era when anatomy was weak. The real revival of innovative practices awaited the 16th C.

Guy's *Chirurgia* is the first didactic treatise (ie see fn.2) and it served as the teaching text for surgeons until the 18th C. This is what Malgaigne wrote in his remarkable *Introduction to the Works of Ambroise Paré*. "I do not hesitate to say that, barring Hippocrates alone, there was no treatise of Surgery, be it in Greek, Latin, or Arabic, that I would set above or on a par with the magnificent *Chirurgia* of Guy de Chauliac."

That Master, he who merits the title as the Founder of Didactic Surgery, must never be forgot, and this large memoir will not be lost, now that a chair of History of Medicine has been installed in the Faculty of Medicine (ie Paris) where the scholarly Professor Laboulbène stimulates the curiosity of his students about the Masters whom formerly they knew only by name, and where new translations have appeared for the works of Hippocrates, Celsus, Paul of Aegina, Albucasis, Ambroise Paré, and others.[7]

Malgaigne's vast project (ie the complete works of Paré) was issued piece-meal. It was his wish that the works that mark every epoch of the Art of Surgery should be published, all the texts—originals or faithful translations—written by the most outstanding authors. That collection should be annotated to include the observations and the doctrines put forth by the lower ranked authors, and such materials should be placed in special introductions accompanying the new editions.

The "Library of Surgery" has yet to include the works from the Middle Ages. That is what led me to provide this edition of Guy de Chauliac's *Major Surgery*. Perhaps it will encourage the issue of other editions; only a small collection will be needed complete the history of Medieval Surgery.[8] But that is a daunting task, to dig through the vast library of sources in which one may not know where to begin. Guy himself wrote that even if we are provided with all the old books, and even if we read all of them, we will need divinely endowed memories to retain all of it.

---

[5]   Primary healing of a wound or incision without suppuration (LDR).

[6]   As compared with the more aggressive ventures of Henri de Mondeville and Jehan Yperman and those of barbers and specialist 'cutters' (LDR).

[7]   Nicaise names the French translations that were published between 1840 and 1878 in Part V of his General Introduction (LDR).

[8]   Nicaise's wish has been satisfied by the appearance in French, Italian, and English of the treatises by Henri de Mondeville, William of Saliceto, Bruno of Longoburgo, Theodoric, Lanfranchi of Milan, Roger Frugard, Roland of Parma, and Jehan Yperman, and Pierre Franco (LDR).

THE MAJOR SURGERY OF GUY DE CHAULIAC:
PREFACE

I have been interested in Guy de Chauliac for many years, and I hesitated until now to undertake the long voyage to the homes of our ancestors. Now that I have completed it, I must admit that it has been a most interesting and useful venture. My horizons are wider. Littré wrote, "After one absorbs the matters of contemporary science, he directs himself to the past. Nothing can better strengthen his evaluation of the new than a comparison with the past; one's spirit is more impartial; one's skepticism is aroused; one seeks facts to bolster authority; one learns this lesson from the agglomeration of philosophies. Try to learn, to understand and to evaluate."[9]

I agree with Littré; one should be au courant with his own times before judging the past. I have tried to keep up with today's science (ie Surgery), and in some ways I have participated in the advances that are in progress: as a teacher in the Faculty, in the medical press[10], in the hospitals, and in the supervisory council for Public Assistance. I have been exposed every day to the facts and concepts of the surgical world.

In this new edition of Guy's *Major Surgery* I have tried to provide a text that is as close to Guy's as I could make of it by collating what I found in the Latin and French manuscripts and printed books. I have inserted notes concerning the accepted medical doctrines of Guy's epoch, and I have cited my sources. That endeavor led me to a much more intensive study of the history of his times than I had anticipated, and I report that in my Introduction to the text. I include a review of the diffusion of Guy's book and of the authors whose works I have cited, arranging them according to their academic sources. And I refer back to Galen whose works dominated the medieval medical world.

My review becomes more complete when we tackle the 14th C, in particular the medical doctrines and instruction, the written works in medicine and in surgical practices. I have leaned heavily on the original documents, and I have dated the Latin translation of the Arabic texts. I have examined the Papal Bulls and other papers that have accumulated in municipal collections at places where Guy lived. I think that my assessments of all those materials are accurate.

I do not doubt that a Reader will find errors and will spot what I have overlooked in the materials that I have studied most earnestly. It is an era that still has many 'unknowns', and I plead that my Reader will pardon my weaknesses.

I have tried to write a biography of Guy, heretofore not available except from data that had slipped into his text. It was my good fortune to have discovered previously unpublished documents.

In the Bibliography of his works, I have revealed the large role he played in his own and later epochs.

---

9   Littré: *Oevres d'Hippocrate,* Vol. I, p. 477 (EN).
10 Nicaise describes his participation in the founding of a new medico-surgical journal (1877) which soon was divided into the two *Revues,* medical and surgical. He shared the editorship of the *Revue de Chirurgie* (EN).

You will find a Glossary at the end of the edition with which to improve your understanding of certain terms in the text that are no longer in use or that have different meanings now[11]. I have appended a complete list of the medications and the instruments that are mentioned in the text. You will find seven miniatures that appear in the various mss. The originals have been copied faithfully by M. Profit.

I have investigated documents from Universities all over Europe, and I have visited in France the collections at Paris, Avignon, Lyon, and Montpellier. In Italy I have studied items at Bologna, Florence, Rome, The Vatican, and at the Museum in Naples. At all of those places I received much assistance: science is an endeavor that has created a brotherhood that has served me well. I thank each and every one, and I ask the Directors of the libraries to share my thanks with their staffs, especially the curators of the manuscripts and rare books in our own marvelous National Library (at Paris) and in the Library of the Faculty of Medicine (Paris).

M. G. Guique, chief archivist for the Department of the Rhône generously provided the precious documents concerning Guy's biography, and he led me to search the archives at Lyon. Also I give special thanks to M. Andre, archivist of the Department of The Lozère, who was a willing helper who sent me several items of interest. Also, I thank M. Bayle for help in my research about Medicine and the Physicians of Avignon.

My co-worker has been Dr. Saint-Leger of Lyon, the Master Botanist who has investigated all the medieval materia medica, and who has allowed me to publish his work without changing a word.

Finally, I admit to having difficulties in translating some of the Latin texts. When in that quandary, I went to M. Person, Doctor of Letters and Professor at the Lycée Condorcet. He was a most patient and valuable assistant.

I bow to the Ministry of Public Affairs and to M. Liard, Director of Superior Instruction, and to M. Delisle, General Administrator of the Bibliotheque National, who granted me access to all the mss that I needed, and authorized my reproductions of the miniatures.

Finally, I offer my warm thanks to my Editor, Felix Alcan, who checked all the details to insure that the Book was worthy of the Medieval Surgeon.

I dedicate this Edition to the Faculty of Medicine at Montpellier, to celebrate Six Centuries of Existence of the University, and to renew the homage of Guy who dedicated his own work to his first teachers, five hundred-twenty-seven years ago.

---

[11]   The Glossary is not included in this English Edition. See Fn. 2 in my Preface (LDR).

HERE BEGINS NICAISE'S LONG INTRODUCTION

CONTENTS OF NICAISE'S HISTORICAL AND BIBLIOGRAPHICAL INTRODUCTION

PART I

THE MIDDLE AGES

The Sciences Of The Times

S ciences always are affected to a degree by current political events; one must study the latter in order to trace the history of the former. Here we will limit our political chronicles to a review of the principal events as they occurred during the epochs defined by the chief medical advances[10]

We define the long Middle Ages as the years that begin with the fall of Rome and extend for eleven centuries until the fall of Constantinople. We divide it into four parts: First is the period of invasions. Second is the Feudal Era, including the Crusades, that extends from the Treaty of Verdun until the 12th C. Third is the 13th C itself, when our Modern Era begins, when civilization was reawakened; the signs of progress mark it as the pre-Renaissance. The Fourth resembles the third but is less explosive. It breeds the Renaissance and the Reformation, and it includes some of the 13th, 14th and 15th Cs. The latter three periods are ours to examine, each separately.

The First Period—From the 5th to the 9th Cs

Those were the times of the invasions and the instability of kingdoms. After the death of the Emperor Theodosius in 395, the Roman Empire was divided in two. The Western Empire was seated at Rome during the next 81 years until, 476. The Eastern Empire, seated at Constantinople (or as re-named Byzantium) lasted 1008 years, until 1453.

By the end of the 5th C, Gaul had been invaded from all sides, by Visigoths, Burgundians, and Franks. The last, led by Clovis, dominated it from 481 to 510.

---

[10]   Nicaise's scheme was based on his own interest in the history of surgery on France. His four-part division before the Renaissance in France is re-worked here into three periods insofar as Medicine and Surgery are involved (LDR).

5

During the 6th C Gaul was continuously involved with invasions or internal conflicts. The 7th C was relatively peaceful, controlled by Dagobert (628-638), then followed by the conquering Merovingian Franks.

At mid-century 8 the Carolingians replaced the Merovingians, and Pepin the Short, supported by the Pope, was named King of Gaul. Pepin then defeated the Lombards in Italy and ceded their territories to the realm of Pope Stephen II. That was the beginning of the Temporal Powers of the Papacy.

Pepin's son, Charlemagne ruled from 771 to 814. He sponsored a brief Literary Renaissance: the lore of Greece persisted here as well as in the East. But that was a brief enlightenment, and the darkness of the 5th C returned and endured until the 9th.

However, we call attention to the introduction of the calendaric reforms in France under Pepin and Charlemagne. The years were numbered from the date of birth of Jesus Christ. The New Year began of March 1 or January 1. Christmas was set at December 25 and Annunciation Day was March 25. A date for Easter was set that lasted until Hugh Capet's reign (987-996), and again to the reign Charles IX (1560-1574).

While civilization faltered in the West, it blossomed in the East. The Arabic Empire began with Mohammed in 622, the year of his Hegira, and the Empire spread rapidly by conquest. By 750 it included North Africa, Spain, Septimania (ie Southern France along the Mediterranean Sea) up to the Loire. The Caliphates of Bagdad and Cordova were established by 755. Haroun-al-Rashid (786-809) was Charlemagnes coeval, and was a protector of Arts and Literature.

Little is known about medicine in the West during the early years of the Era. Some schools persisted at Rome until the mid-seventh century; some towns had medical establishments, notably where the Lombards had lingered. A few studied medicine, and Charlemagne promoted the founding of schools in monasteries and cathedrals. We will return to them in a later chapter.

What is most notable for us is the establishment of a school at Salerno and the Arabian Renaissance; both began in the 9th C.

The Second Period. The 10th to the 12th Cs[12]

This begins at the Treaty of Verdun in 843[13], and lasted until the reign of Philip Augustus (1180). It was the three-century-long Feudal Period.

---

[10] Nicaise's used the traditional dimensions for the term "Middle Ages" to describe the eleven centuries between the falls of the two halves of the Roman Empire, West and East. In more recent times the term "Dark Ages" has been used to describe the earlier medieval epochs, corresponding to Nicaise's First Period, as it described western Europe. His Second Period is that of restlessness—a convenient metaphor—while his Third and Fourth Periods more or less describe what some historians call the "High Medieval" years that blend into the Renaissance and the Reformation in northern Europe. All of these terms are loose generalizations that are

When Charles the Bald[11] died his hereditary fiefdom covered most of France; then began what we call the Feudal Era. No longer did Roman and Barbaric (ie Gothic, Lombardian, etc) codes of law control the Lords and serfs; local customs replaced them. From early on, the feudal Lords defended France from the Saracenic invaders in the south and from the new invaders, the Norsemen (Normans) in the northwest. But the same Lords were not slow to begin fighting among themselves to dominate their opponents, all of whom were established in fortress-castles. By the 10th C the new arrangements were well established. The Carolingians had ceded their authority to the new masters, and were replaced by the Capetans in 987.

The Crusades began at the end of the 11th C; the first took place in 1095. However, the real century of crusades was the 12th, with the principal ones in 1147 and 1189. It was not until the reign of Louis VI (1108-1137) that the royal powers gained its advantage over the feudality. As a result of the Crusades, the gap between the ranks of the feudal hierarchy—the Lords and the serfs—was narrowed; commerce and industry grew; a world of ideas was set in motion and the people of the cities and the countryside were restless. The communes which the Church had organized as parishes took on independent voices. In the cities an aristocracy of the middle class took form, and in some cities in the south the precedents of ancient Rome were seen in the creation of senates and consuls. Elsewhere the artisans came together in corporations.

Family names were adopted and appended to the surnames received at baptism which had been the common usage. For example, in 1171 there were one hundred-ten Lords named William, now they added place-names or professions to distinguish the families. The latter became hereditary while the surname designated a particular member of the family.

During that period the Arabic civilization reached its apogee (more of this later). Their schools were famous and produced works that were translated into Latin, books that breathed life into the schools of western Europe during the 11th and especially the 12th Cs. Salerno's flame was ignited, followed by Bologna after 1119.

Medicine also moved forward. The number of lay practitioners expanded. In many towns the physicians came together in corporations and took on apprentices. In the 12th C the small schools at Montpellier became famous for offering free instruction (more later), and Paris soon followed. During the next century the small schools were aggregated as Universities.

---

restrictive and sometimes pejorative, and they are used as convenient devices for writing and teaching history. However, they should be expressly limited to place as well as to time. The "Dark Ages" in Gaul were bright in southern Spain and in Baghdad, and certainly in China (LDR).

[11]   Charles I of France (The Bald) one of Charlemagne's grandsons via Louis I, was King of France from 840 to 877. He fought and defeated his brother Lothair, King of Germany at Fontenoy and replaced him as the Holy Roman Emperor Charles II with the Treaty of Verdun (LDR).

The Third Period—The Pre-Renaissance of the 13ᵗʰ C

The restless spirits and the emancipated thoughts set loose in the 12ᵗʰ C were accentuated in the 13ᵗʰ. People wanted to be relieved of the heavy taxation and the injustices and the insecurity of feudalism; the Crusades had stirred the thinkers and the translations from the Arabic medical texts by Constantine (Africanus) in the 11ᵗʰ C and by Gerard of Cremona in the 12ᵗʰ added flesh to the 'bare bones' of Surgery in Europe.

The 13ᵗʰ C was the great century of the Middle Ages, the century of Philip Augustus and Saint Louis (IX).[14] Philip (1180-1233) founded the Archives and provided the stautes for the schools at Paris in 1215; the school became the University in 1250.

Louis IX (1226-1270) is the hero of the Middle Ages. He built the *Hospice of the Fifteen-Twenties* and the *Saint Chapelle*. His confessor was Robert Sorbon, who formed a congregation of poor students of theology, which later became The Sorbonne.

Philip III (1270-1285) annexed the county of Venaissin to his royal domain, but ceded it to the Papacy along with Avignon. That allowed the Papacy to leave Rome and settle at Avignon in the next century.

Gothic Architecture reached its triumphant zenith, as seen at Saint Chapelle and at the cathedrals at Paris, Reims, Rouen, Strasbourg, and Amiens.

In Literature and in the Sciences we glimpse the beginning of the Renaissance. After the collapse of the Western Roman Empire the written remnants of science were preserved only by the clerics, and were of little use to them. But in the 13ᵗʰ C the lay schools burgeoned and the range of studies expanded, and national literatures grew (ie in the so-called vulgar languages).[15] Schools other than those in the monasteries were established in the larger cities, such as those at Paris, Angers, Montpellier, Orleans, Toulouse, et al. The corporations referred to in the preceding period now were strengthened. Although books were rare, oral teaching satisfied the needs; wherever there were willing pupils there were willing instructors. In southern France the more educated Jews played an important role.

The separate schools were brought together by the Church and organized as Universities. In the 13ᵗʰ C Universities were founded at Paris (1200), Oxford (1206), Valence (1209), Naples (1224), Padua (1228), Toulouse (1229), Cambridge (1229), Rome (1245), Coimbra (1279), Montpellier (1289), and Lisbon (1290). In the 12ᵗʰ C others appeared at Avignon (1303), Orleans (1305), Grenoble (1339), Pisa (1343), Valladolid (1346), Prague (1348), Florence (1349), Pavia (1360), Angers (1364), Cracow (1364),

---

[14] Nicaise's History focuses on northern Europe, on Gaul-France in particular, and on the world of Medicine. We are reminded that Philip Augustus was coeval with Frederick II, HRE, whose intellectual and political activities in Italy preceded those in the north by many years (LDR).

[15] In France: The Chanson de Roland (1080), The chronicles of Villehardoin (1210) and Joinville (1309). In Italy: Brunetto Latini (1220-1294), Dante (1265-1321), Giotto (1276-1330), Villani (1276-1336)—all were Florentines. In Germany the Nibelungenlied was published, etc. (EN).

Orange (1365), Vienne (1365), Genoa (1368), Cologne (1385, Heidelberg (1386), and Palermo (1394).

The Universities taught Theology (Canon Law), Arts, and Medicine. The Faculty of Arts taught the Trivium (grammar, rhetoric, and philosophy), and the Quadrivium (arithmetic, geometry, music, and astronomy). Moreover, a University could develop around a single Faculty, as it happened when Montpellier began with Medicine alone.

All the Universities were under the jurisdiction of the Church that held full powers; most of the instruction was by priests and all of the instruction was in Latin. Before the Universities held sway, the instruction in the small schools was free and taught only by practicing medical and surgical professionals, whereas in the Universities the curriculum was determined by Bulls, and the books that were selected for reading and commentary were chosen by the ecclesiastical authorities. The content was never practical[16] and it became entirely dogmatic and bound by ancient tradition.

The system of instruction discouraged originality among the students. The philosophy of Aristotle was blindly accepted as it had been taken from the Arabic translations, of selected works that did not emphasize his science and reduced it to scholastic exercizes (ie syllogistic). That had regrettable consequences for Medicine. The material of ancient texts obviated the need for direct observation of nature; there was no progress in that. Yet, despite that restriction, some progress was made when the physicians discovered Greek sources that had been overlooked during the preceding two centuries. That is what brought the school at Bologna to its apogee (see later).

Surgery advanced more than did Medicine. Daremberg described how the surgeons of the 12th-14th Cs escaped the bonds of Scholasticism. Petrarch said the same, and placed the surgeons ahead of the physicians.

In addition to the bonds of scholasticism Medicine was retarded by Astrology, a negative influence that reached its apogee in the 16th C and remained until the 17th C. Also, the alchemists and the sorcerers (ie in witchcraft) increased in number and in their popular appeal. Medicine and Surgery had to compete with such measures that were favored by princes as well as by the masses; all of them were driven by their superstitions.

The Fourth Period (14th and 15th Cs)

The liberating energies of the 13th C continued in this Period, that of the Renaissance and the Reformation, although the calamities of the 14th C dulled its éclat when compared with the 13th. We should not label it as an abased and ignorant century simply because there were no Medical or Surgical geniuses. A large number of people were active, and some new Universities were founded. Villani[17] published his *Chronicles* and Boccaccio his *Decameron,* and some important medical and surgical works appeared. A sign

---

[16] Not concerned with direct medical and surgical applications (LDR).

[17] Giovanni Villani (1280-1348) wrote histories of Florence and others (LDR).

of emancipation was the appearance of Wycliffe's Bible (1384, posthumously) that presaged Evangelism, and came to be called The Morning Star of The Reformation. John Huss who followed was immolated at Geneva in 1415. Gerson (1363-1425) decried the loose practices of the clergy and formulated one of the first doctrinal codes for the Gallic Church. *The Imitation of Christ* is attributed to him (ie rather than to Thomas a Kempis).[18] All these are signs of the ferment, and they signal the modern era just ahead. The urge to learn was as great as in the 13thC, and the treasure troves of manuscripts in the libraries of the monasteries were avidly hunted; however, among the many newly discovered theological tracts there only a few that were valuable medical sources. Nevertheless, those signs of progress were obscured by the major events that disturbed the world.

The Papacy at Rome was weak. In France there was continual conflict with the royal powers. The authors of the *Literary History Of France* wrote that the discipline in the monasteries was passed to the authority of the King. Under Philip the Handsome (1285-1314), Clement found refuge at Avignon, and that was the haven for seven popes between 1309 and 1376.

The Hundred Years War between England and France began in 1337. Philip IV was defeated at Crècy in 1346, where the King of Bohemia was killed. In 1356 the French King (Jean II) was taken prisoner (ie until 1360) at Poitiers. France was laid waste by England and its allies. The Plague of 1348 was terrible. In 1360 The Treaty of Britanny spelled disater for France. Some relief followed under Charles V, but Charles VI remained enthroned even while mad after 1392 (ie during decades of internecine strife).

The early years of the 15th C were similar to the 14th. Charles VII (1403-1461) held court only at Bourges when he took the throne in 1422, after the English had captured Rouen in 1419, and then laid siege at Orleans in 1428. Nevertheless, patriotism survived everywhere in France. The end of warfare and pillage was the universal wish, and when France was at its nadir, Joan of Arc appeared and reanimated the hopes and fervor of the people.

The second half of the century was marked by other major events. Constantinople fell to the Turks in 1453. That caused the Greek scholars and scientists to emigrate to Italy with their books. Other Greek mss were discovered and the Greek language was studied. Even earlier, in 1393, Chrysolore a Greek, came to Italy as the Byzantine Ambassador, and sponsored instruction in Greek. His pupils included Guerin of Verona and Aurispa and Philadelphus who then traveled to Greece and there investigated the monastic libraries in 1423 and 1427.

It was the century of the invention of printing and of the use of gunpowder (ie in rifles). Columbus discovered America. And, at the end of the century, Charles VIII went to war in Italy.

---

[18] Nicaise here backtracks: Gerson was a straitlaced and recalcitrant dogmatist, not a forward-looking progressive (LDR).

Louis XI founded several Universities and Medical Schools. He established a postal system in 1461; it served only the king and the Pope until 1506. Meanwhile the University at Paris had its own system of postal relays along the main royal roads; it allowed students to communicate with their families.

The use of gunpowder led to great changes in Surgery as well as in warfare. The impact was not felt in the 14th C; Guy de Chauliac made no mention of wounds caused by firearms. The first to write about them was Brunschwig in 1497, followed in 1514 by Jean de Vigo.[19] The first use of gunpowder in cannons in France appeared in 1338 at the siege of Puiguillaume, as cited by Du Cange. Marianus (ie Sanctus) claimed that the Moors used it at the siege of Algesiras in 1343. The English used it in the bombardment of Crècy in 1346. In Italy as in Spain cannons were used in the 14th C, but only as elsewhere in attacking fortresses. The arquebus did not appear until the 15th C. The monks at St. Denis wrote about it in their history of Charles VI, reporting its use to discharge lead balls in 1414 (see Malgaigne, p. 69).

The Church During The Middle Ages

The Church, despite its ups and downs, dominated through the centuries. By monopolizing the written word and the teaching it controlled them.

A Note About The Calendar

The Julian calendar timed the years through the Middle Ages. The year contained three hundred sixty five days and six hours. The true solar year had eleven minutes and nine seconds less. The civil year of three hundred sixty five days was adjusted by adding one day every four years. By 1582 the amassed annual excess of eleven+ minutes had accumulated ten days beyond the civil year. Pope Gregory XIII added the ten days to the civil calendar after October 5, 1582. Another correction added three days during each four hundred year-span by eliminating the leap-day during every century that was divisible by four hundred. The Gregorian reform has been adopted by most countries except Russia (ie as of 1890). Today, the difference between the Julian and the Gregorian calendars is twelve days. Therefore, the dates stated by Guy are 'old-style'.

About Clock-Time

Until the 14th C a day had twelve hours of daylight—from sunrise until sunset. Noon came six hours after sunrise. Night began at sundown and also contained twelve hours.

---

[19]   Nicaise recognized his error five years after writing this. In his edition of Pierre Franco's Surgery (1895) he gave first claim to Pfolspründt who wrote about firearm wounds in 1460 (LDR).

Therefore the length of an hour varied with the day-time sun. The hour was measured by sundials, sand-hour-glasses, and water clocks.

Jean de Dondis described a mechanical clock in the 14ᵗʰ C at Padua, his home city. The first mechanical clock in France was driven by weights, installed by order of Charles V, near the end of the 16ᵗʰ C.

Christianity reached Gaul near the middle of the 2ⁿᵈ C, and Lyon had the first church. Constantine (280-337) freed the churches and the physicians of tax-burdens and granted the Church the privilege of receiving donations. In time, the temporal authority of the clergy accompanied the growth of its moral powers, and by the end of the 4ᵗʰ C the towns were governed by their bishops.

Some monasteries appeared in the 3ʳᵈ C; the first was that of St. Anthony in Egypt. St. Martin's was first in the West, at Ligugè near Poitiers, founded in the year 300. By the 6ᵗʰ C there were two hundred-thirty-eight monasteries. Around 530 Benedict of Nursia set down the rules for the monks at Monte Cassino, and those statutes were adopted by nearly all the monasteric orders in 817 at the Council at Aix-la-Chapelle. The rules imposed tillage of the soil, reading and copying mss, and the care of the sick. A small amount of contemporary writing was saved, but for the most part the libraries of the monasteries became the refuge of old books.

During the epochs of the invasions, of mass mortality, of plagues, and of pillages, the cloistered life was sought or planned not only by the faithful but also by scholars and others who looked for safe havens. But as early as 373 the Emperor Valens ordered the seizure and punishment of those who tried to enter the monks' world by false pretense of religiosity.

In the 9ᵗʰ C the Papal Monarchy (ie the Holy Roman Empire) was affirmed and the Church established its own Feudality. In France it owned about a fifth of the domain, in Germany and England nearly a third. But long before that, it had organized a system of ecclesiastical dioceses, dividing them into provinces, and then further into parishes. The terror aroused by the approach of the milleneum (ie the threat of Judgment Day) strengthened the hold of the clergy on the people. The monasteries multiplied: seven hundred two were founded in the 12ᵗʰ C and two hundred eighty seven more in the 13ᵗʰ. Schools were established in the abbeys. In the 13ᵗʰ C the Church created the Mendicant Orders that were responsible only to the Pope, the Franciscans, the Dominicans—Albert the Great (1222-1289) was a Dominican—the Carmelites, and the Augustinians.

The Church then created Universities by putting together the small free (ie not ecclesiastical) schools.

Martin Luther and Calvin instituted the Reformation early in the 16ᵗʰ C and printing furthered the dissemination of all its precepts. The result was a "Third Inquisition", called "The Index" that began in 1542. The inquisitors alone determined what books, ancient or modern, could be printed. In 1540 the Jesuits were founded.

We now know the significant role played by the Church in the teaching of Medicine in the 13ᵗʰ and 14ᵗʰ Cs; less is known about the earlier centuries except what happened at the medical school at Salerno.

However, we do know that some monks did study medicine after the organization and codification of the monasteries by St. Benedict in the 6th C. From that time even until now one monk at each monastery gave medical care at the hostel or infirmary of the abbey, not only to the friars, but also to strangers and to the poor of the region who came to him. Soon those infirmaries were filled with the sick. That served to strengthen the influence of the church, and the numbers of the medical monks increased. Apprentices were accepted and small schools appeared in many monasteries and cathedrals, and Charlemagne encouraged that development. Some abbots and bishops studied and practiced medicine, and the monks and the priests deprived the lay practitioners of much of their activity, and some of the laics sought instruction at the shurch schools

The clerics also practiced surgery until the Councils restricted that activity—at Reims in 1125 and at the Lateran in 1139. The Council of Tours in 1163 proscribed the shedding of blood by priests, and that meant the practice of surgery: *Ecclesia abhorret a sanguina!* During the rest of the 12th C the lay practitioners regained some of their former status. But in the 13th and 14th Cs, the Church took control of the small schools in the larger cities, as at Montpellier.

So we may see that the Monks and priests practiced medicine, but they made no advances in the sciences. The instruction was static; they knew no Greek and the great classics had not been translated into Latin. They had crude formularies, abridgments of the old ones in the neo-Latin of the era. There is no evidence that they used the serious classics other than the few that had been copied in common languages. Nevertheless, the rich accumulations of the great books and manuscripts were preserved in the holy refuges of the privileged monasteries. They were preserved but not exposed. Nevertheless the monasteries did a great service by saving the precious resources, even though the preservation often was careless.[20]

---

[20]   As cited by Malgaigne, p.47: "Benvenuto of Imola wrote that when Bocaccio had visited the library at Monte Cassino, he found the doors boarded up and weeds growing through the windows. The books were covered with dust. He opened some at random and recognized that some were ancient texts in foreign languages. Some were in open boxes and had been opened. Some pages had been cut out for use to bind small psalters for children. Some had the blank margins trimmed for paper on which to write the prayers, which they sold to women."

Pioggio, a pupil of Chrysolore, found a Quintillian in mint condition at the monastery at St. Gall. He wrote; "The books were not in a library as they deserved, but were in a dreadful, dark corner at the bottom of a tower where condemned prisoners were placed to die." (Muratori *Script. rerum. Ital.,* Vol. 20, pp. 161 and 164, cited by Malgaigne, p.108 (EN).

PART II

MEDICINE AND SURGERY BEFORE THE 14TH C

I n this section I shall discuss the ancient medical concepts that persisted into the Middle Ages, some of which have endured until now (ie late 19th C). I shall briefly describe the authors cited by Guy, arranged by epochs, and noting when the Greek or Arabic works were translated into Latin. The history of those works is our prime source for the history of Medicine through the eleven centuries of the Middle Ages.

I shall describe Galen's dominating influence from when he first wrote until the Renaissance, during the seven centuries for which we have pertinent materials. After the fall of the Western Roman Empire, the Greek physicians left the West and the Greek language fell from use by those who remained behind. It was ignored in the monasteries until the first complete Latin translations of Galen's works appeared in the 11th and 12th Cs. Therefore, during the 5th through the 12th Cs Galenic medicine was unknown in the West, excepting a few fragments.

In addition to the mss and their translations, we will seek our history in the Bulls and statutes related to the Universities, and we shall examine documents from local archives.

Guy names more than a hundred authors in his own *Major Surgery,* citing them in about 3,300 places (according to Joubert).

Part A. Medicine Before The Middle Ages—Two Sections

1. From Hippocrates To Galen

We look to the ancient Greeks as the real sources of the medical doctrines of the Middle Ages discovered when the medieval savants were strongly influenced by the Greek philosophers.

Pythagoras (6th C BC), during the 'Century of the Seven Sages'[21] established the theory of the four elements that constituted all bodies: earth, fire, air, and water.[22]

Empedocles in the next century refined that theory by claiming that cold and heat represented air and fire, and that dryness and wetness represented earth and water.

Later in the 5th C BC, the era of Pericles, the philosophers were the savants, and the physicians were ranked among the empirics; Hippocrates was classed between. He drew the line between Medicine and Philosophy (ie the other sciences). While seeking to apply progressive concepts that could affect both fields for mutual benefits. He took Socrates' (470-400) dictum "know thyself" and set down certain principles (ie to use one's senses) in the study of medicine.

Galen revered **Hippocrates** not only as the Lawgiver of Medicine but as the guide in philosophy and science. He saw him as the forerunner of Plato, Aristotle[23] and Theophrastus, all of whom influenced the world of knowledge. Aristotle dominated Medicine and Philosophy in the Middle Ages until the Renaissance, when Plato superceded him as the favorite (ie of the Church). A brief presentation of how they influenced medical authors follows.

The so-called Divine **Plato** was a pupil of **Socrates** and a contemporary of Hippocrates. Born in Athens in 430 BC, he died in 348. He founded the Academy at Athens. His beliefs included the immortality of the soul, metempsychosis, the Deity, and the concept that inborn Ideas are the only realities. The Ideas had divine origin and are perceived by a superior faculty, by thinking, or as memories of previous lives. Our morality depends on the acceptance of the Ideal as it resembles God. He accepted the Pythagorean four elements; and that concept existed in Medicine until the Renaissance.

During the Middle Ages the Platonic influence was displaced by that of Aristotle. Although Guy continued to nod to Platonism, when he used the term Philosopher, he meant Aristotle, an appellation that continues.

Aristotle (384-322 BC) the so-called the Prince of Philosophers, was a pupil of Plato. He established his own school at Athens, the Lycée, known as the Peripatetic School

---

21   Thales of Miletus, Solon, of Athens, Bias of Priene, Chilo of Sparta, Cleobulus of Rhodes, Periander of Corinth, and Pittacus of Mytilene. The story is told that all of them were wise enough to decline to act as a judges in a difficult civil suit that had been assigned by the Priestess at Delphi to 'the wisest man among the Greeks' (LDR).

22   That concept may have been derived from observing a burning log that produced a flame, smoke (air), steam (water) and ashes (earth) (EN).

23   Hippocrates antedated Aristotle (384-322). More recent information places Hippocrates (ca 460-377) about thirty years earlier than Nicaise (LDR).

because Aristotle expounded his views while walking. He was a teacher of Alexander the Great.

He rejected the Platonic doctrine of the Ideal and of inborn ideas. All reality exists in things themselves and can be envisioned in several ways: as the elements that compose the object; as its inner nature or essence; as its cause; as its purpose (ie use). From that we can distinguish matter, form, efficient cause, and final purpose. A Divine presence is perceived in the last, the Goal to which everything aspires.

He rebutted the Platonic concept of Inner Ideas by insisting that it is direct observation via our senses that provides knowledge and the data from which we generate general principles based on which we can logically deduce conclusions (ie by syllogisms). Many conclusions can be derived from a few principal propositions.

During the Middle Ages Aristotle's primary concept, that of Direct Observation, gave way to the intricacies of dialectics, to the subtleties of logic, to what became scholasticism and when that was applied to Medicine it retarded its progress.

Aristotle accepted the four elements and the four qualities. He divided the parts of an animal's body into 1. Basic Simples, groups of similars such as liquids, solids, bones, flesh, nerves, tendons, et al., and 2. Composites of dissimilars such as organs and limbs.

We credit the Arabs and the Greeks that emigrated from Constantinople with bringing the knowledge of Aristotle into the West. Only his *Organon* was known For a long time, a collection of works devoted to Logic. In the 13th C a Papal Bull proscribed it, but Albert the Great had the interdict lifted.

Guy cites Aristotle more than sixty times in his descriptions of the Philosopher's method and in his reviews of the works: the *Organon, the Animalia, On the generation of Animals, On The Soul, on Problems, On Metaphysics, and on Meteors.*

Hippocrates, was the contemporary of Socrates in the age of Pericles. I cannot find a general doctrine that runs through his works. However, what are dominant are the doctrines of the four elements and four humors—blood, bile, pituit, and atrabile. He proposed the concepts of fluxion and of crisis.[24] He was less systematic and logical than Galen, but he emphasized (ie before Galen) the importance of observation as the most certain and the best source of knowledge.[25]

Many of Hippocrates' works were translated from the Greek into Arabic in the 9th C, and from Arabic into Latin during the 11th and 12th Cs by Constantine of Africa and Gerard of Cremona. However, it was Galen rather than Hippocrates who dominated Arabic

---

[24] He established 'diet' (ie meaning the general regimen for a healthy way of life) as the primary therapeutic measure. He elucidated it in his *Treatise on The Regimen In Febrile Illness*. (EN).

[25] Some curious fragments link Hippocrates to Democritus (5th C BC) A fanciful tale has it that Democritus. requested a medical visit by Hippocrates. Diogenes Laertius said that H. went, accompanied by a girl who Democritus. described as a virgin. But when they returned the next day, Democritus called her a woman (ie no longer a virgin). The anecdote is apocryphal. Democritus. had proposed that a swollen neck (ie goiter) was a sign of pregnancy (EN).

Medicine, and that of the West in the Middle Ages. Galen's prestige continued unabated until after the Renaissance.

Hippocrates had written many books of Surgery which were not Latinized in the Middle Ages, and Guy had no access to them. However, he commented in Chapter I of his treatise. "I believe that the orderly arrangement of Galen's books put those of Hippocrates and others after him into a second rank."

The Hippocratic opus with Galen's Commentaries were not published until the 16th C. We are indebted to Pétrequin of Lyon for his excellent translations of Hippocrates' surgical works (*La Chirurgie d'Hippocrate,* Paris 1877-78, two vols.).

Guy cited Hippocrates one hundred twenty times in from his *Aphorisms, Prognostics, Regimen for Fevers, and Signs of Sudden Death.*

Hippocrates died early in the 4thC BC. After him the scepter of Medicine passed from Greece to Egypt (ie Alexandria) where the Ptolemies of Greek origin founded, around 320 BC, a school and a library. The school at Alexandria did not always shine brightly. In the 1st C AD, during the early Christian Era, it yielded its fame to Rome where the Greek physicians came to the fore, after Asclepiades.

The early Alexandrians cited by Guy were **Ptolemy, Heraclides,** and two **Apollonians.** He had no sources of books written between Hippocrates and Galen, and his citations all were taken from Galen.

Ptolemy was a 3rd C BC Alexandrian known as the author of the *Centiloquium.* **Heraclides** of Tarentum (3rd or 2nd C BC) wrote a *Commentary on Hippocrates* that was cited frequently by Celsus and Galen. Only fragments of the originals have survived.

Apollonius of Antioch, 2nd C BC was accredited by Galen as the author of a book titled *Medications Easy To Find And To Prepare.* The book has been attributed by others to Dioscorides. The **other Apollonius** (pupil of Erasistratus) is preserved in a frequently cited work by Oribasius.

Here follows a list of some Roman physicians who preceded Galen:

Asclepiades of Bithynia practiced medinie at Rome during the 1st C AD. A few fragments of his works are extant.

Dioscorides of Anazarbe was a Greek physician during Nero's reign. His chief work was *Treatise of Materia Medica*, a collection to which Greek, Latin and Arabic writers were devoted. It was translated into Arabic the 9th C but not into Latin until much later. It endured as a classic until the 17th C. Guy did not know it. As noted above, the authorship was sometimes attributed to Apollonius.

Damocrates, a Greek physician, was later named **Democrates** by Pliny, and as such is known to us. He wrote about medications around 67 AD and was cited by Galen.

Archigenes of Apamea (fl. 98-115 AD at Rome) was mentioned frequently by Galen. Arabic translations appeared in the 9th C. Only fragments are extant.

Among the Roman physicians was one who most regrettably was not known to Guy. He was **Celsus.**

Celsus worked at Rome during the reigns of Augustus and Tiberius (28 BC to 37 AD). His multivolume *De Re Medicina* was the most important and most intensive text written in Latin; it remains as a precious resource, providing a summary of Medical and Surgical treatises written during the three centuries between the post-Hippocratics and the Alexandrians up to 30 or 40 BC. His works have disappeared, especially those that illuminate the great school at Alexandria.

Celsus' books have an interesting history. I can find no reason why Galen did not mention them. Perhaps Celsus was lost or hidden as was the case for Aristotle whose works were not published until the time of Cicero and Sulla (137-78 BC), after they were discovered by Apellicon.[26] Celsus remained an unknown until Petit Radel (see Malgaigne, p.109) discovered him filed under his surname, Cornelius, in Isidore of Seville's catalog (ca 630) and in those of John of Salisbury (fl. 1176), and Gerbert, who became Pope Sylvester II in the 12th C. Afterward, Celsus was preserved in the monasteries. The Ms I in the Medicean Library at Florence dates from early in the 12th C, and the Ms V was copied from it in 1427. In spite of those discoveries, Celsus remained hidden in the Middle Ages. Only in 1443 did Thomas of Sarzane, who became Pope Nicholas V, find a copy at the Church of St. Ambrose at Milan, and published it. After it was printed, Celsus was famous.

There are several French editions: that of Ninnin in 1753, of Fouquier and Raties in 1824, and Vèdrènes published a bilingual (French and Latin) edition in 1876 that plagiarized Ninnin The edition contains pictures of the surgical instruments found at Herculaneum and Pompey (buried in 79 AD) which were used in Celsus' era.

2. Galen

Galen (131-201 or 210) was born at Pergamon and lived and worked at Rome more than five centuries after Hippocrates. Although he wrote many books, none of them was a treatise of surgery.[27] However, as Guy stated, large parts of several books are devoted to surgery, especially the book *On Abnormal Swellings*.

His works had been translated into Arabic, and from Arabic into Latin by Gerard of Cremona in the 12th C.[28] Guy had available two translations for his use—one from Arabic, the other directly from a Greek Ms, by Nicolas of Reggio in the 14th C. He used a ms sent to Robert, King of Sicily, by Emperor Andronicus.[29]

Guy cited Galen 890 times and mentioned nearly all the places that had surgical context. Peyrille, (*Histoire de la Chirurgie*, Vol. II, p. 507-ff) in his discussion of Galen's works admits using Guy as his resource for Galen's surgery. In all, Guy cited thirty-one of

---

[26]   Apellicon was a wealthy bibliophile who found or purchased Artistotelian mss after Sulla had sacked Athens. He brought them to Rome. See Sarton,Vol.I p.462. (LDR).

[27]   See Hahn, *Dict. et Encycloped. des sc. med.* (EN).

[28]   Gerard of Cremona, 1114-1187, probably an Italian Monk who was enlisted by Raymond the Archbishop of Toledo to work at his famous center for translating (LDR).

[29]   Andronicus III, Paleologos, Eastern Emperor, 1296-1341 (LDR).

Galen's books, many of which have received different titles applied by different translators, and that can be a source of confusion. My list that follows uses Hahn's chronology, and I provide the various titles other than those used by Guy.[30]

1. The Sects (*De sectis*): viewpoints of the three main sects at Galen's time. Dogmatists, Empiricists, and Methodists. Guy, in his own Introduction, was wrong in claiming that this section was part of *The Art of Medicine*. See *2*. Casini titled it *Differentiis febrium*.
2. The Arts of Medicine (*De Constitutions artis medicae)*. Guy called it *de Constitutione artis dogmaticae.*
3. The Elements According to Hippocrates (*De Elementis secundum Hippocratem)*. Two books.
4. The Temperaments. (*De Temperamentis*). Two books. Constantine called it *de Complexione.*
5. Melancholic Humor. (*De Atrabile.*). Guy titled it *The Black Bile.*
6. On unequal tempers (*De Inequale Temperie)*. Guy called it *Unequal Distemper or Unequal Intemperance.*
7. On a Proper Life-Style. (*De Bono Habite.*). Guy called it *Euchymia.*
8. On The Natural Faculties (*De Facultatibus Naturalibus*. Three Books. Daremberg's title *A Defense of the Elementary Qualities Against Eristratus and Aesculapius.*
9. On Anatomic Demonstrations (ie dissections). (De *Anatomicis Administrationibus)* in Fifteen Books. We have only eight and part of the ninth books in Latin translation. An Arabic translation of the remaining books is in the Bodleian Library at Oxford.
10. On the Action of Muscles (De *Motu Musculorum*) in Two Books
11. On the Uses (purposes) of The Parts of The Body (*De Usu Partium Corporis Humani,*) in Seventeen Books. Daremberg's Title is *On the Usage of the Parts.*
12. On The Affected Parts. (*De Locis Affectis*) in Six Books. A very important work dealing with the local manifestations and the diagnoses. Constantine called it *De Interioribus Membris.* Guy sometimes called it (in Daremberg's translation) *The Affected (or afflicted) Places., or Internal Maladies, or Maladies of Internal Organs*
13. On Various Fevers (*De Differentiis Febrium)* in Two Books.
14. The Critical Days. *De Diebus Criticis)* in Three Books.
15. On Plethora (*De Plenitudine),* or multitude, or plethora
16. On Abnormal Swellings (*De Tumoribus, Praeter Naturam)*
17. On The Qualities of Simple Medicaments. *(De Simplicium Medicamentorum, Temperamentis et Facultatibus)* in Eleven Books. Guys's title is *Simple Medicaments*

---

[30] I will use my own English translations where that is feasible (LDR).

18. The Medical Arts, (*Ars Medica*) so-called τεγμη ιατριχη. The arabists called it 'the small arts' (*Microtechn*). It summarizes the Galenic system.It is the treatise of Medical Art. Constantine called it the microtechni. Honein's *Introduction to the Art of Galen* was translated. see Leclerc, Vol. 1, p 46.

19. On Illnesses and Complications (Symptoms). (*Des Maladies et Des Accidents*). Leclerc grouped several independent works under this title of Guy's: *The Differences Among Illnesses; The Causes of Illnesses, The Different Symptoms (Three Books), The Causes of Symptoms.*

20. The Composition of Medicines according to Places (*De Compositione Medicamentorum Secundum Locos*) in Ten Books. The Arabs called it *Miamar.*

21. The Composition Of Medicines according to Types (*De Compositione Medicamentorum Secundum Genera*) in Seven Books. The Arabs called The Catageni.

22. On the Methods of Treatment. (*Methodo Medendi)* in Fourteen Books. Constantine's title was *Megatechni* or the Art of Healing. Guy called it *The Method, or The Therapeutics*

23. Glauco's Methods (*Ad Glauconem Medendi Methodo*) in Three Books. Daremberg's title is *The Therapeutics of Glauco.*

24. The Regimen of Health (*De Sanitate Tuenda*) in Six Books. Constantine called it the *Road to Good Health, or The Preservation of Health, or the Mechanism of Health, or the Keeper of Health*. Guy was wrong in claiming that the Arabs used the title Mechanism (ie Engine) of Health for item 22. above.

25. The Organs of Nutrition (*De Alimenarum Facultatibus).*

Guy also cited the *Arguments about Medicines used by Asclepiades and Erasistraticos, On the Uses of Medicines, On The Voice, Movement of Liquids, and Galen's Commentaries on many of the Hippocratic Books (The Epidemics, The Aphorisms and The Prognostics).*

Galen adopted and copied Hippocrates' ideas and accepted his concepts of the elements and the humors, and he formulated a body of doctrines. Guy said that Hippocrates planted the seeds and Galen cultivated them as a good laborer who added to the garden. But Galen let pass without emphasis on Hippocrates' insistence on the experimental method and on direct observation as the basis for progress. He did not hide them, but they remained submerged until they came to replace scholasticism and science conducted by textual exegesis. Galen dominated Medicine between the 6th and the 13th Cs.[31] So it is that we must understand Galen's doctrines dealing with elements, humors, temperaments and diseases if we are to understand the writings of the Middle Ages.

---

[31] Bishop Eusebius of Caesaria complained in 313 that Galen's authority equaled that of the Lord (EN).

Guy was wrongly numbered as an Arabist, whereas really he was a Galenist. He used the Arabs to supplement and amend Galen, but his basic medical doctrines came from Galen, and he always added the importance of direct observation

The Medical Doctrines of Galen

Galen relied on Hippocrates for his use of the theories of Pythagoras and Empedocles as they explained the four elements, the four qualities, and the four humors.

Health and good habits were the result of the proper balance of the humors. The Temperance (ie Complexion) may not always be perfectly healthy, yet no frankly pathological state would appear if the predominance of one humor was minor. When the predominance was exaggerated or if the humor was unnaturally altered, illness was the result.

All bodies consist of the four elements and their qualities are those of the elements and they vary according to how they are mixed, harmoniously or not. The quality of a body is not of itself but is as derived from each of its elements, which may have qualities that are in conflict with other elements in the mix. So it is that water is cool and moist whereas earth is cool and dry, air is warm and moist and fire is hot and dry.

"In the association of similar and contrary qualities, certain rules govern the accord and the conflict of the elements. For example, one tries (ie when treating) to balance any two qualities with two opposites, and not allow an imbalance of one against three. That is the simplest form of the doctrine of opposites." Deschambre also says, "that is how we deduce the elementary qualities, how they function normally, and how their deviation becomes pathological in man."

Galen established his own Medical Doctrines by applying the theory of the elements to that of the humors which take their qualities in various degrees according to how they are in accord or in opposition.

Galen's Theory of Humors

All diseases come from one or more humors. All abscesses, abnormal swellings (ie tumors), and putrid fevers derive from putrid and corrupt humors. Wounds, fractures and ulcers heal as a result of the nourishing actions of humors in the affected regions. That is why the treatment of wounds and ulcers must include modifying the patient's blood (the whole blood is a mixture of the four humors) to bring it to normal in quantity and quality. Bleeding (ie phlebotomy et al.) corrects the quantity, and purgation eliminated the bad qualities.

A humor is said to be *Natural* when it is just right for health. It is *Unnatural* or *Contrary To Nature* when it is not.

In a Natural Humor we distinguish what is nutritious for the body, aside from its other functions. What remains of the humor after it has served its role as a natural nutriment (ie whole blood) becomes an excrement. Blood is contained in veins and arteries.

How is blood generated? The aliments are attracted[32] into the stomach by its attractive virtue and is retained there by its digestive virtue and is converted into chyle. The chyle is attracted into the slim (ie small) intestine where the liquid part is attracted into the mesenteric veins and then into the portal veins. The digestion of whatever of the aliment that was not converted to chyle in the stomach is completed in the upper intestine and the mesenteric and portal veins before all of the chyle reaches the liver. There the intrinsic heat and special virtue of the liver converts it into blood, that is, all the natural humors: that which is nutritious and that which is not.

Blood is the humor for nourishing, and it can function as such only after its has been purged of two kinds of excrement. One is attracted into the gall baldder as yellow bile, and the other is attracted into the spleen where it becomes black bile (ie melancholy or atrabile). Those two humors are Natural but they are not nutritious.

The purged blood now goes throughout the body and feeds it, not alone as sanguine humor. Blood contains some of all the humors, bile, melancholy and phlegm, (ie pituit) as well as the sanguine, all of them in harmonious proportions for a healthy temperament (ie complexion). When the mixture is not harmonious, the person is sick.

The phlegm that remains in blood is cooked by the heat of the body as a whole, not in a special place (ie like the gall bladder or spleen).[33]

Every humor is formed from the four elements and expresses the element that is overabundant or preponderant in its composition. In the sanguine humor (the main consituent

---

[32]   The distribution of blood and its separation according to the functions of the various parts is explained by each having its own special attractive powers. The vis-a-tergo or pump had little to do with the distribution of nutritious blood from the liver, although the heart pumped the spiritual blood in the arteries.

The blood went where it was attracted, and after it served its purpose the residue was excreted. There is no return or circulation. (LDR).

[33]   Nicaise's brief recital fails to describe the theory of digestion that was current in the late medieval period and was rooted in Galen. A *First Digestion* occurs in the stomach, duodenum and portal venous system to form chyle. The chyle is *Second Digested* in the liver to form the humors that are the blood. The nutritious blood from the liver is delivered throughout the body via the hepatic veins and vena cava, and every structure attracts what it needs and passes on the unwanted or what is needed to other structures including the organs that will excrete the residues (excrement), After the blood reaches certain special organs and nourishes them, and a *Third Digestion* occurs. Bile is formed in the liver, melancholy in the spleen, and in the heart the vital spirit is added as blood passes through pores in the interventricular septum. In the brain the digestion produces catarrh which seeps into the nasopharynx (nb. the pituitary gland). The digestive product of well-nourished kidneys is the urine. The secretions of the intestine are digestive products added to the non-chyle residues of ingested foods. Other digestive products are excrements, such as sweat, saliva, ear-wax etc., representing the end of the line after the nutritious part of the blood has fed the part. (LDR).

of blood) air is preponderant and the humor is warm and moist, which are the qualities of the element air. Therefore, it is best for nourishing muscular parts and is distributed via the veins (ie from the liver) and arteries (ie from the heart) as they spread throughout the body.

Fire is overabundant in the yellow bile (choler), and the bile is warm and dry, the better to stimulate the intestine to move along and expel its contents after it and its related structures have been nourished by the blood.

Earth characterizes melancholy (atrabile) which is cool and dry. It excites the appetite after the spleen has been nourished along with other cool and dry structures such as bones.[34]

Water predominates in pituit which is cool and moist as is its source. It is formed after the brain has been nourished (ie by venous blood from the liver), as are other cool and humid structures. As part of blood, it moderates it and serves to lubricate the joints.

I repeat: Health is the condition of a perfect mixture of humors and qualities. Disease occurs when the mixture is unbalanced and one or several predominate or are unnatural.

According to Galen's Dogmatic School, the qualities of every body depends on which element predominates. Beaugrand and Hahn published a summary of the various qualities; I will insert it here. It may assist the Reader in his comprehension of the intricacies of the Dogmatistic Theory.

1. When Air predominates, the dominant qualities are humid and warm, as observed in temperate climes, during the Spring season, in blood, in children with sanguine complexions and in sanguine illnesses
2. When Fire predominates, warmth and dryness characterize the summer, bile, youth, bilious temperaments, and bilious maladies.
3. Because Earth is cool and dry, it predominates in those climates, in the autumn, in Melancholy, in adults, in melancholic temperaments, and in cachectic maladies.
4. With Water, which is cool and wet, as in those climates, in winter, in pituit, in old age, in those with phlegmatic temperaments, and in catarrhal illnesses.

Dechambre (1886) said that we can still find traces of those views today in what are called constitutions,or seasonal illnesses, or climacteric ailments, the bilious characteristics of Southern peoples, the lymphatic temperaments of Northen peoples, biliousness during the summer and where the climate is warm, the catarrhal disorders during the winter months or where the climates are cool.[35]

---

[34] The connection between the gall balder and the intestine was easy to demonstrate. However, a route from the spleen that could deliver the melancholy was not clearly defined until the pancreatic duct was implicated by anatomists after the 16th C (LDR).

[35] As recently as 1939 my professor of pathology devoted a complete lecture on The Thymico-Lymphatic Constitution which he thought was more common in the North Central American States. (LDR).

As to the dry and wet qualities, Michael Levy demonstrated that Galen implicated the secretions from the skin-dry or sweaty. That was a typical Galenic ad hoc explanation of observed phenomena in the language of temperaments that he accepted.

Galen also accepted Aristotle's division of the animal's substances into Simples (ie similars) and Composites (organs and limbs).

I shall repeat: Health depended on the harmonious relationship of the four humors wherein they temper each other to achieve a proper balance of the elemental qualities. But such perfection is rare, and the imperfections are what constitute the *Temperaments*, which are sanguine, phlegmatic, bilious, melancholic or combinations in which one or another of the qualities predominates: cool, warm, etc. In addition to the general temperament of the body, Galen defines particular temperaments that characterize organs: for example, the brain may be cool or warm, dry or humid.

The Dogmatic School used the term *Complexion* instead of *Temperament*. Therefore, a complexion is a mixture, a harmony, a state of agreement among the four elementary qualities in order to set in balance the contraries. The Greeks called the condition of harmony *Crasis* or *Euchemy,* and the body that was in such a state was *tempered*, that is, a *tempered temperament. Tempery* described a tempered condition. Sometimes the term *temperature* was used instead of temperament.

In pace with the tempered temperament was the untempered state, when the qualities are not in harmony. The body then is said to be untempered, a state of untempery or to be untemperate.

The Predominance of humors varies with the age of the person, the season, the climate, etc. Blood (ie the sanguine humor) predominates in the spring and during adolescence. Bile is more abundant in the summer, in mature adults.

The Dogmatic Theory is most confused when it classifies the human conditions according to what was called Natural and Non-Natural, and their related annexes, and in things that are Contrary-to-nature.

The seven Natural Things in humans are the elements, the temperaments, the humors, the parts (the members) of the body, the virtues or faculties, the bodily functions (operations) and the Vital Spirits. Altogether, the Natural Things are intrinsic, they are the anatomy and the physiology of the body.[36]

The six Non-Natural Things (ie external factors) are called Hygiene. When they are in order, they maintain health; when otherwise, it is destroyed. They are: air, food and beverages, exercise and rest, sleep and wakefulness, excretion and retention, and passions or affections of the soul.

---

[36] A Virtue or Faculty is an efficient cause (ie the direct impulse) depending on the temperament. An action or function or operation is an activity belonging to a faculty. The Spirit is a subtle substance, airy, transparent and clear, made from the most delicate and diluted part of the sanguine humor. The spirit has three forms: animal, vital and natural (EN).

The five things Annexed To The Natural and the Non-Natural are: the weather and the seasons, the region, coitus, the métier (career, profession), a person's habits, including bathing.

The three Things Contrary-to-nature are pathological and are destructive of the natural state of the body. They are illness, their causes and their symptoms (signs).

The Cause of a malady is a disturbance of a contrary-to-nature thing that precedes and brings on the malady. It has two aspects, the internal and the external. The internal is divided into the antecedent and the conjoint. The antecedent had existed before the malady; the conjoint immediately precedes the malady and accompanies the disorder. Any of the causes can be congenital or be acquired during the life of a person.

Guy said, "The surgeon discovers what he needs for treatment first by what he determines to be the things-contrary-to nature, and then the Natural, the Non-natural and their annexes." That method of seeking indications is a reliable way for the surgeon to choose treatments that conserve, preserve and cure.

A default of the Non-Natural things, such as the life-style (habits), leads to maladies. The defaults can be observed in the Simples (the similars) which suffer defective harmony of their elementary qualities in an untempered state, such as when the simples (warm, dry, etc.) are combined in composites that are warm and most.

The humors can cause harm by being in excess or be insufficient, or in their actual makeup. The altered humors appear as untemperate states; the defects of crasis appear only in the non-sanguine humors. Those defects are called *cacochymia,* derived from χαχοσ bad and χυμοσ (juice), and acrimonies.

Most maladies primarily are the results of an excess or a lack of humors, or damaging alterations in one of the four.

A malady due to excess is called plethora, whether of blood, bile, melancholy or phlegm. True plethora is an excess only of the sanguine when the qualities of the others are not altered. However, when there is an excess of any of the other humors, all of the liquids of the body are affected, including blood, and the condition is called a general cacochymia.

Bile, pituit, and melancholy can be damaged during their production, that is, in their crasis, as when they are exposed to too much heat or humidity. They become too warm, too salty, too acid, too bitter, and they disturb, ferment and corrupt the blood. The damage caused by those humors is called putridity.

Inflammation is caused by seepage of a humor into regions where it does not belong. It is phlegmatic inflammation when it is due to sanguine humor alone, but it is gaseous, edematous, erysipeloid, or scirrhous when air, pituit, bile or atrabile also participate.

Galen's treatments were consistent with the Dogmatist Doctrines. One tried to weaken the humors, or dilute or thicken them; to cool or warm them; to purify or evacuate them. The medications had a dual function: One was to purify the humors and restore their crasis; the other was to evacuate the evil or the excessive humors.

One needs many medicines to properly defeat the multiple effects of the altered humors. Galen was bound by a single dominating principle, blindly and uncompromisingly. That precept was to combat an illness with its contraries, and to aid Nature by imitating it.

First one must make the correct diagnosis and determine the intemperences (heat, cold, etc.) so to use opposing medications that can reduce the patient's irritability, cool him, calm him, and temper him. Relax the tissues and the pores or tighten them as needed. Begin with the most gentle medicines. When the malady is complicated, direct your attack first against the principal element. Keep the patient away from sources of additional illness and aggravations. Avoid alimentary excesses. Keep the patient where the air is clean and the temperature is comfortable.

The signs of an impending malady should guide one to special medicines to suppress it. The dosage should match the severity of the symptoms.

Galen's composite medicines were those in general use during his epoch: bizarre formulations. He classified them according to their elementary qualities: warm, moist, dry, etc. One could measure their qualities by their flavors and odors. The warm ones taste salty, the dries caused bitterness, etc.

The effects of treatment were primary or secondary (subsequent), as described by Boyer in his excellent analysis of Galen's doctrines. I have used it in addition to the works of Dechambre, Hahn, Brochin, et al.

1. (per Broyer) The Primary effects may not appear immediately. Fire (ie cautery) is promptly effective whereas castoreum works slowly. The primary effects are Natural (intrinsic) or haphazard. That is, varying from case to case. For example. cool water may be heated. Vinegar is cool but may contain warm elements. All these distinctions and properties, varieties of actions, and degrees of potency are Aristotelian. Broyer wrote as follows:

   "There are four degrees of quality in medicines (see p. 627 in the Text). Chicory is cool in the first degree, pepper is warm and potent, both in the fourth degree; ciguë (hemlock) is cool and potent; fire (actual cautery) is hot; caustics (potential cautery) are hot and potent.

   "The experts could design medications (for their primary actions) to suit the requirements for three different actions, no matter what their contents: poisons, antidotes, or purgatives.

2. The Secondary (subsequent) effects were different. Some caused the pores to open or to close, some hardened or tensed the tissues; some relaxed or softened them; some modified the humors as they were combusted or matured or made ready; others were evacuants and eruptors; others assisted the eruptors; others are suppuratives, expectorants, sedatives, et. al. Several medications could cooperate in a unified action in some organs or against some humors.

Such were Galen's doctrines as they affected all the authors during the Middle Ages, although some had access to them in the 13th C through recent Latin translations of Arabic versions taken from the Greek. One can see Galen even after the Renaissance, in Paré and in the works of the many authors who were followers of Guy.

I repeat my claim that a Reader should know Galen's doctrines in order to understand Guy and the others whom he cited. The informed Reader will be able to grasp the ideas hidden in obscure terms in the writings of the medieval authors. In the Classic texts we should attend the names if only to understand their meanings (so said Guy).

Galen had more influence on Medicine than on Surgery because he did not write a special surgical text. He could have done so, because he was a surgeon in his early years, serving the Gladiators in the Arena at Pergamon. Mark Anthony Severinus[37] chided him for his timidity, because he favored medical treatments when manual treatments were necessary, and in that way slowed the progress of Surgery. That was the case in his own era, and that deterrence continued long after him.

Part B. Medicine In The Middle Ages : Five Sections

Section 1. European Medicine

After Galen, Medicine came to a halt, and, as Daremberg said; an era of conservatism and immobility began.[38] Nevertheless, we can detect some signs of activity until the end of the 5th C. Medical centers existed at Rome, Alexandria, and Athens until the division of the Roman Empire in the 4th C. Then droves of physicians and other intelligentsia left Rome and went East.

A few Latin medical writers bridged the gap to the New Greeks (ie The Eastern Empire) at the of end of the 4th and the beginning of the 5th Cs. For the most part theirs were collections of recipes and superstitious formulas[39] that later gave rise to the Christian Recipe Books of the Middle Ages (see Daremberg. p. 246). He also believed that the classic medical traditions persisted during the early medieval years after Theodosius, and that some Latin medical books were re-edited, collated, or were newly translated (from Greek) between the 1st and 7th Cs, after Galen. He believed, contrary to what most historians have claimed, that the Barbarian Invasions did not destroy everything. The Roman schools persisted until the 7th C and served as models for those of the Merovingian and Carlovingian kings later on. In addition there were two kinds of ecclesiastical schools: some were operated by bishops in the cathedrals and others were in monasteries and churches where the monks were the teachers. The Lombard Code also shows that there establishments for medical care.

The Merovingian and Carlovingian kings had their own physicians. Charlemagne founded royal schools at cathedrals and monasteries where medicine was taught after 805. Most historians have said that his initiatives were not continued, after his death, but Daremberg has disagreed. He discovered a ms at Ravenna dated at the end of the 8th C that showed that public lectures were held, dealing with Hippocrates and Galen. And, at St. Gall, monks were copying medical mss. And at Monte Cassino, at Einselden, and

---

37   5th or 6th C author (probably Alexandria) of a book on clysters (LDR).

38   Dominated by Galenism (LDR).

39   'Superstition' is Nicaise's term, here and later, for 'old-wives' tales, mystical elements, alchemy, astrology, miracles, etc. (LDR).

at Rome there were copyists from the 8th through the 11th Cs. Daremberg and de Renzi collected from various archives the names of physicians who lived in the 8th through the 13th Cs, most of whom were laics; but the clerics shared the medical responsibilities.

From the 6th C, and almost certainly before that, some of the works of Hippocrates, Galen, and Soranus had been translated into Latin, and there were formal establishments for translating.[40] Cassiodorus (480-?575) was prime minister of the Ostrogothic King Theodoric. He retired to a Calabrian monastery and set the monks to copy the ancient mss. Were medical texts included? Which ones? At Paris we find mss of the 7th C that include translations of Oribasius in uncial script and some rather coarse translations of Hippocrates, Galen, and Alexander of Tralles. All of this led Daremberg to conclude that the Gothic (ie Barbarian) realms that replaced the Western Roman Empire did not lack physicians or Medicine or the teaching of Medicine. He continued, "The ancient classic medicine later was tied to the Renaissance by Latin translations from the Greek originals made in the 13th C, in the monasteries."[41]

Much remains to be explained, at least the matters dealing with the more advanced classical schools and the reproduction of their books, other than those that were collections of recipes or were abridgements in Arabic translations. We should know the titles of the mss cited by Daremberg. How true is the claim that the monks knew no Greek? If so, how did they copy the old mss? Or did they simply preserve the old texts ? We must be grateful for that service alone.

Medicine was decadent in the Eastern (Byzantine) Roman Empire during the 8th through the 14th Cs. Little progress can be documented.

Now we come to the school at Salerno (see more below in Section 3) which began as one of the neo-Latin schools noted above. The researches of Daremberg, although inconclusive, have thrown light on the obscurities in the history of medicine between the fall of the Western Empire and the development of the school at Salerno not long afterward.

I shall focus on Surgery, and limit the citations to the centuries after Galen and the origins of Salerno, that is, to Oribasius (4th C), Aetius (6th C), Alexander of Tralles (6th C) and Paul of Aegina (7th C), and Philagrius (4th C). Excepting Aetius, all of them, were translated into Arabic in the 9th C. Oribasius and Aetius had not been re-translated into Latin before Guy's era, and he did not know them. Aetius was not Latinized or even known in the West before the14th and 15th Cs. He was rediscovered in fragments in the 16th C.

---

[40]   Nicaise and Daremberg fail to indicate that although the copyists and the translators served as compilers and preservers, their handiwork lay in dusty piles, awaiting discovery and practical medical application in later centuries. See below (LDR).

[41]   In other words, the earlier Latin translations of the Greek classic authors from Arabic translations were displaced by the fresh Latin translations from the original (perhaps occasional copies from scriptoria) Greek mss. The 'new' and more accurate Latin versions were not available to the medieval surgeons before the late 13th C. That is why the pioneer surgeons of Italy after Roger Frugard were called 'arabists' for want of other sources (LDR).

Guy had **Philagrius**, a Greek physician of the second half of the 4th C who was involved in operations for aneurysms which had been treated during the previous century by opening them.[42]

Alexander of Tralles, cited seven times, was a Greek nonsurgical physician at Rome in the 6th C. He was translated before the 9th C (in Arabic). A Latin translation came after Guy's time, therefore his citation probably were taken from Paul of Aegina's *Pandects*. Alexander's works included a *Practice of Medicine, The Art of Medicine in 12 vols., and The Treatment of Gout.*

Isidore of Seville, a Greek born in the 6th C (d. 636), was a bishop and an encyclopedist, Four of the twenty volumes became a medical treatise. They were published by Lieudemann at Leipzig in 1833, in three vols.

Paul of Aegina, a Greek in Alexandria, wrote a seven volume text (the *Pandects*) of which Book 6 was entirely surgical, and was the only volume available to Guy. Paul was a valuable compiler who provided abridgements and abstracts from his surgical predecessers. His work culminated the era of classic Greek Medicine.

Paul's work is very important because it reflects the status of Surgery in his own time as well as what progress had been made after Hippocrates and Galen. His books were translated into Arabic in the 9th C and thence into Latin. Honein's Arabic version was the *Pandects*.

After Guy, Paul's Book 6 disappeared until the middle of the 15th C (see Malgaigne). A French translation of Book 6 by Pierre Tolet and Daleschamp was published at Lyon in 1539. A new French-Greek bilingual edition by Briau was issue at Paris in 1855.

Now we will interrupt our history of Western European Medicine and we will review what happened among the Arabs.

Section 2. Arabic Medicine

I have taken most of what follows from the seminal works of Leclerc who has enlightened us about Arabic Medicine, and I bow to him in appreciation.[43]

The Arabs were more than compilers; they were scholars and innovators. They rapidly and avidly assimilated the sciences of the Christians and they were leaders during the five or six centuries while their science was transmitted to the West.

---

42   The name of Antyllus escaped Nicaise here. That marvelous surgeon of the 3rd C invented operations for aneurysms of the femoral, popliteal and brachial arteies, and wisely avoided others in the groins, axillae and neck. After ligating the arteries proximal and distal to the aneurysm he opened and emptied the sacs and packed them to control collateral inflow. Nicaise mentions only the last maneuver (LDR).

43   Lucien Leclerc *Histoire de la medicine arabe*. That work reviews the entire collection of Arabic translations from the Greek. It describes the Medical Science of the Orient and the transmission to the West via the Latin translations. Two vols. Paris, E. Leroux, 1870 (EN).

During the 7th C, before the invasions by the Arabs, the School of Alexandria ranked high for studies based on a selected sixteen of Galen's works, as are listed by Leclerc. After the conquest of Alexandria and the arson of its Library, the Arab conquerors resumed the curriculum based on the works of Galen that were translated by Honein in the 9th C. Later they were translated in Hebrew (1322). Long before that late date the Arabs had penetrated Greek science (9th C), and that began the epoch of special importance for the civilized world. Whereas at the end of the 8th C they had only the *Pandects* as translated by Aaron and a few books on alchemy, by the end of the 9th C they possessed the entire body of Greek science and could count among themselves scientists of the first rank. Bagdad attracted the savants of Persia and India. That was the century of their Renaissance.

Nestorian physicians had established Greek traditions in Asia Minor when they formed their school at Gondishaipur, where they taught in Syriac. The Syrians were the innovators in Medicine among the Arabs, and the first translations from the Greek were in Syriac (ie the biblical Aramaic), later to be carried into Arabic. When the Caliphs brought the Greek texts (ie from Gondishaipur) to Bagdad, the translators went directly from Greek to Arabic. Leclerc credits the Arabic translators with more skills and accuracy than accorded them today. He counted about one hundred translators; the best was Honein (see below).

They translated the works of philosophers including some of Plato and Aristotle and they thought Aristotle was the great thinker. Munk said that he dominated their logic and its applications.

In Medicine they translated Hippocrates (*Aphorisms, Epidemics, Prognostics,* and *Fevers)*. Dioscorides was as popular as Hippocrates and Galen, and they based their own pharmacopeia on Dioscorides and Galen (*The Simples*), and they added many items. The translations of Rufus of Ephesus, Archigenes, Galen (they called him 'The Eminent One'), Oribasius, Philagrius, and Alexander of Tralles,. Paul of Aegina was translated as Honein's *Pandects*. Paul's Book 6 (Surgery) was Albucasis' 'Master', although it was not frankly credited. Leclerc said that plagiarism was an acceptable practice among the Arabs.

The 9th C, which includes the reigns of Al Mansour and Haroun el Rashid (Charlemagne's contemporary), was one of the most memorable centuries in history. Its greatness depends not so much by comparison with the decadence of Europe in the same years, but on what a brief glance, as follows, will reveal of what was happening in the East. I quote Leclerc. "Barbarian peoples had invaded the world of the civilized; the invaders were uncultured in matters of the spirit which they disregarded and abandoned to the conquered populations. But among the hordes that conquered the Roman world, the Arabs alone had a traditional intellectual culture. They accepted the conquered peoples as teachers and proved themselves to be the worthy heirs of Greek science, and what became an Arabic culture endured for four centuries. It lost its strength only after the great conflicts had disrupted Asia.

The Middle Ages in Europe owed much to the Arabs. Take away their historical roles and their books and you set back the Renaissance more than several centuries. Humboldt wrote, "The Arabs rolled back some of the barbarisms that had been burning

Europe for two centuries, and they revived the ageless philosophy of Greece. They set no boundaries for their efforts to protect the treasury of knowledge that they had inherited; they broadened the older and provided new ways to study Nature."

During the 12th C the Crusades brought trouble into the Orient but they did not interrupt the culture of science in Araby, and they led to the 13th C which was the most important of all. Europe had struggled to escape the tyranny of barbarism; after two centuries of fighting the Arabs, they had discovered the sources of the science that Europe lacked.

In the 10th C he recognized the superiority of the important schools at Cordova and that led **Gerbert** (Pope Sylvester II from 1099-1102) to translate some of the most famous Arabic books.

In the 11th C **Constantine the African** (1015-1087) translated some Arabic books into Latin. He published some of them under his own name, omitting the those of the real authors. His work led to the rebirth of Medicine in Europe; his was a role of great importance. Constantine spent his later years in Italy at the monastery at Monte Cassino in Italy where he was most productive around 1072.

Before the Crusades, Oriental Jews migrated from their schools toward the West and with the training that exceeded the common lot and that of most laics, they opened the door to Arabic medical practices, and they made translations of the Arabic texts such as Ferraguth's of Galen's *Continens*. When forced to flee from Andalusia early in the Crusade years, some of them sought refuge in the Languedoc.

Arnold of Villanova, (ie a Christian Catalan,1235-1312) was one of the last in that region who made translations from Arabic into Latin. Also in the 12th C Archbishop **Raymond of Toledo** translated Avicenna's *Treatment of the Soul* into Latin. Raymond gathered a group of willing workers, among whom was **Gerard of Cremona** (1114-1187), who lived at Toledo for more than fifty years and made more than seventy translations.

Leclerc claimed that the translations made at Toledo are not as good as those from Bagdad. Moreover, the Latin of the translations from Arabic was not as elegant as the translations made directly from the ancient Greek mss made later during the Renaissance.

Much later, original Arabic works were produced in Spain, and others came to light in the French Midi where the Jews continued to translate from Arabic into Hebrew, and then into Latin. That provided another source for our knowledge of the Middle Ages.

The foregoing is a summary to illustrate how Greek and Arabic Medicine was transmitted into the West. The transmitters began with the Jews and with Gerbert; then came Constantine the African and Gerard of Cremona. The works by Jews continued to play important roles in the 12th and 13th Cs. All the above were sources for Latin translations of the following authors (from Leclerc, Vol. II, p 483).

Among the Greeks, translated by Constantine were the *Aphorisms of* Hippocrates with the *Commentaries* by Galen. Other Hippocratic works translated by Gerard: were *Fevers* with Galen's *Commentaries*. Credit Gerard with other books by Aristotle and Galen.

Among the Arabs, translated by Gerard: Serapion the Elder, Mesûe the elder, Rhazes, Isaac (Ishay ben Soleiman, the Jew), Albucasis, Halyabbas, Avicenna, Canamusali, Jesu Haly, Avenzoar, Ali ne Rhodhouan, and Averroés.

Leclerc identifies no fewer than three hundred translation from Arabic into Latin. Those translations were distributed all through Europe during the 12th and 13th Cs. They promoted and stirred the eager scientific spirit of the 13th. Greek Medicine was represented by four works by Hippocrates and twenty five by Galen. Ninety of the translations concerned Medicine. A dozen of the three hundred had first passed through Hebrew versions.

Guy owned the works of the principle Arabs. He cited Serapion, Mesûe, Rhazes, Haly Abbas, Albucasis, Avicenna, Jesus Ali, ali Rhodoam, Canamusali, Avenzoar, Averroés, Rabbi Moses (ie Maimonides), et al.[44] Of the nearly three thousand-three hundred citations in his *Major Surgery* one thousand-four hundred refer to Arabic authors.

There were two **Serapions. Serapion the Elder** (9th C) also was called Janus Damascenus. One edition of J.D.'s works was published by Albanus of Turin. His small book, *The Pandects* has seven parts. A translation by Gerard of Cremona was titled *Breviarum*. A later version by **Alpago** was titled *Practica*. According to Hirsch, the *Pandects,* later was published as *Aggregator Breviarum*. Guy cited him sometimes as John Damascenus or just as Damascenus. A book attribute to Janus Damascenus titled *Aphorisms,* really was written by Mesûe the Elder **The Younger Serapion** (13th C) wrote *A Treatise on Simple Medications,* that was translated into Latin by Abraham and Simon; both were Jews of Genoa.

Honein (809-873) was also known and was cited as Johannitius in many Latin documents. He made a large number of translations from Greek into Arabic. His Introduction to Galen's *Microtechni* was titled *Isagoge Johanniti.*

John Mesûe the Elder (777-855 or 57) was especially known as the author of the *Selecta artis medicae.* Würstenfeld thought that it was meant to be a reply to the *Aphorisms* of Janus Damascenus.

Mesûe the Younger (11th C) wrote about materia medica, and his work was highly respected, although his name did not appear in the title of the Latin translation, *The Medical Handbook (ie Practica).* That work does not survive intact. Guy cited both the Mesûes sixty times, in his text and in the antidotary.

Rhazes was cited one hundred-sixty times. He was born in Persia about the middle of the 9th C, studied at Bagdad, and died there in 932. He was the first great Arabic physician, known also as an encyclopedist. His largest and most important book is *The Continent (The Haouy).* It is a vast collection of ancient and contemporary Medicine; it is not at all dogmatic, and is marked by the insertion of his own esperiences. The twenty—two volume work was translated into Latin in the 13th C by Ferraguth, titled *Continent.* Joubert, as did Guy, who called it *Helham* in Arabic, said that it also was called *Elhan, Elhandi, Elhangi,* as substitutes for *al Houay.*

---

[44] Nicaise's spelling of his French names for the Arabic authors is not consistent from page to page. I assume that he used them as he found them in Leclerc and other sources and did not exercise editorial prerogatives. Neither do I. (LDR).

After the *Continent,* Rhaze's *Mansoury*[45] in ten books, was his most important work. Book VII is about surgery. It is less detailed and of less historical importance than the *Continent,* which included all of Medicine. However, the surgical part of *Mansoury* has been used by surgeons more than what is contained in the *Continent. Mansoury* was translated by Gerard of Cremona, and Book IX was treated with respect at Louvain until the 17th C. In addition to the above, Guy cited Rhazes' *Book of Division (*ie wounds and fractures) *and Book on Joints.*

Halyabbas was cited one hundred-fifty times. He was Persian of the 10th C who died in the 994. His chief work, The *Kamel, or Maleky the Royal Book,* is a complete medical text, not merely a collection, as is the *Continent.* The material is well arranged and the work was very popular until it was displaced by Avicenna's *Canon,* which copied Halyabbas' orderly arrangement of materials and illnesses.

Constantine translated the *Maleky* into Latin at the end of the 11th C, but did not reveal the name of the original author; he published it as his own *Pantegni.* In 1127 Stephen of Antioch made a new Latin translation under the title *The Royal Disposition,* and that is the edition cited by Guy, although Leclerc thinks that Constantine's version was better. Guy also cited Halyabbas' *Theory.*

The **Isaac** cited by Guy probably was Ishaq ben Soleiman el Israely, called Isaac Judaeus. He was a Persian physician (mid 10th or 11th C). He was an oculist who also wrote about fevers et al.

Albucasis, or Abul Cassim, was cited two hundred times, as Albucasis, Azaram, or Azaran Galaf. The *Albumazar* cited by Guy also could have been by Albucasis. He lived at Cordova in the 10th C and died in 1013. His encyclopedic work in thirty volumes was *The Tesrif.*

The entire *Tesrif* was translated into Latin with the title *Alsaharavius or Açaravus.* Leclerc said that the translation is defective. Another translation, in Hebrew, was published by Chem Tob, titled *Chimouch,* which corresponds to the Latin *Servitor (ie* late Latin for Servant).

During the 13th C at Toledo, Gerard of Cremona translated the Surgical part of Albucasis, Volume XXX of the *Tesrif,* into Latin. In the mid 13th C Chem Tob translated it back into Hebrew.

In his *Surgery* Albucasis reviews what he had called the Division of The Maladies into The Theory and The Practice, which form volumes I and II of the *Tesrif.* Volume XXVIII deals with the preparation of simple medications, and was translated by the Jewish physicians Abraham and Simon of Genoa at the request of King Alfonso X (1226-84) who sponsored many translations from Arabic.

Albucasis' books devoted to therapeutics (medicines) also were translated piecemeal. Guy cites his Antidotary and his Large Antidotary, whenever matters dealing with

---

[45] Better known as *The Almansor,* a title that many later authors thoughtb was the name of an Arabic physician (LDR)

medicinal compounds appear in the twenty three volumes of the *Tesrif.* Leclerc said that those were the sections that were titled *Liber Servitoris,* rather than *A Description of Simple Medications.*

The *Tesrif* was widely used by many medieval physicians and it was well known even in England during the time when Schenk (1609) lived.

Albucasis' *Surgery* was very important. It was the first presentation of Surgery as a separate science, and it was the first book with illustrations of surgical instruments. However, it was almost entirely based on what Albucasis took from Volume 6 of Paul of Aegina's works, taken without due credit.

Roger of Parma[46] and William of Saliceto did the same in their own uncredited use of Albucasis.

His work was a major resource for the development of medieval Surgery, and we see that well in reading Guy de Chauliac. Albucasis divided his book in three parts: a Treastise on cauterization, another on manual operations that showed the instruments, and a third on fractures and dislocations. Channing (Oxford, 1778) published a bilingual Latin-Arabic edition and reproduced Albucasis' drawings. Leclerc produced a French edition in 1861, with etchings.

Guy cited **Avicenna** six hundred-sixty times. That well-named Prince of The Sciences was the greatest physician of the 11th C. He died at Hamadan (Persia) at age 58. He was both a physician and a philosopher, and he wrote major works in both fields.

His treatise on Medicine covers the entire subject, arranged systematically. It is *The Canon*, a Greek word that means The Rule. The five parts are set in good order, better than that of the *Continent* of Rhazes, and is more complete and less terse than the *Maleky*, and it is more analytic in depth.

In addition to the *Canon* guy cited Avicenna's *Anatomy of the Muscles* and his *Book of Naturals. The Canon* was translated into Latin by Gerard of Cremona at the end of the 12th C, and by Alpagus. According to Malgaigne, Arnold of Villanova had made Latin translations of parts of the books, now lost.

Avicenna's influence on Medicine exceeded that of Rhazes, and it endured in Europe for five centuries. That influence also pertained in philosophy, as seen in his success in the medieval schools—books that were translated in Latin at the dawn of the 13th C.

In the 17th C a reading of the *Canon* was part of the curriculum at Louvain. Guerner Rolfinch gave public readings at Jena and Plempius gave public commentaries in 1658. A miniature of the 16th C, as the frontispiece of this volume, shows Avicenna's book more sharply marked than those of the other notables in the picture.

Jesus Ali (also called Ali Ben Issa, and Issa ben ali) was cited by Guy more than sixty time, under the names Jesus, Jesus Hali, and Jesus son of Haly. He was a Persian physician who died around 1010. His specialty was ocular disease; his book *The Tedkirat*

---

46   Nicaise refers to Roland of Parma who wrote his own version of the *Surgery,* of Roger Frugard, the first such work in medieval Europe (LDR).

el Kahhálin (A Handbook for Oculists) was the forerunner of **Canamusali**. Jesus took much of his material from Galen and Honein and he cited Paul of Aegina and Criton. The book had three parts: a description of the eye, maladies with observable signs, and maladies that have no detectable signs. The parts that describe treatments are better than the theoretical. The book was translated in Latin early with the title *Recognition of Ocular Ailments And Their Treatments.*

Ali Rodoam or Ali Rodoan or Ali ben Rodhouan, frequently cited, was an Egyptian physician born near the end of the 9th C and died in 1061. He was a better physician than a philosopher. Guy twice in error (in Ch.1 and in the Antidotary) attributed the *Techni* to him, intending to cite the *Pantegni* of Haly Abbas.

Canamusali or Acanamose or Acanamosul was a famous Egyptian oculist in the 11th C. Guy cited him frequently. Leclerc believes that he really was named Omar ben Ali el Mously. He was credited with inventing an operation for removing cataracts by suction. However, **Salah Eddin,** an oculist during the 13th C claimed that the priority belonged to**Tsabet ben Cora**, and that Canamusali actually had rejected the procedure. Canamusali's monograph appeared in print in the Venetian collection of 1499, filling fourteen pages after those of Jesus Hali.

Avenzoar was cited twenty one times. He was an Arabic physician in Seville in the 12th C (d. 1162). He dedicated his principal work, *The Teissir* to Averroés who was his pupil. Although he accepted the doctrines of the Greeks and the earlier Arabs, he modified them by reporting his own experiences. He was as independent as Rhazes, but he turned away from surgery in favor of medical measures. He was one of the really great Arabs, along with Rhazes and Avicenna, and was dubbed 'The Glorious One'.

A Hebrew version of *Teissir* was re-translated into Latin in 1280. Malgaigne said that the Latin translation was by Jean de Campagnie; another by Paravicini appeared in 1285, assisted by Jacob, a Jewish translator.

Averroés was cited twenty times. He was the most important Moslem author. He lived at Cordova (1126-1198) and was the pupil of Avenzoar, and like him was a philosopher-physician.

His medical treatise was *The Colliget* in which he dealt with and responded to the general questions raised by Avicenna in Book I of the *Canon*. He reviewed what Avenzoar had written about each malady, and reiterated his claim that his master was the greatest physician since Galen. Renan believes that the Latin translation of the *Colliget* came in the mid 13th C. Leclerc thinks that it came from Averroés' original Arabic. Another translation is attributed to Armengaud, and was re-edited by Alpagus. Armengaud was a physician at Montpellier late in the 13th C and was one of the royal physicians of Philip II (The Fair).

Renan, in his work about Averroés and Averroëism, gave a full description of The Great Philosopher's personality.

Rabbi Moses Maimonides or Rambam was better known as philosopher and theologist and Talmudist than as a physician. He was one of the most famous Jews. His book of *Medical Aphorisms* was translated into Latin as *The Regimen of Health*. Guy cited it several times. Dr. Rabinowitz published Maimonides' *Treatise on Poisons* in 1865.

To summarize: During the Middle Ages the Arabic Civilization produced medical works which established their own importance in the world's scientific literature. It served to instruct the West during many generations that followed.

Section 3. The School Of Salerno

Salerno was the only bright spot in the West, amid all the neo-latin schools in the cathedrals and monasteries. Its origins are obscure. Daremberg and de Renzi researched the question in depth, and they tend to tie Salerno to the small Greek schools that flourished in southern Italy and in Gaul until the knowledge of the Greek language disappeared from Western Europe. A small group of 'Grecians' seems to have found refuge in southern Italy. In the 14th C Nicolas of Reggio found a Greek Ms of Galen and translated it into Latin. And that is all we know about Greek at Salerno.

The fame of Salerno was dominant for several centuries, and its name is associated with more than half of the Latin translations and compilations that were the bases for instruction in Italy, Gaul, England, Germany, and even in Spain (ie where Arabic medicine flourished). Until Constantine, Salernitan Medicine, as it was elsewhere in the West, was Neo-Latin. Daremberg believed, however, that it was Greco-latin.

We can distinguish two epochs of Salernitan medical literature. The first specially represented by Gariopontus (before 1040) derived from neo-latin translations of predominantly Methodist authors. The second period (after 1050) during the first two-thirds of the 12th C was predominantly humoristic, notably by Trotula, Cophon, and the Platearis family. Even the authors who wrote in their own names knew little about the Arabs excepting perhaps a few echos via the Jewish physicians. But, as Daremberg showed, they had translations of some of Galen's books—he counted sixteen of them, perhaps the same sixteen in use at the school of Alexandria—some by Hippocrates (*Aphorisms, Prognostics, Epidemics*), some by Alexander Of Tralles, and some of Paul of Aegina. The translations probably were made at Salerno during the 7th to the 11th Cs. At Salerno there were lay practitioners as well as clerical physicians, even as priests, who taught Medicine, and there were trained mid-wives.

Salerno was well known long before Gariopontus. The Salernitan physicians were quoted by the middle of the 9th C, and Salerno was famous by the 10th. We can date texts related to the school and to the 'medical town' in the 10th C. At that time the medical doctrines were more Methodist than Dogmatic, and they did not change until after the translations of Constantine the African appeared during the second half of the 11th C.[47] Among the physicians of that first period we already have cited **Gariopontus.** He wrote his *Passionarius,* a collection of medicaments. He played an important role as source of information about neo-latin Medicine at Salerno before the introduction of the Arabic texts.

---

[47]   Methodists, contrary to Hippocrates and Galen, rejected the dogma of 'essence' (that the body was the source of its ailments), and prevented investigation of primary causes, hidden causes, and organic phenomena. They tried to explain all illness by invoking 'tightness' and 'relaxation' (EN).

That important turn of events began with the translations of Constantine. They broadened the scope of medical studies which had been limited by ignorance of the Greek language and by the poverty of Latin translations—merely a few scraps of Latin. Leclerc agrees with Daremberg that Galenism was feeble throughout the early Middle Ages. On the other hand, Daremberg, as noted earlier, held that there was a time when some of the Greek originals were translated into Latin, and that Western Medicine was not entirely bereft, as is the general opinion today. But those early translations were loose and fragmentary, and for the most part consisted of recipes and superstitious formulas. (ie see fn. 39). That set them apart from what was happening in the School at Salerno, and shows how feeble were the strengths of the ecclesiastical and lay schools during the long period between the early and the later (ie 12ᵗʰ C) Middle Ages. Nevertheless, much is left for the historians to discover about that epoch; to find new documents and new mss before we can provide definite answers.

Constantine left Salerno and went to Monte Cassino and translated at the monastery Among his translations as listed by Leclerc are *the Pantegni,* from Halyabbas' translation of the *Maleky; the Viatique* from Arabic by **Ibn Eddjezzar**; the *Treatise on Urine and The Treatment of Fevers* from Isaac; and *The Commentaries on Hippocrates* of Galen. Those translations led to the shift of emphasis in the Methodism at Salerno; it was wedded to Galenism and to the theory of humors. Salerno then became City of Hippocrates.[48]

Among the Salernitan physicians during the 12ᵗʰ C were **Nicolas Praepositus**, wrongly confused with **Nicolas of Alexandria**. He taught at Salerno where he became Dean (ie Praepositus). His *Antidotary* came early in the century and it was highly regarded during the rest of the Middle Ages. He cited some Greek and Latin authors and some earlier Salernitans, but not once did he mention an Arab. His book was expanded and became more popular, as enlarged by **Nicolas Myrepsos** of Alexandria (mid 12ᵗʰ C).[49] (See below, Jean de St. Amand).

Nicolas P.s *Antidotary* was mentioned by ***Mathew Platearis.*** His *Book of Simple Medication Circa Instans* was better known as the *Circa Instans* (Readily Available).

---

[48] The full list of Constantine's writings from Galen contained in the Basel edition of 1536 follows. Part I: 1. *Signs of Disease in 6 Bks., 2. Treatment of Diseases and Symptoms, 3. On Urine, 4. On Ailments of the Stomach, both the Natural and the Unnatural, 5. The Dietary Regimen for Various Ailments, 6. On Melancholy, 7. On Coitus, 8. On Differences in the Life-Spirit, 9. On Incanatations and Curses. 10. Disease of Women, 11. On Surgery—Phlebotomy, Arteriotomy, Scarification, Fractures, etc. 12. The Degree-Rankings of Simple Medications.*

Part II. 1. *Common Sites of Illnesses. 2. On Human Nature. 3. Elephantiasis., 4. Medications derived from Animals (EN).*

[49] Sarton (Vol II, p.1094) says that Myrepsos means maker of Ointments. His book contained 2566 recipes as compared to Nicolas Praepositus Antidotary with only 140-50, and leaned more heavily on Arabic sources (LDR).

In the 12th C (1134) Roger II, King of Sicily (1130-1154) officially established the School of Salerno as part of his government. Until then it was independent. In 1224, Emperor Frederick II completed the absorption.

It was the school for the earliest real Italian surgeons. **Roger,** so called Roger of Parma, who flourished around 1230 (see fn 51 re Roger Frugard). He described medications and manual procedures. Guy cited him ninety times. Roger's book was printed in the Venetian Collection of 1499: forty-eight pages, titled *Master Roger's Handbook.*[50]

Roland (Capellutti) was Roger's disciple, born at Parma, he worked at Bologna, but truly he was a Salernitan. His *Surgery* published in (?) 1264, the so-called Rolandina, is a transcription of Roger's plus a few additions. It also appears in print in the Venetian Collection, twenty-four pages titled *The Small Book Of Surgery.*

The *Glosses of the Four Salernitan Masters on the Surgeries of Roger and Roland* was cited by Guy twenty-five times. Daremberg published a Latin edition in 1854. (see Roland in the Translator's Bibliography).

A certain **Jamier** was a Salernitan, cited forty times. Guy wrote (see Chapter 1.) "We can identify Jamier who performed some violent operation and made some foolish comments, but in most matters copied Roger."[51]

At the end of the 12th C the School at Salerno was in decline and produced no significant scientific works, but it continued as center for treatment. In 1748 the Parisian Faculty of medicine consulted it in the conflicts between the physicians and the surgeons. On 29 November 1811, the School officially closed its doors.

Section 4. The School at Bologna

Some schools[52] in northern Italy had existed before the high-medieval years. Ravenna was well-known in the 8th and 9th Cs; Padua came later and Bologna was parallel with Salerno. Daremberg found one notice of the school at Bologna in the 12th C. It listed the thirty-one practicing physicians who were the teachers. Another counted forty-seven in the 13th C, when Bologna's fame overtook its rivals. By the 14th C it had equaled Montpellier and Paris, and those were the three schools attended by Guy.

In the 13th and 14th Cs Surgery was the metier of many reputable authors. Their works indicate an advancing scientific milieu and a shared knowledge. We should not hold Guy's

---

[50] Roger Frugard's *Surgery* was written in 1170 by his pupils, led by Guido of Arzzo II. See Translator's Bibliography (LDR).

[51] Sarton (Vol.II, p 634) identifies him as Johan Jamatus, or Jamaticus, or Jammarius, or Jamerius. He worked in southern Italy at the time of Roland. Wrote a compendium based on Roger in nine books (LDR).

[52] The term 'schools' refers to informal aggregations of teachers and students. Some of them, after many years of independence, were congregated by local cathedrals or large churches, and much later became parts of the Universities (LDR).

book to be the only book by a single original author, although it was more complete, and more systematic, and it exhibited a degree of skepticism, and was more thorough in its citations than those of its predecessors.[53] Those remarks illuminate the state of medieval Surgery. As to particulars about Bologna, the following were its great savants and teachers:

Hugh of Lucca was contemporary with **Roland**. He must be regarded as the head of the line, a master surgeon, the first European worthy of that title according to Malgaigne. He died around 1258, and although he left no writings of his own, his teachings appear in the works of **Theodoric**.[54]

Bruno of Longoburgo, was a Paduan surgeon who wrote his *Major and Minor Surgeries* around 1252. Both appear in print in the Venetian edition of 1499, filling sixteen and four pages. Bruno was the literary 'father' of **Dino del Garbo.**

Theodoric was a famous surgeon at Bologna, and was cited eighty-five times by Guy. He wrote his *Surgery* after 1264, and in it exposed the teachings of his Master Hugh of Lucca. He copied much from Bruno. Theodoric cited Roger and Roland (ie but made no mention of Bruno)[55]. His book is printed in the Venetian collection, filling seventy-four pages.

William of Saliceto was cited seventy times. His *Surgery* was published at Bologna and Verona in 1275. William was a priest who became the most famous surgeon of the 12th C. His book is more complete than those of his predecessors, but it is much smaller than Guy's. He leaned on Galen and the Arabs, and he included sections on disorders of women. Malgaigne said he was a more skillful operator but was less well informed than Guy. A French translation by Nicolas Prévost, a physician, was printed at Lyon in 1492, and at Paris in 1506 and 1596.

Taddeo Alderotti of Bologna (1215-1293) taught there, and he enjoyed great fame as a promoter of Scholasticism (ie Aristotelianism a la Thomas Aquinas). Guy cited him frequently.

Authors From Schools Other Than Bologna (as cited by Guy)

Jean of Saint-Amand was the provost Canon of the Cathedral at Mons in Puelle (ie now in Hainault, Belgium). He was a physician in the 13th C, frequently cited by Guy. He served as a priest in the royal entourage of Louis IX, and translated and commented on the *Antidotary* of **Nicolas Myrepsos,** of Alexandria (Guy called it the Antidotary of Nicolas). He was a Greek physician of the 13th C who had absorbed the *Antidotary* of **Nicolas Praepositas**, the Salernitan, and made it part of his own. According to Chèreau, the *Commentary* by Saint-Amand was required equipment in all royal apothecaries in France, so ordered by the Faculty of Medicine at Paris. It was part of the codex of 1649. Guy mentioned another book by Saint-Amand, *The Aureoles*. He also wrote a

---

[53] See below, the text and footnotes re Henri de Mondeville (LDR).

[54] Theodoric called him "my Master". Many claimed that Hugh was Theodorics real father (LDR).

[55] That interesting issue is discussed in Tabanelli's edition of Bruno. See Translator's Bibliog. (LDR).

Concordantiae (Abreviationes) about some of Galen's works. It was expanded by **Jean of Saint-Flour** and titled *The Companion of the Concordances.*

Gilbert The Englishman was physician during the second half of the 13th C. In his book, the *Compendium of Medicine, Both The General and The Particular,* he attacked the minor empirical practitioners who exploited the public. He argued against the application of the scholastic methods both in the theory and in the practice of Medicine. He favored the authors who followed the Hippocratic doctrines rather than those who invented new ones. Guy cited him as 'The Gilbertine'.[56]

Peter of Spain or Portugal was born at Portugal early in the 13th C. He studied at Paris and Montpellier and became Pope John XXI in 1276, and died in the following years. His *Treasury of the Poor* probably was translated into French by Arnold of Villanova.

Lanfranchi of Milan was cited more than a hundred times. He was an Italian who carried the Italian Surgical Science into France. He wrote his *Minor Surgery* at Lyon, and a *Major Surgery* at Paris in 1296. He had been a pupil of William of Saliceto, and his books reflect the erudition of his Master, but they lack William's orderly presentation. Lanfranchi must be ranked ahead of the other surgeons who preceded Guy, and placed second to him. His books appear in the Venetian Edition of 1499, occupying ten *(Minor)* and ninety-two *(Major)* pages.

Anselm of Genoa, or Anserin of The Port, was cited often. He may be the same person as **Simon of Genoa,** (Simon of Janua, Januensis), named after the Ligurian port city. Guy cited another Genoan, **Jean,** nephew of Anselm. Simon (1270-1303) was the physician for Pope Nicolas IV. His book, *Medical Synonyms, or Key to Good Health* was an alphabetical list of simple medications taken from Greek, Latin, and Arabic sources.

Meanwhile, practical operative surgery was slipping away from the educated surgeons. Bruno decried the release to the barbers of small surgical procedures such as phlebotomy and scarification. By the time of Lanfranchi that lapse included the application of leeches and cauteries. Lanfranchi himself avoided tapping ascitic bellies and operations for hernia, cataracts, and bladder stones.

Section 5. Physicians Of The 14th C Cited By Guy

Arnold of Villanova (near Bracelona) lived at the end of the 13th C into the 14th. The exact dates of his life and of his writings are uncertain. Some documents date his death between 1309 and 1313. He was at Montpellier in 1289, when it was officially dedicated as a University, and at other times. He was famous as a physician who wrote much about many topics. One can examine a printed edition of his works collected in *The Literary History of France,* Vol. 28, p. 26, 1881. He discovered Rhazes' *The Spirit Of Wine* (ie the vital essence) and he recommended that a physician should be familiar with the flavors and the

---

[56]   See the discussion of Aristotelianism in Part II, Part A. of this Introduction (LDR).

scents of all the plant materials in foods and medicinal compounds made from their juices. Guy wrote about his use of "Arnold's Water' in treating wounds.[57] A copy of **Mundinos's** *Anatomy,* annotated by Arnold has been questioned because it was not published until 1315, long after Arnold's death. Perhaps Arnold's famous name was inserted to 'advertize' it. Arnold's *Handbook of Health For Use By Pope Clement* (Clement V, 1305-1316) is attributed also to the School Of Salerno, written in Latin verse. However, most historians credit the work to Peter of Spain (ie see above). Guy also cited Arnold's *How Medicines Work.*

Peter of Abano (near Padua), was born in 1250 and died in 1316. He learned Greek at Constantinople before studying medicine and philosophy at Padua, after 1303. Guy cited him as the author *The Conciliator of Various Philosophies and Medical Precepts.*

Henri de Mondeville is cited almost one hundred times as Henri d'Hermondeville. He was a pupil of Jean Pitard and became a royal surgeon for Philip The Fair. He had studied in Italy before he became a professor of Anatomy at Montpellier in 1304. After 1306 he worked at Paris and began to write his *Surgery* and completed Vols. I and II in 1312 (per Pagel). He died between 1317-1320 before completing the book.

Henri has been investigated by Chereau, Corlieu and Pagel. The latter published an edition of Henri's *Anatomy* in 1889. At present, Pagel is working at a new Latin edition of Henri's complete works collated from various mss in the libraries of Paris and Berlin. Pagel's work will fill a great need.[58]

Henri was the first French author of a Surgical treatise; Lanfranchi was an Italian who wrote his books when he was in France. Henri described the many sources, ancient and contemporary. Pagel praises him highly for the descriptions of his own experiences and for his progressive ideas. His book is organized on the same plan as those of William of Saliceto and Lanfranchi, and later followed by Guy, who exhibits his own

---

[57] Here the Catalan-Spaniard expounds a Frenchman's 'nose' for wines, and describes the favorable effects of alcohol in wounds (LDR).

[58] Three years after publishing this edition of Guy de Chauliac, Nicaise published a French translation of Henri de Mondeville (see Translator's Bibliography). He then came to fully appreciate the greatness of Henri as compared to that of Guy. The character of Henri and the quality of his book were exhibited and the interest of medical historians has been awakened during the century after Nicaise's edition. It is apparent that the differences in the style and the contents between those of Henri and of Guy tell us that one marks the end of an epoch and the other marks a beginning. The gap between them was created by the Great Plague in Europe after 1340, and by the factors that kept the work of Henri in Limbo until the 19th C. I am convinced that Henri's openly expressed criticism of the clerical medical establishment led to the disfavor of the authorities who selected the texts for the curricula of the universities. The Churchmen did not condemn him but they ignored him We know that Guy was more tactful, respectful, and liberal with praises. Note how often Guy bows to his Masters, yields to his colleagues, and comforts his Popes. His social graces contributed to his successes and financial security from childhood on (LDR).

regard for Henri. I am astounded that such a work as his has not been published, and we have only Chereau and Pagel as our sources. But that will not be the end of it.[59]

Bernard of Gordon is cited thirty times. He was a professor at Montpellier where he wrote his handbook, *The Lily of Medicine* in 1322.[60] It was the best medical text written in the West until his time. He described the use of a metallic disc held in place by buckles for the treatment of hernias. The *Lilium* was translated into French in 1377 and was printed at Lyon in 1495. I quote a French comment re the translation. "Here ends the handbook, the product of a noble experience at Montpellier, after twenty years, completed in 1322 and translated into French at Rome in 1377, and printed at Lyon in 1495, on the last day of August."

Mundini (de Luzzi) was a professor of anatomy at Bologna, and is said to be the restorer of that science in the West. He died around 1326. His *Brief Anatomy* appeared in 1315. It dealt mostly with the viscera, and it enjoyed great favor. Although he had made dissections of human bodies, his descriptions are those of Galen; such as the liver having five lobes, etc. And Guy agreed! We may ask why he did not use his own eyes? Be not too surprised: behind the retinal eye there is a cerebral eye that dominated it. Even now in the 19th C we can find errors in the anatomical texts.[61]

Dino del Garbo of Florence is cited almost forty times. He also was called Dinus or Dynus of Florence. Born at the end of the 13th C, he died in 1327. He studied Medicine under Taddeo Alderotti at Bologna and acquired his own great fame. His son who lived at Florence until he died in 1370, was equally well known.

Bertrucius died in 1347, after teaching anatomy at Bologna after the death of Mundino. He was one of Guy's teachers and was cited frequently. Daremberg credits him with the invention of a perforated catheter that used a metallic wire or braided cord as a leader.

Albert of Bologna is cited a few times. He may have been Albert Zancari (or Antoine Bumaldi). He was a well know physician early in the 14th C. Cellarier, accepting the work of Garzonius, thought that Albert was a professor at Bologna who was the teacher to whom Guy pointed as his surgical master ("My Master"). He died in 1348.

John of Gaddesden worked at Oxford around 1320 and died around 1350. He studied Medicine under Gordon at Montpellier and Surgery at Paris during Henri de Mondeville's time. He was a priest who wrote *The Rose of Medicine,* the English Rose.[62]

---

[59] Nicaise inserts here bibliographical references to the works of Chereau, Corlieu and Pagel. In lieu of that I refer to Nicaise's edition of Henri and to the new English edition in the Translator's Bibliography for this volume. Many of the references will be found there (LDR).

[60] Lilium (lily) connoted purity, and Rosa connoted floridity (LDR).

[61] Nicaise's tongue-in-cheek comment reminds us that Galen ruled that brains ruled the eyes, even of those with brilliant visual senses. The marvelous anatomical drawings by Leonardo da Vinci show us how Galen also ruled that artist's eyes (LDR).

[62] It had four parts: particular maladies, fevers, surgery, and a pharmacopoeia. It was printed at Pavia in 1492 (EN).

Guy wrote, "Finally have inspected a faded English Rose sent to me by a pupil, and I believe that I detected a gentle scent. The volume in hand also contained some Spanish fables and some of Gilbert and Theodoric." Malgaigne said that John copied Gilbert and Henri de Mondeville freely.

Dondi (also Jacques de Dondis, 1298-1359). He was called the Aggregator because of his large compilation of remedies. He was a Paduan by birth.

His son **Jean de Dondi**, lectured on Medicine at Naples from 1317 to 1345. He translated many Greek works into Latin at the request of the Angevin princes, of Charles II and of King Robert II. He made a new translation of Galen; Guy owned a copy.

Raymond of Molier was Chancellor of the University at Montpellier in 1334. Guy referred to him many times as 'my Master'. (See Albert, above).

Chalin of Vinario (Raymond) is also Raimond Rinaldi of Vinario. Marini said that he was one of Pope Clement VI's (1346-52) physicians. He wrote a work on the plague of the 14th C which was published by Dalechamp.

Peter of Dye King Jean II whom he attended during the wars, dubbed him his physician-surgeon.(ie before his exile in England).

Jean of Alais a contemporary of Arnold of Villanova at Montpellier, had a long life and was a physician for Clement VI at Avignon after having served Clement V.

Bienvenu or Benvenutus wrote about ocular maladies. His book was printed, and copies are still in our National Library, undated, but printed around 1474. The author's full name wa Benevenutus Grassus Hierosolimitanus. Probably he was Jew who practiced at Salerno and Montpellier.

Etienne Arnauld of Montpellier wrote the *Isogoge on Hippocrates' and Galen's Physiology and Anatomy*. It was printed at Paris in 1587.

Jean Jacques was a colleague of Guy at Montpellier around 1364. He left two works: *A Medical Thesaurus* and *On The Plague*.

Guy also cited lay 'cutters', the non-clerical surgeons: **Nicolas Catalan** at Toulouse; **Bonet**, Lanfranchi's son at Montpellier; **Perrerin and Mercadent** at Bologna; **Peter Argentiere** at Paris; **Peter Arelata** at Arles—not to be mistaken for Peter Argelata; **Peter of Bonant** at Lyon; and **Jacques**, the Papal apothecary at Avignon.

In addition to the above, I mention *The Lives and Morals of the Philosophers,* which probably was another title for *The Lives fo Philosophers and Physicians* writte by **Leon The African,** mentioned by Leclerc (Vol. I, p.338). And, finally, Guy cited: **Alcoatin, Americ of Alais, Master André, Jean** (Anselm's nephew)**; Bernard of Metz, Criton, David, Jean of Crepatis, Jordan, Macrobius, Odet of Lyon, Master Paul, and Paul of Orlhac.**

TABLE I: AUTHORS AND FREQUENCY OF CITATIONS BY GUY (from Joubert)

Acanamose	12	Alcoatin	28
Albert of Bologna	4	Alexander	7
Albucasis	175	Alexander the commentator on sects	3
Albumazar	2	Americ of Alais	7

Master André	1	John of Crepatis	1
Anselm (Simon of Genoa)	6	Jen of Saint-Amand	8
Apollonius	3	John of Parma	1
Archigenes	6	Jean Jacques	2
Aristotle	62	Jean, Nephew of Anselm	2
Arnold of Villanova	8	Johannitius	2
Asclepiades	1	Jesus Haly	62
Avenzoar	21	Jordan	2
Averroés	29	Isaac	1
Avicenna	661	Lanfranchi	102
Bienvenu	4	Macrobius	1
Bernard of Metz	1	Mercadent	1
Bertrucius	14	Mundini	6
Bonet (Lanfrtanchi's son)	1	Nicolas the Catalan	1
Bruno of Longoburgo	49	Nicolas Praepositus	11
Platearius	1	Odet of Lyon	1
Commentator (? Averroés)	9	Ovid	1
Jean de Saint-Flour	6	Master Paul	2
Criton	1.	Paul of Aegina	1
David	1	Philagrius	1
Damocrates	1	Peter of Argentier	4
Democrates	1	Peter of Arelata	3
Dioscorides	2	Peter of Bonant	15
Dino del Garbo	36	Peter of Dye	1
Dondi	1	Peter of Spain	6
Etienne Arnold	2	Peter of Orlhac	3
Gaddesden	2	Plato	2
Galen	890	Ptolemy	1
Gilbert	2	The Four Salernitan Masters	25
Gordon	26	Rabbi Moses Maimonides	12
William of Saliceto	68	Raymond of Moliere	3
Haly Abbas	149	Rhazes	161
Haly Rodoan	5	Roger	92
Mesûe	61	Roland	4
Henri de Mondeville	68	Serapion	9
Heraclides of Tarentum	1	Taddeo Alderotti	4
Hermes	1	Theodoric	85
Hippocrates	120		
Hugh of Lucca	1		
Jacques the Apothecary	1		
Jamier	36		
John of Damascus	3		

PART III

AN ESSAY ON MEDICINE AND SURGERY IN THE 14TH C

I n this epoch there were three major theaters of science: Bologna, Montpellier, and Paris.[63]

Early in the century Bologna enjoyed the greatest fame. Yet, in spite of the panache of Mundino, it declined with the years. It failed to prevent its students from carrying off its books; dire threats and punishments were ineffective. One hundred-thirty years later, Petrarch bewailed the loss of Bologna's splendor (Malgaigne), and that of Salerno had long since faded. What remained was Montpellier and Paris

l. The University at Montpellier

While most of Europe was in the hands of the barbarian invaders, Montpellier was favored by its proximity to Spain and its Arabic culture, which had reached north to include Septimania as far east as the Rhone. The Arabian march northward was halted by Charles Martel in the victory at Poitiers in 732. What remained in Spain founded the Caliphate at Cordova in 755. In 860 the newly formed Kingdom of Navarre set itself between the Midi of France and Arabic Spain in the in the south. In 1035 the Navarene reign was divided and the Kingdom of Aragon formed south of the Pyrenees included the Balearics and the territory of Roussillon north of the mountains[64], especially the town of Montpellier. In turn that was retaken by the French under Philip II in 1349. However, from 1204 until then it was ruled by the Kings of Aragon and Majorca. The Arabic rule elsewhere in Spain was cast out in 1492, when Granada was retaken. The Arabs (Moors) had ruled for eight centuries.

During the 9th C, before the Arabic 'Renaissance' had reached full flower, southern France (the Midi) really was separated from the Arabs. Leclerc says that is why there are

---

[63] 'Science'. Nicaise's usage is close to the ancient word that meant knowledge in general, sometimes equivalent with 'philosophy. Here we will accept Nicaise to mean Medical and/or Surgical Science (LDR).

[64] Roussillon was the part of ancient France north of the Pyrenees as far as the Languedoc. Now it is settled by the French Basques (LDR).

so many more citations of Arabic Medicine (ie Avicenna, Rhazes, et al.) from the East (Bagdad) than from Spain (Cordova). Nevertheless, some Arabic material filtered up through the Pyrenees during the centuries of Moorish reign.

Another circumstance that favored Montpellier was the large Jewish population in the Midi. Even while suffering persecution from both the Christian and the Arabic Spaniards, they had held on because their commercial activities were successful. Among them were many well-educated physicians who knew Arabic and whose medical education had been much better than that in the north, where Latin translations of the Greek and Arabic books were lacking until the 12th C.[65]

Many documents attest to the large role played by the Jewish physicians in southern France during the Middle Ages. That role continued under the Popes at Avignon who protected their own Jewish physicians. Although the edicts of expulsion forbade Christians from dealings with Jewish physicians (Council of Aix, 1338), the actual times of the separations were not long, because their influence persisted, supported by their greater knowledge and continued availability as practitioners. The princes and the towns sought their services.

M. Bayle, (*The Physicians of Aragon During The Middle Ages*) describes their important function in the towns. M. Vidal of Perpignan in his equally interesting book[66] devoted an entire chapter to the Jewish physicians in the larger cities, and the large number of books in their collections. Many of them came from Spain after having studied at Cordova. Vidal cites an edict of King Ferdinand I of Aragon (23 July 1415): "No Jewish man or woman may practice medicine or surgery or be an apothecary for a Christian."

Many medical mss in Hebrew still exist.[67]

During the last years of the Middle Ages the level of intellectual activities in the Midi was higher than in the rest of France. Vidal lists the titles of the medical books found in the collections of the 14th C, and the list is long. Most of the items are Latin and Catalan translations; the Catalan language was the common tongue in Catalonia, Roussillon, and Majorca. He found translations of Guy, Lanfranchi, Theodoric, Roger, The Four Salernitan Masters, and William of Saliceto—all in Catalan. The large number show us how well Medicine and Surgery were cultivated, not only by the Jews, but by all the local savants, and Montpellier stands out clearly at the center of the group.

We should add that the Papacy at Avignon attracted to the Midi and to Montpellier many foreigners, including savants, and the Popes looked on the University at Montpellier with favor during later centuries (after 1309).

---

[65] Renan wrote: *French Rabbis in the 14th C* in *The Literary History of France,* Vol 27, pp 431-753 (EN).

[66] P. Vidal, 1888 *The Jews in the Ancient Counties of Roussillon and Cerdagne*, Paris. p. 60 (EN).

[67] The Bibliotheque Nationale at Paris owns one hundred-ten volumes of medieval Hebrew medical mss; books by Italian and French physicians and surgeons. The original Hebrew was by Jewish physicians. Many of the translations were from Latin and Arabic. None were directly from Greek. Many of the mss were written at Beziers, Carcassone, Lunel, et al. (EN).

From the 12th C on, the Medical education at Montpellier had a fine reputation.[68] and a single Faculty did not monopolize the teaching; the special schools taught freely. Each Master had his own paying students. In January, 1180 Count William VIII of Montpellier granted them the rights to teach Medicine to local citizens and to foreigners.

By 1220 the numbers of teachers had increased, and as a group they could define the scope and the duties of both sides of the 'contract'.[69] In 1220 Cardinal Conrad issued the statutes of the free schools and placed them under the jurisdiction of the Bishop. I quote, "For many years the science and the practice of Medicine have shed shining lights and have flourished under the banner of Montpellier, and have spread throughout the world the healthy abundance and the invigorating yields of its fruits."

The statutes were not meant for a Faculty; they permitted the special schools to come together as an association, a sort of University with a single set of rules for all. That charter—for want of a better term—was confirmed in 1239 by a Legate of Pope Gregory IX, and in 1258 by Pope Alexander VI himself. That grant endures.

The group of free schools were called The University of Medicine and it could award three degrees: bachelor, licensee (ie the right to practice), and Master. However, one could practice medicine if he had only the first and second degrees. Every town had its own special medical arrangements, and it could award the license after an examination, which we will describe later. A physician who left the university with a bachelor's degree or a license had certain advantages over others in obtaining the necessary privileges. The few who had become Masters became the physicians for Popes, Kings, Princes, and dignitaries of the Church; they could practice anywhere.

The rank of Bachelor at the University of Montpellier came after three and one half years of course work there or at another acceptable university. A student could be awarded the bachelor's diploma after two and one-half years if he was certified by a Master at Montpellier or if he already was a Master of Arts at the University of Paris and had some practical experience as a physician for at least six months in a town, Montpellier or nearby. Furthermore, the candidates had to be able to 'read' the required texts. That was required of every Bachelor who had to read and discourse on a list of authors designated by the authorities before obtaining the required practical experience in Montpellier or nearby. That meant that he had to apprentice himself to a practicing physician in town.

A bachelor who sought a license was presented (ie certified as ready) by his Master for an examination, as he had done before his baccalaureate, and after he had read an

---

[68]   Jaffe's doctoral dissertation (1853) cites a text from Montpellier, in 1337 (see *Rev. Therap. du Midi,* Montpellier, 1885): "One of Saint Bernard's letters mentions an Archbishop of Lyon who fell ill at Saint-Gilles en route to Rome in 1153. He returned to Montpellier where he consulted physicians as to what was his diagnosis." (EN).

[69]   See Germain, *Cartulary of the University of Montpellier,* Vol. I, p 18, 1890. His book was issued as part of the celebration of the Sixth Centennial of the University. In it I have found most of the documents that I have used here (EN).

additional book of medical theory and one on practice while in the school of one of the Masters of the University of Medicine (ie during his apprenticeship).

By 1309 the applicant for a license must have completed six years of study. He had to agree not to leave Montpellier and to work toward the degree of Master by studying for at least two years, unless receiving special permission from the Bishop of Maquelonne.

The constitution for the University as awarded by Nicolas IV on October 16, 1289 restricted the award of degrees to the Masters of only one school, and that became the Faculty of Medicine. Although it removed the privilege from the 'free' schools, they could continue to teach. That is how the Faculty of Medicine became the University of Montpellier, and was called the School Of Montpellier.[70]

What was taught? The oldest document, a Bull of Pope Clement V, dated 8 September 1309 (already at Avignon), states that: "he had consulted his physicians and chaplains—William of Prixia and Jean of Alais as well as the physician, Arnold of Villanova. He then decreed that every bachelor who wished to become a Master in the Faculty in The School of Mantpellier must study these books: By Galen: *On Complexioms, On Crises and Critical Days, On Illnesses and Complications, On A Healthy Way-of-Life;* and books by Avicenna, Rhazes, Constantine, and Isaac. Then he had to expound on the two books of *Commentaries* by Hippocrates and Galen, and on the *Techni;* on Hippocrates' *The Prognostics and the Aphorisms,* and *Fevers*; on Joannitius' *Isogoge;* on *The Fevers* by Isaac; and on the *Antidotary* of Rhazes.

The statutes of the Faculty of Medicine were reviewed and edited in 1340 based on the Bull of 1309. The Masters must have read: Vol. I of *The Canon, Illnesses and Complications, Crises and Critical Days, Complexions of the Maladies, Simple Medications, Complexions, Aphorisms, Regimen for Acute Diseases, The Prognostics, On Joints, On Internal Organs, The Book of Dispositions (Life-Styles), Ad Glaucon, The Four Salernitan Masters,* from the *The Four Canons* (either I or II), *The Pulses* by Joaninitius, *The Urine* by Theophilus, *and the Tegni and Prognostics, The Regimen for Acute Maladies,* and *The Regimen for Good Health* (ie Of Salerno), and *The Natural Virtues.*"

The Masters also could read Isaac's *Fevers, and Universal Diets, Parts I and II of the Fourth Canon, or The Third Canon,* or other books by Galen.

In regard to Anatomy: The statutes of 1340 order the Chancellor to provide a complete dissection every two years. In 1376 Duke Louis of Anjoiu authorized an operation on a cadaver of a criminal. The dissections at Montpellier antedated those at Paris.

Certainly, Surgery was taught elsewhere than at the Faculty, by the free surgeons, as it had been taught earlier at the free schools. While working his way through the grades leading to Master In Medicine at Montpellier, Guy also studied Surgery, perhaps with one of the free surgeons. Even as late as his era, some Masters of medicine continued to study surgery. Raymond of Molieres was a Master in Medicine attached to the University and he openly taught Surgery. And there were still a few successful Jewish surgeons at Montpellier.

---

[70]   Therefore, the University of Montpellier was founded on only one Faculty (LDR).

However, that official tolerance of surgeons by the University did not last. When the Popes left Avignon, an interdict forbade members of the University from practicing surgery; that was fifty years after a similar injunction had been issued at Paris.

An official establishment for teaching surgery had been mentioned in a letter from Charles V. (3 June 1399), but it did not take effect until 1597. According to Symphorien Champier the Faculty opened a course of Surgery limited to barbers, controlled by a chancellor. The first professors were Griffin and Jean Falcon. The University did not allow reading of texts in French[71] and the students knew little if any Latin. The commentary was in a barbaric half-latin-half-french. We have a copy of such a commentary given by Falcon (Vol. 28, etc.). A permit of 31 March 1490, cited by Germain, granted the candidates their rights, and named them 'barber-surgeons'.

Henry IV in 1597 created a chair at Montpellier for a professor to teach Surgery and Pharmacy to students enrolled at the University.

To sum up: In the 14th C the Faculty of medicine at Montpellier recognized no special degree of Master of Surgery. There were practicing surgeons and teachers, including some clerics and Masters of Medicine, as was Guy. But the latter group disappeared after the interdiction of the Faculty. What remained were the laics, including the Jews who were educated in the Arabic lore.

Also there were barbers who, as noted by Guy, had formed their own corporation in the 14th C. Some say that was what they had wanted since 1088, but Malgaigne said that was not the case. According to Germain, the barbers' statutes were not set down until 1252, but it was simply a matter of them using their shaving razors for treating the sick and the wounded. In Guy's epoch the barbers were performing minor procedures that were disdained by the surgeons, such as extracting teeth, etc. And in the 15th C the barbers at Montpellier referred such cases to lowly specialized 'tooth-pullers', supervised by them. As noted, the Faculty of Medicine provided a course in surgery for barbers in 1490, but not until 1597 did they offer a course for their own enrollees.

2. The University of Paris

We are uncertain about the origins of the University at Paris; they do not go back as far as Charlemagne's Royal (Palatine) Schools, nor do they go to the 10th or 11th Cs as was claimed by Jacques Mentel and Gabrielle Naudé. Probably Medicine was one of the subjects taught at the free schools, as at Montpellier, founded by physicians or surgeons, alongside the schools at the monasteries and cathedrals. Daremberg believes that those began early in the Middle Ages. The term 'institution' is an exaggeration; most of the schools (ie that taught Medicine) consisted of a single Master and a few apprentices. Early in the 12th C, according to Hazon[72], the practice of medicine was controlled, and the

---

[71]   The professor read to the assembled students from the book, perhaps the only one in the library, and then discussed the material in the common (vulgar) language (LDR).

[72]   J.A. Hazon, *A History in Praise of The Faculty Of Medicine At Paris*, Paris, 1793 (EN).

practitioners had to meet the standards set by the Masters. In 1332 an official document reiterated what had been the custom for a century.

The Literary History of France (Vol. 24, p. 469) states that there was no formal course for teaching Medicine before 1160. Before that the instruction was by monks and by Jews. A concordat of 23 March 1268 describes how the physicians (ie physici) formed a separate corporation. There was no Faculty of Medicine, just the corporation; many of the members probably had apprentices and associates, as at Montpellier.

It is certain that at Paris at the beginning of the 13th C there were many schools for the study of theology, canon law, arts, etc. In 1215, during the reign of Philip Augustus, Robert de Courçon, the Legate of Pope Innocent III, gave the schools their first legal status, just as Cardinal Conrad would do at Montpellier five years later. Around 1250, the Schools of Paris took the name of University. At first Medicine was included in the Faculty of Arts; it separated in 1280 (Chomel). Ch. Jourdain claims that some documents show that the Faculty of Medicine existed as early as 1261, at least by 1268.

Therefore, a Faculty of Medicine existed during the second half of the 13th C; it was entirely ecclesiastic. Its regents could be widowers but they could not be married. (see Chomel, p.160, and Malgaigne p. 45). Later, in 1452, Cardinal Astouteville disposed of that restriction. Nevertheless, the Faculty remained under the stewardship of the Church until 1595 (Henry IV).

The Faculty conferred Degrees. At the beginning they had no meeting hall of their own and the ceremonies took place in the churches selected by the leader. In 1492 a building was assigned, on the Rue de la Bucherie (ie Butchers' Street).

In 1348 the Faculty gathered to discuss the Great Plague that was invading France. The document that reports that meeting is very interesting (See Doctrine II, Ch. 5, in Guy's text).

In 1352 King John I limited the right to practice Medicine to Masters and Licensees of Paris or of other major schools that had systems of medical instruction similar to that of Paris (Described by Jourdain).[73]

It would unfair to compare the medical education of that epoch with what we have today. The medical Faculty consisted of Masters of Arts chosen from their own group until 1634. Only two of them taught the medical course, and the assignment lasted for two years. One professor taught *The Natural and The Non-natural Things*—anatomy, physiology, hygiene, and dietetics. The other taught *Things Contrary To Nature*—pathology, material medica, and therapeutics. The Master sat on an elevated Professor's Chair and commented, assisted by some bachelors who read aloud from the texts.

In 1634 the Faculty created a course in Latin for the students at the University, twenty-seven years after Montpellier had done the same.

---

[73]   Jourdain *History of The University of Paris Until the 17th and 18th Cs* 1866. see Vol. I (of two) (EN).

   Nicaise quotes in full the document in Vol. I of Jourdain's book, in Latin. The summary above is adequate (LDR).

At first Anatomy was not taught. After the 14th C a cadaver, the victim of hanging, was dissected three to five times a year. The dissections were celebrated events that lasted up to seven days. The scene at Bologna is pictured in the miniature in Ch. 1. of Treatise I in the text. A student had no formal bedside instruction at the University until the end of the 18th C. That experience was provided during his apprenticeship to a physician, a surgeon or a barber-surgeon of Saint-Côme; the apprenticeships antedated the Faculty.

These are the books which were used by the Faculty at Paris in the 13th and 14th Cs.[74]: by Hippocrates: *Aphorisms, Diets, Fevers, Prognostics* by Joannitius: *An abridged version of Galen's Art*; by Philarete:*The Pulses*; by Isaac: *The Viaticum, Fevers, Diets—General and Special, Treatise on Urine*; by Theophilus: *Treatise on Pulse, Urine*; by Gilles de Corbeille: *Treatise on Urine, Different Pulses*. Chomel wrote, "These were the books explained in the readings (ie lectures) to bachelors who were under oath not to pass on what they learned to others, and no others books were to be read or explained. Nothing was changed in 1350."

Much later, additions included the books of Avicenna, Rhazes, Averroés, and Albucasis, and the more recently discovered (translated) works of Hippocrates and Galen, adopted as rare copies became available. That was the total until the epoch of Fernel (1496-1558).

The Commentaries of The Faculty of Medicine of Paris 1395-1786, Vol I. p. 2., describes the contents of the library of the Faculty in 1395. I have taken the list from Corlieu's *The Ancient Faculty Of Paris,*1877. p.148.

1. Owned by Pierre Desvolées: *Abreviationes synonymorum and Treatise on Theriac* by Januensis (Simon); by Averroés: *Colliget, and An Old Edition of Avicenna's Anatomy.*
2. and 3. Avicenna's *Canon* bound with Saint-Amand's *Concordances,* Mesûe the younger's *On Simple Medicines,* Nicolas Myrepsos' *Antidotary.* In another single large tome, many books by Galen.
4. Owned by William Boucherii: Peter de Saint Flour's *Concordances;* Albucasis'*Antidotary*, and the complete *Continens* of Rhazes in two volumes.

We have noted that there was no Surgery in the curriculum of the Faculty of Medicine. However, at the end of the 13th C there were a few surgeon-clerics and Masters of Medicine who solicited Lanfranchi, newly arrived, to offer a course in Surgery. We know that Pitard and Henri de Mondeville attended.[75] But, after a few years, the Faculty at Paris returned

---

[74] Chomel, 1762. *An Historical Essay On French Medicine,* Paris, pp. 117. 124, 150. And Malgaigne, Vol. I. p. 42. (EN).

[75] Another attendee, almost ignored through the centuries, was Jehan Yperman, the first great Flemish surgeon (LDR).

to its primitive intolerance of Surgery, and in 1350 it forbade its bachelors from any manual surgical activity,

Therefore, for a long time, the Faculty's only surgeons were physicians, while two other groups practiced Surgery: the barbers who were everywhere in France, and the surgeons of Saint-Côme who worked only at Paris (ie and Luzarche). In the 14th C their importance was greater than before as they set themselves in opposition during the conflicts between the Faculty, the surgeons and the barbers of the 15th C.[76]

In the 13th C the free surgeons, as granted by the provost of Paris (1254 and 1258) came together as a Confraternity at Saint-Côme. In 1268 they pledged obeisance to a designated official, and in 1311 a charter from Philip II delineated the rights of their Master Surgeons. In 1355 the Corporation had nine members. Their statutes were amended in 1370, and the Corporation could include licensees and bachelors. Yet they were not a school or a college. The first Master of the College of Surgeons was identified by Malgaigne in a Royal Ordinanace of 1533, but actual teaching at the College came much later. The various steps toward the goal are described by Corlieu (*Teaching at The College Of Surgeon From The Beginning Until The Revolution*, Paris, *1890*.[77]

In addition to the surgeons were the barbers. They performed minor surgical procedures and wanted to expand their scope and be independent. That was conferred by Charles V in 1371. On 11 January 1494 the Faculty of medicine, as a thrust against the surgeons at Saint Côme, voted to permit the Faculty (one professor) to teach the barbers in a course of surgery. He read aloud in Latin the pages in Guy's book (ie as required in the University) while others commented in French.

The conflicts between the surgeons, physicians and barbers lasted for four centuries. Finally, the surgeons and the barbers were united—at least for a brief period, The College of Surgeons and the Faculty of Medicine remained apart until all the Universities and associations were suppressed in 1793.

3. Medical Instruction In The 14thC and The Degrees Awarded By The Universities

Before the 13th C there were no general systems for teaching Medicine other than what was provided without much supervision in separate towns to suit the local needs. Early in the Middle Ages the Greek and Roman schools were succeeded by the neo-Latin schools—as was Salerno when it began. When many monasteries and cathedrals established small schools, Charlemagne improved the situation by establishing them as Royal (Palatine) Schools. In

---

[76]   I refer the Reader to Nicaises' edition (1895) of Pierre Franco's *Surgery* for a complete and detailed description of the conflicts. See Bibliography. (LDR)

[77]   Until the 17th C the insignia of the Barbers was a white metal receptacle, to distinguish them from the yellow (gilt) metal of the surgeons. See Brau of St.Pol, Lias, *Rev. Scientific,* 1890, p 143 (EN).

addition there were a few 'schools' taught by lay practitioners. In the 12th C corporations were conceived by the lay practitioners as they had been formed by other artisans (ie guilds) to bring together members of the various crafts. The teachers became known for what they taught, and the municipalities favored them; in that way some important schools had their beginning. Then the Church intervened to obtain the right to award privileges to practice and placed the schools under its jurisdiction. At Montpellier the separate schools had existed from the beginning, and the Church there issued only general regulations. Later on, the Church rejected the right for any of the schools other than one to award privileges, and that one became the Faculty of Medicine, the members of which were nearly all clerics.

The major reformation of the 13th C was the formal organization of Universities under the jurisdiction of the Church, included in which were the Faculties of Medicine.

The degrees previously awarded by the free schools, the corporations and the municipalities now were given only by the Universities. Nevertheless, the lay schools continued to teach and to issue license to practice medicine and surgery.[78] The Faculties did not provide enough practitioners: their course of study was too long, the costs were burdensome, the instruction was entirely in Latin, and Surgery was not part of the curriculum. Soon we shall describe how physicians were recruited other than from the Universities by relating what happened in an important town (ie Avignon).

First let us describe the Medical degrees awarded by the University. The full story was revealed by a long an tedious search; I gathered information from mss, Papal Bulls, and official documents from the Universities of Montpellier and Paris, in order to produce a clear exposition.

What was a University? In the 13th C the word was always followed by a modifier, such as 'University of Masters' (ie teachers), or 'University of Scholars'. In 1220 Cardinal Conrad began a 'University of the Schools of Medicine' at Montpellier. Later on, the second term was implied and the word University was used alone. But the Papal Bulls did not use it; instead we find the terms 'studium' or 'Parisian studium'. In the Bull issued by Pope Clement V, dated 8 September 1309, we read "The Bachelors who wish to become Masters in the Faculty of Medicine of the Studium of Montpellier, must . . . ."

The Titles: 'Medical Universities' were used in Papal Bulls. Those Universities conferred degrees of three ranks: bachelor, licensee, and master—there were no other grades or titles.

The title Doctor of Medicine was not used by the Faculty of Medicine, although in the faculty of Laws a Doctor of Laws was considered to be equal in rank with a Master of Medicine. However, some authors and chancellors in towns, especially in Italy, used the term Doctor instead the more customary 'physicus' or 'medicus'. Later, when Robert Saint Germain was Dean of the Faculty of Medicine at Paris (1413), the term 'Doctor' was adopted to replace 'Master'.

The terms physicus or medicus were used for practicing physicians who had been trained at a University or had been admitted as a member of a corporation or was admitted

---

[78]   The licenses were granted by municipalities. See below (LDR).

to practice by a municipality. The terms were current both in speech and in writing, but were not used in the University, as today we say 'physician' or 'surgeon'.[79] Guy de Chauliac was qualified as a physicus in official documents, Papal Bulls, etc. where he is mentioned.

During the 14th C in the Midi physicians were designated not only as physici but with the terms 'meges' (or melges). Canappe in the translation of the *Prologue* (see Nicaise's bibliography of printed editions of Guy, #37). wrote "Misters Meiges of Montpellier, Bologne, etc". It was a proper term for a physician. Later on it was reserved for charlatans, and even today in Switzerland and other countries it persists as such. Peyrillhe was wrong in demeaning Guy's background, because he had found the term meges used in some of the mss of Guy's book, where it was used instead of medicus or physicus.

There was no university-based title for surgeon in the 14th C. A few, including Guy, were clerics and Masters in Medicine, but that title given to surgical practitioners soon disappeared completely.

We have noted that many physici were not from the universities, and the same is true for the surgeons and members of the corporation at St. Côme at Paris, or of the barbers' corporation.

Also, there were many 'specialists' who treated various medical and surgical maladies; as dentists, otologists (as was Guy), etc. There were specialist 'cutters' for certain operations, as for bladder stones, hernias, etc. Among that crowd were some who had no licenses, whereas others were authorized to practice after passing an examination. Here we will deal with the latter group.

Some valuable documents were found in a folder titled *The Medieval Physicians of Avignon,*[80]. Bayle had great curiosity about the history of his native citty and he has been generous with his assistance in my own researches at Avignon.

Before any of our Universities were founded, the towns took certain measures to assess the qualifications of those who wished to practice medicine. That led Charles II of Anjou, in a letter dated June, 1297 and sent to all the seneschals of the Provence including Avignon, to declare that no physician could practice unless he had passed an examination performed by the city council and the royal court. That body also could take away a privilege.

An article about a regulation set forth in Avignon early in the 15th C by the presiding judge of St. Peine describes the scope of practice for barbers and lay surgeons: Article 103: Barbers and Surgeons: Every barber and surgeon must declare to the office of inquiry of the civil court the extent of his practice: wounds, contusions, fractures bones, torn tissues, etc. that he will be called on to treat on the same day when he is called. In default he will be fined forty pounds, ten of which will go to the accuser. Every barber

---

[79]   The use of the term 'Physic' to mean Medicine has an ancient origin. Physic was part of Aristotle's philosophy. It included more than the study of the body, but also the study of various phenomena of air, earth, and water. Medicine was part of those studies, hence medici were physicians. 'Physic' remains as standard usage today in England (EN).

[80]   G. Baytle; Avignon, 1882. (EN).

and surgeon who has not taken an annual oath not to practice before taking the official oath in the civil court will be fined forty pounds."[81]

Surgery was not taught in the early years after the Universities were founded. The Towns had to continue their own surveillance of the abilities of the practitioners. Here is an example.

On 10 August 1460 an assigned court examined Salomon de Vetri, A Jew, who wished to practice surgery at Avignon, after several years of study (ie apprenticeship). The examiners were Michel Piandi, Master of Arts and Licensee in Medicine, a physician for the legate at Saint Siége in Avignon[82], and William, and Anguithilli, a barber surgeon and Jury Master of the city, chosen for his well known abilities.

On 21 August, in the presence of Master Stephen Posieux, Notary Public; Sir Clavaire of the Civil Court; and certain witnesses; Master Michel, a physician, and Master William, a Master Surgeon, The jury examined Salomon Moses of Vetri in an uninterrupted session regarding the art of Surgery. They reported the examination to the court in writing and all signed it. To Wit: As required by our commission, given by the magnificent Sir Montdragon, Seneschal (ie Viguier) of the city of Avignon, we have examined Salomon Mossé a Jew, on the theory and the practice of Surgery, according to the ancient modern canons, and we affirm that we recognize his sufficient aptitude, especially in the theory, which is the basis for the practice. That is why we declare him capable of performing surgical operations on the human body. Inasmuch as Surgery is the last-ditch treatment in Medicine, as was taught by Joannitius and Avicenna (Section IV of Book I, Ch. 1) we presume that this use of potions and diets according to the precepts will be suited to the maladies in the majority of the cases. It will be his responsibility to operate only when his interventions will be necessary, and he will consult a physician in those decisions."

The report was approved by the Seneschal who gave Master Salomon the authorization to practice the art of Surgery in Avignon and its environs. "The Seneshal, having seen, read, and understood the report of the said Physician Michel and Master Surgeon William, has judged the request of Salomon Moïsé de Vetri to be just, etc. He concedes and awards to the said Mosse the right to practice the art of Surgery in the city of Avignon and environs, and requests that he place his hand on the book of the law of Moses written in Hebrew, and take the oath to practice his art faithfully, legally, and without falsehood or fraud. From what he has requested, the said Mosse de Vetri will be permitted to dress (ie to treat) one or many maladies as named, and authenticated by my Notary Public[83].

---

[81] Box 11, Piece 15, Municipal Archives. Cited by Bayle (EN).

[82] Here is a physician for a Papal Legate who also is a dignitary who was licensed only as a Master of Arts, a valuable title. The Master's degree was sufficient to grant him a license to practice. However, at Montpellier a candidate (ie not a Master of Arts) needed six years of study before obtaining his license (EN).

[83] The applicant's list defined what he believed he was capable of treating. The examiners quizzed him on those topics, The Notary made the list and the proceedings a matter of record (LDR).

Signed below.[84] Made at Avignon in the Palace of the Civil Court, near the Archives of the Treasury of the Court, etc."

Many Papal Bulls decreed the addition of a Faculty of Medicine at the University of Avignon, but nothing came of the matter. First was a Bull from Boniface VIII (1303). On 13 Nov., 1441 the Bishop of Avignon issued a ruling to include some Masters of Medicine. On 11 January 1459 Pius II added three regents to a Faculty of medicine. Despite all, it did not happen, and the city council tried to provide medical instruction by attracting foreign physicians. In 1467 the Council set aside one hundred ecus a year for a newly arrived physician.

In 1480, the Council discussed the issue. "We the Councilors of Avignon humbly offer Master William Imberti, Master of Arts and Bachelor of Medicine at Montpellier as follows: Although we have here a Faculty of Medicine, the study and the practice of the science has been entirely neglected. We offer him, on behalf of the City of Avignon, as a teacher of Medicine and philosophy the following contract: 1. He will assemble a corporation of physicians similar to what was approved in the University of Montpellier. 2. He will enjoy all the privileges of judging as The Head Master the awarding of degrees of Master, Licensee, Bachelor, and Student.

He will teach for one hour daily after two P.M., if the present Dean Master Pierre Rodin and Professor Master Jean Guillermi wish to conduct their readings from eight to nine A.M. Professor Guillermis is here pro tem, coming from Montpellier at the invitation of the Council, to teach for three years, to be paid one hundred florins per year. Professor Imberti was hired with the same terms."

Those documents reveal that there was a municipal medical organization in an important medieval city. Probably that is what happened in other places.

Alongside the 'regular' physicians and surgeons there were practitioners with licenses or university degrees who were allowed to function, as we see in this document cited by Bayle (p.44):

On 12 October, 1444, Pierre de la Thouroye, sergeant at the civil courts and his wife Catherine appeared before Gilles Rastelli, Notary at Avignon, to notarize a contract between them Master Guido Rastelli, a tool maker. Catherine is afflicted by a serious malady in a breast and Master Guido agrees to operate and totally remove it, with the Lord's help. Pierre Thouroye and his wife agree to pay him twelve florins after the treatment is performed, as witnessed by some physicians and other competent persons.[85]

The barbers also practiced surgery at Avignon, as is attested in the acta of notaries (contracts, wills, inventories). Bayle cites the following (Rastelli, again): "In her will of

---

[84]   The Notary was Anton Bonaud at Avignon. See Bayle p.39 (EN).

[85]   Minutes of G. Rastelli for 1441-442, p.46. (EN)

  This contract is cited again in Nicaise's edition of *The Surgery* of Pierre Franco (1895). Nicaise suggested that the notary may have been related to the 'cutter', and that the charlatan was free to act. 9 (See Bibliography (LDR).

July 17, 1452 Dame Léonarde Pachaude, widow of Master Mangin Guarin, a barber and resident of Avignon, in recognition of the services rendered every day during her illness and of the expenses he paid on her behalf, she bequeaths to Master Pierre Theurot, a barber of Chalons-sur-Saon, also living in Avignon, all the instruments used in a barber's shop, including basins, mirrors, pots, wash-basins, sharpening stones and wheels, vials, braziers, boxes, lock-boxes, chairs, razors, scissors, gowns, surgical books, and everything in general that pertains to the work-place of a barber and surgeon."

In addition, there were many 'irregular' medical practitioners in Avignon; a regulation set forth by Pope Gregory XIII (21 November, 1577) lists them, and they probably had been there for a long time. Article IX of the regulation: "No one other than a licensee of another academy or of the Faculty of Medicine at Avignon, shall practice pharmacy, surgery, perfumery, barber-surgery, massage-therapy, dressings (ie wounds) obstetrics. All empirics are banned."[86]

Women Physicians

We should add them to the list because many of them acted as apothecaries, physicians and surgeons. Beaugrand has published a good brief history of the matter.

From the most remote antiquity women were known as healers: the Egyptians and the Jews had mid-wives. In Greece obstetrics and certain medical interventions were consigned to women; they are mentioned in the Hippocratic corpus. In Rome they could be general practitioners; Pliny and Galen mentioned some of them by name. During the Middle Ages women continued as such. Albucasis, writing about bladder stones in women, said. "You must refer them to the trained female physicians. However, there are only a few of them."[87] Avicenna cited women in reference to some ocular maladies.

Women played a considerable role at Salerno. The' matrons' or 'wives' were famous. The earliest mention is by Orderic Vital a Benedictine monk who wrote an ecclesiastical history during the first half of the 12th C. The most famous was Trotula of Roger. Both J. and Mathew Platearius, and others cited the women of Salerno, who treated diseases of both sexes. They prepared medicines (ie their own inventions) that were used by other physicians during that era. Some of the recipes had mystical properties and were used when accompanied by kinds of incantations. However, most of the physicians were not

---

[86] All prior attempts to restrict charlatans had failed. What follows is a copy of a curious document attached to an English ms of Guy's *Surgery*. I found it in our National Library. It was attached to the third blank leaf at the end. It is an ordinance by Henry V (1420), published as an act of Parliament. It was translated as a French MS, but it is not mentioned in the copy of the ms published in *French Manuscripts,* 1840, Vol. III. 1840. (ie an abstract is appended in Old French that describes an act of Parliament to curb the medical and surgical charlatans who were not trained or do not meet the standards of the times.) (EN)

[87] Leclerc, p. 150 (EN).

believers. After Trotula's book, *The Suffering of Women*, appeared, some Arabic women physicians came to practice at Salerno.

The women continued to practice in Italy and some of them were known by name and had fine reputations and were mentioned as late as the 15th C.

There were no such famous women in France, although many of them practiced in the cities, after passing the usual examinations by the Master Juries. An edict of 1311 forbade them from practicing in Paris unless they had been properly examined. An edict of 1352 confirmed the earlier one. Du Bouley[88] reports another edict that voices some complaints from the Faculty at Paris, about the question of female physicians. "There are some women who are practicing surgery without legal authorization, and women without special knowledge participating in that practice."

Guy wrote about women who practiced surgery. They formed the fifth and the last group of the operators of his era. "A sect of women, most of them idiots, who refer their sick patients with all kinds of illnesses to the Saints basing their actions on this prayer. "The Lord has given what he was pleased to give, and the Lord uses me as he pleases. Bless the Lord's Name. Amen."

By the 16th C, as observed by Pasquier, the practice of medicine by women had almost entirely disappeared, and their numbers were rarer as the years passed to reach what we have today (1723). "There still are some savants with special interests in natural sciences and medicine, but few are practitioners."

In our time (ie 1890s) the study of Medicine by women is increasing, especially in some countries (ie other than France).

As to mid-wives: Until the 17th C they assisted women in labor, sometimes simply reading to her. In the 17th C a course of instruction was arranged for them by the Faculty of medicine and The College of Surgeons (ie at St. Côme).

4. Medical Doctrines In The 14th C

A precise description of the doctrines is difficult because the 14th C was a time of change and turmoil and curiosity, when scholasticism no longer reigned supreme. However, many old prejudices prevailed among such leading physicians as Arnold of Villanova and Bernard of Gordon who held on to some mystical concepts: a belief in astrology and talismans, etc. They were led to apply mystical interpretations to observed natural phenomena in the belief of invisible forces—earthly and sidereal. That strange mixture continued what had existed through all the preceding epochs. We can detect that strange mélange of concepts, the reasonable and more advanced side-by-side with the superstitious; and criticism of scholasticism (ie the syllogistic rather than the phenomenological) found in works of physicians and Masters of the school at Montpellier.

---

[88]   *History of the University* Vol. IV, p. 672 (EN).

Medicine lacked a solid foundation; anatomy was cultivated by very few; physiology was reduced to vague notions; and anatomical pathology awaited Benivieni of Florence for its birth in the next century. Nevertheless, Medicine did advance as a result of the study of the ancient authors and by increased attention to the manifestations (ie symptoms and signs) of illness, leading to some amazing hypotheses that would stand the test of time.

A physician and surgeon of the 14th C who wanted to be current with the most advanced concepts had to know more than the old principles; he was better when he was aware of 'arts' other than medical. Guy wrote, when commenting on Galen, that a real physician would be better than the cooks, carpenters, smithies and others who also claimed to be medical practitioners, if he knew something of geometry, astronomy (ie astrology), dialectics, or other fields of knowledge. A physician should know the different branches of Physic (natural sciences) as defined by Aristotle, if he deserves the title of Physician.

The medical doctrines of the century were taken from the Arabs, who in turn had learned them from Galen. The Arabic works by Joannitius (Honein), Avicenna and Averroés were available to most students. Astruc wrote, "Arabic Medicine was in great favor at the school of Montpellier, and equally so in other universities. Ermengault Blasius, a physician for Philip the Fair, wrote commentaries on the works of Averroés and on the Cantica of Avicenna. He was proud of his ability to diagnose an illness simply by studying the appearance of the patient. It was called the Sphenic Art."[89]

Jean de Dondi and Jean de Parma accepted the Arabic doctrines, and Petrarch reproached them vehemently.

Everyone was not equally carried away, Guy among them, as we shall see later when we discuss his guiding principles, which were Hippocratic-Galenic. As I have explained them, they dominated the late Middle Ages and Guy.

The educated physicians in the towns tried to formulate a system of health (ie hygiene) and of disease (therapeutic pathogenesis), and they were dominated by the concepts of humors in general, their alterations, and their influences. From that they deduced their choices of medications. Thus, Jean de Dondi, Petrarch's friend, insisted that there were six matters to be considered at the onset of a disease: the use of salted meats, of salted fish, of raw vegetables, fasting, pure water as a beverage, and fruits.[90] One sees that he was guided by a well-conceived system (ie a Regimen of Health) during the time when two diseases predominated in medical practices: the Plague and leprosy. And even today, when microbiology is our guide, we have nothing better than Dondi's diets in treating the same diseases.

---

[89]   Jean Astruc (1684-1766) was French Physician-Historian. 'Sphenic': Nicaise guessed it derived from 'sphedoni', meaning agile or prompt. I think it comes from the sphenoid bone, having to do with the facies of the patient (LDR).

[90]   In his insistence that diet be the first therapeutic measure, Dondi was Hippocratic (LDR).

Urinoscopy

In retrospect, I find nothing more crude than the empirical use of urinoscopy as a guide. That measure also was exploited by the medical charlatans, even as it continues to our time. The ancients had deemed it important, and by the 14th C diagnosis by urinoscopy was universal, as we illustrate in the following document. The town of Collioure had a hospital as early as the 13th C, and in 1372 had hired a physician, Albert del Puig, with a contract that he examine patients according to standard medical practices of the time, specifically by urinoscopy.

Although Quesnay was not very through in his selection of documents for his *Critical and Historical Review of The Origins Of French Surgery*, he did cite an Act of Henry II, an order that physicians taste the excrement of their patients. That was satirized by Moliere, who had a Minister Of Justice edit the order (ie on stage).[91]

Astrology

Astrology was important in the 14th C: the art of predicting the future by studying the stars.[92] It was accepted by the common folk as well as by the lords, kings, and educated persons, and was authorized even by Thomas Aquinas. Gerson wrote, "This is a true science that has degenerated and should be restored. Everyone must bow, more or less, to the influences of the constellations of the zodiac, to the Egyptian days and to happy or unhappy days." See Gerson, pp. 559 and 566.

The Faculty at Montpellier taught astrology. Gordon (*Lilium Medicinae*) advised physicians that the alignment of the planets be consulted when treating every patient. They should keep an accurate calendar of the lunations and conjunctions, and they should know the aspects and the complexions of the stars, and they should tell their patients what the stars are saying. That information will determine when to bleed.

Gordon wrote, "The hour of a person's conception will determine his shape and form as influenced by the constellations and by the entire universe. That influence is dominant from the moment of his birth and controls the conduct of his entire life." Gordon also claimed that chronic illnesses were governed by the sun's course, and that febrile illnesses were governed by that of the moon.

---

[91] Quesnay, p.57:"Nothing is more singular than the regulation of Henry II. His edict #209 dealing with the pleas of the survivors of patients who had died as the result of the failures of their physicians. He ordered that they be considered homicides, and that the mercenary physicians be forced to taste the excrements of their deceased patients and to give sympathy as required. If they do not comply, they will be held guilty (ie of murder) (EN).

[92] See Nicaise's Bibliography. The French Ms No. 1357, written in the 15th C by several scribes. It describes a collection of documents about medical astrologers, made by Symon of Pharos, during the reign of Charles VIII, relating to the 11th through the 15th Cs, until 1494 (EN).

Guy admitted to some astral effects among the causes of the Plague of 1348. He tells the Reader of his treatise on astrology that he, too, should be a bit of an astrologer (Bk. VII, Section VI). But he was far less intense than the avid astrologers, Arnold and Gordon.

Witchcraft

As to *Sorcery* (ie witchcraft)[93] which claimed so many victims in Europe during the 16th and 17th Cs, and *Magical* charms[94] and *Incantations*[95] he was not at all convinced of their influence. However, just to keep things in balance, he remembered the advice of Hermes to wear a belt of seal-skin or sea-lion skin on which was engraved a gilt image of a lion, to be worn when the sun was in the sign of the lion and the moon is not in opposition with Saturn, and is leaving it. It seems that those belts were much favored in Avignon where they were taxed as haberdashery.

Gilbert the Englishman recommended for the cure of sterility and impotence that a man wear a parchment strung to his neck on which was written, with the juice of the large consolida, the following: "The Lord has spoken. Swell up (UTHOTH), and multiply (TABECHAY), and populate the Earth (AMATH).[96]

Gordon recommended repeating the following, three times, into the ear of an epileptic. "Gaspard brought myrrh. Melchior brought frankincense, and Balthazar brought gold. Carry these three things in the name of the Lord. He will cure you. Let Christ have pity on a fallen man.[97]

Let those examples illustrate the expanded status of astrology and the faith in talismans. I shall end this discourse and bow in appreciation to E. Begin, a first-rate military surgeon, for his study of Guy's use of astrology. The following passage is taken from an unpublished work by Paré; I am indebted for my copy to Baron Larrey. "The influence of charms and

---

[93]   The sorcerers laid claim to a pact with the Devil, and to their control over his evil doings, leading to a 'witches sabbath' (EN).

[94]   Magic claimed to create effects contrary to Nature (EN)

[95]   Incantations were medical prayers such as "The Lord makes you sick (cross yourself); The Lord determines your fate (cross) and finally he takes you away (cross) and removes you from the world of the living (cross). In the Names of The Father, The Son and The Holy Spirit, Amen). Saint Nichasius, Martyr, had tumors in his neck and body, and he implored our Lord Jesus Christ to take them away (cross four times), Holy Mary (cross), Saint John (cross), Saint Sebastian (cross,) Saint Roch (cross), Saint Blasius,(cross), Saint Catherine (cross), and all the Saints as well as our Holy Lord to intercede for me with the Lord Jesus that the tumors will do me no harm. Amen (cross). Christ reigns (cross)." From Ms 1076, folio 56-58, in the library at Lyon. Cited by Guigue(EN.)

[96]   Are these the divine message to Noah's family to 'Go Forth, Multiply and Fill the world ? Genesis, Ch. 8, Verse 8 (LDR).

[97]   Grand mal epilepsy was called The Falling Sickness since the time of Hippocrates (LDR).

incantations seemed to be inadmissible, but he did not formally reject the use of astrology. Indeed, you can see that he found a relation between the celestial motions and what happens on earth; the living spirit can sense it. Yet even in setting aside most superstitious beliefs, a judicious person, as was Guy de Chauliac, could find something useful in it."

Alchemy

Aside from astrology, magic and witchcraft we must question how alchemy was practiced by such savants as Arnold of Villanova, as well as by large numbers of ignoramuses who claimed to be alchemists, who exploited the credulity of the masses. Philip the Fair called them charlatans in an edict of 1311, and much of Alchemy hung on far into the 15th C.

Alchemy was the chemistry of ancient and medieval times. Instead of being applied to an investigation of the body, it sought to transmute metals as a way to prolong life. It was mixture of astrology and magic. The work of the alchemists, however, led to some discoveries.

King Siphoas II, the second Tut of Egypt credited the art to Hermes (the Greek Mercury) a God who lived nineteen centuries before Christ; the practitioners were called Hermetic. By the middle of the 7th C their numbers had increased. The Philosophers' Stone was first mentioned by them in the 12th C (see Chereau).

Alchemy was taught at the School of Alexandria until the Greeks lost interest. It was soon restored by Arabs who applied themselves, and gave it the name *Alchemy* meaning the best of chemistry. The books were translated into common languages at the behest of Gerbert, who became Pope Sylvester II (930-1003). It found its way into the West in the Arabic texts, having expanded with the passage of time. In searching for chimeras, the alchemists came upon some interesting new substances.

Géber (8th C) invented the Strong Water (nitric acid), Aqua Regia, the infernal stone, corrosive sublimate, etc. Rhazes discovered eau de vie in the 9th C (later attributed to Arnold of Villanova), and invented several medicines based on it. He also invented orpiment, realgar, and borax. Albertus Magnus (13th C) made caustic potash (lye) and quick-lime. Roger Bacon commented on the role of air in combustion. And I will add the name of Raymond Lull to other alchemists of the 14th C.

Chereau lists the principal alchemical writings after the 8th C in the *Dict. Encycl. des Sc. Med.)* Vol II, 1865, p.568.

5. The Practice Of Surgery In The 14th C

Let me repeat what Guy wrote concerning the surgeon, and what is required of him before he undertakes operations (see Chapter 1).

"There are four requirements: First, he should be educated. Second, he must be deft. Third, he should be ingenious. Fourth, he must be indulgent (obliging). In the first, he should know more than just the principles of Surgery learned from books; he also must know Medical theory and practice.

"As to Theory, the educated surgeon (ie a Master in the University or a Master from the College at St Côme) should know about the Natural and the Non-Natural Things and what are Contrary to nature. Among the Natural, *Anatomy* is foremost, as I will show later on. He should understand *Complexions* and how that varies according to the nature of the body, and how that leads to the choice of proper medicines and the assessment of the patient's innate vigor. Among the Non-Natural things (ie hygiene) he should know about air, food, and beverages, etc, because they are fundamental in health and disease. He should know what things are Contrary to Nature (ie pathology): to recognize a disease and what special actions should be taken to cure it. If one does not know its causes, he will not know what to do, and simply act by guessing.

"In the practice of Medicine he should know how to adjust the patient's habits (ie Diet and Life-Style) and to prescribe medicines. So much for the Art of Medicine.

"As to an educated surgeon's manual skills, he must have learned by watching other surgeons at work (ie a long apprenticeship).

"As to his ingenuity, I refer to good judgment and a good memory.

"As to the fourth requirement, I mean that he should be bold only when he is sure of what he does, and be wary of the risks of them that may lead him to refuse to engage in the harmful treatments and practices. (see fn. below). He should be chaste and sober and be amiable. He should be circumspect with his predictions (ie assurances of success). He should not be greedy or extortionate for his money; his fees should be moderate, set according to the amount of work entailed, the seriousness of the illness and the operation, and the ability of the patient to pay."

Any further comment from me will be to no avail. I can add nothing to what Guy wrote about the need for the surgeon to know the medical concepts of general pathology and pathogenesis in order to successfully treat medically before applying the surgeon's skills. However, I cannot avoid his comments about 'harmful treatments'[98] The physician should tell the patient when he cannot cure him medically, and that he, as a surgeon, may be able to salvage him. Guy's mention of 'bad treatments', which were called such in the works of medical authors, who openly recommended patients to avoid operations. That statement probably is a concession to a fearful public whose opinions were labile and suspicious of all surgery, who were aroused by the ignorant physicians and charlatans who outnumbered the few who were well trained. Guy demands a surgeon's devotion to his task, as was demonstrated by his work during the Plague, when most of the physicians fled from their duties. He insisted that a surgeon be sympathetic and merciful.[99]

---

[98] The Bad (harmful) treatments refer to all surgical operations, resorted to only after diets and medications failed (LDR).

[99] The Plague of 1348 was followed by several recurrences when the physicians abandoned their stations. The towns recruited physicians to care for their residents; their contracts required them to remain in the towns during episodes of the plague (EN).

We add that the physicians' situations sometimes were dangerous. Jean of Amand, physician and barber for Pope John XXII (1316-24) was accused of a plot to kill the Pope by witchery and poisons, of having devised small wax figurines on which he cast spells; he was burned alive. The same Pope immolated as a witch Francisco de Stabili (also called Cecco of Ascoli) at Florence. He was a long-time retainer who was accused by Dino del Garbo, another Papal physician. John XXII believed in astrology, sorcerers and sorcery. He had some interest in Medicine, and he wrote a Thesaurus Of Poverty, published at Lyon in 1525.

King John of Bohemia, for whom Guy wrote a small work on cataracts, had a French physician tied up in a sack and cast into the Oder River because his guaranteed cure of the King's ophthalmia failed.

Such events partly explain the fears of surgeons for the consequences of a patient dying after an operation. On the other hand, Malgaigne described the precautions taken by Roland(ie of Parma) when he was faced with operating on a patient with a herniated lung, after other famous surgeons of Bologna turned away. Roland first asked the Bishop for his permission; then he obtained a formal (ie witnessed) consent from the patient, and from his feudal lord and from about thirty of his friends who would attend the procedure.

That explains the existence and the success of the nomadic operators, the 'runners' who, said Guy, fled from the scene after an operation at the first signs of failure or of impending death. Mingelousaulx described a nomadic lithotomist of Bordeaux with similar actions. (Ming. Vol. II, p. 739).

Furthermore, the lack of anatomic knowledge, the fear of hemorrhage, the failure to ligate the main vessels during amputations even though ligation of arteries during treatment for wounds had been performed for a long time—all explain the hesitation of surgeons.

We can see analogous behavior in our own times. During the last thirty years, only the surgeons at major centers dared to perform important operations that were attended with high rates of mortality. Others, lacking the authority that could support their efforts and its consequences, turned away from such 'bad treatments'. However, today, as the risks are decreasing, thanks to antisepsis, surgery is becoming less centralized, and the 'bad treatments' are being performed.

Nevertheless, according to Petrarch, Guy, whose reputation and authority were considerable, he who had been part of the medical staff for three Popes, Clement VI, Innocent VI and Urban V—those physicians had great influence—performed many of the operations that he described in his book.

Guy divided Surgical Doctrines of the 14th C into two large groups: those of the logicians (the Dogmatists) who were Galenists, as was Guy, who were educated in both Medicine and Surgery; and the Empirics who had no use for theories to explain and to search for causes; they applied themselves only to what they observed on the surface. Guy called them 'mechanics' and distinguished them from real physicians and surgeons. The first of the latter were Roger and Roland and the Four Salernitan Masters.

As we have noted, Surgery was practiced also by various kinds of workers whom Guy divided into five types. The first treated all wounds with cataplasms, and tried to induce suppuration. That group included Roger, Roland and The Four Salernitans.

The second group tried to keep wounds dry (ie free of pus) and they dressed wounds with wine, as did Bruno and Theodoric. They were the antiseptic surgeons of their epoch.

The third group used measures half-way between the two above. Their dressings used ointments and mild plasters.

The remaining two sects kept their distance from the others. The fourth belonged to the Teutonic Knights, to armies and fighters. They used a mixed brew—oils, lanolin, and cabbage leaves, and they dressed all wounds believing that God favored good prayers, herbs and holy stones.

The fifth group consisted of women, and, according to Guy, of idiots who placed their patients in the hands of titular saints, basing their doctrine on the concept that "The Lord has given what he pleases (ie the disease) and He will use me when He pleases. Bless the Name of The Lord, Amen.

Indeed, in the hands of all of the above, surgery was active in the 14th C, when there were plenty of war-wounds to treat. The results of the treatments by the second group were excellent (ie as compared with the others). They used pure wine as a kind of antiseptic, and they heated it with herbs and balsams, and they alternated that with eau-de-vie (ie brandy) and brine. They used rain-water to wash sick eyes, sometimes after boiling it. Guy's chapters on wounds suggest that he aimed for healing by first intention.

When suppuration occurred, he tried to get rid of it before it stagnated, sometimes using counter incisions with solid drains or hollow tubes. Surgeons operated on some tumors of the breast, on scrofules and on sebaceous cysts.

Guy seldom amputated except for gangrene; a policy that endured for two centuries until the era of Paré. Malgaigne wrote (Vol. II, p.232), "Amputation at the thigh frightened surgeons in Paré's time as it did his predecessors, and the same was true for amputations in the arm. In his works about the history of the great surgeons, Dalechamps said that they did not go beyond what Albucasis had taught: in cases of gangrene they amputated through the joints of the fingers or wrist or elbow, or the toes and ankle or knee. When the gangrene extended beyond those joints into the thigh or arm, he deemed it fatal. Albucasis also cited the ancients who amputated in continuity, but distal to the elbow or knee. He proscribed disarticulating the ankle or dividing the lower leg, and he doubted that it had ever been practiced."

Fractures and dislocations often were referred to 'dressers' during later centuries when surgeons refused to treat them. Guy disagreed with that practice because it put the patients at risk. He favored treating eyes, teeth, hernias, hydrops (ie tapping the ascitic belly), bladder-stones, all of which had been relinquished to 'specialist cutters'. However he referred phlebotomy. He disdained the 'runners' and he refused to turn away from extracting teeth.

The foregoing shows how important surgery was in the hands of the few skilled surgeons in an era when anatomy was deficient and when ligation of the great vessels during amputations was not practiced. I will not delve into surgery as practiced by the charlatans.

I have discovered little information about surgeons' fees. An article by Bayle refers to a record of Dulceline de Sade. She was a sick woman of that family who was treated in1348 by three physicians, of whom two were Jews and the third was a Christian. Each was paid a half-florin per visit. During the reign of Charles VI (1380-1422) a tournois pound equaled one florin plus sixteen pontifical sous, then current at Avignon. That corresponds in today's France to twenty-seven francs and three centimes. A half-florin, or twelve sous, would equal eight francs and seventeen sous today.

6. The Pharmacy And Materia Medica Of The 14th C

Guy's profuse list of medicines numbered about seven hundred-fifty items. In the chapter that describes their degrees he lists the three hundred that he used most frequently, and he describes their qualities according to Galen, Serapion, and Avicenna, as well as his own assessments. The other four hundred-ninety are mentioned here or there in the text.

When I asked Professor Saint-Leger to comment on the botany of the 14th C, he replied, "My summary comment is simple: there was no botany, just a list of medical substances. Botany as a discrete science did not exist after Dioscorides. In ancient times, Aristotle and his pupil Theophrastus were alone in the study of animals and plants, from which they drew some useful (ie medical) applications.

"The materia medica gathered by Galen and Dioscorides were the objects of special studies by Arabic physicians, especially Serapion, Mesûe and Isaac in Amram.

"That heritage in turn was collected by the Masters of the second period of the School of Salerno: Constantine, Platearius, and Matthew Sylvaticus. In that transition, their Materia Medica lost its formerly exclusive oriental character and a large number of remedies was added from plants that grew in Italy. That tendency was even more important as the instruction of Salerno spread across Europe. I can safely say that in the 14th C the Materia Medica was what was taught by those three authors of Salerno."[100]

The medicines used by Guy differed little from those used by Galen, Serapion, and Avicenna.

In my footnotes and Glossary I have placed the botanical names of all the substances used by Guy, and I repeat the Latin terms used by Guy, and the French names used in my translation. Prof. Saint-Leger has helped me in that onerous task. His vast knowledge increase the value of the notes; they can be useful when applied in other medieval medical works.

During the 14th C the physicians were busy preparing medicinals, and they wrote a large number of antidotaries, and they were added to all important treatises; Guy's appears as Treatise VII in his *Surgery*. Guy insisted that it is important for physicians, especially for those who were surgeons, to know how to create and compose remedies, and how to administer them to patients (ie frequency, dosage, etc.) Often he must attend patients in remote places where there are no apothecaries, or the one who is there may not know how to prepare what the surgeon needs.

---

[100] Dr. Saint-Leger is the author of an article in which he emphasized the influence of Salerno: *Research On The Ancient Herbaria*; Paris 1886 (EN).

We know that apothecaries who could prepare medicines were available only in the larger towns. Frequently the physician had to prepare and administer his own. The frequently used medications could be prepared in advance and be stored, such as the clays, and the pills and the troches that were made from medicinal powders and kept as solids, or could be crumbled or easily dissolved, such as gums, bread crumbs, et al.

We lack information about the arrangements in a medical pharmacy. The so-called official (ie standard) preparations also were sold in grocery shops (ie spiceries), as is shown in a statute of Avignon (No.1242, Art. 130) which says that spice-sellers should be independent of the physicians. An edict of the synod of Avignon (15 April 1341) allows Christians to purchase medicines from Jewish apothecaries and spice-dealers.

Early in the 14th C we still find spice-dealers acting as apothecaries and preparing medicines at Avignon. That led to a ruling by the Seneschal (Art. 19) forbidding spice-dealers from preparing medicines with fraudulent medicines and dosages.

At first the pharmacists sold only the preparations prepared in advance and had names standardized by long commercial usage, with formulas found in books. That is what gave them the name 'apothecary', from αποτηχη, meaning storehouse (magazine or depot).

Even in the 14th C their attributes were not clearly defined—that awaited the 16th C, according to Graves. He said that the apothecary had a longstanding relationship with spice-dealers and aromatists. "Certain merchants, especially in the Orient and in Italy, monopolized the sale of spices, drugs, confection, and many other popular compounds. But by then there were some apothecaries who prepared medicines as ordered by physicians." Guy's text is evidence for that, as is shown in the miniature at the beginning of his Antidotary.

Jean de Jandun in 1323 (*Treatise on The Pleasures of Paris*) wrote, "The apothecaries who prepare medicines and who make the endless varieties of aromatic spices are located in the vicinity of the famous Petit Pont and in other busy places. They are furnished with beautiful jars that contain the most sought-for remedies."[101]

The statutes of the University of Montpellier (1340) say that unlicensed apothecaries cannot sell laxative medicines unless ordered by Masters (of Medicine) of the University.[102]

The medicines of the 14th C contained many subtances brought from the Orient on Venetian ships which had a monopoly for carrying cargoes between the East and Europe. Grave said, "Venice easily introduced all the medicines from its markets and its immense storehouse. A flotilla left the Arsenal every year and carried afar the products it had assembled; they landed in Africa, Spain, Netherlands, and England, and every ship was loaded with spices, drugs, and aromatic substances. That traffic continued until the discovery of the New World."

During Guy's era many of those substances came to Avignon where the Popes were surrounded by the crowds of people who were engaged in religious rites. The salt-tax

---

[101]   Graves, *Paris and Its Historians of The 14th and 15th Cs*, 1867, p.43. (EN).

[102]   Laxative potions were among the very few oral medications prescribed by surgeons without having to consult a physician (LDR).

records at Avignon for September 1397, under the Category *Spice Shops,* list one hundred-forty-five medicinal substances to be taxed as imports. The Glossary contains a summary of that tariff.[103]

The medicinal substances in the 14th C were seldom used singly; just as it was for Galen, the Arabs, and the Salernitan. The compounds were complex, as is seen in Guy's recipes. The complicated formulas of the Arabs often contained foul matter, and Guy often acceded to that potent polypharmacy, although the formulas in his times were somewhat less complicated than the Arabic originals. Yet everywhere superstition and ignorance attributed curative properties to the unusual and fetid substances prescribed by the Arabs. In the magic of the Middle Ages the reptiles and horrid animals, the philters, the disgusting recipes, and the bizarre ancient formulations played an enduring role. The ignorant patient believed that his physician's prescription was impotent unless it contained some of those items, and the physician had to comply with the feelings of the masses, and he added them to his own recipes to assure that they would be accepted, and in so doing gain the confidence of his patient.[104] Even today one may find similar substances in prescriptions used in China.[105]

---

[103] In The Glossary that I have omitted from this Edition, Nicaises lists over three hundred-sixty medicines and spices found in a tax-book for Avignon in 1397 written in Catalan. All the hundred-forty-five medicinals in the list are in the Antidotary (LDR).

[104] A table of equivalents for measures and weights is placed in Treatise VII (LDR)

[105] Dr. Bilance, a physician at Shanghai, reported this in the *Semaine Medicale* of Paris 21 May 1890. Here is a brief extract.

"Sir Robert Hart, Inspector-General of Chinese Customs, has published *A List Of Chinese Medicines* (493 pages), It includes all the medicines used by the Chinese. They have many of our medicinal plants such as aconite, gentian, artemisia, datura, malva, et al. They also have a large number of strange and repellant drugs: dried silkworms, scorpions, millipedes, the scales of cicadas; kidneys and penises of seals, donkeys, dogs, and deer; tiger bones and teeth, human excrements; snake-skin; cow and donkey dung; dried placentas; the droppings of cicadas, rabbits, and goats; bear bile. The collagen of donkey is a kind of glue used as a tonic, made from evaporating water from a spring in the Tung-O district of Shantung Province. The animals' hides are macerated in the water before it is boiled down (EN).

PART IV

BIOGRAPHY OF GUY DE CHAULIAC

A lthough many authors have searched for more, what little we know about the life of Guy de Chauliac comes from bits in his *Major Surgery.*

Gobet was the archivist of the Count of The Provence, and he was a corresponding member of the Royal Academy of Sciences at Toulouse. He addressed the Academy in 1783 with a paper titled *Research on The Life and Works of Guy de Chauliac.* Peyrilhe studied it and Chereau owned a copy that I have not been able to find. According to Peyrilhe at contained little more than what was already known.

Cellarier (1886) published an interesting article, *An Introduction To The Study of Guy de Chauliac.* It is a study of the man and his work, but does not provide previously unpublished material about his life.

We quote (see my Bibliography) a brochure by Moulin about the conference at Follin.

The historians have called him Guido de Cauliaco, whereas his real name was Guigo de Chaulhaco. Most of the documents from the 14th C that I have been able to examine used the name Guigo, and he is called such in all the acta of the chapter at St. Just. (see below, Pieces Justificatives). and in the Bull of Innocent VI, dated 16 April 1353, when he made Guy a Canon of the Church at Reims. Several official items there beg the question. The name Guigo is found in my Introduction, in the description of the Latin Mss of his treatises, numbers 5, 14, 15, 16, and 17. It appears in only one printed edition—now lost—of Nicolas Panis, the conclusion of which begins "here ends the book named Guido, of the practice of surgery by the Master of Surgery named Guigon de Calliac."

The given name Guigo was added to his place of birth, Chaulhac; place-names were the common practice before the use of a family name supplanted it. The village of Chaulhac keeps that spelling even now, and the locals pronounce it 'tchauliaq". How does one write the words phonetically, when etymology as the basis of spelling was not introduced until the 17th C? We can understand why (ie hearing the natives speak it) the name Chaulhac was written in various ways.

Thomas, who published the Bull of 1353, also believed that Guigo was correct, and Guido was used as a popular nickname for Guigo in the Gevaudin, Lyonnais, and Dauphiné. However, Guy became the customary name through later centuries, and we will use it here.

We know little about his early years. The opening chapter in his book tells us that he was born near the border of the Auvergne, in the diocese of Mende, in the hamlet of Chaulhac, which lies on the Gevaudan (ie the stream) on the plateau of Mount Morgerine. It sits at one thousand-fifty meters elevation, sixty-one kilometers from Mende. Chaulhac was and is the chief place of a commune that consists of three hamlets: Chaulhac with one hundred-forty residents, Nozercille with one hundred, and Paladines with fifty. The last-named is the birthplace of General Aurelle of Paladines, a hero in the battle of Coulmiers in 1870. The church (ie at Chaulhac) is Romanesque, restored in 1868. The commune of Chaulhac is a dependency (ie civil) of the canton of Malzieu, in the arrondissement of Marvejols in the Department of the Lozére. The territory consists of gentle hills and valleys. Its people are a sturdy folk, as I can attest, after I visited Guy's birthplace on May 22, 1890. The entire region keeps its old conservative traditions.

In the 14th C, the Parish of Chaulhac was a dependency of the feudal barony of Mercoeur, and Guy remained on good terms with the lords all through his life, as we shall see in a document of 1367. It also confirms Guy's birthplace, and lends consistency to tales about his youth, and explains his good fortune to have those lords as his sponsors. "It was an ancient, powerful, and illustrious feudality in the Auvergne, equal in age and nobility to that of the early royal families. The palace of Mercouer was leveled in 1567 on orders from Charles IX."[106]

The date of his birth remains unknown, but we can safely place it during the last years of the 13th C. In 1325 he was qualified as a Master, indicating that he had to begin his arduous studied when he was at a proper age. In his Introduction (1553) he wrote, "this hard work of my old age." All of it lends credence to my belief that he was born in the 13th C. Peyrilhe and Malgaigne agree.

We have nothing except legends about his youth.[107]. Tradition has it that he began as a simple farm-boy, and that he was known for a certain operation even before the fateful one that changed his fate.[108] The niece of an old gentleman (ie perhaps one of the Mercouers) had been unseated from her horse during a hunt, and had fractured her leg. The local talents could not set it properly, and the family had consulted a witch who told them that she could be cured by a local peasant. It is implied that she meant the son of a local farmer at Chaulhac. He was brought to the patient's castle, and two days later, the nobleman went to church to thank the Holy Virgin, his patron, for the cure. It was said that the family then undertook to educate the young healer. So we may believe that Guy had rendered some service to the Mercouer family, the lords of the region, and that the episode was the beginning of his fortunes.

---

[106]   Bouillet. *Nobiliare d'Auvergne*, Clermont-Ferrand, 1851. Vol. IV, p. 113. (EN).

[107]   L.D. Mirbel, *Biographic Galleries,* and A. Harr *Illustrious Peasants,* 1841, and Moulin *Guy de Chauliac,* 1884. (EN).

[108]   This implies that he had some kind of apprenticeship with lay practitioners, perhaps nomadic, while yet a teenage farm hand (LDR).

We are wrong to assume that he, a completely illiterate peasant was placed in the school of the Cathedral at Mende before he could enroll at Montpellier; the school at Mende was not founded until the 14th C by Urban V. Therefore it seems more likely that he received his basic education at the local church. He became a cleric, a necessary condition for anyone in that era who wished to continue his education.

Did Guy begin his medical education at Toulouse or at Montpellier? On page 163 of his book he wrote, "my Master of Toulouse", and when he listed the surgeons of his epoch, he named Master Nicolas of Toulouse, a Catalan. Perhaps he began as an apprentice to a local (ie Tolosan) physician before he went to Montpellier, or he returned as an apprentice to complete the necessary phase leading to a license. Cellaries believes that it is possible that Guy actually began his medical studies at Toulouse.

At Montpellier his Master was Raymond of Moliere (see Guy's Introduction) who was Chancellor of the University in 1334. Later on he probably was apprenticed to a surgeon in town to complete his training. The Faculty itself did not teach Surgery, and Guy had to learn from one of the 'free' surgeons who had set up schools long before the Universities existed. I suppose that Master Raymond, a Master of Medicine, did not disdain Surgery.

Guy left Montpellier to study at Bologna after 1326, under Master Bertrucius, who succeeded Mundino who died in 1326. Bertrucius taught anatomy, and Guy described the Course (Tr. I, Ch. 1). Cellarier is correct in his disagreement with Malgaigne, whose use of a defective copy of Guy's text led to his debasement of the role of Bertrucius.

In addition to the course in anatomy, Guy probably had teachers other than Bertrucius for Medicine and Surgery. Guy speaks of one of them, but does not name him. He mentioned Albert of Bologna whose course on the *Aphorisms* he had attended. In fact, when Guy was at Bologna, a famous Master Physician, Albert, was mentioned by Boccacio (*Novel X*), "Not many years ago, there was a very great physician in Bologna, famous throughout the world, perhaps he still is alive. His name was Albert."[109] Also, according to Cellarier, Master Albert must have been Albert Zanari, who was mentioned by Antonio Bumaldi (*Writers About Bolognese Things*) who also claimed he had earned the praises of Boccacio.[110]

Albert of Bologna (ca 1270-1348) read and commented on the *Aphorisms* of Hippocrates.

Finally, in his preface Guy cited two operating surgeons at Bologna during his term: Peregrino and Mercadent. Cellerier correctly notes that they were not Masters; rather they were reputable lay surgeons (cutters).

Guy went to Paris after Bologna, judging that he listed chronologically the names of the physicians to whom he dedicated his book. First come those from Montpelllier, second from Bologna, later those from Paris and Avignon. He arrived in Paris after the deaths of Lanfranchi, Pitard and Henri de Mondeville. The last named died beween 1317

---

[109]   Taken from Cellarier, 1556, p. 55. (EN).

[110]   Cellarier, edition of 1849. Later, in the edition of 1856, p. 56, Cellarier wrote: Johann Antonio Bumaldi was pseudonym of Ovidius Montalbinum, almost an anagram. See Muratori, Vol.21, and Johan Garzoni *De Dignitate urbis Bonon.* preface and p. 1162 (EN).

and 1320. He cites no surgeon living at the time when he arrived at Paris, excepting Pierre Argentiére. Cellarier again disagrees with Malgaigne who claimed that Guy attended lectures. But Guy himself wrote that his first stay at Paris was very brief. However, he did establish good relations with the Masters at the University, an important matter, as he describes it in his Introduction. He wrote about it again in the chapter on the plague, where he describes his own electuary, derived from the Masters at Montpellier and Paris. We are reminded that Philip VI's edict of 1348 ordered the physicians of the School of Paris to convene and discuss measures to combat the plague. At other times he describes the surgical practices at Paris and mentions a minor incident that probably occurred during his first visit. He said that a shoemaker shaved off and dug out one of his corns, without permission.

Malgaigne asked whether or not Guy traveled in Germany. He thought yes because he wrote about a Bohemian who awakened his interest in wounds of tendons and nerves, and he mentioned the Germans of Prague who splinted a fractured limb with a mechanical device (glossocome). In his Introduction Guy mentioned the treatment of war-wounds by the Teutonic Knights. However, in the 14th C, Germany was not a scientific center of the sort that would attract foreign physicians. It is likely that Guy obtained his information from some surgeon who was an attendant of Jean of Luxemburg, King of Bohemia. Malgaigne thought that was the case.

What were Guy's degrees (ie titles)? He called himself a Master in Medicine of the University of Montpellier, and that was the highest of all medical degrees obtained after years of study, first having passed the baccalaureate's and the licensee's examinations. He would have lectured (ie reading and commenting) to students. Cellarier assumed that Guy substituted some of the apprenticeship in Montpellier and spent the time as an apprentice to a Master at Toulouse.

I have shown that the titles of physicus and medicus were not university degrees; they were used by municipal offices and by the public to describe both a Master in Medicine or a lay physician. The medical authors used them indifferently, as did Guy. He also used 'treatise of physic' and 'treatise of medicine' when applied to an ancient author. Furthermore, he named someone who practiced surgery indifferently, either as a 'medicus' or a 'chirurgeon'. Yet he made a clear distinction between the barbers and the physicians and surgeons.

Many authors have called Guy a 'doctor of medicine', although that title did not exist in France during the 14th C. The are wrong also in calling him a Professor of Surgery at Montpellier, because Surgery was not taught there as a special subject in that era. He had lectured there on medical subjects before obtaining his license and the degree of Master, as was obligatory, and that was the extent of his teaching at Montpellier.

Peyrilhe (Vol.III of the ms of his *History of Surgery*) stated that Guy simply was a 'mege surgeon' or barber. But in the 14th C in the Languedoc, the term mege denoted a physician, a Master of Medicine, as confirmed by a ms at Montpellier, and the barbers were not called surgeons, as was the case later on at Paris and elsewhere. Their role was much more modest: extracting teeth, performing phlebotomies, et al. Guy never was a barber; he was educated as a cleric, became a Master of Medicine, and practiced Surgery.

In Guy's introductory chapter he states that before Avicenna all surgeons also were physicians; after him it was vanity that led to the separation of surgeons who were demeaned as mechanical handworkers, and as such they included Roger, Roland. et al.

He praised William of Saliceto and Arnold of Villanova for being both, physicians and surgeons, as was Galen. And they did not demean themselves when they prepared their own medications. In that way Guy separated himself from the role of a mechanic, and he aligned himself with the few physicians who also were surgical specialists.

In the capitularies (records of the chapters of monks and clerics at churches) of St. Just, Guy is called a physician (Piece 1, p. 171). And in the Bull of 1553 he is included with the physicians of the Popes Clement VI, Innocent VI, and Urban V, and not with the barbers and the lay surgeons.

What were Guy's activities after he obtained his degrees and completed his studies at several universities? Peyrilhe, who probably got his information from Gobet, said that Guy was less than twenty-five years old when he received his degree in 1325. In that year he was a witness at an obit (death service) at a chapter at Langeac.[111] In the land-rolls at Chaulhac, still extant in 1784 (when Peyrilhe wrote), he was listed as Master Guido of Chauliaco. We lose contact with him in the years between 1325 and 1344.

We pick up the trail in a capitulary of The Church of St. Just at Lyon. On May 17, 1344 he served as a junior canon in the distribution of the estate of Jean de Chatelet, the recently deceased canon and provost who had died two days earlier. The document states that Master Guido Cauliaco will have, as a bachelor in the chapel, fourteen sous and four pence; in that of St. Baldomarin, four sous and four pence; in that of Gresciacin, three sous; a total of twenty sous.

I believe that Guy visited the University of Bologna and Paris after he received his degrees at Montpellier. He wanted to practice surgery, and for that he went the University of Bologna that enjoyed a great reputation for Surgery in the 14th C, where Mundino added fame for his work in anatomy, and was succeeded by Bertrucius.

After he had learned his skills, did he travel from town to town, or did he settle in Lyon as his base for visiting other towns as a surgeon? I believe he did that. In 1344 he participated in the chapter at St. Just in Lyon where he had lived for a long time. I have traced two of his brothers and a nephew in Lyon, where his family came to live. He practiced there quite a while, until he was called to Avignon in 1348 or before. Therefore, we may assume that he had settled in Lyon around 1330. Pernetti said[112] that he was born at Lyon. Of course, he was wrong in that; Guy himself said that he was born in the diocese of Mende.

However, in his introductory chapter Guy wrote that he had worked in many places and that he went from town to town. Can say that he was a nomadic surgeon? I think not. He lived in Lyon before going to Avignon, and that he went to nearby villages from time to time. That would have fulfilled a need when so few real surgeons were available. Guy held a high rank and he was sought for from afar as a consultant.

---

[111]   A commune near Chaulhac in the Haute Loire (EN).

[112]   In *Catalogue des Lyonnais dignes de mémoire,* vol.I, p.142. (EN).

Finally, during his years at Avignon he often went to Lyon where he was a Canon at St. Just, and that required a residence in town. He continued there as a Provost until he died. Think how difficult such travel was in that era. There were no well-kept roads and no carriages; you walked or rode a horse, preferred by Guy. During his trips Guy stopped for while in the towns en route and earned his keep by practicing his art. He could have traveled between Lyon and Avignon by boat on the Rhone River, as was described by Petrarch who wrote about his own hesitancy for travel, and who requested special arrangements be made for him to travel to Vaucluse by boat or by horse.

When did Guy go to Avignon? He was there when the Great Plague struck in 1348, and he served Pope Clement VI. How long had he been there? A fair guess is that he was called when Clement VI became Pope in 1342, when he was a monk at the Benedictine Abbey of Chaise-Dieu (near Brioude) where later on he was entombed. Guy was an emeritus physician, a protégé of the lords at Mercoeur, and probably was well-known. Chereau (*Dict.encyc,* p. 680) wrote that the Pope became Guy's Maecenas. There is no documentary evidence for that claim. Perhaps Guy had treated the monk before he became Pope.

In fact, Petrarch said (in his letters) that Clement VI had under gone a trepanation. Much later, in 1709, a permit for that operation was found when the Pope's body was disinterred and examined. It was reported that the Huguenots did not profane his tomb when they savaged the abbey, as had been claimed.

In his *Memoir on Literature and History*, P. Desmolets recalled that Jacques Boyer, a Benedictine of the Congregation at St.Maur, had mentioned, according to Petrarch, the trepanation of Clement VI. A note written by that priest in which the same event is mentioned was found at the National Library (Paris), in Ms 12664, Folios 102 and 103.[113]

---

[109] A letter is attached to Desmolet's *Memoir* of 1729, was labeled *Historical and Critical Comments about the Dioceses of St.Flour.* P. 188 deals with the body of Clement VI. "In 1709, when the flooring of the choir of the church and monastery of Chaise-Dieu was lifted,the bones of Clement VI were found, along with bits of deer-hide in which the body had been wrapped before transport from Avignon. On March 19, Barthemi Pissavin, Master Surgeon and Anatomist, was called, and he found evidence of a trepanation in the Pope's cranium. He had not read Petrarch, who was certain that the Pope's memory of events and of what he had learned from books was good. And Petrarch added that it (ie the trepanation) was not a gift of Nature, but was the happy outcome of a blow to his head suffered in childhood, from which he carried the scars that Pissavin recognized at once, without any knowledge of what to expect." (EN).

(Pissavine recognized that a depressed fracture had been elevated by means of a trepanation. (LDR).

A note from J. Broyer: "One can see flaking off of some of the outer table of the left parietal bone, about three fingerbreadtgs long and two wide. At the front of that defect one across the coronal suture and a thick-fingerbreadth from the sagittal suture is the site of a trepanation, confirming Petrarch's statement" (EN)

Therefore, in 1348, Guy was in Avignon in the service of Clement VI where we see his devotion to duty during the time of the terrible plague. He describes the event faithfully in Tr. II, Ch. 5 of his book. That is the period of his relationship with Petrarch, which was said to be that of enemies. I now believe that there is no real basis for that tradition. Certainly they knew each other, perhaps they met when Petrarch was a student at Montpellier in 1319. After that time, the poet had long stays at Avignon and Vaucluse until 1353. Laura was a victim of the plague and perhaps she had been treated by Guy. That was when, so it was said, that Petrarch's dislike for Guy began. Yet there is nothing tangible to justify that claim, and in Petrarch's letter about the death of Laura he makes no mention of physicians or their treatment.[114]

But Petrarch disliked physicians in general, as we see in his letter to Clement VI (March 1353) and in his *Invective Against Physicians*, written in reply to the Pope's letter in which he said he spoke of "a toothless old man, born in the mountains" giving no name. Many have claimed that he referred to Guy. However, I agree with the Abbe de Sade who thinks the remark refers to Jean of Alais, another of Clement VI's physician who was very old, whereas Guy was not much older than Petrarch who was born 1304.

Clement VI died (December 6, 1352) and in 1353 Petrarch left Avignon, and returned only once, in 1360, when he was sent from Paris on behalf of King John who wished to be released from his imprisonment in England, to return to France.

At the accession of Urban V to the Papacy in 1362 (following Innocent VI) Petrarch turned away the Pope's invitation, unwilling to return to the papal court, although he was a friend of Urban V. and a compatriot. The Pope's invitation seems to deny any animosity between Petrarch and the surgeon. Petrarch died at Arqua near Padua in 1374. I will add that what we know of Guy's character lends weight to that story.

Guy was Clement VI's[115] physician and was with him when the plague arrived at Avignon in 1348. Gaetano Marini[116] lists the papal physicians chronologically: For Pope Clement VI (1342-52): Stefano Seguini, Giovanni da Firenze, Stefano Ancelini, Raimondo Raimaldi di Vinario, Guelmo de Lavetagio, Lorenzo dal Biarz, Giovanni la Marescala, Guidone de Chauliac, Pietro Angerii (surgeon), Giovanni da Genoa (surgeon, Giovanni Gabrielli (surgeon), Alberto da Erbipola (physician of the family), Giacomo Capelluti (dead), Giovanni d'Alais (dead). According to the list Guy was the last of the physicians,

---

[114] These letters express the regrets but not the pain (ie of a poet), and they justify what has been said about Petrarch's relationship with Laura, that she was only an imagined beloved; Petracrch kept his real heart for others. Albert Tassoni (*Memoirs of the Life of Petrarch*. Vol. II, p. 476) in re Petrarch and Laura, said, "Petrarch enjoyment of her was like that of a rat in an apothecary's shop, taking only what he can get by licking the surface of a well-sealed jar." (EN).

[115] Clement VI (born Pierre Roger) came from a noble family in the diocese of Limoges. He was a Benedictine monk at Chaise-Dieu, a Doctor at Paris (ie Theology), then Archbishop of Rouen, then Cardinal at Rouen (EN).

[116] G. Marini, *The Papal Physicians*, Rome, 1784, p. 78 (EN).

not classed as a surgeon. He had been in the Pope's service since 1348, but did not receive the title of physician until 1352, as stated by Marini. In fact, in the book titled *Officials of Clement VI* on the date January 2, 1356, we find (in Latin) "Guigo de Caulhiaco, the Sacristy Viennen., Master of Medicine, was received by order of the Pope, in the dining chapel". Marini commented, "One would write Medicus D. PP. as for Lorenzo and Giovanni and not simply doctor of medicine if he had been elevated to that high degree". Marini concluded that Guy became the chief physician during 1348.

Guy served Innocent VI from 1352 to 1362 in the same role.[117] The Papal physicians (from Marini) were: Lorenzo di Biarz, Guidone de Chauliac, Pietro Pestagalli, Guglielmo Ghezzi, Giovanni Gabrielli (surgeon).

On April 12, 1353, a few months after his elevation, Pope Innocent VI awarded Guy a Canonship with a prebend, when the opening appeared at the Cathedral of Reims, at the death of Etienne Chaulhaguet. The appointment was announced in a Bull published by Thomas.[118]

"To our dear Master Guy de Chauliac, Canon of Reims, our Chaplain, Greetings: "The loyal personal service you have provided until now at our Apostolic Court, ceaselessly, and wisely; for your honorable conduct and character, your good common sense and abilities, all have been revealed to me personally, all are sufficient reasons for us to show our favor and our appreciation with liberality. Until now the canon and the prebend of the Church at Reims has been Etienne Chaulhaguet, and he was the chaplain of our chapel in that church; they are vacated by the death of Etienne, who belonged in our Holy Seat. Therefore, nobody but I can fill his positions[119], and in consideration of all the services listed above, I wish to give a special thanks, and I shall confer on you the Canonate and prebend with all the rights and prerogatives therewith. This is my own action, and is not a response to a request from you or anyone else. It reflects my freedom of action in the scope of my Papal Authority.

Given at Villeneuve-lez-Avignon, April 16, 1553."

Ch. Cerf, the present (ie1890) Canon of the Church at Reims gave me that Bull of which a trace was found in the list of prebends of the canons drawn up in 1784 by Canon Lecomte, and kept in the Archives of the Archbishop. On page 89 one reads about the prebend no. 69. "On the left side, Stephen of Chautalaqueto, 1349" Then we see Guido de Chauliaco, 1553." We also found it in a work by H. Weyen, *Ecclesiastic Worthies of Metropolitan Reims,* a ms. in the library of the city of Reims, folio 351. The only difference is Weyen's prebend is number 68 where Lecomtes's is 69. "Guido de Chauliaco is authorized by Pope Innocent VI on Nov. 5, 1553 to fill the canoncy of the

---

[117]   Innocent VI, born as Etienne d'Albert in the Limousin, He taught civil law at Toulouse (EN).

[118]   *The Letters of the Papal Court,* extracted from *A Miscelany Of Archeology and History.* Published by the French College at Rome, 1884, p.70 (EN).

[119]   I think that statement defends Pope Innocent's prerogatives as the 'True Pope" during the long strife between the popes at Avignon and their opponents at Rome (LDR).

dead Stephen Chaulaqueto. The aforesaid Guy has received it with the defined rights of the Church and all others."

Canon Cerf told me that Guy accepted the position. Did he actually go to Reims or was it delivered to him more than six months after the Bull? The notes in the Archives at Reims suggest that it was the latter.[120]

The distant benefice forced Guy to travel. Did he practice surgery en route? There is no evidence except that he remained as Canon for six years, keeping the prebend until 1350, when he was named Provost of the Chapter of the Cathedral at Lyon. That promotion probably explains his resignation at Reims. His successor as Canon was Johan de Rupe, by a procuration of April 6, 1359. It is so stated in the Archives at Reims.

Guy became Provost and on August 18, 1359 we see him as such, bowing in homage to the lord of the region, William of Turey, Archbishop and Count of Lyon. He is named such in the act of homage (see Piece de Justification #2). M. Guigne of Lyon, in his book, *The Late Arrivals,* indicated that Guy had been Provost at St. Just as early as January 15, 1359, in a document in the Archives of Lyon.[121]

After Innocent V died in 1362, he was succeeded by Urban[122], and Guy became his premier physician. Marini listed the others chronologically: Raimondo de Salaironis, Guglielmo Ghezzi (died), Giovanni Giacomo, Robino da Singallo (surgeon), Gandolfo da Cremona (surgeon).[123]

Guy remained as Canon and Provost at St. Just until he died. We see that on June 25, 1366 he opened a new secular registry of justice where the chapter's acts were transcribed. On the blank opening page is written "These secular papers are the concerns of the man, Guido de Caulhiaco, Provost of St. Just at Lyon."

The feudal Lord, Charles of Alençon, had replaced the previous Lord William as Archbishop of Lyon, and Guy as Provost of St. Just, renewed his homage around January 16, 1367, repeating the oaths he had offered to the predecessor on August 18, 1359.

---

120 The Archives include a chapter labeled Guigo Cauliaco which states "procured on Nov. 5, 1553". A list of prebends of the church at Reims is kept in The Red Book, folio 782. It is the 'secret' book of accounts (EN).

121 G. Guigne, *Stories of the Hundred-Years War; The Late Arrivals,* Lyon, 1866. "On December 4, 1358, armed bands advanced on Lyon, and a tax was levied to pay the costs of defense. On December 27 Archbishop William of Turey convened a council together with the Chapters of the Churches of St. Just, St, Paul, St. Irenius, and St. Nizier; and the Abbot d'Arnay; and the Masters and Councils of the local trades. A tax of two deniers was imposed on nobles, priests and citizens to start on January 15, 1359. The Provost of St. Just protested taxing the ecclesiastics as beyond the civil authority." (EN).

122 Urban V was born Guillaume Grimoard in the Gévaudan. After he became a Benedictine monk he taught law at Montpellier, Toulouse, and Paris. He founded the Cathedral at Mende and a college for twelve students at Montpellier (EN).

123 The Reader will note that nearly all of the medical staffs for the three popes were Italians; Guy was one of the few exceptions (LDR).

On September 30, 1367, an edict of the Chapter said that Guy, The Provost, was named by the feudal Lord Jean Quartier as inn-keeper and receiver of payments in coin He was allowed fifteen days in which to decide whether to keep the position of inn-keeper for himself—as was a long-established custom—or to pay in coin (or something else) to a substitute, as was permitted. That replaced what he had earned at Avignon in the Pope's service.[124]

I am obliged to M. Andre, the archivist of the Department of the Lozère for this important item. We know that in 1367 Guy also was a Canon of the Church at Mende—see above, where we described Guy's life-long good relations with the lords of Mercoeur who had favored him since childhood, and without a doubt had contributed to his good fortunes. As a result he was included in the solemn oath of fealty made by Beraud, Lord of Mercoeur to Pierre, Bishop of Mende, who also was Count of Chalons, for all the castles and villages that he held in the region of Chalons and for the diocese at Mende. The ceremony was at Avignon on December 19, 1367, in the Consistory of the Papal Palace, in the presence of Cardinals, Bishops, and Guido de Caulhaco, Canon at Mende.[125] This piece indicates how well Guy stood with the Pope.

In the same year (1367) Urban left for Rome on April 30, and Guy did not follow him. The Pope returned to Avignon on September 24, 1370. In that interval, Guy de Chauliac died.

We do not know the exact date, and there are different opinions as to that question. One claims that it happened on July 23, 1368. Another celebrates the anniversary on July 13, and another says July 17. The celebration of an anniversary did not require an exact date, as the last two claims indicate. The best guess is July 23, as is seen in a note in the Archives of the Department of the Lozére (register G 2728), which states that he died on the Monday after the festival of Ste. Mary Magdalene, which was July 23. In the same register one may read in the 'accounts-received' that the Lord Montterreuse paid one franc for the funeral of Guidon de Chaulhac, Canon of Mende; and Sir B. Angelart paid forty florins for the obit of Guy de Chaulhac, Canon of Mende.

A capitulary of St. Just (July 13 1369) states that Guy de Chauliac, former Provost, had bequeathed to the Church ten florins for a High Mass to celebrate his birthday in the church. That amount was insufficient, and Etienne de Chauliac, Guy's nephew, pledged sixty gold florins to meet the required funds for a celebration mass in the church every year on the birthday, July 13. Until Etienne had completed the donation, he would pay one sou per book to the total of three florins for the birthday mass, to be divided among the members of the church as follows: fifteen sous to those who attended the morning prayers, forty sous to those who attended and partook of the anniversary, and five Vienne sous to the sacristy for ringing the long toll with the bells.

---

[124] A hospice was a house near a monastery where a needy person or a traveler could be taken in, either gratuitously or for payment. In the Middle Ages nearly every monastery had a hospice, and one or another member of the chapter could serve as inn-keeper, that is as the supervisor of the place. If he could not do the job, he either had to pay somebody to substitute, or he had to be replaced. As it was in Guy's case, he was allowed to choose what he wanted to do. (EN).

[125] Here the town is spelled as it is today (1990) (EN).

The ledgers for the receipts for birthday masses between 1428 and 1479 place Guy's on July 17 instead of July 13 which is the date in the capitulary number 10. The expenses for the ceremony were paid by strangers in the chapter and by the family.[126]

Where did he die? Tradition has it that an old lady said he that he wanted to see again the region where he was born, and that he was buried in the chapel of a nearby castle which later was destroyed in the religious wars of the 16th C. But all that is myth. One document, a letter of 1368, allows us to fix the place near Lyon, and not at Chaulhac or Avignon.

That important letter was addressed on September 15, 1368 by Jean, Bishop of Châlons to the Obediencer[127] and to the chapter at St. Just, in response to a request by a sacristan that he have the seat once owned by Guy. We learn that after speaking with Jean Quartier, Guy had wished to be buried at St. Just and that Cabacet (Etienne de Chauliac) had agreed, and that two or three days before his death Guy had mentioned that he was a deacon, and that he wanted to be buried in the tomb of the priests. However, there is nothing in writing to that effect.

The Bishop, in consideration of Guy's wishes, authorized the release of the seat of the dead Provost, and that included the hospice among his benefices In view of Guy's request to be returned to St. Just after death, he grants the seat to the sacristy and approves the entombment of Guy at St. Just.

Therefore, Guy did not die in the Abbey at St. Just, not in Lyon proper, but nearby. That is attested by the date of the disposal of his prebends.[128] In fact, Guy's earnings were divided on July 25, two days after his death on July 23. The brief interval indicates that Guy died near Lyon. Other testimonials about the day of his death are hearsay, and are worth less than what are in real documents.

In *Gallia Purpurata*, a life of Clement VI by Simon the Monk, it is said (p. 70) that Guy died at Avignon in the term of Urban V, and was buried in the Cemetery of Flowers, near the Rhone.[129]

In Vol. III of *The Annals of History of Avignon* by Polycarpe of Riviére (a Carthusian who lived early in the 17th C) there is an addendum written in a hand different from that of the author. In it we find a note about Guy. It claims that he died of the plague at Avignon and is buried in the Cemetery Of Flowers in a grave reserved for the victims of the plague

---

[126] Usually the capital designated for a celebration of anniversaries was converted into land rented to farmers. The heirs did not have to appear at the rite (EN).

[127] The Obediencer was the highest rank in the Chapter (EN).

[128] The usual procedure at St. Just and at the Cathedral of Lyon for the disposal of the deceased benefices of the titular canon took place the next day after the interment, and sometimes on the same day (EN).

[129] Simon the Monk, *Gallia Purpurata*, Paris 1636. In the table of contents we read "Guido Chaulia, surgeon of note, died". And in a note opposite his name appears 'nauclerus' (ie a ship's captain!) (EN).

that had been established by Clement VI in 1348. Although the note is full of errors, I believe it should be reproduced here in full, as my token of respect for the diligence of M. Barrés the Librarian of the town of Carpentras (ie who discovered it).[130]

Bouch also said that Guy died at Avignon.

A Latin Ms in the National Library (Paris) has three epitaphs written in honor of Guy. In the *Catalog of Ancient Mss*, published in 1744, they are judged to be about Guy. But How else can one identify a Guido Calliensis other than Guido de Cauliaco? I think we can identify Guido de Caliac as our man, who in the past had been called Guido de Calliac.

I have spoken frequently with Michel Deprez, Conservator of The Department of Manuscripts at The National Library, who has graciously examined those documents for me. He thinks that the three names belong to the same person (ie but not Guy). Deprez believes that they refer to an Italian physician, perhaps from the town of Cagli; hence the Latin Callinus, who lived at the time of Cosimo Medici, the father of his country (Florence), or to his grandson Lorenzo the Magnificent. Although I think that Deprez's opinion is not well-based, the epitaphs are curious enough to be offered to a Reader, who may form his own opinions.[131]

---

[130] "Guidon Chaulia or Calliaco, deacon, physician, and surgeon, died at Avignon (Fol.s. 129 and130). In 1348, during the papacy of Clement VI, the City of Avignon was the Apostolic Seat. He (ie the Pope) believed that God was punishing his people for their sins and brought the plague, a blight over all Europe for three years, when almost all the world was attacked and almost lost, especially in Avignon. The disease was so terrible that it did great harm and killed thousands. Pleading to St. Peter, overtaken with piety and charity, he sent Guy de Chauliac into the town—a deacon, physician and surgeon—a sturdy man, to treat the poor who were victims. He performed nobly in treating the sick. He came from a place in nearby France, called Gabalin, on the border of Auvergne in the diocese of Mende. He was a celebrated Master of Medicine and a surgeon of merit second to none, and his reputation endures as the author of a book he wrote in 1363, titled *The Major Surgery* of Guy de Chauliac, the famous surgeon of Montpellier. He deserves eternal praise for great and faithful service to the victims of the plahgue in Avignon and at the hospital at Champ-Fleury, a place for treating the sick and burying the dead, one of whom was Guy. He was more than just a physician, he treated the spirits as well as the bodies of the sick. Those qualities are rarely seen in a person. He was God-lke in his dignity, both consoling the sick and in treating their wounds, abscesses and ulcers. After serving Avignon so honorably in those perilous times of contagion, he died at the plague hospital and was buried at the Chamflory in 1362 . . . in the month . . . . Every surgeon who passes through Avignon stops to pay homage at the grave of Guidon Chauliac" From Vol.III of Polycarp de la Riviére's *History* . . . . *)* that is Ms 5031 in the Library Inguimbert at Carpentras (EN).

[131] "Here is an epitaph for Guido Calliensis, the most eminent of all physicians.

I

I. Here lies one who deserves to be celebrated for his talents through hundreds of centuries, now covered by a soft earthen bed.

(Nicaise prints the original Latin and a French Translation made by M. Person, Professor at the Lycee Condorcet. My English translation from the French is in the footnote. LDR)

I would like to have had a portrait of Guy for this book, but I found none that was authentic. Those that exist are fantasies, full of anachronisms.

Ranchin, Chancellor of Montpellier in 1612, had portraits made of all the celebrated physicians and professirs of the Faculty of medicine, among which is one of Guy robed as a professor, showing him with a large blond beard. Below it is a label stating 'Guy de Chauliac, 1361" Spanish editions (1574 and 1596) of his *Major Surgery* also contain portraits that are without value. A portrait published by Dugés in 1827 was part of a collection of A. Tardieu. It was a copy of a portrait owned by the Faculty Of Medicine of Paris; the original no longer exists. The latter shows him with a pink beard, an anachronism (see the miniature at the beginning of Treatise I, here). The portrait published by Figier in 1867, a copy of Ranchin's, is worth no more than the original.

Guy de Chauliac's Successors and His Family Inheritors

Two days after his death his prebinds and his positions as Provost and Canon were distributed among his colleagues on July 25,1368. The capitularies of St. Just provide the names of the recipients. The initial list states that Guy's revenues came from sources owned by the chapter: farm-lands, income-properties, services, and rents at Brignais,

---

"Guy, who saved his Prince, had been celebrated from his earliest years, also was a resource for the common people.

"Long ago Pallas taught him and then Apollo taught him Medicine, a noble art, all agree.

"The entire senate wept for him, and the people and his Prince showered him with his own tears.

"Happy is he whose funeral merits such bereavement. If he could be saved by such he would be supplied for another term even more intense.

"I would be happy if I could fill these elegies with joyous praises for the living, instead of sad verses dedicated to his ashes."

II

"Guy Speaks For Himself: "Attention, Ye who pass here, and do not disdain the words of a dead man. Pay attention to what follows.

"Guy was my name. Minerva and Apollo were my teachers. Now I am deep in eternal sleep.

"I, Guy, have learned about Nature and The Stars. My skilled hands have been able to defeat Death.

"I was able to restore life, much to the joy of the citizens, to the Great Prince who was the father of his country."

III

Another Self-Spoken. "Here is immured Guy, and with him his piety, knowledge, loyalty and science of the heavens and of Medicine." (EN).

Chambert, la Chapelle, St. Vincent, Dagny, Dardilly, Dargoire, Escully, Francheville, Grézieux, Marols, "Meons", Quincieu, Rochefort, St. Galmier, "St. Gildas", St. Foy, St. Martin-Lestra, Strata, Valonne, Vercieu, and Careysieu. The total revenues from them came to Eighty-one silver pounds, three sous, and one penny. A pound equaled twenty sous, and one sou equaled twelve pence.

A second list, dated September 18, 1368, gave Guy's participation in the revenues from eighteen sources to total seventy-nine pounds and three pence, mostly from Chambost, St. Barthelmy-Letra, Chauczans, Sagon, Macherel, la Chevre-de-Gilmel, and at Vilars, properties owned by a certain Jolletus who took no share of the distribution. The last items on this list also produced four pounds, ten sous, and six pence.

On July 25 the revenues were divided among the Obediencer of Brignais, the Sacristan, the Choir-Master, nine Canons, nine Bachelors (ie newly appointed canons), five five-year veterans priests, and the ruling Abbot.

The final accounting of the financial relations between Guy and St, Just was not concluded until 1639. The documents indicate that Guy did not share in the ownership of the lands and castles owned by the church, and the lay heirs of his estate were pursued by the chapter, especially because the property at Brignais had been taken by the late arrivals in March, 1362, after his estate first was divided; and the property had deteriorated without care.

On August 14, 1369, the Eve of the festival of The Assumption, the Chapter appointed three commissioners—two canons and one legal consultant—to examine the payments from the obediances and other places that Guy had shared. A few days later, August 17, the Friday after the Festival, the Chapter assigned the Sacristan and two Canons to examine the accounts of the Choir-Master regarding his shares of the earnings from the land-rents Guy had received from properties owned by St. Just, and to review the final closure of the account.

The commission of August 14 concluded that twenty gold francs should be paid to Guillot de Chauliac, Guy's brother, to pay for Guy's share of the rentals of properties owned by the church, and that Guillot would then pay the twenty francs to the Choir-Master on April 1, 1370.[132]

M. Guigue kindly showed me a curious document about the birthdaymasses after Guy's death, taken from account ledgers at St. Just. The expenses were paid by some of the tenant farmers (see fn 123).

On July 17, 1428, the Mass was paid for by the heirs: Jean Claron (called Bretel) who pledged sixty sous on behalf of the family of the castle at Brignais, and by his daughter who lived with Colin Estaiet of the Parish of St. Paul at Lyon, for which the girl pledged thirty sous as the sister and the co-inheritor. Four years later, in 1432, the mass was endowed by the son (ie Etienne) and the sister.

---

[132] I assume that Guillot was entitled to the earnings that Guy would have collected before the day of his death and would remit to the Chapter a year later that part of the rents he had collected that had become the Choir-Master's share after the benefices had been reassigned (LDR).

In 1463, ninety-five years after Guy's death, we note that a mass was endowed by Thomas Joyeux and Jean de Mercoeur of Brignais in the name of the heirs of Etienne Bretel (called Claron) of Brignais, Jean Lynea (called German of Brignais). Benoît Leretier was the tenant of all the benefices at that time.

From 1477-79, the day is not given, the masses were endowed by the tenant Leretrier instead of Joyeux and Lynea.

From these data we can conclude that Jean Cleron (Bretel) had two children, a son Etienne Bretel (called Claron)—name-changing by transposition as common in the 15th C)—and a daughter who was with Colin Estaiet, They, brother and sister, paid in 1428 and 1432.

Etienne Bretel (Claron) died, and the mass for Guy was endowed by his heirs, Thomas Joyeux and Jean Mercoeur, and by Jean Lunea. Between 1477 and 1479 they were replaced by Benoît Leretier of Brignais.

The Family of Guy de Chauliac

We know that Guy was born in the Gévaudan, that he had lived in Lyon for a long time, and although he moved to Avignon, he maintained his relations with Lyon, where he was considered in-residence as a Provost of St. Just. It seems clear that he attracted his family to Lyon, and after his death we find there his two brothers and a nephew. One of the brothers, Guillot de Chauliac, had a son Etienne, called Cabasset. The other brother was Bernard de Chauliac.

Guillot de Chauliac

The first mention of him that I could find is dated July 13, 1369. He had complained to the Chapter on behalf of his son regarding the finances of the Canonate and the prebends. He agreed to accept the decision of the two clerics acting for the city of Lyon. We have seen that on April 1, 1370, he paid twenty gold florins to the Choir-Master to pay the earnings from Guy's former prebend.

Guillot was a merchant, and that he had business relations with Pierre Boyer of Mende, involved in the construction of the Cathedral there. In the account-ledgers of that Cathedral, Guillot is mentioned several times. On March 16, 1373 Boyer sold Guillot a drape of Venetian silk, a half-drape of gold cloth (lamé) from Lucca, four black mantles, and one each green and red, all for eighty-four francs. He had owed Guilhot six florins for a piece of yellow sendal (Chinese silk). He had bought from Guillot twenty—six jugs of wine at eight francs each, and he accepted the fee for delivery by pack animal (donkey or other beast of burden) to carry six measures (ie saumades) of wheat.[133]

In May, 1374, Guillot again appears in the ledger, and again on May 2, 1376 for Burgundy wines sent to Boyer's cellar. A final entry is in dossier no. G-395 states that

---

[133] The ledger (in Latin) stated that the balance due Boyer was three florins and one gross (EN).

Boyer bought fifteen saumades of cheese from Guillot, each weighing ten pounds, and fifteen saumades of oats, each weighing twelve pounds. Each saumade was priced at three and three-quarters florins.

A review of the records of St. Just reveals that on June 18, 1379, The Chapter gave five asinées of cheese to Guillot as payment for a horse that a priest had borrowed amount and lost. On July 16, 1879 the Chapter ordered payment to Guillot of eight florins by the Choir-Master for things that Thomas of Guy was in default.

An acta of St. Just dated April 9, 1383, announced the death of Guillot, fifteen years after Guy's. The same acta noted that Etienne, Guillot's son, owed the Chapter two hundred-forty gold florins, a carry-over of his father's debts dating back to April, 1382.

Etienne de Chauliac, Called Cabasset

Guillot's son and Guy's nephew was dubbed a cleric after a term of service at the Chapter of St. Just in August, 1369, when he was a witness. On July 13, 1369 he pledged payment of anniversary services for Guy, his uncle and probable testator; both were accepted as part of Guy's will.

Etienne was a merchant and a farmer as was his father. On August 17, 1369 he sold the Chapter a silk cloth for use in the church, priced at 42 gold florins.

On May 13, 1370 the Chapter assigned Etienne one fourth of the annual revenues from the produce at Brignais, one hundred-sixty-four florins. But then he had to pay for the message-service from Brignais about one-fourth of that, about thirty-one and one half florins. However, he could use that fund to pay for upkeep of the castle. On September 7, 1370, Etienne again was charged with the necessary repairs, or be held in debt for all the rents.

On February 25, 1374, he was named Porter (for the Chapter, and verger (ie beadle) for the Cloister of the Church. On January 22, 1375, he swore his oaths for the two jobs.

On March 15, 1376, as verger, the Chapter sold him, the house of a former obediencer and canon, for eighty tournois pounds, worth eighty gold florins. Etienne could not raise the money and he contracted to pay the Church and the Chapter an annual and perpetual rent of five florins until he paid off the original price of eighty florins.

On July 16, 1379, he received various sums for supervising the properties at St. Foy, Darzilac and St. Galmier.

On April 11, 1383 the Chapter tried to recover interest for the amounts that Etienne owed, to be used for a birthday mass for Jean, Bishop of Châlons, but was willing to accept the unpaid principal of two hundred-forty florins.

On April 9, 1383 he paid off some of his debts and sold to the Obediencer and the Chapter at St. Just for forty Florins the vines that he owned in the Parish of Brignais, in the wine growing region of Monessoblium. That sum was deducted from the two hundred-forty. On the same day the Chapter rented to him for the rest of his life all the vines that he could attend for a fee of four florins a year, payable by quarters, as was usual in the Church.

The acta of October 21, 1385 announced his death.

Bernard de Chauliac

An act of St. Just of February 24, 1375[134] announced a new Canon, Bernard de Chauliac, to replace the deceased Guy. In September, 1376 he shared in the disbursement of the revenues of Girerd, and received fifty sous as a bachelor at St. Foy and the same from St. Just.

Bernard died on October 18, 1381, thirteen years after Guy and before Guillot and Etienne. The division of his benefices that amounted to nineteen pounds, seven sous and five pence was made the following day, paid to all the members. In April, 1382 the Chapter ordered that the benefices enjoyed by Bernard, other than what he had inherited, be paid to the Church.

Guy de Chauliac's Written Works

Guy wrote several books, which are the chief sources about his own person. What has come down to us is the *Major Surgery (ie Grande Chirurgie)* and small essays on astrology, cataracts and hernias. Several other books are anonymous and may be his: a Formulary (called The Minor Surgery), a Treatise on the Plague, and a Regimen of Health For Popes.

The Major Surgery[135]

That work is his claim to historical fame. He wrote it when there were no complete surgical treatises suitable for teaching, and there was no surgical instruction in the universities. Although there were some remarkable books, none of them dealt with the entire subject. Guy tried to summarize them and he titled his book *A Surgical Collection.* The title *Grande Chirurgie* was given it in the early Venetian printed editions, to distinguish it from a *Chirurgia Parva* which was attributed to him.

Several medieval surgeons had written pairs of treatises, Large and Small, such as those by Bruno of Longoburgo and Lanfranchi. So it was that the Venetian printers simply followed a tradition by publishing a small surgery, as if it was by Guy.

Guy wrote his book near the end of his career at Avignon, after retiring from active practice when he was about sixty-five years old; he died five years later, in 1368.

Therefore the author's long personal experience as a surgeon is reflected in nearly every page. He comments on the works of other authors, and he evaluates them. In his Prologue he wrote, "I have added nothing of my own except perhaps a bit here and there that I thought was worthwhile." However, his book is much more original than he admitted, and his critical comments (ie about what others had written) often are lively.

---

[134] Nicaise has converted the dates as given for a year that began 'old-style' at Easter, to that which begins, as now, on January 1. (EN).

[135] See The English Translator's Preface concerning the Titles Major and Minor (LDR).

He was a Master of Medicine, after completing a long course of study at the University of Montpellier, the degree was the highest that one could receive in that era. He went further, seeking exposure to famous teachers in other schools. Books were rare and information did not spread rapidly; and oral teaching was the usual means of transmission, and he had to travel to find his teachers. He went to Bologna where anatomy was foremost, and then on to Paris.

During those twenty years when he was a physician for the Popes at Avignon, and afterwards until his death, he retained his vigorous mind. His favored situation near the Popes allowed him to collect as many books as he could find, including those that were just coming to light in his own epoch, especially the Latin translations of Galen by Nicolas of Reggio and the treatises by Gilbert the Englishman, et al. One can judge from the list of works that he cites that no physician of the 14th C had a richer library at his disposal.

When Guy undertook to write, there still was a large gap between the surgical treatises of the ancient world and those just coming forth from the newly awakened medieval surgeons. Dezeimeris (1835) wrote, "In order to equal the high level of the ancient books, he needed the impetus, the love for what he undertook, and the persistence to surmount the difficulties in that era of rare books and the need for his tedious study of them in depth. When we consider Guy in the milieu of the century that he personifies, we must acknowledge that Guy possessed all those qualities. Endowed with a happy intelligemce, a straightforward spirit, and a demanding common sense, he persisted in his labors and in that way studied and extended his own scope to include the works of his predecessors. His ambition was to trace the evolution of surgical science. Ackerman, a wise and prudent historian, wrote, "Guy's *Surgery* could rank with anything that was written before him. If that is the case, this surgeon of the 14th C could be the only author that deserves such praise. At least he created a work infinitely superior to all the others that had appeared in the same era and for a long time afterwards. Posterity has judged that so, and for three centuries it was the classic par excellence. He made study easy and profitable, and all other countries are indebted to our fatherland."

Ackerman's statement is not entirely correct. Guy had no access to Hippocrates' surgical works, or to the works of Celsus, Oribasius, and Aetius. He did have some of the works of Galen, Paul of Aegina, and the Arabs, and those of the earlier medieval surgeons.[136]

We can understand what his world was like when Guy began to write his book. As to his guiding principles and to his interpretations of the materials that he selected for comment, he wrote, "I was amazed by how they followed like a flight of cranes. One repeats what another said before him. I do not know whether it was due to reluctance or to admiration that they turned away from anything that was not established doctrine that had been approved by the authorities. They had misunderstood Aristotle's *Metaphysics*

---

[136] When Ackerman wrote (1792) he lacked the works of Henri de Monedeville, not available until Pagel's edition (LDR).

when he showed that two things impair one's vision and knowledge of the truth; they are bias and fear. Although Socrates and Plato are friends, the real friend is truth. Truth is holy and deserves our respect. Whoever follows Galen's Dogmatic doctrines, which are based on logic and experience, should look for meanings and not words. I do not say that he always had good (ie written) evidence for his claims, because Galen invoked evidence in addition to the use of common sense and personal experience, which are the tools of good judgment. Therefore, at the beginning of his *Miamir* he wrote that the citation of things that we read (ie as citations in our writing) means that we agree with the person who wrote them. Therefore, I will proceed with my work, with God's help" And that is what Guy did.

What language did he use for his *Surgery*? He had been a cleric and had studied Medicine at the University of Montpellier, where Latin was the only language in use, as was the case in all Christian universities. It was the every-day language of clerics and it was used in all scientific works.[137] Later, when Guy was a Canon and a Provost at St. Just at Lyon, it was spoken and it was the language of all legal documents. Therefore, Guy knew the language and used it in his book. But the Latin of that era was a debased language—a mixture of classical Latin with words from French, Provençal, and Arabic to which were appended Latin endings.[138] We find few Greek words in Guy's text. That language was not known by the medical clerics, and most of the Latin translations of the Greek classics were from Arabic texts derived from the Greek originals. The few Greek words in Guy's text came from Latin translations of Galen by Nicolas of Reggio.

---

[137]   Again I remind the Reader that the word 'Science' as carried into 19th C French meant all fields of study except Literature, Theology, and Law, which in medieval years included philosophy, alchemy, astrology, physics, optics, medicine, et al. The Faculty of Arts could dabble in any of them (LDR).

[138]   *Nicaise's Note About The French Language:* The Reader will better understand the Latin of the 14th C if we briefly review the changes that it underwent before Guy's era. In that way we also will see the origins of our French language, Therefore, I shall insert here an abstract taken from M. Petit de Julleville' book (*General Concepts Regarding the Origin and the History of the French Language*).

The Gauls of 50 BC spoke Celtic and Iberian, and no Latin. After Caesar's conquest, Latin spread, but the imported language was not classical Latin of the Augustan century; rather it was that of the common Roman folk who came into Gaul, and it became the 'low gallo-roman Latin. Before the Franks invaded Gaul from the north, that low Latin was spoken from Brittany to the Pyrenees. When the Franks came in later, it became the 'low merovingean Latin' spoken everywhere in the Frankish realm.

After three centuries of decadence, classic Latin disappeared; by the 6th C AD it survived only in some literary works. Then the rustic Latin replaced the classical except where it was less debased in books written for general audiences. But it was the rustic Latin of common speech, not that of the written books that gave rise to French, two centuries later.

In spite of some crude elements, Guy's style is neat and concise; he frequently used aphoristic phrases and there are many elisions. His picturesque descriptions are interesting, and there are some phrases that the modern French reader cannot capture. Most of them in French faithful to the Latin originals.[139] That is why I based this edition on the old French. Many of Guy's terms are difficult to translate. In those places I have inserted the original Latin into the text and in footnotes and in the Glossary. The Reader may use them to adjudge my interpretation.

Most of the writers on the subject agree that Guy wrote in Latin: Joubert, Peyrilhe, Malgaigne, and I. Indeed, did not Guy examine some of the many copies and translations of his *Surgery* that appeared after he published the original, during the five years before his death? His reputation certainly drew attention to his book, and the still available mss show that it was translated into several languages almost immediately. I believe the answer to my question is Yes.

A comparison of mss and printed editions reveals differences, but not as many as Peyrilhe claimed. They refer especially to long citations of items from Galen, and do not alter the sense of the citations. I find little difference from the original, although there are several wrong words; I blame the copyists. That led me to search through many mss and editions and to correct most of the errors.

In the French mss and printed editions there are differences explained by the evolution of the language during the 14th and 15th Cs, and that led to variations in many translations. Furthermore, a copyist took it upon himself to change words and to use what seemed to be more suitable.

By the 7th C the Latin had become another language, by chance and by irregular usage. The low Merovingian became old French, the language of the 'oïl', after many centuries of long and obscure labor.

Latin continued in use by medieval clerics, but it was a barbaric Latin, spoken not just in France. It was the cook-book Latin as can be seen in some of the Pieces Justificatives at the end of this section. Guy's Latin is better than what was spoken in the Chapter at St. Just.

An interest in 14th C Latin leads one to learn what every-day French was like at the time, and in the centuries that followed, because the French mss translated from Guy's *Surgery*, in his own era, and later published in printed editions, is the language that I have used here.

French was born in the 9th C, after a slow but spontaneous transformation from the low Roman Latin that had been spoken in Gaul for four centuries. At first the changes came about in the streets, and the vocabulary of low Latin became French, and that continued, whereas the literary and written language was Latin. Cultured French came later, beginning in the 11th C. Therefore the questions remains: Was the French of the oïl also spoken in northern Farnce during the Middle Ages? And was a companion tongue, the languedoc or provençal also spoken in the north? Both versions evolved from the same low Latin of Roman Gaul (EN).

[139]   The old French MS itself was translated for the original Latin (EN)

Did Guy himself write another version, not Latin? He knew the language of d'oc (Provençal) as well as French and probably Italian. During his time there were numerous similarities among the Romance languages. In *Memoirs on the Life of Petrarch*, Vol., p. 76, it is stated that in northern France the common language of speech and writing contained many more Latinisms than in other provinces—indeed, Provençal also was romance language.

Desbarreaux Bernard believed that Guy wrote in the Roman dialect of the Languedoc. There are no documents to support his claim, although we know that there is an ms of Guy's book in that language. We know for certain that such translations existed in the 14th C and that the mss were available for sale early in the 15th C, along with the *Surgeries* of Lanfranchi, Theodoric, Roger, Platearis, William of Saliceto, and of the *Treasury of The Poor*, all in Provençal.[140] Vidal cites an abridged version of Guy's *Anatomy*, as *Hun Libre Appelat Nathomia abrevjada di Gido*, written on paper (ie not parchment).

Germain wrote that the French Ms at Montpellier could have been by Guy himself or a contemporary, as it was thought to be when it was published. But, in truth, that Ms is a copy.

Pulin Paris wrote that the English Ms was written during Guy's life, or shortly after his death. In sum: all the translations and copies come from Guy's Latin original in the 14th C.

In his clinical chapters, Guy followed the order he set down at the beginning when he stated that Surgery included three parts. First, (ie the Where) know the Anatomy, where the ailment lies. Second, (ie the What) know what is needed for planning treatment by looking for the cause, so you will know what to treat. That material occupied five parts of his treatise: Aposthems, Wounds, Ulcers, Fractures and Dislocations, and Particular ailments. Third, (ie the How) Know the instruments to use to achieve a cure, and the antidotary. The entire book is arranged on that plan.

So it is: his book is not only more complete than any that had preceded him, it has an admirable arrangement, both for the general plan and for the specifics. Guy had been infused with Aristotle's logic, and that explains his plethora of citations, and why he his book played such a large role in medical education, and that is why Guy is called The Founder of Surgical Education.

I have omitted further analysis of the treatises. The Reader soon will recognize and discover how Guy followed his rules, as can be seen in my arrangement of treatises, chapters, and paragraphs, and in the use of Italics.

I must comment: Guy rarely said that he did something or other, and he never claimed priority over another author's.

Guy has accumulated vast numbers of encomiums. Fallopius compared him with Hippocrates. Jean Calvo of Valence who translated the *Surgery* in Spanish, said that

---

[140] The great but incomplete *Surgery* of Henri de Mondeville is not mentioned. It fell from favor among the Church-controlled Universities, especially after the Plague. Its re-discovery in the 19th C led to a general re-evaluation of Guy viv-a-vis Henri (LDR)

Guy was the first Lawgiver of Surgery. Friend in 1782 called him the Prince of Surgery. I have mentioned Ackerman's praise. Peyrilhe said the the *Surgery* was the wisest and the most complete work of its kind since Hippocrates, and that even on our era (1784) a Reader will profit. Bein, in an unpublished essay about Paré said that Guy had written an immortal source for French Surgery. Malgaigne (*History of Surgery, Part I. of his edition of Paré,* said. "I believe that, excepting Hippocrates, there is no single treatise of Surgery—Greek: Latin, or Arabic—that I can set above or even equal to the magnificent work of Guy de Chauliac."Daremberg was less elegiac, but he said "Guy appears as an erudite, skillful, but not radical (ie venturesome) surgeon. What is special is his method of exposition, his middle-of-the-road position between timidity and recklessness, leading him to choose the better way, almost as it was if Ambroise Paré was in the 14th C, even if he lacked Paré's originality."

I will conclude this section with an opinion expressed buy Verneuil at the Conference of Educated (ie University) Surgeons: "The service rendered by the *Major Surgery* was immense. It was the beginning in France of an era of splendor. It is justifiable for us, his posterity, to give Guy de Chauliac the title of Father of French Surgery."

Lesser Works By Guy de Chauliac

The Treatise on Astrology

Guy mentioned (Doct. II, Ch. 5., and in several other places) that he had written a small book on astrology, at a time when every physician was a bit of an astrologer, and as Guy wrote, he had to know the material. Guy probably first learned his astrology from Ranchin's *Apollinare Sacrum,* and then from Arnold of Villanova. Guy propably dedicated his book to Pope Clement VI. No Ms or printed edition is extant. An Ms at Avignon *Astronomia Guidonia, was* claimed by Malgaigne to be by Hoenel. Another ms at Avignon is in leaves, on one of which is written the name of the author, Guidonis Bonati, Astronomer, who was a well known astrologist from Forli, whose works was translated from Latin into French. I believe that he was the source of Malgaigne's erroneus attribution to Hoenal.

Treatise on Cataract

Guy stated (Doct. II, Ch. 2) that he had written a treatise on cataract for King John of Bohemia, later called *On Liquid Diets* by some of the king's biographers. The historian Dubrawius, cited by H, Molliére, said that the king had lost his right eye in1337 due to excessive humidity (ie probably ophthalmitis with lacrimation). He was nearly blind by 1340, and he traveled to Montpellier for relief, and probably that was the incident about which Guy wrote. Later the king went to France where he was killed at the battle of Crécy (1346). At that time, Guy was at Lyon, and already well known as the consultant for so famous a person, and probably, Guy already was the protege of Clement VI. Not a trace of Guy's treatise remains.[141]

The Treatment of Hernias
Mentioned by Guy, but there are no extant pieces.

Books Attribute to Guy

A Formulary for the Minor Surgery

I have found six mss and nine printed editions of the Minor Surgery, and two of them have formularies. 1. "medicines used by Guidon de Cailhat for treating aposthems and pustules, in 1340". 2. "medicines for wounds and ulcers prescribed by Guidon de Cailhat at Avignon, where he was the phyasician of Pope Clement VI in 1340'.[142] The two were not always in the same printed editions. In the editions 0f 1500 and 1559 give them as medicines for wounds and ulcers; the editions of 1482, 1493, and 1533 give both as above. The contents are not as given in the *Major Surgery,* rather they are the concoctions of some abbreviater; the compounding is not Guy's. Peyrilhe agrees. I have abandoned any desire to make a new edition of what I believe to be an apocryphal entity, not worth the effort. The Mss of the *Minor Surgery* all are from the 15th C; all in France, a total of nine between 1482 and 1559; four in Latin, three in French, and one in Dutch.(ie Flemish.)

A Treatise on Plague

Guy gave a good but brief dissertation of the plague of 1348 in his *Surgery*, and he made no mention of another essay. Yet, there is one attributed to him. The first mention of it that I could find is in a small work of 1538 in our National Library with a collection of diverse works. The Table of Contents at the beginning indicates a small work titled *The small tract on the Plague by Guy, not previously published by Squironis, the Most Excellent Translator . . . .* I could not find it in the compilation, and no pages are missing.

William of The Innocents (*Surgical Method)* mentioned a book by Guy. "When I was a student at Montpellier under Master Heroard in 1566 I saw a Ms in old Latin titled

---

[141] Humbert Molliere (1888), *The Care of the Wounded before the era of Standing Armies.* "King John was at Breslau and had been treated for an ophthalmia by a French physician, who had assured him of a cure. When the right eye was lost, the physician was wrapped in a sack and thrown into the Oder. An Arab then was called to Prague, and he was not happy at that, and he probably would have met the same end if had dared to promise success." In 1340 John went to Montpellier for consultation. Molliére believes that the king had cataracts. His age in 1337 and the progress of the disorder all suggest that diagnosis." (EN)

[142] The Pope in 1340 was Benedict XII (EN).

Treatise on The Plague by master Guido de Cauliaco. Because it was very hard to read, I began to transcribe it and translate it into French. When the troubles (ie religious wars) intervened, I gave it up, and I had to attend the wounded who survived at the church of St. Peter. Therefore, I asked the son of Professor Heroard, a very well educated and agreeable young man[143] to protect my book until I could publish it, in honor of the original author and of Surgery."

Desbarreux (*Bull, of Bibl.* Vol. X, p. 835) acknowledged the existence of a book by the Father of French Surgery after he read the above note by William of The Innocents. He added, "That book was in the library of Urban V at Montpellier. It then was seen at the home of Simon Cortaud at Avignon in 1640, nephew of Heroard, and cited in a letter to Charles Spon from Cortand, who indicated that he wished to publish it.

Guy himself does not mention a second treatise on the plague in his book of 1363. Probably the attribution is wrong (ie it was a segment of the Major Surgery). Many of his contemporaries wrote about the same subject, among them Chalin de Vinario, and perhaps the book attributed to Guy is by one of them.

According to Jean Simler, Dresse owned a ms containing works attributed to Guy on stones, and herbs, etc.. And Jean Schenkins, a physician at Freiburg in the Brisgau, owned another attributed to Guy titled *Consilia*.

William of The Innocents attributed a *A Papal Guide To Health* (Regimen) to Guy. Indeed, Guy himself often spoke about a memoir of *A Regimen for the Pope*, containing prescriptions for his general health, and for treating ailments, but he never named the author. Judging from those comments, that memoir existed before Guy became a papal physician.

---

[143]   Jean Heroard, of Valgrigneuses, who later became the Royal physician, in 1597 (EN).

PART V

A BIBLIOGRAPHY OF GUY DE CHAULIAC

H ere I shall name and briefly describe the Manuscripts and the Printed Editions of Guy's *Surgery* which appeared after 1363. That will show how well his work was received and its important role in surgical education.

I shall add notes about the *Minor Surgery* although Guy's authorship is far from certain. Many authorities have attributed it to him, and that is why I deal with it here.

I have searched the great libraries in European cities and Universities. In America I consulted the librarians in Washington and Boston to the same end. In all cases I came away pleased to know that the scientific confraternity is real; my correspondence with the scholars who supervise the libraries has been more than simply agreeable and instructive.[144]

On my own, I have searched all the libraries in Paris, Lyon, Montpellier and Avignon, and in Italy the libraries of Rome, Bologna and Florence.

There are many mss and printed editions; multiple editions can be found in some libraries, whereas others own one or two. Only one or two editions have been lost. I have tried to identify all of them, their dates of publication, and the libraries where they may be found. When an edition is not in a French library, I will name where it can be found. Therefore, when I cite only a foreign library, that will mean that there is no copy in a French library.

I list the editions im which Guy's work was published as part of a compendium with other works. That, too, will allow the Reader to assess the progress of surgical practice and instruction. Therefore, I hope to make of this Section a 'History Of A Book', supplemented with many bibliographic notes in the text.

First, let us go to the years before the invention of printing, to the era of the mss. Then we will go on to the printed editions.

The *Surgery* was translated before the end of the 14th C. in nearly all the languages that were read by scholars. I have found thirty-four complete mss: in Latin (22), in French (4), in Provençal (1),in Catalan (1), in English (3),in Dutch (1), in Italian (1), and in Hebrew (1). Sixteen mss contain only parts or fragments, or are commentaries or abridgments. There are six mss of the *Minor Surgery*.

---

[144] I note with interest that Nicaises did not mention German, Russian, Austro-Hungarian, Danish, etc. libraries (LDR).

The National Library at Paris is rich with with printed editions; in addition it has fourteen mss of the *Major Surgery*: ten in Latin, two in French, and one each in English and Hebrew.

The existence of so many mss after more than five centuries tells us how many more must have been made of the work of just one author. Guy came at a very favorable time. The 13th C had awakened restless and curious minds, written works were vulgarized, and the discovery of cloth-fiber paper, all led to an increase in the number of non-secular copyists[145]. Of the thirty-four mss which I found, nine were written on parchment, one of them a palimpsest. Nineteen are on paper, and four are mixed—pages of parchment or paper.

A. The Manuscripts From 1363 to 1478

A brief History:

During this entire epoch Guy's Treatise was a principal source of instruction when Surgery was still accepted in the School of Medicine at Montpellier. However, after the Popes left Avignon, Montpellier copied what had happened at Paris fifty years earlier,

---

[145] According to Champollion, the surface for writing during three thousand-five-hundred years was papyrus. It remained in Egypt until it went to Greece in the 7th C BC. The papyrus sheet was rolled on a cylinder and became a volumen, and it us said that Herculaneum had two thousand of them. They were hard to read and were easily damaged (ie by unrolling). Even before 197 BC a substitute writing-surface was in use, parchment from Pergamon, made from the skin of goats, lambs, sheep, and deer. The best parchment was from calf-skin ('de veau' or vellum). Parchment was used regularly until the 10th C AD. At times during the middle ages, parchment was rare and expensive. People scraped away the writing on old parchments and wrote anew on the bare surface—called palimpsests—and economical copyists used abbreviations. The Library at Florence has a Ms copy of Guy's *Surgery* that is a palimpsest. In the 10th C the Venetians brought paper made from cotton into Europe. Recent examiners have said that it was coarse linen. In the 12th C even that was rare, and it did not come into common use until the 13th C, during the reign of Philip the Fair (Philip IV).

Therefore, the 14th C was favored by the use of paper for writing in local languages just when the ferment of the 13th C encouraged the writing of mss outside the scriptoria of the monasteries, other than in those of the Carthusians and Celestines which were set up for secular copyists. Copying translations, pictures, and drawings was completely secularized during the 14th C.

I have written this story of papyrus, parchment, and paper because the successive discoveries of those products have played important roles in the advances of human culture. Knowing when they happened helps one to understand many historical facts (EN).

and forbade its members from practicing and studying Surgery. Although Balescon of Tarentum published a Treatise on Medicine (Montpellier,1418) titled *The Pharmacopoeia and Surgery of Philonium,* it offered little of interest as an instructive treatise of Surgery (see Malgaigne).

Therefore, we must go to Italy to look for surgical texts in the 15th C, and there were only a few: Nicolas of Florence (called Falco) (d. 1441) wrote a large book on medicine and surgery, which was a compilation based on Arabic texts. It made no mention of Lanfranchi or Guy. Peter of Argelata, a professor at Bologna (d.1433) wrote a *Treatise of Surgery* in six volumes, in which he copied much from Guy without attributions. Leonard of Bertapaglia (d.1460), a professor at Padua from 1424 to 1429, wrote a surgical text based on Avicenna. He promoted astrology beyond that of the 14th C.

Therefore, there were few books, and none of them equaled Guy's, although few mentioned him, and most were slavish copies of the Arabic texts.

1. Manuscripts Of Guy's *The Major Surgery*

(Nicaise's diligent examination and detailed descriptions of thirty four mss fill many pages. Here I will limit the following to a simple list with few comments, and I offer a bow to Edouard Nicaise, the meticulous historian of French Surgery. LDR)

Guy's original Latin Ms. has been lost. Thirty four copies and translations are extant, distributed in many collections.

1. Twenty-two are in Latin
2. Two are in Provençal
3. Four are in French
4. Three are in English
5,6,7, One each in Italian, Hebrew, and Dutch.

2. Fragments of Mss, Commentaries, and Abridgments

Seven Fragments: One in Latin at Basel; One in French at Paris (anatomy); One in English at Cambridge (anatomy); One in Latin at Paris describes Guy's degrees; Three in French at Paris :Guy's Prologue, Part of the Antidotary, and Synonyms for Guy's Terms.

Four Commentaries: Four in French at Paris, by various later authors.

Five Abridgments: Four in French at Paris, and Dijon, one or more chapters; One in Latin at Cambridge.

3. Manuscripts of Guy's *Minor Surgery.*

Six Copies:: Five in French at Paris; One in French at St. Genevieve.

B. Printed Editions From 1479 To 1890

A Brief History 16th and 17th Cs:

Early in the 16th C most of the surgical publications came from Italy. As it was in the 15th C the Renaissance in other metiers barely touched Surgery. That awaited the arrival of Ambroise Paré. The invention of printing opened the way for common-language editions of the works of ancient surgeons, and medical education was broadened. That meant that Vesalian anatomy soon was well known.[146] The use of fire-arms in battle increased the demand for surgeons, and when the nineteen-year-old barber (ie Paré) suppressed the use of boiling oil in the treatment of wounds caused by arquebuses at the battle of Pas de Suze in 1536, and when, fifteen years later, at the siege of Damvilliers, he adopted for the first time the method of Celsus and he ligated the arteries during amputations. That began a revolution in the practice of surgery.

If the 15th C had few notable works, a groundwork was laid for a robust output in the 16th C, vulgarizing the matters to be printed. Chrysolore taught Greek to a few scholars after 1393. In 1453, after Byzantium fell to the Ottomans, the Greek savants emigrated to the west and carried the new knowledge in their mss. The works were translated into Latin and printed. And America was discovered in 1492.

Those great happenings stirred the spirits and fed the enthusiasm for exploring the ancient works. At first the authors simply adopted the old ways and offered less original material than in the13[th] C. Because Guy's work was the only real academic surgical treatise; it became increasingly important. It had summarized in marvelous fashion the surgical books of the ancient and medieval epochs. At the birth of the renaissance his book was ready with a synthesis of what was known to date. We can understand why its influence endured until the 18th C, until the analytic era of the 17th C led to progress in all parts of Surgery. Guy's book promoted the emergence of other general texts that went beyond recitals from the past. The principal surgeons, the authors of important works in the16[th] and 17th Cs were as follows:

Jean de Vigo (1460-1520, who published his *Practica Copiosa* in 1514. It was the first Italian work that dealt with fireworks and the French Disease (ie Syphilis). No completely surgical text had appeared during the century after those of Peter of Argelata and Bertapaglia. Vigo enjoyed a great success for thirty years. However, we see that it was simply an abridgment of Guy. Vigo's book was translated into French by Godin (Paris, 1530 and Lyon, 1514).

Marianus Sanctus, a pupil of Vigo, published (1514) a compendium (abridgment).

Berengaria of Carpi (1470-1550) published his *Skull Fractures* in 1518. He favored trepanation.

Michelangelo Biondi (1497-1565), was the last surgeon of the Roman school. He wrote on amputations in 1542.

---

146   Vesalius' works were written in Latin. Many copies and extracts soon appeared in common languages, including a generous bite taken by Paré (LDR).

Fabricius of Aquapendente's *Pentateuch of Surgery*, (1592) was the sign of the rebirth of surgery at Padua, while Bologna remained in the doldrums.

In the North, German Surgery was still raw, but it soon sparkled.

Jerome Brunschwig of Strasbourg published his *Book of Surgery, and Practice of Military Surgery in 1497.* It was the first book to mention fire-arms, anticipating Vigo by seventeen years.[147]

Hans von Gersdorf published his short *Handbook for Military Surgeons* in 1517. He copied Guy in nearly everything. Both his and Brunschwig's books had pictures. Brunschwig's early editions mimicked the style of manuscripts in format.

Paracelsus (1493-1561) published his own *Major Surgery* in 1536. It had a French translation by Dariot on 1608.

Conrad Gesner (1516-1565) published a single-volume compendium of ancient and modern surgical works at Zurich in 1555.[148]

Guido Guidi (Vidus Vidius) (1500-1569) published his *Greco-Roman Surgery With Translations and Commentaries Of Galen,* (1544). It contains treatises by Hippocrates, Galen, and Oribasius.

Even when faced with the treatises by Guy and Vigo, the French physicians refused to practice surgery. And although there was a little progress in surgical theory, the practice of surgery, the hand-work, was abandoned to the barbers and to charlatans. Three men of the 16th C, well taught and experienced operators, elevated Surgery and began a new phase in the history of Surgery: **Paré, Franco, and Würtz.**

Ambroise Paré (1517-1590) followed the path opened by Guy and Vigo, especially by Guy, whom he did not often cite by name, and for that I chide him. As did Guy's, Paré's books began a new era in Surgery. Guy founded academic Surgery and Paré founded surgical practice, and in 1545 his little book on wounds caused by arquebuses established his name and transformed the practice of Surgery. The first edition of his collected works was printed in 1575, followed by many others in many languages until the end of the 17th C. In 1840-43 Malgaigne published a fine edition of the works by the greatest surgeon of the 16th C.

Pierre Franco published his small *Treatment of Hernias* in 1556. It was re-published in German in 1881-82, and in a re-edited French text by Verneuil in 1884.[149]

---

[147] In his edition of Pierre Franco, 1895, Nicaise corrected his anachronism. First priority belongs to Heinrich Pfolspündt, before 1460. (LDR).

[148] My copy of the book spells the compilers' names as *Andreas and Jacob Gessner.* I cannot find the source for the confusion with Conrad Gesner who was called the German Pliny, who was a compiler of other works (LDR).

[149] Nicaise did not mention Franco's major work published in 1561. Nicaise himself went on to publish a complete edition of Franco along with a long history of French surgery, in 1895. See Translator's Bibliography (LDR).

Felix Würtz (1518-1575) was a surgical reformer who worked at Basel, His large book is a remarkable product. *The Practice of Wound-Surgery, condemning many harmful Practices.* It had many editions including one in French (Savin, 1672-89).

Gabrelle Fallopio of Modena (1523-62), surgeon and anatomist died at age 39, a follower of Guy.

Jacques Dalechamp (1513-1588) published his French Surgery at Lyon in 1570. It is little more than a translation of Paul of Aegina.

Fabricius of Aquapendente (1537-1619), another great surgeon-anatomist. He published his *Surgical Pentateuch* at Frankfurt in1617; and his *Operative Surgery* in two parts at Padua in 1617. The *Pentateuch* was translated in French in 1649 and 1674 at Lyon.

Therefore, from the surgical point of view the 16th C was very active. In its early years, Champier, Falcon, and Canappe published new editions of Guy. Translations of Vigo, Paul of Aegina, and fragments of Hippocrates were published. Instruction for barbers was provided, and from them came Paré and Franco. The schools of Vesalius and of Paré propelled Surgery, and the instruction mostly was in common languages. But the operators were timid and applications of topical medicines continued to predominate surgery. There were many more practicing barbers than 'true' surgeons. The basis for academic Surgery in the 17th C was didactic and learned from books, in particular from Guy's *Major Surgery.* Editions of his book multipled prodigiously in the 16th C.

In the 17th C Surgery continued to expand although not much new appeared. Many good surgeons surfaced in Italy and France, and a few daring surgeons did the operations, when most of the surgeons were timid. It was a half-way epoch: The Anatomy of Vesalius the impetus of Paré, the discovery of the circulation (1626) all led to changed concepts. There were new ways to interpret facts, but before revising the science of Surgery inherited from Guy and the ancients, the new facts had to be accumulated[150] and studied. The 17th C was an era of analysis that continued well into the 18th. New books for teaching were produced and formed the basis of surgery of the 19th C, until our era of antisepsis and dry (ie no pus) wounds. And today we again find ourselves in an era of analysis, of accumulating data, of producing new monographs that will lead to new general texts.

Let us return to the analytical 17th C that began with **Fabricius Hildanus.** He published his *Six Hundreds (The Centuries),* his cases. That led to other similar productions as in the works of **Scultetus, Purmann, Bonet, and Wiseman.**

Fabricious Hildanus (1560-1634) may be called the restorer of German Surgery. He published many fine books including the *Hundreds,* the first of which came out in Basel in 1606. Most of the observations in the six sets were translated into French by Bonet (Geneva, 1669).

---

[150] 'Facts': Nicaise uses the term later on to mean accumulated and published reports of clinical cases. Here, I believe he refers to raw data (LDR).

Charles Guillemeau (1544-1612, set the basis for a new obstetrical science.

Pierre Pigray (1532-1613), like Guillemeau, was a student of Paré. A prolific author. His *Epitome of Medicine and Surgery* copies Guy de Chauliac.

Other French surgeons of the era were **Vigier, Mery, Dionis and Duverney.**

Marcus Aurelius Severinus (1580-1656) was the leading Italian surgeon.

In Germany, Surgery was studied intensively by **Johan Scultetus** (1595-1645) A famous practitioner his *Surgical Armamentarium published in 1652.*

Matthew Purmann (1648-1721) was another famous German surgeon. He was the first to try blood transfusion in Germany, in 1668. He was skilled at trephination, both as a civilian and a military surgeon. And like others in his century, he collected data, and published them. We see another analyst in **Jean Bonet** of Geneva (1620-1689). His rich collection of data in *Observations and Surgical Histories* (1670), and *An Anatomic Repository* (1679).

In England we find **Richard Wiseman** (1625-1687). His *Several Surgical Treatises* are great collections of observations and data (1676). He was called the English Paré.

And, finally, we must mention **Mingelousaulx** (fl. ca 1670) of Bordeaux. He published a new edition of Guy. Until that one, none of the printed works were complete editions, although they were useful for teaching. A professor, that is, a Reader could fill in the details that were missing, and that was his commentary.

Printed Editions of *The Major Surgery*

In all, there have been one hundred-twenty-nine printed editions. I have found sixty; another nine have disappeared without traces. Sixty others are partial (ie fragmentary), or abridgments and commentaries.

Repeating what I did with the manuscripts, I searched through European and American libraries and universities by correspondence, and I searched in person in the libraries of France and Italy. Of the seventy-four editions of the *Major Surgery* sixteen are in Latin, forty-three in French, five in Italian, four in Dutch (Flemish), five in Catalan, and one in English. Fourteen were issued in the 15thC, thirty-eight in the 16th and seventeen in the 17th.[151]

There are extant sixty editions of commentaries and of the *Minor Surgery.* Four came in the 15th C, thirteen in the 16th, twenty-seven in the 17th, ten in the 18th. Most of them are in French, a few in Latin, Italian, Flemish and English.

The 'disappeared' editions include two by Nicolas Panis (1478, 1490). the first Latin-Venetian edition of 1490, another Italian in 1493, one not dated and unlabeled, one Catalan (1498), two by Champier (1498, 1508), two English (1541, 1579), and one by Canappe (one chapter of the *Flowers*) 1591.

---

[151]   I regret that I must charge the Reader with doing her/his own sums when matching the numbers in these paragraphs (LDR).

The first printed edition appeared at Lyon in 1478, five years after printing was introduced there.[152] That first edition was a French translation by Nicolas Panis. In 1470 an Italian edition was published at Venice. The first Latin edition also came from Venice in 1490.

(Translator's Note)

Nicaise filled many pages with his detailed descriptions of the sixty editions that he knew. Here I will limit the following abridgment to naming them by number and source, excepting a complete description of the edition that he used for his own edition, that of Joubert, Number 44. LDR.)

1,2,3. Three French editions, 1478, by Nicolas Panis, Lyon and Paris. One is lost.

4. A French edition,1503, by Symphorien Champier, Lyon. One is lost.

5,6,7. Three Italian editions, 1480-1521. One is lost

8,9,10,11. Four Latin editions, Venice, 1490-1546. One is lost

12. Latin edition, Bologna, 1499

13,14,15. Three Latin editions, Venice, 1513.1519,1546

16-20. Five Latin editions. Lyon, 1537-1572

21. A copy of No. 20, by Symphorien Beraud. Lyon, 1572

Six editions of,Tagault, all abridgments, Paris, Venice, Lyon, Zurich,etc.

22,23. Two editions in Catalan, Barcelona 1492, and Seville 1498. One is lost.

24-27 Four editions in Dutch (Flemish), Ghent, 1507-1646

Nine Editions by Falco (28-36

28, 29, 30. Three Glosses in French without text, Lyon,1515

31-36. Six editions of text and glosses, French, Latin, Spanish, 1520-1537

Ten Editions by Canappe, most designed for Barbers (37-43)

37-43 Seven editions in French (three incomplete, see below),Lyon, 1538-1609

---

[152] Wood-cut engravings were introduced in the 15th C; letter-printng followed. Gutenberg began using movable type in 1436. The technique was introduced at Rome in 1465, at Milan-Venice in 1469, at Lucerne and Paris in 1470, at Bologna, Ferrara, Florence, Naples, Pavia, and Treviso in 1471. Then came Russia in 1486, Sweden in 1493, Scotland in 1508, Turkey in 1726, and Greece in 1821. At first the printed books resembled the mss in lacking title pages, places, colophons, and pagination. Spaces for capital letters were left open to be filled by brush-painters called rubricators. There were introductory and terminal statements. Slowly the patterns changed and books acquired titles, dates, colophons, and pagination, and the rubricatrors vanished. Later on we find headings, prefaces and comments (ie foot-notes and marginations) (EN).

Seventeen Editions by Laurent Joubert 1579-1659

44. French edition, Lyon, 1579. *This is the edition used by Nicaise. Long notes are below, after the descriptions of the editions).*

45. Another French edition by Joubert, containing most of the annotations added to the edition #44. These notes appear in Nicaise's edition

46-56 Editions based on #44, in Latin and French, Lyon, Tourinon, and Rouen 1585-1659. Four editions are entirely of notes for the other editions.

Two Editions by Mingelousaulx 1672 and 1683

57. In French, Bordeaux, 1672

Master Simon Mingelousaulx was a jury physician at Bordeaux. His two volume edition of Guy came out in 1672. In the advertisement promoted by the printers Joubert was reproached for separating his annotations from the text in his edition (#44) of Guy. That led Mingelousaulx to undertake this translation and to place clear and brief versions of the notes Joubert had omitted.

Mingelousaulx' translation is in 17th C French and it supplants Guy's terms with French interpretations, different from what Guy had meant. The Reader of Mingelousaulx' text may not be reading Guy's. The old French, a cousin of the roman latin (of the south), is what Guy spoke, and it needed only a literal translation (ie not an interpretation).

57. Another version

Five Portions (ie Fragments)

58. Prologue only, Lyon, 1542

59. Prologue and One Chapter, Lyon, 1552

60. Guy's History, Toulouse, 1556

61. The Anatomy, Flemish (with Lanfranchi's) Louvain, 1481

62. Variola and Morbillis (from Arabic), Jena, 1790

Thirteen Commentaries

These are the substance of what a Lecturer (Commentator) said, often in the common language, after someone read aloud from the the text, often in Latin[153]

63. Three by Innocens of Toulouse, 1595-97

69-69. Six by Ranchins of Montpellier, Paris, Lyons, Rouen, 1600-1628

70. By Courtin, Paris, 1656

71. By Abeille, Paris, 1696

---

[153] The Reader often was an 'undergraduate', usually a cleric. Today he would be called a 'teaching assistant (LDR).

Two Annotations on Ttreatise VI of the *Major Surgery*

72. By André du Laurens, Paris, 1639
73. By East, Baker and Clowes, London, 1579

Twenty-two Abridgements
The Flowers of Guy Twenty-two Items, 1549-1705

74. In French,Two issues by Canappe., Paris, 1595
75. Seven issues in French, by Raoul, Paris and Rouen, 1549-1671
76. Two issues by G.S.D.M, Paris and Brussels, 1659, 1680
77. Five issues by Lazare Meysonnier, Lyon, 1650-1686
78. French, by Raymond Dares, copy of #77, Bordeaux, 1694
79. Two issues, no attribution, Paris, 1677-1678
80. In Flemish/French bilingual, Rotterdam, 1650
81. In Italian, Theobaldo Marmier, Rome, 1652
82. In Spanish, Antonio Juan de Villa Franco, Valencia, 1705

Twelve Called *The Master Surgeon* By Laurent Verduc

83-87. Eleven issues in French, Paris, 1691-1744
88. In Italian, Florence, 1697

C. *The Minor Surgery*

None of these are authentically attributable to Guy

89. Two Formularies in French, Paris, 1482
90. A Formulary in Flemish, Delft, 1482
92. A Formulary in English, 1542
93. In Latin, combined with Albucasis, and Jesus Haly, Venice, 1500

D. A Chronological List of Sixty-nine Printed Editions of The Minor Surgery

(The List contains all the Items described in depth in Part C. It is not necessary to reprint it here LDR)

E. Nicaise's Discussion And Transcription (In Old French) Of Joubert's Notes Concerning Item #44, the basis for Nicaise's Edition (and Mine)

Laurent Joubert was born at Valence in 1529; he died in 1583. He was a well-known physician in the 16th C. He had attended the Universities at Paris and in Italy, before he received his degree as Doctor from Montpellier in1558. He remained there, first as a

Professor of Anatomy and then as Chancellor (1574) He wrote several books, two of which were *A Treatise on Fire-Arm Wounds*, Paris 1570, and *Common Errors In Medical Practice*, Bordeaux, 1570. In this section we have special interest in his two editions of Guy's *Major Surgery,* 1579 and 1585, one in French, one in Latin. The French translation is better than all the others that have been published to date; it went through seven editions. The French annotations were added in of these after 1584, one year after his death. The Latin edition with notes appeared in 1585.

The French edition is not a translation of his Latin edition; the text is different, and that version is almost a verbatim translation of an early Latin Ms of Guy's text. The French book seems to me to be from another French ms; and that also was Peyreilhe's opinion. It has many points in common with the French Ms. owned by the Faculty of Medicine at Montpellier, and Cellarier thinks that Joubert did a fine job with it by collating several mss. Nevertheless, many obscure passages remain, and some are incomprehensible. I believe that it is closer to the original Latin edition than to the early French translation.

(Nicaise submits a detailed bibliographic notice about the book, after inserting Joubert's own Preface. LDR)

Joubert's Introduction

To My Kind and Scholarly Readers

I wish to assure you that I have diligently and curiously researched as well as I could, all the materials listed in the catalog of authors who have written on this subject, and I have cited many of them for their opinions about the operations and the observations of Master Guy. I have had two reasons, both based on need and gain. First: to obtain a more faithful understanding of what the good Doctor meant. I think I have been successful, and I have left some of his statement intact. Second, for the benefit of the curious 'cutter' who is a student of surgery, who may seek help in his performance, that he may more

---

[154] Joubert repeats what Henri de Mondeville said about the importance of identifying the source of the citations for the sake of the student who can more easily find and read the original His statement is in Notable XI, p.220 of my edition, as follows:

NOTABLE XI. *The Advantages of Precise Bibliographic Citations*
"Several of the well-known physicians and surgeons to whom I have given my book for proof-reading and to edit especially for its prolixity have criticized me for my insistence on stating precisely the sources of my citations of other authors, by giving 'chapter and verse'. The answer gives my two reasons. 1. To make it easier for scholars to track down the sources. 2. As a result their comprehension of my text is better and clearer.
The first reason is supported by Galen (*On Internals,* Bk. I, Ch. 7, where he discusses this topic), stating that one should use as facile and as compact a style as he is capable of.

easily find the cited passages and explanations of what may be obscure.[154] A too brief sentence is more difficult when one does not know what comes before it or follows it. Indeed, I have been able to annotate nearly all the passages without the help of other authors. There are a few that I have not been able to find anywhere even after a thorough search, although I may not have been far from answers that were lost or hidden in the worm-eaten or mildewed collections of Alcoitin, Henry or Jamier, or in the Concordances of Bienvenu (did he not write about hidden causes and miracles in certain diseases and treatments ?), or of Tadeo Alderotti of Bologna, and others. I have consulted the Library of the laborious Gesner to see if something resembling what I seek is mentioned in any of the books in his collection. He does not mention Alcoitin, Henry or Jamier. He cites Tadeo of Florence but not of Bologna. I sought out many physicians and surgeons who were known for their writing, but I found none whom he had mentioned. Perhaps Guy had read someone before his time and he did not say where, at least for my sake. I was loaned a copy of the Four Masters by Master Philip Guillen, Doctor of this University, who now practices at Avignon, his home-town, a fine man, curious, diligent and wise and kind enough to help. Therefore there are many passages that I have been able to annotate, many of Guy's citations of Hippocrates, Galen, Paul of Aegina, Avicenna, Rhazes, Averroes, Avenzoar, Rabbi Moses, Azarum, Haly, Halyabbas, Jesus son of Haly, Acanamusali, Mesûe, Dionysius, Gordon, Arnold and other princes of medicine. Add to them the principal surgeons: Lanfranchi, Theodoric, William of Saliceto, Bruno, Roger, Roland and The Four Masters, and others who should not be omitted. All of the citations fill the margins of the pages. That has not been my labor; to clarify all the obscure passages I have used Doctors and Scholars at the University, and I name François Humeau, my doctoral son who now is at Poitiers for his labors with six or seven hundred citations of Avicenna. Humeau now is famous for his lectures and writings, subtle and erudite. His diligence has allowed me accurately to cite all the passages from Avicenna. François de S. Vertunian, known as Lanau of Poitiers, also erudite and well known for his talents and depth of knowledge and for his exquisitely focused work on Hippocrates' *Treatise on Head Wounds,* has helped me to identify many passages and in making my index. I must confess and admit that a proper reward

---

Why not make it easier for students by leading them directly to the locations of cited sources rather than make them spend days and nights hunting for a single reference, and then another and another? In similar fashion, my second reason, is supported by The Philosopher (ie Aristotle, *Ethics,* Bk. I) who explained that we must understand the meaning of words (ie statements) when they are placed in different contexts, either used before or after various differing statements. A perfect comprehension of a proposition or a citation is impossible if it is taken out of context, and the same is true for single words, leading to misunderstandings. Therefore, my second reason is valid. By giving a precise citation I encourage students to seek out the sources and to avoid being deceived (ie by out-of-context remarks)" (LDR).

for their work is beyond my means. And the many offers of help in the writing of this book were more than I could accept.

I offer this to my Readers: it is important to locate a citation. The initial search and the quotation, such as from *The Prognostics* of Hippocrates should provide more book and chapter, but should add the aphorism. *The Techni* of Galen, also called *The Microtechni*, which means The Small Art, was originally in three parts, and was cited in that way by Guy, whereas today it is separated into chapters; my note follows the latter citation. In many other Galenic books I use the example of Fiellon, because the treatises are more clearly defined than those by Froben and the Uintas. The books classifying the medications according to the places of their origin and their therapeutic types (Guy called them *Miamir* as did the Arabs, but in the *Catageni* the Greek names are used) have several sections, and I follow that pattern. There are three books of topical medications: the first is legitimate; the others are only attributed to Galen. One of them is dedicated to Solon, the Prince of Physicians[155] and contains many items found only in the Greek texts. Master Guy did not read them because they were not available to him in his era. We are aware that the Galenic texts he quotes were the old-style translations, not those made by the moderns, as we can tell from his use of the titles *Miamir, Techni, Catageni.* Also, he always uses the term Therapeutics, whereas we say The Method of Treatment. There are many other examples that, God Willing, I shall annotate in respect of Guy's Latin original Ms. To locate his citations from Rhazes I have made good use of the work by Jerome Surian, printed at Venice in 1542, and not the older edition called *Elham* or *Elhandi*, or the *Continent* by Rhazes. When Guy cited Arnold of Villanova he did not specify that it was from *Some Considerations of The Work of Medicine.* Jean St. Amand wrote *A Commentary on the Antidotary of Nicolas Prevost* which is not clearly divided into sections or chapters. I cite him (the author now is called Mesûe) by page and column because now it is printed that way in folio (a large volume), and still it is not divided in chapters, etc. It was printed at Lyon in 1525. Bruno and Lanfranchi each wrote two books, a Major and a Minor, and Guy himself also is said to have written a Minor Surgery, of which no mss are available. In Lanfranchi's and Bruno's books, the smaller books are bound with the larger ones. In Bruno's there is a rubrics section at the beginning of several chapters, which are labeled with the syllable *rub.* The same is true for Roland's *Surgery.* There are two books of Roger: one deals almost entirely with maladies arranged head-to-toe, including abnormal swellings, fevers, and some medications. That book was printed in 1549 by Bernard Venette, along with other books not relevant here. The second book of Roger is purely surgical, printed at Venice in 1546; it is bound with other surgical treatises. That was the book (ie by Roger) cited by Guy. The items from The Four Masters, which is a commentary on Roger, are difficult to locate because they are

---

[155] I assume that Joubert was poking fun at the misdirected title. Solon's medical fame relates to his use of honey and his abomination of quack abortionists (LDR).

unlabeled. One cannot find treatments in that text because, as far as I know, they never have been printed.

F. THE NEWEST EDITION (Nicaise's, Paris, 1890)

A Detailed Analysis By The Author

Although this new edition is published under my name, I believe that the text truly is Guy's: the words, the style, his special terms, and even the obscurities. Other editions, such as those by Falcon, Joubert's Latin, and Mingelousaulx have altered the text in order to explain it. In so doing they depart from what Guy wrote.

However, mine is a French edition, a choice that came easily, because Guy's language was a bastard Latin, a mixture of Latin, langue-de-oïl and langue-d'oc with some Arabic words. On the other hand, the French of the 15th C was very close to the roman Latin of the street, and we can see that in the early French mss of the *Major Surgery* (ie translations during the same century from Guy's original Latin Ms). Mingelousaulx wished to translate it into 17th C French, and he made a bad product, often far from what Guy wrote. Therefore, I decided to use the old French which is what I find in Joubert's excellent translation.

There have been six French translations in printed editions: those of Panis, Champier, Falcon, Canappe, Joubert, and Mingelousaulx. Joubert's is the best. Peyrilhe and Cellarier believe that Joubert collated ancient copies of Guy's mss, and I agree, even if that does not increase its value. The edition used here is better than Joubert's own Latin edition, although it has its own defects. In some places it wanders afield when Joubert's collation was incomplete. Some special terms are wrong, as in the titles of physicians used in the 14th C—Joubert used those of the 16th C. Some terms used by Guy are replaced by the original Greek words, whereas in the 14th C Greek had not yet entered the language of science. The Reader will encounter terms used by Guy that came from translations of Galen's Greek by Nicolas of Reggio. In some passages Joubert is incomprehensible, and that seems to have been his intent. He explained that he had discussed the question with physicians and surgeons, and he wrote, "It is regrettable that good books dealing with serious matters are published in some strange language. One may speak as Aristotle did in responding to questions put forth by Alexander about matters he heard in the Master's pari passu lectures (acroamatics). The answers may or may not have been forthcoming. It is not always necessary to go to school to listen to a teacher's spoken remarks. Joubert followed Aristotle and Averroes in making some passages obscure, especially when dealing with doctrines, in a way that is not very clear for an average surgeon. Peyrhile criticized Joubert for such actions, and we may not recognize it for what it is, and now the practice has faded.

In my edition I have restored Guy's text using the 14th C terms, titles, etc., and contrary to the policies of Averroes and Joubert, I use punctuations, indentations, and italics, and

I have added explanatory notes. For that purpose I have followed the Latin texts word for word, using Ms #9 in the National Library and the printed editions of 1499, 1537, and 1559. For the reference French mss, I used #25 at Montpellier and #29 at Paris, as well as those of Canappe, Joubert, and Mingelousaulx.[156]

To continue: I have inserted in footnotes some of Guy's own Latin text when the translations have been too difficult and are not the same in various earlier editions. Also, I have inserted notes to explain certain old French words in use today that had different meanings in the 14th and 16th Cs.

(The long note that follows may be of special interest only to Readers of Nicaise's French edition, and may offer little to Readers of this English edition. Nevertheless, I place it here to illustrate some of the translator's problems in making sense of Joubert-Nicaise's old French text. LDR)

The French Language of the 16th C

The language underwent many changes, and it was not the same everywhere. Every Province and every writer invoked the privilege of speaking and writing the local idiom. The Latinisers abused that privilege, and even the minor clerics used words taken directly from Latin and Greek sources.

There was no standard orthography. Claimed to be based on etymology, certain letters of no use were inserted, such as s*cavoir*(savoir), frui*ct* (fruit), de*b*voir (devoir), chevaul*x*, (chevaux), e*s*tang (etang), sein*g* (sein), il veu*l*t (veut). The added g in speech was pronounced as *ng,* such that the written word also was spelled with an *ung.* The *h* usually was silent, but it was restored as in huit.

On the other hand both the *I* and the *J* were used as I, and *u* was sometimes a *v.* The various intonations of e, é, and è were not marked with accents except at the end of a word.

Grammar was even more haphazard: frequently the subject followed the verb, which often fell between an auxiliary and a participle. Sometimes an attribution or a classifier came before the verb. Every kind of verbal construction was acceptable without concern for the clarity of the phrase, even when obviously they were obscure.

The articles often were omitted and some contractions of the articles and prepositions were used: ou meant en le; es meant en les. They said l'endemain for le lendemain, l'ierre for le lierre, and l'uette for la luette.

Descriptive adjectives were not clearly distinguished from pronouns: icelui or chacun which today only are pronouns were then used as adjectives, and the comparatives had disappeared before the 16th C.

---

[156] The numbers are those in the preceding Bibliography for Manuscripts and Printed Editions (LDR).

Numerating adjectives were taken directly from Latin and were made of the number with an added iéme.

Personal pronouns were used as subjects with self-understood antecedent, especially in negative or subordinate phrases, and often were separated from the verb by many words. The antecedent pronoun *ce* often was omitted before its relative *qui* or *que*. Dont meant not only quoi, but d'ou as well as de qui.

In interrogative phrases they wrote chant-il but pronounced it chante-t-il. Infinitives were used as substantives. Present participles were used as adjectives without clear distinction. The rules for past participles were very loose. Adjectives frequently were used as adverbs: premiere for premierement. Adjectives had only one gender ending. The feminine *e* was just coming into use in the middle ages,

Prepositions often were omitted: Si Dieu plaist. At times *de* replaced *que* in a comparison phrase. *A* replaced *que* as an indicator of possession, and it replaced many prepositions that indicated ends or means, such as avec, dans, en, par, pour, selon, sur, and vers.

Many conjunctions then in common use now are lost: *jaçoit qu*i meant bienque. Pourtant meant pour cela and not toutefois as it does today.

(*I omit two sentences that deal with French usages and to the Glossary referred to in my Preface. LDR*)

Also I have reproduced part of Joubert's notes, keeping what has permanent implications as well as some that will be of interest to students of 16th C Medicine.

I repeat: My intention has been to avoid replacing Guy's explanations, yet to offer the Reader the means to form his own explanations for obscure passages. I have added this Introduction that includes an Essay about Medieval Surgery and a Biography of Guy.

Exhaustive research has allowed me to relate his life and his works to his time and to describe the important roles his works have played. I have tried to illuminate the milieu in which he worked and the written materials that he consulted, and the Doctrines dominated medical education. Only by understanding the differences between our own doctrines today and theirs can we comprehend a book written in the Middle Ages.

I have reproduced for this edition seven miniatures which I found in the Mss of Guy. There is no need to describe what the Reader will see for himself. They are interesting historical documents from an era before printed pictures, engravings and drawings were in use. Miniatures were windows in which to see the scenes, and they are especially valuable for the history of anatomy. The course (ie a dissection) at Bologna lasted four days, and followed a strict routine involving the master and his assistants. The miniatures show the costumes that distinguish the wearers. Two miniatures picture a teacher and his pupils. They are particularly curious as to their dimensions: the Master is gigantic as compared with the small pigmy-like students. I must thank M. Profit for his good work: the reproductions are perfect.

PART VI

SUPPORTIVE DOCUMENTS THAT ARE CITED BY NICAISE

(Pieces justificative)

(Nicaises transcribed 33 documents, all in Latin, dated from May 17, 1344 to 1367. They were found in archives of Municipalities, Monasteries and Cathedrals, and were written during Guy's life. One set, in French, from 1428 to 1479, attests to memorial Birthday Masses after Guy's death. I will not reproduce them here. LDR).

PART VII

BIBLIOGRAPHY OF WORKS THAT REFER

TO GUY AND HIS WORKS

(This List of Authors, Titles and Dates, illustrates the extent of Nicaise's scholarship LDR)

Abbe de la Sade, *Memoire pour la vie de Petrarque,* Amsterdam, 1764

Ackerman, *Instit. historiae med.,* Nurmburg, 1792

Allut, *Les Routieres au 14ᵗʰ C,* Lyon, 1859

Amoreux, P.J. *Notice hist. et bibliog. sur la vie et les ouv. de L. Joubert,* Montpellier, 1814

Arbaud,(Damasse), *Hist. med. de Montpellier,* Montpellier, 1839

Astruc, *Mem. pour servir a l'histoire de la Facut. de med. de Montpel.,* Paris, 1767

Barjavel, *Dict. historiq et bibliogr. du depart. de Vaucluse,* Carpentras, 1814

Bernier, *Histoire chronol. de la med. et des med.,* Paris1695

Baluze, *Vitae paparum avenionensium* Paris, 1693

Bayle, *Dict. hist., et. crit.,* Amsterdam, 1734.

Bayle et Thillaye, *Biographie medicale,* Paris, 1855

Billings, *Index Catalogue Of The Surgeon General's Library, Washington, 1884*

Black, *aesquisse d'une hist. de lamed. et de la chir.,* Paris, 1798

Bouche, *Histoire de Provence,* Aix, 1664

Boyer, L., *Dict encyclycoped. des sc. med.*, Paris, 1876

Breghot du Lut et Pericaud, *Catalogue des Lyonnais etc.,* Lyon, 1839

    Notes et docum. pour servir de la ville de Lyon depuis 1350, Lyon, 1839

Brunet, J., *Manuel de libraire,* Paris, 1860

Ch, Drs. Al. et Ch.-Ed, *Le secret des dames deffendus a reveler,* Paris, 1880

    (The two Drs. Ch. are anonymous, to avoid censure re the contents LDR.)

Cambis-Valleron (marquis de), *Annales de Avignon (ms).* 1706-1772)

Campbell, *Annales de la typographie néerlandaiseau 15ᵗʰ C,* Paris,1874

Cellarier, *Introduct a l'etude de Guy de Chauliac,* Montpellier, 1856

Chereau, *Dict. encycl. des med.,* Paris, 1886

Chevalier, Ulysse, *Repertoire des sources historiques du moyen age,* Paris, 1877-83

Chomel, *Essai historique de la med.,* Paris, 1762

Classen, *De medicis primorum medii aevi,* Breslau, 1856

Chrispohe (Abbe), *Hist. de la Papauté 14th C*, Paris, 1853

Desbarreaux, Bernard, *Bull.de Bibliophil.*, Paris, 1852

    Les medecins plus célebre Lille, 1852

Delorme, Raige,*Dict. de med.*, Paris, 1839

Dezeimeris,*Dict. history. de la med.*, 1834-35

Didot, *Nouvelle biographie*, Paris, 1862

Duges, *Notice sur Guy de Chauliac*, Montpellier, 1827

Eloy, *Dict. histoire. de la med.*, Mons, 1778

Fantone, *Istoria della citta d'Avignone*, Venice,1678

Figuier, V*ie de savants illuistre de moyen age*, Paris, 1866

Fischer, G.J., *Ann. anat and surg.* Brooklyn, NY, 1880

Folin, *Conferences historique*, Paris, 1866

Friend, *Hist. de la med.*, Leyden and Paris, 1748

Frère, *Manuel de bibliography normand*, Rouen, 1860

Germain, *De la med. et de sci. occult a Montpellier*, Montpellier, 1872

    La med. arabe et la med. grecque a Montpellier, Montpellier 1879

Germain, A. *Les Maitres chirurgiens de l'ecole de chir. de Montpellier*, Montpellier,1880

Germaine, A., *L'ecole de med. de Montpellier*, Montpellier, 1880

    Le anciennees ecoles de Montpellier, Montpellier, 1881

Gesner, Conrad, Zurich, *Bibliotecha instituta,*1574 *(See Fn 1, LDR).*

Gobet, *Rech. sur la vie et les ouvrages de Guy de Chauliac*, a lecture, unpubl. ca 1782

Graesse, *Tresor de libres rares*, Paris, 1861

    Hist. litteraire de la France, Paris, 1862

Gregory, *Storia della Vercellese lett. ed arti*, Turin, 1819

    Biogr. du Dict des med. sci., Paris, 1820-1825

Guardia, J-m., *Gz. med. de Paris*, Paris, 1865

Haeser, H., *Biographisches lexicon der hervoraragenden Aerzte*, Jena, 1884

Hain, *Repertorium bibliographicum*, Stuttgart and Paris, 1826-1838

Haller, *Bibliotheca chirurgica*, Berne, 1774-75

Häser, *Lehrbuch der Med. und der epidemische Krankheiten*, Jena, 1875

Havard, *Le moyen age et ses institutions*, Tours, 1878

Heister, *Inst. de chirurg.* (a transl. of Paul of Aegina), Avignon, 1770

Henschel, *Biograph.litterar., (*Janus), 1847

Jaffé et Potthast, *Romanorum pontificium regesta manuductio*, Rome, 1884

Karr, K., *Les Paysans illustres*, Paris, 1841

La Croix du Maine et du Verdier, *Les Bibliothec. franc.*, Paris, 1773

Laval, *Cartulaire de l'universite d'Avignon,*Avignon, 1884

Lemoine, Simonem, *Gallia purpurata*, Paris, 1638

Malgaigne, *Introd. d'A. Paré*, 1840

Mangeti, J.J., *Bibliotheca scriptorum medicorum veterum*, Geneva, 1731

Marini, Gaetano, *Degli archiatri pontifici*, Rome, 1784.

Michaud and Poujoulat, *Mem.* pour servir a l'hist de France depuis le 13th C, 1836

Mirbel, L de., *Premieres galleries biographiques* (cited by Moulin)

Montfalcon, *Precis de bibl. medicale,* Paris, 1827

Morerii,L. *Grand diction. history.,*Paris, 1759

Moulin, *Gui de Chauliac,* Lozere, 1884

Neubauer, *Juifs au 14th C,* Paris, 1877

Pauly, A. *Bibliog. des sc. med.,* Paris, 1872

Pulin, Paris, *Des Manusc. francais de la bibli. de Rois,* Paris, 1840

Petit-Radel, *Reche. sur les bibl. anc. med.,*1819

Peyrilhe, *Hist. de la chirurg.,* Paris, 1784

Pericaud, *Bibliograph. lyonnaise,* Paris and Lyon, 1851

Policarpe de Riviére, *Annales de Avignon,* 17th C ms.

Portal, *Hist. de l'anat. et de la chir.,* Paris, 1770

Prouzet (l'Abbe), *Annales du Gévaudan,* Mende, 1843-44

Quesnay, *Recherches sur le orig. e les progress de la chir. en France,* Paris 1744

Rauchin, François, *Apollinaire sacrum,* Montpellier, 1880

Renouard, *Hist. de la med.,*Paris, 1846

Riolan, *Curieuses recherches,* Paris, 1651

Rodrigues, *Notice sur Guy de Chauliac,* Montpellier, 1841

Rousssel, Th., *Le pape Urban V et Petrarque, Lozere, 1858-9,*

Sprengel, *Hist. de la med.* Paris, 1815

Thomas, A., *Les letters a court de Papes,* Avignon, 1884

Tiraboschi, *Storia della letteratura italiana,* Modena, 1772-1782

Tourtelle, *Hist. de la med.,*Paris, 1804

Zambrini,(Fr)., *Le opere vulgari a stampa dei secoli XIII e XIV,* Bologna, 1878

HERE ENDS NICAISE'S GENERAL INTRODUCTION

THE
MAJOR SURGERY

In The Name Of The Lord

HERE BEGINS THE ENTIRE TREATISE

OF THE SURGICAL AND MEDICAL TOPICS

COMPILED AND COMPLETED IN THE YEAR OF THE LORD MCCCLXIII

By **GUY DE CHAULIAC,** SURGEON, MASTER IN MEDICINE
IN THE VERY ILLUSTRIOUS STUDIA OF MONTPELLIER

Figure 2
A Lecture in the 15th C
A Miniature As A Frontispiece For A French Ms of Guy de Chauliac
Ms 396 Bibliotheque Nationale of Paris Fol. 1

GUY DE CHAULIAC'S PREFACE

Having first thanked the Lord who has given me a long life and has spared my body and granted me peace and has kept me active, who has given me knowledge of the Art of Medicine and of the way to heal with divinely inspired efforts, now I will try hard to discuss what I have garnered. Before I comment on the collection of matters dealing with the Art of Surgery and Medicine, I must thank the living and true Lord for giving me all the materials without which this kind of introduction would be impossible. I will give him my heartfelt devotion here and throughout this work, and elsewhere, and plead for His continued assistance and for the protection of his Son. They gave me a happy start and provided the milieu in which to accomplish something useful, and by his command I will bring it to a good ending.

What led me to this collection was not the lack of books, but the need to unify and find a useful concordance. None of us has at hand all of the books, and none of us has been able to read them all cover-to-cover and to remember everything he has read. A lecture may serve a good purpose, but it has limited value, as are models or short essays; they provide only temporary benefits. Knowledge (ie science) has been an accretion from its very beginning. We are like children perched on the shoulders of giants, and we can see what they see; and yet somewhat farther. Therefore, our hypotheses and constructions may be valuable, and, as noble Plato wrote, even brief written notes may be less obscure than a long lecture, and they can be re-read and remembered.

And for the sake of my old age and to keep my wits alive, this book is for you, the]Masters of Medicine at Montpellier, Bologna, Paris, and Avignon, and especially for those who served the Popes and have been my companions in that service, with whom I have been nourished, and have studied, read (ie lectured), and operated, trying to avoid mediocrity. Here I will try to re-state and to modify to a degree the principles set forth in the speeches and writings of the sages and savants as they have put them in their surgical treatises. That is why my book should be called an Inventory, A Surgical Collection. I have added almost nothing of my own other than a bit here and there that I think may be worthwhile. Certainly it is not perfect, and doubtlessly it will be superfluous in places and obscure in others. I ask for your emendations; and I ask that you pardon my ignorance.

A SINGULAR CHAPTER

Here are the Premises:

The Matters To Be Known By Anyone Who Wishes To Avail Himself Of The Art Of Surgery[157]

D ear Readers: You will see that this Chapter (ie Commentary) is arranged similar to the Inventories of civil wills, where the most common and the most valuable items of an estate are listed first. This Singular Chapter describes how I have done that here. First come the usual (ie common) and the most important matters needed by anyone who wishes to avail himself of the Art of Surgery. That is what the Philosopher (ie Aristotle) wrote in his *Physics*, "We know that the proper process of science is to go first to the more general and then to the particulars."

---

[157] In the Venetian Editions *The Singular Chapter* is called *The Universal Chapter.* In it Guy states the Generalities applicable to the entire work; they contain material s needed by whoever wants to profit as a *surgeon, et al.* The meaning of the term Singular has been debated. I will say only that the word derives from the Latin 'singulus, meaning 'unique'. This chapter stands alone, and really is unique, whereas all the other chapters are part of the Doctrines and Treatises. In other words, the Singular Chapter has its own rare excellence. Unlike Galen, Guy defined Surgery more than simply a method of treatment. He makes it a part of Medicine. He blames Avicenna and the physicians after him for their disdain of surgery. He praises William of Saliceto and Arnold of Villanova as shining lights in the faculties of medicine where they performed both as physicians and surgeons. After Guy, however, the separation between Physicians and Surgeons was even more accentuated.

In this Chapter Guy provides a brief and interesting history of Surgery, and he offers us the gift of his erudition, a review of the books that existed in his epoch, and he shows us the methods of his study. He provides a masterful and honest description of the conditions faced by the true surgeons. At the end of the chapter he outlines the plan that he follows in composing the book. As an aside, he tells us something abut himself. He says that he practiced at Lyons for a long time, during which he period he operated at many placed (EN). This Singular Chapter appears in many mss and printed editions. Nicaise describes them all in detail, and I shall not reproduce that bibliographic material (LDR).

The Definition of Surgery

Let us begin by asking what Surgery means. Although there have been many definitions, they all take root from our father, Galen, in his *Introduction to Medicine,* where he wrote, "Surgery is part of the curative art that makes use of incisions, cauteries, and the setting of broken bones." In Book I of his *Treatment of Fevers,* He added this to his definition, *"and other manual operations."*

That is a good definition of what now is considered a third category of Medicine; it is broadly defined as the science of treatments of maladies that require manual procedures which are not to be used before the other measures, which are food, beverages, and restoration of a good life-style (ie and the applications of medicines). Therefore we agree to define Surgery as follows: *Surgery is the science that teaches the methods and the arts of closing wounds, making incisions, and performing other manual operations for curing people whenever that is possible.*[158]

The science of Surgery is part of the general therapeutics, and no one can deny that it is called an Art in many places. The word *Science* has a broad connotation that is not limited to pure Medicine. The mind considers them together, and the term covers both of them.

Indeed, there are two kinds of Surgery. One is academic, and it deems itself alone to deserve the term 'Science", and that kind of surgeon need never to use his hands. The other is surgical practice and is what commonly is called Surgery, the handiwork that deserves the term Art. That person may know nothing except how to operate. That kind of Surgery was included in Aristotle's category of Mechanical Arts. Galen, in Book 1 of his book *On Diets* said, "Someone cannot become a seaman or have another similar career simply by reading book. Only by diligent practice can we become adepts and artisans, etc. etc." We must agree that his statements are true.

Then we should add. "for treating people when that is feasible", because, as my Master Raimond of Montpellier said, nothing is absolute; while some things are quite certain, a physician cannot always offer relief or achieve a cure. To demand certainty from a physician is like asking one who stammers to deliver a sermon, both use imperfect instruments. The subtle doctor (Averroes) told us to act as our art dictates.

That is why the Art commands us to strive for a cure, with three exceptions: 1. When the malady is known to be incurable, as is leprosy. 2. When the malady is curable but the patient will not accept the recommended treatments or he cannot tolerate the suffering caused by them, as in the treatments for a necrotic ulcer on a limb (ie requiring amputation).

---

[158] Joubert's own definition: "Surgery is skill or a science adopted by someone who usually or occasionally is a physician. He is responsible for teaching not only surgeons, but also apothecaries, each of whom has his own particular skills for providing what a physician orders. Surgery, narrowly defined, is what a surgeon does, but in a broader sense, yet what he does is a physician's responsibility." (Joubert)

And 3. When the treatment itself will engender complications that are worse than the primary malady, such as inducing a virulent fever (ie mal mort) or by over-treatment of chronic hemorrhoids (ie mild bleeding or discomforts). Hippocrates taught that treating such may result in hydrops or mania.[159] Galen said the same in his *Therapeutics*, Book 4. Good treatments should not cause pain or suffering; they should preserve and not destroy; they betoken a good physician rather than a bad one. In Book 12 of the same he wrote that one cannot promise the impossible simply to obtain a fee. Avoid harmful treatments and false promises, lest you incur a bad reputation.

The word *Surgery* derives from *cheir*, meaning hand, and *ergeia* meaning operation. So it is that Surgery is the science of manual operations.[160] Let us proceed:

Surgery includes opening the body to treat a malady that surgery alone may cure. And it can prevent an illness and preserve one's health whenever that is possible. Those are its goals.

The Parts Of Surgery And The Operations

Surgery, according to Joannitius (ie Honein) has two parts: treatment of soft tissues and of hard. In action it does five things: operations on swellings (ie aposthems), treatments of wounds and ulcers (ie unhealed wounds), the restoration of fractures and dislocations and a miscellany of maladies for which it is suited. There are three types of procedures: The separation of continuity (incisions), the restoration of continuity, and the riddance of redundancies. Phlebotomy and scarificatioin are in the first category. Closure of wounds and the reduction of fractures and dislocations are in the second. Ablation of tumors such as aposthems and glands are in the third.

A Surgeon's Instruments

The surgeon's actions vary. Some are routine and others are special to a case at hand. The common measures involve what the physician routinely employs, and the special measures may require metallic tools. The medical treatments include attention to general regimens: foods, beverages, phlebotomy and leeches, ointments, plasters and powders The metal instruments are for cutting: scissors, razors and lancets; those used for removing things: large and small forceps. In addition we use cauteries (olivaries, cultelaries); tools for probing: sounds, and tubes; and catheters, dilators, and needles. And there are special instruments for trephining the cranium, and scythe-like lancets for operations at the anus.

---

[159] Mild bleeding from hemorrhoids was accepted as a form of phlebotomy, a kind of prophylaxis that taps off the humors that could cause ascites. Hemorrhoidal veins were thought to be emunctories of the liver; they could be dammed back and even offend the brain. (LDR).

[160] Nicaise notes that Guy knew no Greek. In his Latin Ms he wrote *ciros* and *gygos*, which are Arabic versions. (EN).

That list tells you that a surgeon's work is artificial[161]. He should always have at hand five kinds of ointments: the Basilicon to wall off an abscess; the Apostles' to mondify; the White Oinment to consolidate; the Golden to generate flesh; and the Dialthea to soften the scar. His portable equipment should include five or six instruments: scissors, forceps, probes, razors, lancets and needles.

A well-armed surgeon should be able to perform useful operations, provided that the patient has been informed as to the nature of the procedure. Galen (*Therapeutics)* said that the surgeon should first treat things that are contrary to Nature, then go after Natural things and the Non-Natural with their annexes.[162] Galen also said (*Therapeutics,* Book 2) that we should begin our treatments by attacking the primary ailment and then go after the secondary ones that derive from the primary, and persist until we reach your goal of curing the malady. We follow those principles: first make a diagnosis of the malady and assess its nature, and then proceed step-by-step until we know everything, the obvious and the obscure.

Next, after determining the indications as stated, Galen said that we should determine what we should do, and what is or is not achievable. Then decide how to attain our goal. He continued (at the end of Parts 3 and 7) noting that if our goals are not unlimited yet are consistent, as for treating an ulcer or an uncomplicated wound, our choices will be easy. But if they are manifold and contrary, as for treating an undermining sordid ulcer or an aposthem in noble region, we must be aware of complications.[163] First determine if the patient's life is threatened. Then find the causes for the threat and determine that it is impossible to treat the (ie the primary cause) first because the threat is imminent; it is a time for a quick decision. When we decide that our treatment should aim at the ailment,

---

[161]   Artificial meant art-ful, the work of an artisan with special skills (LDR).

[162]   The seven Natural things are a person's innate endowments: According to Guy they were: the elements, the temperaments (ie complexion), the humors, the limbs and body parts, the virtues (ie faculties), the body-functions, and its spirits. Four things are Annexed to the Naturals: age, color, dietary habits, and sex.

   The six Non-Natural things (not part of the body) (hygiene) are important for health, and health will suffer if they are not considered next: air, food and drink, exercize and rest, sleep and wakefulness, excretion and retention, and emotions. The five Annexes to the non-natural are the temperature (ie weather), the season, the geographic region, coitus, personal habits (bathing, etc.)

   The three things Contrary To Nature (ie the pathology) destroy the Natural dispositions: disease, their causes, their symptoms. (EN)

[163]   Joubert's Note: An ulcer is a double malady: it is an excavated wound, and it may be a triple-threat if it is sordid or sloughing, or it may be inflamed (a fourth element). You must begin your treatment by cleaning away the slough, and finish the treatment by promoting the closure of the defect by cicatrisation. In the 'triple' case you begin by allowing the ulcer to fill with granulation tissue. Galen describe all of them at length in Books 3 and 4. (J).

so be it. In other words determine the order in which we should deal with the problems. As Galen put it in Books 3, 4, and 7 of his *Therapeutics,* and he said that it in all three places search for a cause, or find the reason sine qua non, or what is most urgent[164] At times the need for haste is such that you must treat the threatening complication first as when a nerve has been pierced, or a vein is hemorrhaging or a muscle has been torn or a displaced bone appears in an ulcer.

How To Perform An Operation With The Instruments Cited Above

Arnold of Villanova said that there are four considerations. 1. The operating surgeon should know what operation is to be performed on a human's body. 2. What good reasons lead him to an operation? 3. Is an operation necessary and feasible? 4. How does one perform it?

The first is answered by considering all of the operations as described above and selecting the correct one. The second is answered by faith in the surgeon who dares to operate on a human, that he will know that what he is doing, that he will be successful, and that he will be confident of its safety. The third considers the operation as it affects the various parts of the body. The fourth is knowing that the operation will be tolerated by the body, that is, if it is done well and is acceptable to the patient, and that the surgeon has experience in performing it, and in preparing the patient before and caring for him afterward.

As an example, when we wish to eliminate the water of ascites (hydrops), we should decide that an operation is necessary, and we decide that among all the methods available that we should use the knife. Then we consider why we should operate and we agree that surgery is a good way to treat ascites: at least it will relieve the discomforts. Third: is an operation necessary? We say, "yes", because there is no other way to treat real ascites. Can the patient tolerate the operation? Yes, if he not too debilitated and we do not drain all the fluid at one session. Fourth: let us consider the technique. We shall lay the patient supine, to expose the skin of the abdomen below the navel, on the right side if the patient's discomforts are mostly on the left, and on the opposite side if the symptoms are on the right. We will pinch up the skin (ie and panniculus) and incise it (ie simply a nick) where there are no exposed blood vessels. Then we will thrust in a cannula and drain as much fluid as the patient tolerates. We withdraw the cannula and release the skin that will cover the opening in the abdominal wall and stop the leak. When you wish to tap off more fluid, repeat the maneuvers and abide by the patient's tolerance. So be it.

A History Of Surgery

I shall describe in order all the surgeons whose works I have studied, and have cited in my book. I think that is a better way to distinguish each from the many others.

---

[164] Mingelausoux explainedf this passage: that Guy referred either to a direct complication of the malady, or to a separate threatening accident that demanded treatment first (M).

First came **Hippocrates** who towered above all the others. He was the first Greek to bring Medicine out of obscurity, as Macrobius and Isidore wrote in their *Etymologies,* and as is quoted in the prologue of Rhazes' *Continent.* Medicine had been dark during the five centuries after **Apollo and Aesculapius,** they who invented the art. Hippocrates lived for ninety-five years and wrote many medical works as are decribed by Galen in Book 4 of his *Therapeutics* and elsewhere. I believe that Galen preserved Hippocrates and many others from obscurity.

Galen was the disciple who farmed the seeds taken from Hippocrates, and he added to the crops. He wrote many books in which Surgery is mixed with Medical topics, especially in his *Book of Tumors Contrary-to-Nature* where a summary of Surgery appears, and in the first six books of his *Therapeutics*[165]. In the last two he deals with aposthems and other maladies that require manual operations. Seven other books which made up the *Catageni* listed and classified the medications; I regret that we retain only a summary version. Galen was the premier scientist during the reign of Marcus Aurelius Antoninus, about one hundred-fifty AD. He lived to age 80 as testified in *The Lives and Conducts of the Philosophers.* According to **Avicenna** in Book 4 of his *Fractures,* and in glosses of the same, the long interval between Hippocrates and Galen lasted three hundred-twenty-five years,. But a truer chronology measures the interval at five hundred-eighty-six years.[166]

After Galen, we cite **Paul of Aegina** who made much of Surgery. He devoted his entire Book 6 to it, which was used by **Rhazes** in his *Continens* and by **Halyabbas** in his *Royal Dispositions.*

After them we meet **Rhazes, Albucasis** and **Alcaran.** The claim that all three names represent one person is not born out by the *Almansor* and its Divisions[167], by Albucasis in his *Divisions, Surgery),* and by Halyabbas, who described them as separate persons. He quotes Rhazes often, from his *Continens*—also called Helham in Arabic—in which he collects all the works of his predecessors. However, the exact citations are not stated; and a tedious and thankless search through that long Helham now seems unnecessary.

Halyabbas was a great Master, and his 'seed' is in his *Royal Disposition,* in which he devoted the ninth part of the Second Sermon to Surgery.

---

[165] Wounds, ulcers, and fractures are discussed in Books 3, 4, 5, and 6, but not in Books 1 and 2 (EN).

[166] Joubert claimed that the numbers were added to one of Guy's mss. He placed the correct figure in his Venetian edition of 1559. It is not in the Montpellier Ms, but it appears in several earlier Venetian editions (EN)

[167] 'Almansor' is the title of Rhazes' great work, it does not name a person, However, Al Mansour was a highly respected mathematician at Cordova. In his General Introduction Nicaise implies the Rhazes named his book to honor the Second Caliph of Bagdad who antedated Rhazes by at least one hundred-fifty years. The name Mansour means 'victorious' and it may have been a boast by Rhazes rather than a tribute (LDR).

Avicenna came next, the honored 'Illustrious Prince'. The fourth book of his *Canon* is a nicely arranged surgical text, as are all of his works.

We know that before Avicenna all medical practitioners were physicians as well as surgeons. After him, however, whether due to timidity or to the increased complexities of treatment, Surgery was separated and relegated to the hands of 'mechanics'. The first to appear in Europe were **Roger, Roland** and **The Four Masters**. They wrote books entirely devoted to surgery and included much that was empirical (ie Anecdotal rather than Methodistic). Jamier followed: a brutal surgeon full of meaningless twaddle. All of them followed Roger's lead In most matters.

Then came **Bruno** who very carefully provided a summary of what he had learned from Galen, Avicenna, and Albucasis, even though he did not have a complete translation of Galen's works. He completely omitted anatomy from his book.

Theodoric followed. He plagiarized Bruno and added some 'fables, about **Hugh of Lucca** who was his Master.

William of Saliceto was a valuable resource; he produced two treatises, one on Medicine, the other on Surgery. I consider them both well made.

Lanfranchi also wrote a work which included many of the things he had learned from William, although his book is arranged differently.

In the same era **Master Arnold of Villanova** flourished in another Faculty.[168] He wrote many fine books.

Henri de Hermondaville began to write a very powerful treatise at Paris and was noted for his melding of the methods of Theodoric and Lanfranchi. His premature death prevented him from completing the work.

During that era **Nicolas of Reggio** worked in Calabria and being proficient in Latin and Greek, he worked at the behest of King Robert, and translated many of Galen's books with which we now are endowed. They are the best and most elegant translations, better than those that had been made from the Arabic versions.

Finally, I shall mention a faded English Rose that had been sent to me and which I scanned; where I had hoped to find a delicate aroma, instead I found fables from Spain, from Gilbert and from Theodoric.[169]

In my day I have met good practicing surgeons; Master Nicolas Catalan of Toulouse; Master Bonet, Lanfranchi's son from Montpellier; Masters Peregrin and Mercadant of Bologna; Master Pierre de l'Argentiere of Paris; Pierre Bonant, a veteran practitioner at Lyon; Master Pierre of Arles; and my associate Jean of Parma.

Last, I list myself, Guy de Chauliac, both a surgeon and a physician. I come from the border of the Auvergne in the diocese of Mende. I am a chaplain and a physician and

---

[168]   This is Guy's first mention of a French surgeon. And the first to be titled 'Master', that is, one who was the product of a University, and the first from Montpellier. Guy tips his hat to an alumnus of his own school (LDR).

[169]   The reference is to John of Gaddesden's *Rosa Anglica*, full of mystical nonsense, ca 1310 (LDR).

a friend of our Noble Pope. I have witnessed many operations and I have read treatises by the old masters, especially those of Galen which I have in translations. I studied them diligently and at length at the many places where I have found them. Now I live at Avignon in this year 1363, during the first year of the Papacy of Urban V. In this year I have completed the assemblage of this work, with the help of my friends, and with the good will of our Lord.

The Sects

Two chief sects of operating surgeons: persist to this time: the Logical or Rational[170], and the Empirics, of whom Galen disapproved in his *Book of Sects* and in his *Therapeutics*.

There have been five groups of the Empirics. The first group included Roger Roland and the Four Masters. They believed that suppuration was important in all wounds and aposthems, and they used poultices and bouillies based on what they read in Book 5 of the *Aphorisms* where it says, "The Gentle is Better and the Harsh is bad" [171]

The second group included Bruno and Theodoric who tried to treat all wounds without pus, using wine as the only substance. That method also was Galenic (*Therapeutics*) where he stated, "The dry is healthy, not so the wet."[172]

The third sect included William of Saliceto and Lanfranchi who tried to steer a middle course, using ointments and soft plasters based on Galen's *Therapeutics:* he advised following his simple dogma: treat safely and cause no pain.

The fourth sect are the Teutonic Knights and policemen, and other warriors. They treat all wounds with magic potions, with oil, wool-fleece, and cabbage leaves, believing that the Lord endows words, herbs and minerals with curative powers.

---

[170]   A few commentators have read 'Laics' for Logicals', and they are wrong, because the laics were the empirics, with no rational (ie book-taught) bases. Many of them are illiterate. The physician-logicians (ie the rationals) were declared the best by Galen (J).

[171]   They used plasters which today are called cataplasms made from flour, roots, leaves, fruits, seeds, and flowers, all cooked, peeled and strained through cloths to make the pastes. A poultice-bouillie has the consistency of a cataplasm, but is made of wheaten flour wet with some liquid and thickened while heating and stirring to become what we call bouíllee d'armotte. We surgeons call it a triarpharmac mol when it is composed of wheaten flour, water and oil (J).

Much of this and the ensuing paragraphs is an almost verbatim transcript of what Malgaigne wrote in the Introduction to his edition of Paré. (See pp. xxxix ff). I cannot say who came first: Guy, Joubert or Malgaigne. Certainly Nicaise followed Joubert nearly always and did not compare him with Malgaigne word for word in making this edition (LDR).

[172]   Malgaigne wrote: "One makes a serious error in attributing this to Bruno and Theodoric (who copied Bruno). Bruno cited the *Aphorism* "Laxa bona" and credited it to Roger. This is not meant to detract from Guy." (Malgaigne ibid) (EN).

The fifth sect are the women and the idiots who relegated all sick persons to Saints. Their only treatment was based on this prayer:" The Lord giveth as he pleases, and he deals with me as he pleases. Bless His Name, Amen."

These sects will be refuted in the text that follows, and I will not carry on with it here.

However, I am amazed that people follow leaders like cranes in flight; they repeat what someone else has said. I cannot decide if it is fear or love that makes them listen only to what they are accustomed to and has been approved by some authority. They misinterpret what Aristotle wrote in Book 2 of his *Metaphysics* where he demonstrated that when there is a choice between two approaches to the truth, one should ignore biases and fears, and even if Socrates and Plato are our friends, the truth is even more a friend.[173] Truth comes first, it is holy and honorable. That characterizes the Dogmatic Doctrines of Galen as expressed in his *Book of Sects* and especially in the *Therapeutics* which is entirely a composite of reason (ie logic) and experience, where one seeks the facts and ignores words. The same is taught in the epilogue of Chapter 7 of his *Constitution of the Dogmatic Method;* he said it again in Chapter 10 of his *Natural Faculties:* "He who will be ahead of others in knowledge cannot become so of a sudden. He must be born intelligent and be well taught from the first lessons on. When he reaches boyhood and then puberty his restless spirit will love the truth. He will study diligently night and day and he will learn everything that was taught by the most renowned Ancients, and when he reaches full manhood he will be endowed with the ability to judge for himself what he examines at length and with due care, and accordingly he can advise what he agrees with and what he rejects." He added, "I hope what seems relevant to me will avail the Reader. However, to some it may seem superfluous, or a fable told to a fool."

I do not claim to have experienced everything that is written here. Galen wrote in many places that there are two sources from which to make a decision. Yet, there is a third source, other than what is based on personal experience and logic (ie reason). It is dependence on what others have witnessed. He said (early in Chapter 3 of the *Therapeutics)*, and in the *Miamir* or *The Compositions According To The Places),* that one increases his faith in what he reads if he has confidence in the authors. And that is why he repeated all the medications that he took from the works of expert physicians. And I will do the same here, with God's help.

The Requirements For All Surgeons

Now I shall discuss the conditions to be met by all surgeons who use their art on human bodies, and I will describe how to operate. Our forefather Hippocrates commented with this subtle conclusion to his introduction of *The Aphorisms: "*Life is short, Art is complex, The Time and the Occasion may be demanding and occur suddenly, and

---

[173] I interpret this passage as a reflection of the disputes between the Platonists and the Aristotelians in France, especially at Paris at that epoch, as influenced by Thomas Aquinas et al. (LDR).

experience may be misleading and perilous, and judgment difficult. Your actions should involve not only your own experience in such actions, but also the consideration of your patient, your assistants, and your control over the external circumstances.

According to Arnold of Villanova, that eloquent Latinist, there are four sets of requirements: One set involves the surgeon, a second involves the patient, a third deals with the assistants, and the fourth concerns the externals.

The four requirements for the surgeon are: 1. His education. 2. His skills. 3. His ingenuity, and 4. His personality and conduct. Therefore the surgeon must be well educated, not only in the principles of surgery, but also in the theory and practice of medicine.[174]

As to Theory: he must know the Natural things and the non-natural and the contrary to nature things. Of the Natural Things, Anatomy is the most important, a sine-qua-non for surgery, as will be seen in the text. Also, he should understand Complexion and know that the varieties thereof will affect his choice of medications (see *Versus Thessalus*[175], *and* in all parts of *Therapeutics)*. The same holds for knowing the Virtue, which means the life-force (ie spirit).

The Non-Natural Things concern the Air, Foods and Beverages, etc, all of which affect the health of the patient. The Things Contrary-To-Nature are the malady itself which he intends to cure, and its causes, which he must know if he will treat it. His success will depend on that rather than on lucky guesses. He must be aware of the complications in order to attack their causes and forestall them and interfere with the treatments of the malady. Galen repeats all that in the *Glaucon*.

As to Practice: The surgeon should know how to manage the life-style (ie the regimen) as well as the medicaments, because operative Surgery is the third instrument of Medicine. Galen said that in his *Introduction*. The use of medicines (pharmacy), follows the properly managed regimen, and surgery comes third.

Therefore, the surgeon who operates should know the principles of Medicine and have some knowledge of the other arts. Galen (*Therapeutics*) said in opposition to Thessalus, "Physicians should know something about geometry, astronomy, and dialectics (ie rhetoric), and other good doctrines that will affect the conduct of the scrapers, carpenters, merchants, and others who would change their careers and take up Medicine.[176]

The second requirement for surgeons is Skill. He should be an expert when he operates. The wise Avenzoar quote, "Every physician should know things (ie been taught) and then learn from his own experience with them." Rhazes said the same in his *Almansor* (Book 4) and Halyabbas agreed (*Commentary on Hippocrates' Theory)*.

The third requirement is Ingenuity and Common Sense. Haly stated (Book 3 of *Techni).* "The surgeon must have a good memory, common sense, proper motives, good

---

[174] Arnold's term was 'Physic', meaning all of medicine (J).

[175] Thessalus of Tralles, a Greek in Nero's Rome. He was a leading Methodist. See Sarton, Vol. I, p.262 (LDR).

[176] A bitter jab at the mpirics and Thessalus (LDR).

vision, and be healthy". Later he added that the surgeon should be well-made for the job, with slim fingers, strong hands, clear eyes, and lack tremors.

The fourth requirement is a good Presence. The surgeon should be daring when the outcome will be assured, but he should tread carefully in face of peril, and he will avoid harmful treatments[177] and other such actions. He should be gracious with his patients and generous with his friends, and be wise with his prognoses. He should be chaste, sober, and be sympathetic with those who suffer. He should not connive and seek extortionate fees and he should willingly state in advance fees that are commensurate with his work and the patient's ability to pay, in view of the results of his treatments and in view of his professional status.

The three conditions required of the patient are: 1. He must follow the surgeon's instructions, much as a peasant obeys his feudal seigneur (Galen, *Therapeutics*). 2. He should trust the physician (Galen, *Prognostics*) and 3. He should be patient, because patience overcomes fearful thoughts (also Galen).

The four requirements for the assistants: Be gentle, be gracious (agreeable), be faithful, and be discrete.

In regard of the externals, there are many elements that must be arranged for the sake of the patient, as Galen wrote in his *Commentary on the Aphorisms of Hippocrates*.

How I have Arranged My Text

After this Singular Chapter, I will follow a plan which simply is what Averroes suggested, at the beginning of his *Colliget*. He wrote that every Artisan's practices should have three rules. 1. Know your subject. 2. Know what your goals are. 3. Know your tools and how to use them to obtain your goals.

Because our Art is practical and operative, the treatises must be at least of three sorts. However, in our book I have set them into seven Treatises. to cover the three parts described in the *Colliget*): 1. Anatomy, where maladies occur 2. Treatments and 3. Usage.

In summary: The seven treatises are: 1. Anatomy. 2. Aposthems. 3. Wounds. 4. Ulcers. 5. Fractures and Dislocations. 6. Other surgical maladies, and 7. The Tools, including the Antidotary.

Each Treatise will describe two doctrines, and each doctrine will have about eight chapters. Each chapter will have three parts, as in Galen's *Therapeutics*, which will refresh the physicians who wish to follow the Dogmatic methods. 1. The malady and its causes. 2. The symptoms that will lead to your diagnosis and to decisions about treatments and curability 3. The treatments: why and how to operate. That is the order I will follow throughout the book, with God's help.

---

[177]   Again we remind the Reader that 'harmful treatments' in general were open surgical operations (LDR).

HERE BEGINS THE MAIN TEXT OF THE MAJOR SURGERY

TREATISE 1 THE ANATOMY

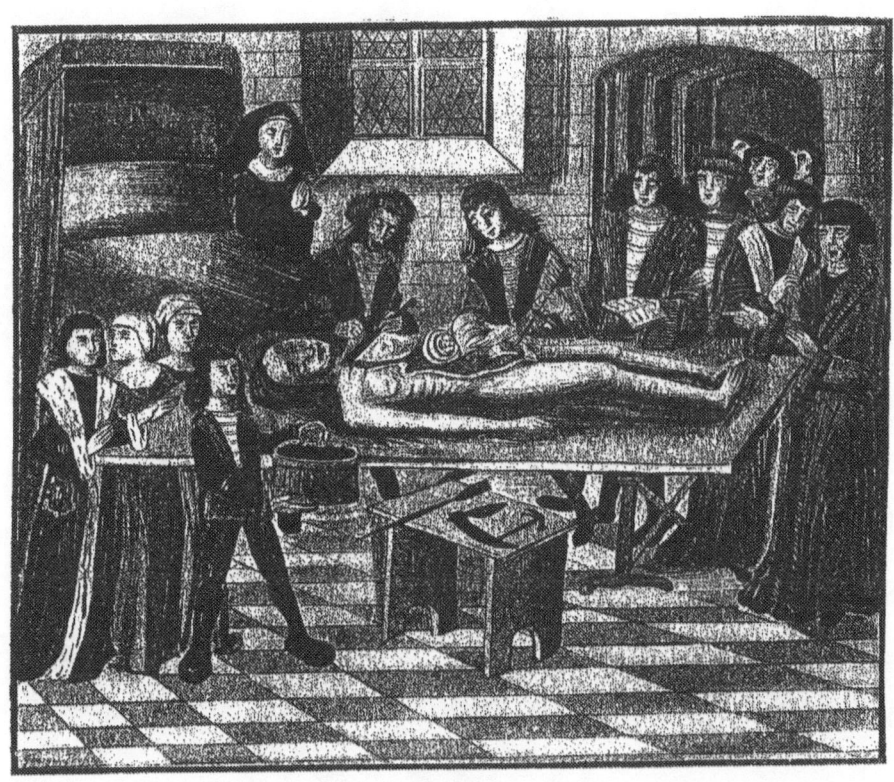

Figure 3
AN ANATOMICAL DEMONSTRATION IN THE 14 C
A Reproduction of a Miniatuer in a 14th C French Manuscript of Guy's Treatise of Anatomy
Ms 184, Library of The Faculty of Medicine at Montpellier

Courtesy of The Surgeon General's Library, Washington, DC
7. Paris—a commentary (French.

RUBRICS FOR TREATISE I THE ANATOMY[178]

---

[178] See the note on Rubrics in the Section on Contents at the beginning of Volume I. (LDR).

ANATOMY: DOCTRINE I

CHAPTER 1

General Matters and The Nature of The Parts

Nicaises Note on The Anatomy

*G*uy owned only a few resources for this Treatise other than a translation of Galen's de Usu Partium (The Purpose of the Parts of the Body) and some Arabic pieces. He did not have Galen's Anatomic Demonstration; it was not yet translated in Latin, although, as Guy knew, it was cited by both Halyabbas and Avicenna. Today,(1890) only nine of the fifteen sections of Galen's book are extant.

Guy completed his anatomic studies at Bologna, taught by "my Master Bertucius". Guy said that his course consisted of four dissections with lectures. Guy had seen Henri de Mondeville's anatomic drawings and paintings.[179] Although he had not performed a dissection on his own, he advised such for his students, both on humans and on animals, at a time when cadaver dissections were uncommon.

Some Comments About The History Of Anatomy In The 14th C

The Hindus had cultivated anatomy more than a thousand years BC, as is described in the Ayurvedas, a textbook of Medicine that was considered to be holy. A Latin translation from the Sanskrit was made by Francois Hessler at Erlangen in 1844.

In the Iliad and Odyssey we find many anatomic references that are more precise than those in Genesis. Hippocrates wrote nothing anatomical. The first honor was Alexandrian, upheld by Erasistratus and Herophilus (330-290 BC). But it was Galen who was the first to make anatomy a prominent feature, and his work dominated the field until Vesalius in the 16th C

---

[179] Mondeville's parchment Ms (Bibliotheque National, #2030) is titled *The Practice of Surgery Fortified by Theory,.* and was published at Paris in 1306. The lettering is gothic-style in two columns. This copy was published on October 14, 1314; it consists of one hundred-eight pages. Only the initial twenty-three deal with anatomy, the remainder are surgical. It contains fourteen small anatomical drawings, none of which merit reproduction here. I think that Henri used larger paintings in his lectures (EN).

THE MAJOR SURGERY OF GUY DE CHAULIAC:
TREATISE I, THE ANATOMY: DOCTRINE I

Daremberg[180] called Galen a great anatomist, although he dissected only animals, especially monkeys and apes, according to Cuvier and Plainville.

Galen, therefore, was the oracle of Anatomy and Physiology for many long centuries. The Arabs copied him, but only in parts and with many alterations. Those translations were the only sources for the medieval physicians. Even in Guy we find many Arabic terms, such as: mirach (the lower abdominal wall), syphac (peritoneum), zirbus (epiploon, omentum), canna (trachea), and mery (esophagus).

An awakening began at the end of the 13th C when Mundini (1250-1326) appeared as an anatomist and famous teacher at Bologna, where he wrote a small work in which, like Galen and Avicenna, he used no drawings. The first printed edition (1476) was titled Mundini's Anatomy, From Head To Feet.

Mundini's original work was followed by many editions and commentaries. Arnold of Villanova's Ms was printed in 1528. As noted in my Introduction, Arnold's authorship has been challenged. Bologna continued after Mundini as a center for anatomical studies, and Guy attended the school where Bertucius, Mundini's successor, conducted the dissections on cadavers. The luster of Bologna faded when religious obstacles were raised against dissections.

An edict (1230) of Frederick II, the Holy Roman Emperor of Germany and the Two Sicilies, required physicians to attend one human dissection every year. That edict was not very effective until Mundini's time. In France, the first dissections were performed in 1376 at Montpellier.

This brief summary reveals the status of anatomy in Guy's time, when he wrote the Major Surgery. He well knew the importance of anatomy for the surgeon, and he placed a remarkably neat and precise summary at the beginning. It reflects what he had learned from lectures during his travels. I think that he could have augmented his limited descriptions if he had wanted it for his book, and such was not his wish. He deemed it better to state only what was strictly necessary. At the end of the book and at other places in the text he added anatomic material in the descriptions of lesions and operations. An example: When he discussed wounds of the abdomen he wrote this about the regional anatomy, "If the syphac (peritoneum) is not sutured with the rest of the abdominal wall (mirach), the healing (ie incarnation) will not be good." That concern, whether or not to close the peritoneum during laparotomies, is debated even today (1890). The wide use of Guy's Anatomy gives us some idea of the state of the science before Vesalius, as testified by the large number of mss and printed editions instigated by it. The Treatise on Anatomy was reproduced apart from the entire work, as I have listed them in my Introduction.(EN).

Here, as before, I shall abbreviate Nicaise: There are seven Mss: 1. Basel (Latin). 2. Paris (French. 3 London (English). 4 London (English). 5. Paris (French). 6. Paris (French),

---

[180]   *The Anatomical, Physiological and Medical Works of Galen.* Paris, 1854-56.(EN).

There are ten printed editions that either reproduce Guy's Anatomy, or are commentaries, or contain drawings, or are coupled with others, including works of history published before Vesalius (1542-43)(LDR).

Return To Guy's Text

According to Galen, who is our Sun in Medicine, in Book 17 of his *de Usum Partium*, in the next-to-last Chapter, there are four uses for the science of Anatomy. The first and most important is the demonstration of God's wonder-work. The second is to recognize any abnormalities (ie afflictions). Third is to anticipate what will happen to them (ie prognosis). The fourth is to support the treatments of maladies. Therefore, all physicians should know anatomy. Galen, at the beginning of his *Treatise on Internal Maladies and The Affected Places*, wrote, "The ancients, as we do today, studied to learn the parts of the body and their symptoms (ie passions), so to tell how to treat each on its own. Although we recognize what we can see and feel on the surface, we must know our anatomy and physiology to diagnose what is hidden from view and how it functions. That will lead us to the source and to the correct method for treatment. Galen also wrote (idem.) that a physician should identify the source of symptoms when they appear, and that information is much more important for the surgeon who accepts his Doctrine (ie Methodism) as fostered in Book 16 of his *Therapeutics,* which, as translated from the Arabic, is titled *On Health.*

Surgeons who ignore anatomy err by cutting nerves and ligaments. A surgeon should know what every part of the body is like, and what lies within, and he can see if a nerve or tendon or ligament has been cut by looking in a wound. Henri de Hermondaville[181] reasoned as follows[182]: "Every artisan is expected to know what he is working with, lest he commit errors when at work. The surgeon is an Artisan for the health of the human body. Therefore he should know its Nature and its composition. And that is Anatomy. For example, if a blind man sets about trimming a tree, he fails because he cannot envision what he is working at. Similarly lacking is a surgeon who is ignorant of anatomy. That is confirmed by simile". That is the case with the malfeasant 'butchers' as described by Galen in Book 2 of *Therapeutics*. They cannot find the joint when amputating, they fracture, they break apart, they tear. I repeat, all physicians, especially those who practice surgery, should know anatomy and they should know how the parts function. Those are the three bases and the elements for every treatment. That is what we can deduce from Galen's *Internal Medicine.* I will avoid a tedious citation and simply say that Galen repeated such comments in the 17 Books of his *de Usum,* and in the fifteen books of his *Anatomic Demonstrations (*ie not available to Gu*y)* as mentioned by Hallyabbas in his

---

181   One of several ways that Henri de Mondeville's name was spelled after his death (LDR).

182   'Reasoned': Here is Guy's first description of the syllogistic method of Aristotle so popular in Paris in his times. He repeats Henri de Mondeville's syllogism! 'Simile' is step two of the system(LDR).

Techni in the section on the minor Arts, and in his *Book of Royal Dispositions*. And the same work of Galen is cited by Avicenna in Book 1 of the *Canon*.

In the following we will place at least the more important anatomical matter that will suffice to guide an operating surgeon in making his incisions and reducing fractures.

Definitions and Scope

Anatomy is dissection to determine the locations and the dimensions of the parts of a normal body, especially that of human, because that body is the focus of our Art. The term is derived from 'ana' meaning correct, and 'tomos' meaning taking apart (ie division).[183]

It is studied in two ways: First is what can be learned from books which explain the functions (ie uses) of the parts. But that is not enough to explain everything that is not obvious to direct observation (ie sight and touch), as stated in Galen's *The Uses of the Parts (de Usum Partium)* Chapter 8. Averroes wrote in his *Collection (Colliget)*, "Our anatomy is limited in respect of what one cannot directly observe".

The second way is to learn from cadavers. We examine the bodies of recently decapitated or hung persons, especially their internal regions, their muscles and pannicules (membranes), their many veins and nerves, especially their origins. We follow the instructions of Mundinus of Bologna.

My teacher, Bertrucius, did it in this way many times. He dissected a supine body on a table in a course of four lectures. First he demonstrated the organs of digestion before they stank. Then he demonstrated the spiritual organs (ie the heart etc,). Next came the animal organs (ie other viscera), and Fourth were the extremities. As Galen suggested n his commentaries on the *Sects,* Bertrucius pointed to nine things in each part: the situation, the substance, the complexion, the quantity, the number, the contours, the neighborhood, the actions, and the uses.[184] Then he listed the maladies to which each part was subject. In that way Anatomy lent credence and assisted a physician in diagnosis, prognosis, and treatment. We also studied bodies that had been desiccated by exposure to the sun in open air, or had been buried or drowned in running streams or had been boiled to prepare them for the study of the bones, cartilages joints, intact nerves, tendons, and ligaments.

Those are the two ways in which we learn the anatomy of the bodies of humans, monkeys, pigs, and many other animals, not just by the study of the thirteen pictures used by Mondeville in his lectures.

---

[183]   Joubert wrote that this statement indicates how deficient was Guy's knowledge of Greek. "Although he was well versed in other languages, he was blind in Greek."(J.)

[184]   Galen's term 'usum' meant purpose, and reason (rationale), and reflected his teleology. Every anatomic structure was created to serve one or more divine purposes. Therefore, 'use' and 'purpose' expressed the 'reason' for its existemce and function. I will employ the terms interchangeably in this edition. Galen's teleology is nicely explained by May in her translation of his *de Usum partium* (LDR).

What constitutes a human body? It is apparent that it is the product of purposeful design, composed of many different items. Every member, as Galen said in his *Usum . . .* , is neither completely separate from nor completely adjoined to other parts. Some are larger, some are smaller, and some are indivisible. Avicenna wrote (*Canon, Book 1)* that the members are initially engendered from combinations of humors and they are either Simple or Composite. Medicine mostly deals with the latter.

Simples

Simples are composed of one indivisible substance that is obvious to our senses, and everywhere in the body it has the same name and purpose. There are ten Simples: cartilage, bone, nerves, veins, arteries, skin, and flesh. We add fat, hair, and nails which are not proper members because they are superfluous, yet they have special uses and are engendered by humors, as are the other parts of the body, as stated in the *Techni*.

Some parts of the body are sanguine (the humor of origin) and can regenerate losses and can heal, such as flesh and fat. Others are spermatic, that is, they arise from semen, and they cannot regenerate or heal other than by cicatrix, as are bone and others. Some simples are warm and moist, others are cool and moist or cool and dry. If they contain a warm humor they may be warm and dry, only as is skin, which is its nature, and as such it is the standard for comparison of all parts.[185] Furthermore, the skin is the standard not only for the particulars of the body, it serves also as such for anything that can be regenerated or corrupted, as Galen stated in the final chapter of his *Complexions)*. Warm and humid members replete with blood (ie the sanguine humor) as are spirit and flesh, where Averroes (*Colliget, Book 2)* said that humidity dominates. Phlegm and fat and marrow are cool and moist. All others are cool and dry in various degrees: bone, cartilage, hair, cords, ligaments, nerves, veins and organs.

The foregoing material really is a sea in which a physician can navigate, and we relinquish to Medicine the task of assessing the natural complexions of the various parts.[186]

Composites

The composite members contain simples. Some of them are complex forms that can be divided into other composites. Although all the parts may not function alone as if they were the whole, some of them may perform such. They are called organs and instruments,

---

[185] Joubert explained: All parts of the body are compared to skin which is the standard, and all others are more or less temperate, and are qualified as the degrees of coldness, warmth, dryness, humidity, and combinations. The heart is mostly warm and, according to Galen, is softer than skin, therefore it is more humid (J).

[186] Mingelousaux wrote. "A physician should should not embark on that sea without the aid of Medical Science (ie Physic) . . . . etc. (EN). This literary 'gesture' repeats Mondeville's, who

and all are controlled by the vital spirit, as are the hand, the face, the heart, and the liver. Galen (ibid,) said, "Nature made every part obey the spirit that empowers the body." Some parts are principal, others are not. The principal organs are the heart, the liver, the brain, and the testicles. All the others are not principal and they vary in importance; the lesser ones include the eyes, the nose, and the hand. More important are the head, the face, the neck, the shoulder and the other eight members, listed here for the surgeon. So too, we shall discuss how the members are composed, their actions, and passions, how their roles are played, their quantity and quality, their conformations, the essences of the simples and similars in them, and their principal functions and their purposes. Some are made for specific functions; others are substitutes; some are designed to protect all the others. Examples of the first are the hands, as explained in Parts 1 and 2 of Galen's *Usum . . .;* examples of other categories are in six other parts of the book. Part 4 applies Aristotle's claim, and all testify to the fact that no part of body was created in vain and does not fulfill a need. Each part has its own complexion and composition and is endowed with its own virtue by Our Creator. Since every composite is made of simples, what holds for the composite, as to its elements (ie complexion, etc.) holds for the simples.

The heart, the most important organ, is said to be dry because its constituent ligaments, membranes, and fleshy muscles are dry. But the multitude of spirits in it which animate (ie provide energy) for the entire body allow the academic physicians to say that the heart is warm and dry. However, the Philosophers of Montpellier say that it is somewhat tempered toward warm and humid.

As to the liver: It essentially is warm and moist because it is mostly fleshy coming from sanguine humor, and it contains many arteries which carry the spirits to it.

The brain is cool and moist by virtue of its marrow-like substance, although it distinctly different from bone-marrow. The latter is engendered from humors, whereas the brain forms from semen. Galen said it is almost warm (Part 2 of *The Parts of Animals).* Therefore we may syllogize our deductions of the complexions of the other composite members, drawing conclusions from the complexions of their constituent substances.[187]

---

passed on to the physicians the chores the academicians liked best, debating the intracies of categories and terminology (LDR).

[187] See Fn. 182, above (LDR).

CHAPTER 2

The Anatomy of Skin, Fat, Flesh, and Muscles

W e shall begin with the skin inasmuch as it covers the rest of the body. It is a tissue containing networks of nerves, veins, and arteries; it was designed to protect and to provide sensibility. There are two types of skins: The first covers the outer surface of the body and acts like leather; it has five varieties (Galen, *Usum . . . .*). The second type of skin covers the organs, a membrane, called a pannicle, such as the membranes that cover the brain, the pericraneum over the skull, and the periosteum that sheathes the bones. Other pannicles are the peritoneum, the pericardium, and the coverings of all the viscera.

Fat is next. It is an oily substance that insulates (ie warms) and lubricates the parts. Here, also, there are two varieties: the exterior fat under the skin is what we call 'fat', and the other sort which is internal, that of the abdomen and around the kidneys, which we call 'lard'.

Next we shall deal with flesh, of which there are three types. The simplest is pink as found in the glans penis and in the gums around the teeth. Another type is glandular or nodose, as is found in the testicles, the breasts, and the emunctories. The third type of flesh is in muscles or lacerti; a large amount of it is distributed throughout the body; it is responsible for voluntary and involuntary (ie manifest) movements.

All the muscles are described in *Usum.* and in Galen's *Voluntary and Involuntary Motion.* Although muscles in a way are Simples, really they are composites of nerves, arteries, and ligaments which thread their way through the flesh, and, a membrane that wraps all the rest within. All this is described by Avicenna in Book 1 of his *Canon.* Muscles and lacerti are the same except for the descriptive terms. A muscle is shaped like a mouse (ie mus) and a lacertus is shaped like a lizard. They have slim ends and plump centers. I disagree with Henri de Mondeville who said they were different in substance.

Galen (*Usum,*) said that a composite muscle gives rise to cords and tubular ligaments that widen as they approach a joint and attach to the pannicle (ie periosteum) and can move the bone. After it passes through the muscle it leaves at the other end, again as a ligament or cord. Then it may reenter another muscle, and again leave as a cord to attach where it can move a bone. The chain continues to the end of the limb. The transition from a rounded ligament to a broad attachment begins before the muscles reach the joints which they will move. We can see it in the arm. The nerves from the vertebrae of the neck pass to the arm through muscles in the neck. When they reach the shoulder they become cords about two or three finger-breadths before they dilate and go over the shoulder attach to the shaft of the humerus and to move it. As a cord it passes two or three finger-breadths before it enters a muscle at the middle of the arm and leaves it two or three finger-breadths above the elbow where again it rewidens and covers the joint and attaches to bones, the same distance below the joint. The same process of widening, cord formation and interposed muscles continues through the forearm into the hand, inserting as flat structures at the ends of tendons, on down to the fingers, so to move them. This explains why wounds that occur within two or three finger-breadths of joints are dangerous, where the bare cords lack the fleshy protection of the muscle. When they (ie the nerve components of cords) are penetrated they may cause spasms and convulsions which may be lethal. Galen explained it in *Techni,* Part 3, and *Therapeutics,* Part 4.

Halyabbas (*Royal Book,* Part 3) said that muscles and lacerti differ in five ways: quantity, shape, location, composition, and where their tendons attach. Galen (*de Usum* . . . Part 6) said that muscles are arranged in four ways: vertical (ie straight), transverse, and oblique in 2 directions. Avicenna counted five hundred-thirty muscles (*Canon,* Book 1).

CHAPTER 3

Nerves, Cords, and Tendons

M uscles contain nerves, ligaments and flesh, and I will deal with them here. A nerve is a Simple created to provide motion and sensation in muscles and other parts. Galen said (*Usum.* Part 4) that Nature had three uses for nerves: 1. For sensation. 2. For motion in parts that move, and 3. To anticipate and defend sensible structures.

As to sensation: cartilages receive no nerves, nor do most glands, but the teeth are innerveated (see Ch. 17 ibid.). All nerves arise from the brain itself or from its extension as the spinal cord, as taught in the School of Philosophy and Medicine (ie at Montpellier). The nerves which leave the brain are softer and mostly serve sensation, whereas those arise from the spinal marrow are tougher and mostly serve motion. Can the same nerve serve both sensation and motion? Galen believed (*Internal Medicine,* Part 1, and *Illness and Symptoms*, Part 4) that the same nerve could carry both but not simultaneously, and that is the belief of the School at Montpellier. Indeed it is a difficult question, and even more difficult is how the messages are carried. By transmission of a substance or by radiation of an aura? It seems to me better simply to leave the matter rest unanswered. We do know this: there are seven pairs of nerves from the brain and thirty pairs from the spinal cord plus one unpaired nerve from the lower end. All of this is described by Halyabbas in the second sermon of Part1 of his *Royal Book*

The ligaments resemble nerves in appearance, however they arise from bone. Some bind the bones together and others attach to the outer surfaces at joints (Galen, *Usum . . .* Part 12) The outer surfaces of the joints are covered by ligaments that can stretch.

Cords or tendons (they are the same) also resemble nerves even more than do ligaments, and like ligaments may lie between nerves and bones, and also they may lie between nerves and ligaments. Tendons arise from muscles and they receive nerves for both sensation and motion. When they leave muscles they are round (ie cylindrical) but they splay at the joints where they attach to bones to flex or to extend them. When one tendon flexes its opposite relaxes. When one or the other is divided, flexion or extension is lost (Galen, Idem).

CHAPTER 4

Veins and Arteries

T he veins function differently from the arteries. The veins arise from the liver, the arteries derive from the heart. In places they are separated, as in the armpit and in the rete mirabile. However, you rarely find an artery that is not accompanied by a vein; almost everywhere in the body they are distributed together. The surgeon needs to know that fact, if nothing else (Galen, idem.). The vein alone carries nutritious blood and the artery carries only spirited blood. As the veins leave the liver and the arteries leave the heart, they divide in two; one branch goes upward and the other branch goes down. Each gives off branches until they reach the extremities, and en route they nourish and animate all the parts. The veins are larger and are at greater risk for hemorrhage. We will deal with particular vessels when we discuss the important organs.

CHAPTER 5

Bone, Cartilage, Nails, and Hair

O ur dissections finally reach the bones in the depths of the body, and they are the hardest parts. By virtue of that quality they defend the contents of cavities such as the chest, the cranium, and the spinal canal. Avicenna (*Canon,* Part 1.) counted two hundred-eight bones other than sesamoids and the L-shaped bone (ie hyoid) at the base of the tongue. Some bones contain marrow, some bones create a fac;, others establish the height of the body. Some function alone, others participate in joints but each is different from the other. Some are straight, others are bent. Some are large, others are small. Those that participate at joints are wider there than in their shafts. At joints they differ: one of them may lodge in a cavity of the other; sometimes they are bound to each other; some joints are joined end-to-end, and some are inserted into another, as are the teeth, driven in like nails. Some joints are serrated as are the cranial sutures, and some bones, as in the forearm and lower leg are bound to each other at both ends resembling tongs. Some joints are compounds, as in the fingers. Some joints allow no movement and are solid, whereas some bones have rounded balls at the ends set into hollows of other bones, and are prone to dislocation, or to complete separation.

Cartilage resembles bone but is softer. It may replace a bone when the latter is absent, such as in the eyelids, nose, and ears, and in certain joints. It allows for freer motion than that of bare bone, as in the chest, hips, and extremities.

The nails at the tips of digits help grasping. Hair is decorative and it is a purge.[188]

---

[188] The meaning is obscure. I assume that removing hair could be a form of purgation (LDR)

DOCTRINE II

The Anatomy of Composite Structures

CHAPTER 1

The Cranium And Its Parts

We have discussed the Basics of Simples and Composites; now we will describe the anatomy of the individual composite structures, both large and small, in eight chapters. Each chapter will have three sections: a description of the part, its uses, and how it affect the different treatments.

We begin with the Head, especially the cranium itself and the brain; the first is the location and the second is the source of the mind (ie reason). I cite Galen *Usum* . . . , Bk. 9, Ch. 4, and *Conservation of Health,* Ch. 9, and in both the *Internal Maladies* and the *Commentary on the Management of Acute Fevers.*

We will describe the nine items concerning the cranium as named by Alexander in his *Commentary on The Book Of Sects*: its purposes, location, neighborhood, quantity, shape, substance, complexion, components, and its ailments. According to the Philosopher (ie Aristotle), the cranium is the hair-bearing structure that contains the animal spirit, which is its Purpose. It is the uppermost part of the body and that is why the eyes are there. Its relation to the rest of the body—the face and its parts, and the muscles that move it that are attached to the neck—is easy to see. Halyabbas (third *Sermon of the First Part*) said," There are two sets of muscles that move the head. One group moves only the head, and they attach behind the ears and to the clavicles. The second group move both the head and the neck. We will describe them later.

The cranium is filled mostly by a large inner cavity that can contain the human brain, which is larger than that of any animal of a man's size. Its shape is that of a ball slightly compressed at the sides, causing the bulge of the forehead and the posterior bulge of the occiput (ie so-called bosses or gibbuses). Galen describes it in his *Techni*, Part 2., and in *Usum* . . . Bk. 8, final Ch., When compared with faces of other animals, it is the least offensive and the largest, and is the most expressive of all round faces.

The substance of the cranial 'bowl' is bony and membranous. Its complexion is cool, derived from its many parts. The components of the cranium are ten or eleven bones, as counted by Avicenna; five are inside and others make up the outer shell.

The Container: outermost is the hair, then the muscles, then a thick membrane, and then bone. Just beneath it are the dura mater and the pia mater, which continue beneath the brain, and beneath them we finally encounter the rete mirabile and more bone, through which pass the nerves.

We need not add to what we wrote about the hair, the skin, and the muscular tissues. The thick membrane that covers the bone is the pericraneum, (ie periosteum); it is nervous in character. It arises from the dura mater to which it is attached by ligamentous tissues that pass through the commissures between the bones. The cranium itself is not a single bone, but is an orderly aggregation of seven contiguous bones. That multitude is a source of protection; injury to any one may spare the others. They are joined by serrated connections which allow the escape of vapors emitted from the brain.

The most anterior of the seven bones is the coronal; it extends down from the dome to reach between the eyes. Behind it is the coronal commissure that crosses the dome transversely. The coronal bone has openings for the eyes, the 'sieves'[189], the nostrils, which are separated by bony tissue like a cock's comb that contains cartilage[190] Sometimes the coronal is divided by a commissure down the center, more common in women.

The second bone is at the rear, called the occipital or lambdoidal, bounded behind by an L-shaped commissure, the lambdoidal. It is hard and has an opening below through which the spinal cord descends from the brain through all the vertebrae the full length of the back.

The third and fourth bones are at the sides, between the coronal and lambdoidal and separated from each other in the center by a commissure in the long axis of the cranium. They are the square parietal bones and meet the transverse coronal suture in front and the lambdoidal at the rear.

The fifth and sixth bones are called the petrous (ie the temporals) because they are stony hard; their sutures are called false because they lie against the parietals[191] and attach over their surfaces. The ear canals pass through the petrous bones as emunctories, over the mammilary projections (ie the mastoid processes). In the rear they reach the lambdoidal sutures and extend to the temporal bones.

The seventh bone is the paxillary (basilar) shaped like a quince over the palate. It has many openings, like a sponge, and allows the superfluities to be purged. The bone substance itself is vary hard.

To sum up: The bowl of the cranium consists of seven bones which can be see in dissection of heads prepared by boiling. Galen confirmed the count in *Usum* . . . Bk. 6, Ch. 15. In addition there are the two bone of the bregma, held in the commisures between the hard bones in front at the rear.[192] The paxillary (basilar), over the palate is considered by some to be part of the upper jaw, whereas others say it is part of the cranium as it lies,

---

[189]   The olfactory cribriform plate (LDR).

[190]   The turbinates (LDR).

[191]   The petrous bones are extension inward of the temporal bones, which do not form serrated commissures with the parietals which they overlie. That is why they are called false (LDR).

[192]   The bregmas here are the parietals, and at the junctures of the sagittal commissure with the coronal and the lambdoidal were the soft fontanelles. Did they contain bits of bone before their openings were completely closed by bone? (LDR.

shaped like a quince, between the two parietals. Here is where William's, Lanfranchi's, and Henri's anatomies are defective. They claimed that the paxillary bone (ie shaped like a seat), which lies beneath the lambdoidal is a cervical vertebra[193]. Also they claim that the petrous bones are part of the overlying parietals and are not in direct contact with the brain, and therefore are not part of the 'bowl', whereas the opposite is true.

Therefore, seven bones contain the brain. The smaller bones are not principals, such as the bone of the crest (crista galli) that divides the cribriform processes (ie the sieves) of the nostrils, or the Bones of Paris that are part of the face[194] and the aigulheus (ie needle-like) that are the styloid processes of the temporal bones, and the clavals (ie mastoid processes) behind the ears, where muscles attach. All are included in Halyabbas' count (*Royal Book,* Part 1, Sermon 2) where he said that there are fifteen bones in all in the cranium, whereas Avicenna counted only five, including the three that meet at commissures and the two applied as flakes at the sides (ie the temporals), lacking the true serrations. They extend into the Bones of Paris, etc. More about them later. Now we go to the contents of the cranium

The internal parts are visible after sawing the circumference of the dome and lifting it off. First we encounter the dura mater and the pia mater. They are the two membranes that are richly endowed with veins: one set for the outer (dura mater) and the other for the inner pia mater, which distributes the veins over the entire brain. The dura mater extends through the commissures to engender the pericranium. The veins and arteries reach the membranes through openings in the cranium below and up through the cranial commissures. Within the membranes lies the brain itself, a soft white substance. The surface is thrown into gently rounded folds, as if added one atop another. Paired (twinned) nerves of sensation exit from beneath the center of the brain. Galen said that they are paired; if one side cannot function the other can serve for both. (*Usum . . .* Book 8).

There are three ventricle within the brain, directed in the long axis; each has two parts, and each part has its own function. The forward parts of the anterior ventricles serve for common sense. The second part is for imagination. The middle ventricle is for deeper reasoning and thinking, and the rear part is devoted to memory and recall. You can infer that the anterior parts of the ventricle are the largest, the middles are smallest, and the posterior are of middle-size. The ventricles communicate openly and allow free exchanges of the spirits. Up front are the mammillary processes which serve olfaction. The rest of the brain gives rise to seven pairs of sensory nerves that go to the eyes, the ears, the tongue, the stomach, and elsewhere. As noted above, we can see where the

---

[193]   The confusion of terms and the sources (man or other animal) concerns the basilar portion of the occipital bone which has a separate center of ossification and does not fuse with the occipital until late in childhood. its spongy character is that of incomplete ossification. Furthermore, what Guy describes as overlying the palate is the extension of the basilar process that meets and joins the sphenoid (LDR).

[194]   The Bones of Paris are the zygomatic processes (LDR).

nerves emerge from the brain and the openings through which they leave the cranium, and how they are covered by extensions of the same membranes that cover the brain. There is a place in the middle ventricle called the Fossette, or opening, and a worm-like region (twisted) created by humps of glandular tissues. Beneath that is the rete mirabilis (ie Circle of Willis) made of arteries alone that come from the heart and provide the spirit that is combusted to create the animate substance for the body.

Finally, you will observe how the spinal cord leaves the parencephalus (cerebellum), the posterior part of the brain, It's bare surface is covered by the same two membranes that cover the brain.[195] It descend through the center of the vertebrae to the lowest part of the back, and it gives rise to all the nerves of motion. It resembles the brain's substance and really it is part of it, and its symptoms are the same. Galen noted that in *Usum* . . . , Part 2, Ch. 12.

That ends our description of the contents of the cranium, including the nine items as planned.

Now we will mention the maladies affecting those parts. The cranium can suffer wounds, aposthems, and sick complexions. Wounds that penetrate the bone are serious, even moreso are those that damage the membranes, and worst of all are those that injure the brain.

The operations that involve the commissures risk allowing the dura mater to collapse onto the brain. All surgical procedures should follow the direction of the hair lines, and they will cut across much less muscle. Those incisions will be easier to close by bandaging, an important consideration for the rounded surface of the head.

---

[195] Here Guy describes the medulla oblongata rather than the cerebellum, called parencephalon in Greek (LDR).

CHAPTER 2

The Face and Its Parts

The parts of the face are the forehead, the eyes, the nostrils, the ears, the temple, the cheeks, the mouth, the jaws, and the teeth.

The forehead consists of skin and muscle lying directly atop the coronal bone, where it bulges slightly, and the outer table (ie of two) is elevated by an increase of the internal spongiosity, appearing as if it was double-thick.[196] The forehead shapes the eyebrows which are there for the sake of appearances and to protect the eyes. Note this: make incisions there in the long axis of the body (ie unlike elsewhere) lest you cut across the muscles that raise the brows.

The eyes are the organs of vision, placed in the spherical caves called the orbits, which are formed from the coronal and temporal bones. Galen (*Usum . . .* , Bk. 10, final chapter) wrote. "The optic nerves are hollow, enabling them to carry the visual spirit (ie the pneuma) from both sides of the brain, where it is united within the cranium and passed to the eyes on the same sides as its origin from the brain. The spirit does not cross to the opposite side, as it has been claimed by some authors."[197]

The eye has seven tunics and three or four humors. The outermost tunic is the conjunctiva, white and thick. It covers the entire eye except the cornea. It is an extension of the pericranium.[198] The other tunics surround the eyeball itself. The various colors of the eye are provide by the iris which surrounds the changing diameter of the center of the eye.

---

[196] The frontal bosses were more prominent in the northern Europeans, as can be seen in paintings by their artists, and are not seen in the Italians' unless the artist's model was from the north! Various explanations have been offered: the cause may have been rickets in countries less favored by the sunlight that producedof Vitamin D in the skins of the southern Europeans (LDR).

[197] The phenomenon of vision was explained thus by Galen. The Pneuma, or Spirit departs the brain, passes through the chiasm and the optic nerves to the retina which receives the image. Then it returns from the eyes, nerves, et al. to the brain where the image is realized. We can conceive it as a built-in flash-bulb in a cerebral camera! Galen thought that all the sensory cranial nerves carried some pneuma, but not to the same extent as the optic nerves which were hollow (LDR).

[198] I think we should read this to mean that he conjunctiva was an extension of, the periosteum lining the orbit, to be part of the outer tunic around the eyeball (LDR).

Three of the other six tunics are extensions from within the cranium; the others are completely external. The first continues the dura-mater which becomes the sclera, meaning tough, and the cornea. The second derives from the pia mater. Within the eyeball it is called the placenta (ie 'secondine') except where it is exposed and is called the uvea, which has an opening called the pupil. The third external tunic extends the optic nerve and lines the eye as the retina; where it crosses in front of the lens it is called the aranea (the spider web)[199]. Therefore, of the seven tunics of the eye, three are continuations within the eyeball of external material.

There are three humors: First is the crystalline (the lens), placed in the center of the eye. It is slim and colorless, and it captures whatever is seen Behind lies the the second, the vitreous which holds the lens in place. Both of the two humors are surrounded by the derivative membranes from the brain (the optic nerve) (ie retina and aranea). The third humor is up front, the albuginea, set between the aranea (ie the iris) and the outer layers. Galen thought that there was a fourth humor, described in the fourth chapter of *Usum . . .* it is completely transparent and is set within the pupil.

Leaving the eyeball itself we encounter motor nerves derived from the second pair of cranial nerves, the six muscles that move the eye, veins, arteries, spongy flesh, and the sources of the tears. All are within the orbit and surround the eyes. The eyelids in front contain cartilages and lashes, long and short. The upper lid is closed by one muscle and opened by two that cross it. Its benefits and its performance are described in detail in Jesu's *In The Eyes,* and by Alcoitin and in the books that are special for the eyes. The above material should fulfill the needs of surgeons.

The nose is shaped by flesh, bone, and cartilage. The fleshy part is skin and two muscles near the tip. Two triangular bones which meet under the nose are joined by a 'rib' down the center between them; they are set at the sides, adjacent to the cheeks. There are two cartilages: one is between the nostrils at the lower end; the other is inside, between the nostrils.

The nostrils are channels that extend upward to the "sieves", on which rest the mammilary processes of the brain, and through which one sniffs the odors. The nostrils open in back of the palate, beyond the uvula. Air is inhaled through the channels to reach where the fumes are evaporated (ie for olfaction), and to pass in and out of the lungs, and they are vents for superfluidities from the brain.[200]

The ears are cartilaginous and can resist fracture. They lie on the petrous bones and they are the organs for hearing. The fifth pair of cranial nerves, the auditories, pass through pores and tortuous channels in the petrous bones. The glandular tissues below the ears are emunctories for the brain, The glands lie close to veins, which, according to Lanfranchi, carry some spermatic matter to the genital regions. If you remove them (ie the glands) that person becomes infertile. That is contrary to what Galen said, and to Avicenna statement in his discussion about phlebotomy.

The temple, the cheeks, and the jaws are parts of the face. Within them are muscles, veins, arteries and bones. There are many muscles, especially the seven that move the

---

[199]   The iris was called 'prunelle' (LDR).

[200]   Including catarrh (LDR).

jaws which attach to the clavals[201] and below. Halyabbas said that there are eight that operated the lower jaw: some for opening it, which attach to several points on the clavals and under the lower jaw. Those that close it pass from the petrous bones under the zygoma (ie to the mandible) and are called the temporals. They are very strong, prominent and sensitive. Damage there is a serious matter, and Nature provided the zygomatic arches (ie ansa) to protect them.

There are other muscles for biting and chewing. They arise from the bulges of the cheek bones (ie maxillae).

All the muscles receive branches from the third pair of cranial nerves. In addition, there are many important veins and arteries in the temple, and behind the corners of the eyes and lips.

There are many bones in the face. Galen counted nine. The most important are the two upper jaw bones that meet and fuse beneath the nostrils. The two bones of Paris[202] (ie zygomas), coming from the temple, form part of the orbital rims and the cheek-bulges. They join the maxillae and meet an extension from the temporal (ie petrous) bones to form an arch under which pass the temporal muscles.

The lower jaw-bone: Galen (*Usum* . . . Bk.2, next-to-final Ch.) said. "the lower jaw bone is invisibly divided. The halves meet under the lower part of the beard. At the other ends are protrusions covered by temporal muscles that hold the jaw in its socket. The protrusions are called rubriform because they resemble small breasts."[203]

Next come the five parts of the mouth: lips, teeth. tongue, palate and uvula (ie luette).

Teeth are bone-like, excepting that they are sensitive, innervated by branches of the third cranial nerve. Usually there are thirty-two of them, although some persons have only twenty-eight. The thirty—two include sixteen in each jaw, upper and lower. The two in the center are incisors. All are rooted into the jaws: some with one root, some with two, and others have three or four. We know what teeth do.

The tongue is a muscular structure, soft and spongy, including many nerves, ligaments, veins, and arteries. It has two functions: speech, and movement of food within the mouth. The fourth and sixth pairs of cranial nerves provide the sense of taste. Its nine muscles are divided down the middle by a band, and they form an inverted V origin at the rear. The muscles attach below to an L-shaped bone. Beneath the tongue are glandular tissues that form the saliva and deliver it through two openings.

The palate includes the roof of the mouth. It is covered, as is everything in the mouth, by a membrane that derives from the lining of the stomach and esophagus.

That is enough said about the parts of the face. It is the site of many maladies, and this chapter should be helpful with treatments and prognoses.

---

[201] Clavals: variously defined, but include the mastoid processes (LDR).

[202] The zygomas shaped the cheeks of that handsome Trojan prince (LDR)

[203] The rounded condyles. The term is a scribe's mispelling of 'ubriform' from 'uber' the Greek word for breast (EN).

CHAPTER 3

The Neck and Upper Back

E veryone knows what the neck is and where it is attached to the rest of the body. Its important parts are the throat and the pharynx (gullet) wherein lies the tracheal artery (a single structure) and another passage for particulate matter. Galen describes it in *Usum . . .* , Part 8. The neck can be considered as a container and contents. The former consists of skin, flesh, muscles, ligaments and bones.

The contents are the trachea and the esophagus, the epiglottis (ie the larynx), nerves, veins, arteries, bones, and a segment of the spinal cord. The principal structure is the trachea. Before describing it, let us say that it divides the neck in its long axis; it is placed in front as the conduit for air into the lungs.[204]

The trachea begins at the epiglottis and consists of almost circular rings of cartilage, open at the back where the circle is completed by a dense elastic membrane that is shared with the esophagus.

The esophagus behind the trachea is the conduit from the throat that passes through the diaphragm into the stomach. It has two tunics, villous and fibrous. The inner villous one continues the membrane that lines the mouth. The outer tunic is muscular. Both continue with the abdominal portion of the esophagus into the stomach.

The two channels begin above the gullet, at an open space behind the mouth (ie the pharynx).[205] We call one the gullet, or epiglottis, a partly cartilaginous structure. It is part of the instrument of the voice, and it functions as a lid when one swallows. It functions somewhat like the tongue, which also is part of the swallowing act. The epiglottis (larynx) consists of three cartilages surrounded by five muscles which move

---

[204] Nicaise notes that Roger Frugard called the neck the region which contained the upper vertebrae, the throat, and the lateral structures which included the organic veins (the jugulars) which also were called the cervicals (EN).

[205] Joubert explained: The two are very different. The gullet really is a passage for food and drink. From Aristotle on it has been called esophagus, but the barbariuans called it 'mere'. The 'gosier' (glottis) is at the upper end of the trachea, called 'larynx' by the Greeks who called the gullet the 'pharynx'. At the upper end of the trachea the cavity is shared by both. The epiglottis is properly named as the lid of the larynx (J).

its parts, raising and lowering them, etc. Galen describes them in *The Book Of The Voice And Its Movements.*

Now we shall describe the double-nerve: part of it descends into the abdomen and is sensory for the stomach and the intestines. The other part is recurrent, going back up alongside the epiglottis (ie larynx) to serve the voice.[206]

You can see the large veins and arteries that divide at the fork of the clavicles. The main branches travel at the sides of the neck. They are called the guides, or apoplectics, both deep and superficial. Cutting them will cause life-threatening hemorrhages.

Before describing the 'container', the skin, flesh, muscles, ligaments, and the seven vertebrae I will offer here a general description of the back-bones.

The vertebrae are bones that form the back. Their centers are open to contain the spinal cord, and there are lateral openings through which transmit paired nerves. Many have additions (ie appendages)[207], and they (the spinous processes) line up and down the center like a hedge-row to form a keel. The back with its several kinds of vertebrae was designed to protect the spinal cord from the head down to the bottom.

The back has four main sections (Galen, *Usum*, Chs 12 and 18): the neck, the shoulder, the kidneys, and a single fused bone called the sacrum.

The neck has seven vertebrae; the shoulder section (ie thoracic) has twelve, the renal (ie lumbar) has five, and the sacrum has four. Therefore there are 24 'true' spondyles (vertebrae with spinous processes) alike in structure and placement. The top three cervical vertebrae are very sturdy but have no lateral appendages or lateral openings. There are many cartilages, especially at the lower end where they make a kind of tail. All together, there are thirty vertebrae. For each of them there is a pair of nerves coming from the spinal cord, and one alone at the lower end. If we count the seven paired cranial nerves, there are thirty-eight in all, as we described them in the chapter on the head. Alongside the spondyles there are long muscles in which the nerves are hidden and protected. In common language they are called the 'long' muscles. The bones are covered, as are the cranium and other bones, with a dense membrane that ties the spinous processes together. Therefore, the neck has seven vertebrae that emit seven pairs of nerves at their sides, coming from the spinal cord in that section of the back. They carry sensation and provide motion in the shoulders, the arms, and to parts of the head and neck.

There are three kinds of flesh in the neck: The 'longs', the cervical muscles that lie against the vertebrae, the twenty muscles and tendons which move the head and the neck (Galen), and the tissues that fill the empty spaces.

Many ligaments bind the head to the neck, and the neck to the shoulders (ie thorax). In addition are two large muscles in front that go from behind the ears down to the clavicles.

---

[206] The recurrent laryngeal nerve from the vagus was Galen's discovery. He proudly described it in several places. See May's edition of *Usum* . . . (LDR.

[207] "additions': the transverse and the spinous processes (LDR).

In back are other large muscles which go between the ribs and the spinous processes. Other larger muscles go from the shoulders to the spinous processes. Those binding miscles, tendons, and ligaments surround the neck and allow it to bob, and bow, and turn the face, neck and head. Without them the joints would fail, as Galen stated.

Having dealt with the six or seven items that relate to each part[208], now we can mention the maladies. The neck as a whole and many of its parts can suffer many ailments: wounds, dislocations, aposthems, all of which are dangerous. It is obvious that incisions should be in the long axis, parallel with the important organs. And that will affect how we close the incisions, as we shall describe later.

---

[208] Joubert asks,"Why, six or seven, after Guy had twice said nine (J).

CHAPTER 4

The Shoulders, Arms, And Hands

Now we approach the thorax, but first we are concerned with what is planted on it: the shoulders and their attached arms.

The Spatula (ie scapula) and the Humerus (ie the arm) are obvious, and we need not say what they are, where they are, and what they are attached to. They are the instruments for grasping and for defense, as described in *Usum* . . . Bk.1. Our maker provided and defended us with hands and intelligence (ie minds). Galen cited Aristotle who said that the hand[209] takes precedence over all instruments, and the mind over all the arts. They (ie the upper extremities) consist of skin, flesh, veins, arteries, nerves, muscles, cords, ligaments, membranes, cartilages, and bones.

First come the shoulders: The skin and other soft tissues are apparent. The muscles and cords[210] that move the arm come from the neck and the chest, passing the shoulder and enveloping the joint between the upper arm-bone (ie humerus or adjunct)[211] to which they are attached. The nerves come from the spinal cord in the neck. The veins and arteries give off branches as they descend, but few of their major branches are in the shoulders.

The shoulder has two bones; the spatula (scapula) in back and the clavicle atop the thorax.

---

[209] Galen's term for the upper extremity as a whole was the 'Geat Hand' (EN).

[210] Note again that a 'cord' usually is a composite of tendon and nerve as it enters or emerges from a muscle. Sometimes the cord included an artery and/or a vein (LDR).

[211] Joubert was wrong in calling the forearm the 'adjutoire', whereas that was the name for the upper arm. Galen called the entire upper extremity the Grand Hand with three divisions: the adjunct (also called ulna), the small arm and the small hand. In the Latin editions the humerus sometimes was called ulna or uvula. The last-named probably was a copyists error. The Montpellier Ms uses both adjunt and ulna when referring to the humerus.

The term 'adjunct (ie adjutoire)' nowadays means assistant, but I believe that it was used by Guy to mean "the bone joined to the shoulder", and it was the way the Montpellier Ms used it. Pliny always used 'ulna' to mean the forearm, whereas Virgil used it to mean the upper arm. 'Ulna' was the source for 'aune' which meant the span of one's outstretched arms.

In order to match the Latin text as closely as possible, I will translate both adjutoire and ulna to mean the humerus (upper arm bone)(EN).

155

The shoulder blade (ie the scapula) resembles a spade in size and thinness as it curves over the back. Near the middle of the bone a thin ridge appears and rises leading to the joint. The ridge is rather like a handle with three projections at the end. The first of them has a hollow to receive the head of the humerus. Another, above it, is curved and pointed, and resembles a crow's beak. The third is on the open side and is curved like an anchor.

The clavicle is tubular and one end lies in an indentation at the upper end of the breast-bone. It has two attachments: one is at the shoulder and the other joins the clavicle of the other side. At the shoulder it is bound to the two appendages where the head of the humerus joins the scapula. There are no other bony appendages there, as Lanfranchi and Henri claimed; direct observation establishes the truth.[212] Galen said (*Usum . . .* , Part 3, Chs. 11 and 12) that the acromion, at the upper end of the scapula, coupled with the end of the clavicle, covered the joint and prevented an upward dislocation of the head of the humerus. Also, three large ligaments pass from the edge of the scapula to the humerus, and the joint is surrounded by many strong muscles that come from the thorax and scapula that hold it in place, allowing one to flex, extend and rotate the arm.

The axilla is the region under the shoulder; it is filled with glandular tissues which are the emunctories of the heart.

Now we go the arms. Galen called them The Great Hands, with three parts: the arm (ie ulna), the forearm (ie the small arm or acrochiron) and the hand (ie the small hand). Those parts have other names, too; see above.[213]

The surface arteries and veins are dealt with as one (ie named the same).[214] They are branches of the main vessels in the axilla which divide there in two and begin to ramify. One branch tends toward the surface, and soon ramifies: one branch mounts toward the head over the posterior surface of the sholder. Two other branches pass down the arm; one of them has many branches and is called the cord of the arm, the other passes more directly on the outer surface of the fold of the elbow where it is called the Cephalic. It passes onto the dorsum of the hand between the thumb and the index finger where it is called the Ocular-Cephalic.

The second axillary division descends within the soft tissues of the arm and surfaces at the elbow where it is called the Basilic. It goes on to the hand between the fourth and the auricular [215] fingers, where it is called the Salvatelle. Where the cephalic and basilic

---

[212]   Joubert said that Galen had written that a bone, called the catacleis, was lodged between the clavicle and the scapula. Lanfranchi and Henri seem to have accepted that error (J).

[213]   Nicaise quotes Canappe with a more modern anatomy; "One is the arm, the second is the ulna, and third is the acrocheiron." (the last is Greek for 'hand at the end'(LDR).

[214]   The detailed descriptions of the veins of the arm and hand were especially important for the phlebotomist (LDR).

[215]   The Romans named the fingers. The fifth was the auricular, used to palpate or scratch within the ear canal. The index was second, the third was impudicus (for impolite gestures), fourth was annularus (for rings). Pollex was the thumb, as a shortened finger (ie polled) (LDR).

veins come close at the elbow they communicate via a short vein called the Corporal or Median. Therefore, in the arm are at least four large veins, and arteries, that bleed copiously. In addition there are many smaller ones that simply trickle and are not causes for worry for a surgeon (ie phlebotomist).

There are four paravertebral nerves from the spinal cord that go to the upper arm: one behind, one in front, one above, and one beneath. All of them divide and send branches that carry sensatioin and motion. All of them lie beneath the surface and become parts of muscles by joining cords and ligaments.

The nerves, flesh, and membranes contribute to form the four principle long muscles. Those that move the forearm lie within the arm. Another four in the forearm move the hand, and four more in the hand move the fingers. In the hand we can see the cords (ie tendons, et al.) where they are most exposed to injury, and always at serious risk.

Also, there are many ligaments attached to bones, or crossing over the joints, or as parts of cords, where they, too, are more exposed to injury.

The bones: the upper arm bone has been called the adjunct as well as the ulna (see fn. 211). It is a tube containing marrow, rounded at both ends. The upper end is a simple sphere that fits into the scapular cave to form the shoulder joint. The lower end is a double knob, between which (in front) is a channel that allows the cords that attach to the ulna to act as a pulley, strong enough to lift a pail of water. Behind there is deeper hollow that accepts a beak-like end of the larger focile when the forearm is extend.

The forearm begins at the elbow, at the cubital joint between the humerus and the fociles. The larger focile lies below the smaller and is longer because of the beak-like extension described above. It reaches the hand at the side of the little finger. At the lower end there is a projection like the malleolus at the ankle. The lesser focile lies in front of the other and approaches the hand at the thumb. It has hollows at the end where it meets the rounded ends of the humerus, and fits against the side of the larger focile. At their lower ends, where their hollows meet the hand, the fociles widen and meet each other in a joint. In the region between the elbow and the wrist the fociles are slim and are separated. The intervening space contains nerves, vessels and muscles.

The lower ends of the fociles form a joint with the bones of the hand, which are arranged in three rows. The rounded surfaces of the lower rank meet the hollows of the second rank. The first rank has three bones and we add the malleous of the larger focile as part of the joint with the hand. The second rank has four bones, one of which meets the first bone of the thumb. The bones of those two ranks are small, whereas those of the four of the third rank are larger and longer. The lower ranks are called the wrist, or carpus. The uppers are called the metacarpus or palm.[216] Each of the fingers has three bones.

Therefore, we count 29 bones in the upper extremity: fifteen in the fingers, eleven in the hand, two in the forearm, and the single humerus.

---

[216]   The Galenic anatomy described four metacarpals, the radial metacarpal was considered as the first phalanx of the thumb (LDR).

As to the maladies, we name aposthems, wounds, dislocations, fractures and paralysis.

As to surgery, we can deduce from the anatomy that incisions should be made in the axis of the limb, and in the skin-lines to avoid transecting muscles.[217] As to dislocations, we can see why those at the elbow are the most difficult to reduce, and those at the shoulder can be more easily reduced with simple manual maneuvers. You should now know the parts and the places where to make your incisions. Furthermore, consider the importance of applying topical remedies around the vertebrae of the neck, where the nerves emerge, in cases of paralysis in the arms.

---

[217]   A reminder: a muscle was its flesh as well as its tendons and nerves (LDR).

CHAPTER 5

The Thorax And Its Parts

The thorax is the Ark or the treasure chest for the spiritual organs. The container consists of skin, muscles, breasts and bones. The contents are heart, lungs, membranes, ligaments, veins, arteries, and the esophagus. I need not describe again what you can see, the skin and the flesh under it.

The breasts are on the surface; they consist of white, glandular, and spongy tissues that contain arteries, veins, and nerves. They are closely related to the heart, the liver, the brain, and the genital organs.

Now a brief note about the eighty or ninety thoracic muscles that Avicenna counted. Some of them are shared with the neck, others go to the arm and the shoulder, and others include the diaphragm and those that are entirely within the chest, such as those of the ribs and the back.

There are three sets of thoracic bones. The first set of seven in front are strong cartilages joining the Bone of the Thorax (sternum). The upper most is where the clavicles rest on shallow hollows.[218] Below, where it lies over the stomach and is called the forcelle, there is a cartilage attached to the Bone of the Thorax that is called the Scutiform (ie shield).

In back there are twelve vertebrae that contain a section of the spinal cord that gives off twelve pars of sensory and motor nerves.

On each side there are twelve ribs; seven are true and five are false in that they lack the full lengths of the seven. You can see how they are shaped.

When you cut through the ribs to inspect the contents, be careful in front where you will encounter the mediastinum and its structures. The principal organ there is the Heart. The Lord placed it at the center of the body where it does not favor either side (Galen, *Usum*, Part 6), although the lower end tends to the left to avoid the liver and the upper part leans to the right, to make room for the arteries.

The heart is shaped like a pine-cone, pointed downward; its large base is above. The material of the heart is firm, almost muscular. Within it are two ventricles, right and left,

---

[218]   Nicaise explains: The furcule (fork) is where the clavicles meet the manubrium; Guy considered the two clavicles as parts of a single bone. The forcelle is the shield of the stomach (boucliere) (EN).

and between them is a cavity in which Galen said that the alimentary blood from the liver is digested to become very thin and spiritual. Then it is sent forth through the arteries to all the vital parts of the body, especially to the brain. There it undergoes another digestion to become the animating spirit. In the liver it becomes natural blood (sanguine). In the testicles it becomes regenerative semen. Every where else the members are restored and invigorated. The spirited blood is the instrument of all the sensory faculties and it is the perfect link between the body and the soul.

The heart has two orifices. On the right, as part of the great vena cava that ascends from the liver, a large branch, the arterial vein, carries blood to the lungs and nourishes them. All the rest of the blood is delivered through the body by branches of that great vein (ie vena cava). From the left side emerges a pulsating vein part of which carries vapors and air to cool the heart; it is the venous artery. At those orifices we find three small flaps, that control the inflow and exit of the blood and spirit between the heart and the lungs, as required.

A sturdy cartilaginous bone (ie sternum, the bone of the thorax) protects it, and the entire heart lies within a tough membrane called the pericardium, which receives nerves, as do all the organs.

The heart is bound to the lungs by ligaments in the mediastinum which hold them firmly. We now can appreciate that the heart is intimately related to all parts of the body. And we can understand its great dignity and that it can suffer and endure pain for a long time.

The lungs abut the heart to cool it. Its pale substance is soft, filmy, and spongy. There are three kinds of vessels in it: branches of the arterial vein from the right ventricle of the heart; the venous artery from the left; and branches of the trachea which bring air for the sake of the heart. Those three vessels ramify throughout the lungs, which consist of five lobes, three on the right and two on the left

The esophagus passes behind the lung at the level of the fifth thoracic vertebra and the ascending vena cava. Both of them pass through the diaphragm. The aorta from the heart lies close to the trachea to form a dense mass with an accumulation of ligaments, glands, and membranes rising to the gullet in the neck.

The thorax has three membranes. First is the lining that coats the inside of the ribs, called the pleura. Second is the mediastinum which divides the thorax down the middle. Third is the diaphragm that separates the spiritual organs above from the organs of nutrition below. It consists of pleura above and siphac below which sandwich between them a spread of tendonous membrane. It carries nerves from the ribs in back. The muscles of the diaphragm are attached to the ribs. It is obvious that the diaphragm functions as a muscle that acts in breathing and in straining to expel superfluidities (Galen).

CHAPTER 6

The Abdomen And Its Parts

'Venter' has two meanings. First it is used as a translation of the Arabic word for stomach, because in the Latin translation of the Greek, the word is meri or esophagus. But in Arabic, 'stomach' is called venter. A second meaning for venter is the entire abdominal cavity, and that is what it will connote here. It is the region in which we will discuss the nine items dictated by Mundinus.

First: The General Region. It is situated beneath the spiritual region (ie the thorax). We enter it near the xiphoid (see fn. 218), in what the ancients called the precordium, what we call the orificial. That is where the stomach lies, down to about three finger-breadths above the umbilicus. The umbilical region is called the sumenial.[219] From the umbilicus downward it is called the lesser (ie lower) abdomen.

The hypochondrium means beneath the ribs, and the flanks (ie the isles) are above the hips. A good anatomical dissection to expose everything inside is best made through cruciate incisions.

The abdomen is enclosed within the mirach, the five lumbar vertebrae, and the paravertebral structures. The mirach is the abdominal wall, consisting of four layers: skin, fat, a fleshy membrane (ie fascia) and mucles with their cords. The siphac is a membrane within and adherent to the mirach. However they are separate structures.

Within the abdomen are the omentum, the intestines, which come after the stomach, the liver, the spleen, the mesentery, and the kidneys. We will discuss the bladder and its attachments when we describe the pelvis.

Eight muscles serve to protect the belly and to aid in evacuating the superfluifities from the organs. The eight are described in Part 4 of Galen's *Usum . . . .* and in Part 6 of his *Therapeutics.* Two long muscles go from the xiphoid to the pubis. Two broad ones go from the back to meet the long ones at right angles. Four cross them, two going from the ribs on the right and two on the left. They cross to the the bones of the pelvis and the pubis and form an X.

When dissect the abdominal wall muscles we encounter the siphac beneath them, called the peritoneum in Latin by Galen. "Peri' means surrounding, and 'tento' means to

---

[219] 'Sumeniale' derives from 'subminia' a kind of woman's dress (EN). Did it cover or did it expose the navel?(LDR).

hold the contents. It is a nervous membrane that intervenes between the viscera and the muscles. It can stretch and can shrink to accommodate the varying sizes of the organs, but it can easily be torn and allow the viscera to escape, as what happens in hernias. It suspends the intestines from the back and helps the organs to reject what they must be free of.

Also apparent (ie in the open abdomen) is the arrangement of the organs within, and you can see why wounds in the middle are more threatening than others at the sides (Galen, *Therapeutics*, Bk. 6), and why closing them is more difficult. Besides, in that part the intestines will eviscerate more readily and be more difficult to replace. Also it is apparent that a wound closed by suture will heal with a stronger scar when the siphac is closed together with the mirach.

Now to particulars: On opening the abdomen you first will encounter the omentum (zirbus or coëffe), also called the epiploon in Greek, derived from 'epi' meaning on top of, and 'ploon' meaning preeminent or 'first-encountered. It is a membrane that envelops and covers the stomach and the intestines, consisting of two tough but thin plies enclosing many arteries and veins, and a lot of fat. It serves to warm the viscera (*Usum . . .* , Bk.4, Ch. 5). It derives from the peritoneum at the back, and you can see that its fatty matter makes it subject to alteration when it is exposed through an open abdominal wound. And you can see why the retained stump must be ligated securely after you excise the altered portion, to prevent serious hemorrhage.

Next we will describe the intestines, and get them out of the way before going to the other viscera. They are a tube with a double-ply wall. They perform the primary digestion and send chyle to the liver via the mesenteric veins, and they carry away the left-over fecal matter, their superfluidity.

There are six parts, all connected end-on, but each segment is different and has a different function. Three segments are slim, others are larger. You can find complete descriptions in Galen's *Usum . . .* , Part 5, Ch.3. The first part begins at the ventricle (ie pylorus) and is called ecphysis or duodenum, meaning twelve. The second is jejunum, meaning empty; the third is the ileum, the narrow-calibre section; the fourth is a blind-end sac; the fifth is the colon; the sixth is the straight segment, called rectum, placed where there are muscles to expel the feces. The best way to anatomize the intestine is to begin at the rectum (also called the longaon), and to avoid spilling feces, doubly ligate it and divide it between the ligatures. Remove the intestine by cutting alongside along the flanks (isles) begin at the colon. which is large in diameter and fatty; it has many saccules that collect feces.[220] It mounts toward the spleen about two fore-arm-lengths before it makes a sharp turn at the left kidney. Then it. It follows to the right under the ribs; to the right side of the stomach, and the third lobe of the liver (ie the pennon) from which it receives a dose of bile to stimulate peristalsis. It turns downward at the right kidney to the hip where it meets the blind-end segment, called the borgne (cecum).

---

[220]   Diverticula in the colon (LDR).

because it seems to have only one on-end opening., although it has an other opening on its inner-side where the contents of the of the intestine enter to be passed on to the colon. The sac resembles the stomach, but is only about a hand-breadth in size. It lies near the groin where it is not securely tethered. Therefore, when one strains too much, it can be pushed into the scrotum, as noted by Avicenna. The ileum (the narrow segment) meets it at the pelvis. It is seven or eight arm-lengths long (ie fore-arms, or cubits), and it twists and turns in the flanks and in back. It meets the jejunum, the segment empties into the ileum the superfluidities carried to it by a large number of mesenteric veins and its share of the bile from the duodenum The jejunum begins at the duodenum, so-called because it is about twelve finger-lengths long; also it is called the porter because it carries matter from the stomach into the intestines, what the esophagus carried into the stomach. Now you can deduce where the clysters function in treating cramps, and where to place the topical medications.[221] For colic apply them over the right and left sides; for iliac passion (ie small-bowel obstruction) apply them on the upper abdomen, near the ribs. Know, too, that wounds that enter the small intestine are not curable because they are membranous (ie thin-walled). Wounds in the thicker colon can be cured, because they are fleshy.

To improve the exposure of the other viscera, doubly ligate the duodenum and remove all that we have described. First examine the remaining mesentery and note that it consists only of countless veins that are branches of the portal vein of the liver that pass to the intestines. All are imbedded within fatty and glandular tissues called rodal, or red, where it was detached from the intestines. The covering membranes function as ligaments to suspend the intestines from the back. After you remove all of it, continue the dissection with the stomach.

That organ is the instrument of the first digestion that generates the chyle. Just as the mesenteric vessels prepare the chyle en route to the liver for its second digestion, the mouth prepares the food for the stomach by chewing it before delivering it to be swallowed by the esophagus (Avicenna). So, too, the residues not fit for chyle are delivered by the stomach into the intestine. The stomach serves as a larder in the center of the body; it will dispense chyle for the use of the entire body (Galen, *Usum* . . . Part 4, Ch.1). It sits centered beneath the thorax, its upper end tilted to the left near the twelfth thoracic vertebra at the very back of the diaphragm. The lower end tilts to the right. Its digestive process is fueled by its own natural heat supplemented by heat from neighboring organs (Avicenna). The lobes of the liver extend like fingers on its right side, and the spleen leans against it on the left, and with its fatty flesh and veins it generates the melancholic humors that stimulate the appetite. The stomach is invigorated by its close situation beneath the heart and its vessels, and it is innervated for sensation by branches of nerves from the brain. Behind it pass the vena cava and the descending aortic artery. Many ligaments attach it to the lumbar vertebrae. So

---

[221]   The medications mostly were topicals, placed on the skin over the region of the malady (LDR).

much for its location, its functions, and its neighborhood. Now we will describe its composition.

It is composed of two layers, continuing those of the esophagus. Its fleshy tunic covers the rest and is a fibrous web, some fibers of which are directed along its length and serve to attract matter into the stomach. The transverse fibers hold the matter until it is ready for expulsion. The body of the stomach is shaped as an elongated zucchini melon which is gently bent so that both its orifices are above the bottom of the sac; that shape favors the retention of food (ie during digestion).

You can see that it is large enough to hold two or three pints of wine. The stomach can suffer many ailments which are best treated by applications of topicals near the twelfth thoracic vertebra in back and between the xiphoid and the umbilicus in front.

Now the liver: It is the organ of the second digestion which generates the blood. It lies in the right side just under the false ribs. It is shaped like a moon; its hump fits against the ribs and its underside with its five finger-like lobes is against the stomach. As are all the abdominal viscera, the liver's flesh is covered by a membrane which is supplied with sensory nerves. That membrane continues under the diaphragm, and attaches to it with tough ligaments; the same sort of extensions connect it to the stomach, and the intestines. It relates by proximity with the heart and the kidneys, and via the latter to the testicles and to all other viscera (ie that are covered with peritoneum).

The flesh of the liver is red, the color of clotted blood; in it are veins and arteries that we will describe. The flesh itself is uniform composite of a multitude of the clusters and veins that engender blood. Galen described (*Usum* . . . Part 4, Ch. 3, and *Natural Virtues*, Part 2, penultimate Ch.) how the chyle is cooked within the vessels to become three substances. Two are superfluidities and the third is a natural liquid, the basis for the humors. Blood (our School calls it the sanguinary mass) contains four natural and nutritious substances, as perfectly described by Galen in his *Elements*. Those humors as formed in the liver are of two kinds: the Natural which are nutritious, and the Non-natrural.

The Natural ones are delivered as blood to regenerate and to nourish the rest of the body. The Non-naturals are separated and delivered to special places for special needs, or they are discharge from the body. For example, Choleric humor goes to the gall bladder as bile; Melancholy goes to the spleen; Phlegm goes to lubricate the joints. Watery superfluidities go to the kidneys as urine. The rejected substances are carried in the blood[222] and sometimes become pus and cause fevers, or rise beneath the skin and resolve without causing much discomfort, or appears as sweats, or be voided in the urine, or appear as pustules or aposthems. Therefore there are four Natural and four Non-natural humors. The liquid humors were named by the ancients as blood (ie sanguine), bile, melancholy, and phlegm. All are generated by the liver and are distributed as follows:

---

[222] Joubert wrote: These (ie the watery substances) are useless or corrupt and can easily be restored, as Galen noted in *Treatise II, Doct. I,* in the Chapter on Special Causes of Aposthems (J),

Within the liver is a kind of vein, the Portal, which divides into many mesenteric veins that are planted in the stomach and the intestines. They attract and send on to the liver the liquid part of chyle. The small branches from the portal vein go everywhere in the liver. Another vein, The Vena Cava, comes out of the dome of the liver, a gathering of many small veins in the liver. They attract the blood engendered within. The vena cava ramifies after it leaves the liver, and its branches carry blood everywhere to nourish the body where third and fourth digestions occur.

The liver also contains a system of tubules and channels which carry the superfluidities of its digestion to their proper destinations.

In sum: We have described the liver as to its actions, its situation, its substance, its relations, etc. Now we will discuss its maladies. Its particular disorder is the impaired formation of blood. From that stem cachexia and hydrops. Galen said that hydrops is a result of faulty digestion in the liver (*Natural Virtues,* Part 2, and *Different Symptoms.* It is apparent that you should apply medicines to treat diseases of the liver on the right side, and that they should be astringent in view of the nature of hepatic flesh.

Now we will describe the organs which receive the hepatic superfluidities. First is the Gall Bladder. It is a purse-like membranous sac, placed under the liver near the middle lobe (ie pennon), available to receive the bilious superfluidity, a special creation of the liver. It has two openings, not far from each other, as stated by Mundinus. One comes from the center of the liver and receives the bile. The other goes to the intestines, passing beneath the stomach, to cleanse them before they evacuate it. We also have mentioned the situation, the arrangement, the actions, the form, the substance, the relations of the gall bladder. It is large enough to contain a full glass of wine (Joubert estimates that at five ounces). You can see why it can suffer disorders. It can be obstructed at its neck or in the channel that empties it. When such occur, no bile can be attracted or expelled, and the bile remains in the blood to be discharged in the urine. The body then is discolored orange-ish. When the gallbladder itself is diseased it cannot function normally, and its neighbors suffer complications. Galen describes them in *Causes and Symptoms,* and in *Applications*, Part 5.

The Spleen receives the melancholic superfluidity from the liver. It is placed to the left of the stomach which it embraces. Its substance is not solid, but is spongy; it is more red than that of the liver. Its shape is that of an elongated rectangle, and its hump is snug up against the ribs to which it is attached by reflections of its covering membrane. It is linked to the posterior surface of the stomach and to the omentum. It has two conduits: one draws in the superfluidity (ie the splenic vein) and the other sends it to the stomach to enhance its functions.[223]

The spleen suffers when the bile flow is obstructed at the liver. That should be treated early by purging the liver when suspected because of loss of body-weight and

---

[223]   I think that Guy described the pancreatic duct and its connection with common bile duct at the duodenum. The pancreas itself was not recognized for what it is, but was thought to be fatty tissues to support the great vessels passing through (LDR).

discoloration of the skin. At that time the spleen cannot provide its usual assistance to the stomach and the patient's appetite fails. Open wounds are not as serious as those in the liver, and the patient can tolerate more potent medications than the liver, and you may use special stomach-purges. Topicals should be placed on the left side (Galen, *Therapeutics*, Part 3).

The two kidneys—left and right—serve to clear the blood of its liquid superfluidities. The left kidney is lower than the right. Their flesh is firm and tough. They are rounded ovals, slightly flattened. Each has a central cavity to receive what is collected there. Each has a double channel: one is a conduit that attracts what comes via the vena cava, and the other carries urine to the bladder. The kidneys are supplied with veins, arteries and nerves which are part of the covering membranes, which attach them at the back alongside the lumbar vertebrae, where they are cushioned.[224] The vena cava and the aorta pass between them near their lower ends. Some veins near the kidneys descend from the vena cava as the origins of the spermatic vessels. The kidneys suffer many ailments, principally obstruction and stones. Treating them is difficult.

After inspecting these organs, remove them all except the stomach when you want to anatomize the upper abdomen, and the kidneys when you dissect the lower abdomen.

Now you can see the size and the number of the lumbar vertebrae, and note that they are larger than the others, and that five pairs of nerves coming from the spinal cord enter the abdomen to serve it as well as parts of the hips and the thighs.

---

[224] Nicaise notes that the perirenal fat is mentioned only in the Montpellier edition (EN).

CHAPTER 7

The Pelvis and Its Parts

The pelvis includes the lower abdomen as far down as the thighs and the reproductive organs.[225] There are three parts: the 'container', the contents, and the external appendages. The outer part consist of mirac and siphac, fat and bones.[226] The contents are the urinary bladder, the spermatic structures, the female organs, the rectum, nerves, veins and arteries. Some parts leave the interior and penetrate and lie outside the mirac: the didymus (two layers of siphac, called the twins), the testicles, the penis, the groins (ie the lymphnodes), the perineum, the buttocks, and the muscles that move the thighs.

The 'containing' tissues are the same as in the upper abdomen. The bones are of two kinds. The three or four vertebrae that form the sacrum with its attached tail of two or three cartilaginous segments. The upper sacrum is very large, and it tapers down to the 'seat' at the lower end. The openings which transmit the paired nerves face forward instead of at the sides as in other vertebrae.

Attached at the sides of the sacrum are two large bones which are the sides of the pelvis; they meet in front to form the pubis. Because they are large and support the ileum (of the imtestine) they are called the ilia, or the bones of the ilia. Below their center there are cavities, called the little containers (ie the acetabulae) which receive the the heads of the femurs. Following the bone downward toward the seat (ie ischium)we find large openings, described by Galen in Part 6, Ch. 9 of the *Usum* . . . as follows: "Between the femurs and the pubis there must be a large opening as a passage for the nerves, muscles, veins and arteries to go down." Also, the ilium narrows to join the other side. The hip really is only a single bone, but it has been given three names by some: the ilium, the pubis and the ischium.

---

[225] A kind of medieval prudery imposed the term 'honteuse' or shameful, the equivalent of the modern 'unmentionable'. The author uses that term especially when he discusses the female pudenda. The entire female reproductive system, from the ovaries to the vulva and clitoris, was called the matrix, although the Reader will note when the word intended only the uterus or the cervix or the vagina (LDR).

[226] Guy's term for the fat was coéffe, which usually meant the omentum. Here he uses it to connote a fatty layer in the abdominal wall (LDR).

The contents: First is the urinary bladder, the sac or basin that receives the urinary superfluidities that come from the kidneys. It is a membranous (ie fibro-muscular) structure with two layers. It forms a sphere that can hold one pint and is set just behind the pubis. Two long tubes which come from the kidneys and enter it an angle are called the urinary pores or ureters. There are muscles and other flesh that close and open the bladder neck, which cross the perineum and encircle the bladder neck at the base of the penis in men. In women they go straight and insert into the vulva for a distance of two finger-breadths. Those muscles control the outflow of urine.

It is clear, in light of the foregoing, that the bladder is subject to obstructions caused by the tissues at its neck or by stones that form from sand retained from the urine. We can deal with that with a syringe, but there are times when stones must be removed by cutting. The incisions should be made in the bladder neck, exposed through incisions in the perineum. More on that later.

The spermatic vessels are the veins that come from the vena cava near the kidneys, and the arteries from the descending aorta. They carry blood to the testicle in both sexes; the blood becomes sperm after a final digestion. The sperm is the seed, the germ, of human nature. In man it goes from the testicles outside the body. In women it leaves the female testicles and remains within.

Now it is clear that by virtue of its origin in the spermatic vessels the sperm is influenced by the heart, the liver and the kidneys. Furthermore, it is affected by the brain via nerves which communicate with the entire body. Therefore, the sperm is an issue from the entire body, not only in its quantity but in its vigor, as is maintained by the worthy Conciliator. (ie Peter Abano)

The lower abdomen in women contains the matrix, where humans are engendered after receiving the semen. It lies between the bladder and the intestine. Its substance is membranous, in two layers, shaped as a sphere with two horns, or hollow arms, each of which suspends at its end a small testicle. At the other ends the tubes enter a large channel. The arrangement is as if it was a male apparatus turned inside-out, as described by Galen (*Usum* . . . , Part 14).The two cellular arms embrace the testicles, analogous to the scrotum Between the them is a 'stomach', resembling the male structures at the pubis. At the neck there is an opening like the end of the penis. There also is a vulva (ie vagina) analogous to the foreskin and glans (ie clitoris) and prepuce which is very sensitive. The vagina (ie the inside-out scrotum) is about as long as a penis, eight or nine fingerbreadths. The uterus has only two cavities which they pair off with the breast, despite Mundinus' claim that there were three pockets (cellules) on each side plus one in the middle, seven in all.[227] Her matrix is linked to her brain, heart, liver, and stomach. It is suspended from

---

[227] This strange translation of a pigs reproductive organs and breasts into a human's allows a bifid uterus (as Galen's claimed to correspond with her breasts. All of the curious anatomizing was accepted when the female organs were shameful and were delicately shunned until two more centuries of ignorance were superceded by Vesalius et al. (LDR).

her back, and the milk veins from her breasts connect with the veins of menstruation. Both Galen and Hippocrates believed that milk and menstrual blood were kin, because sometimes during menstruation the breast may produce milk.[228]

The matrix suffers many ailments, most of which are treated with suppositories inserted far inside.

We now return to the rectum, as it was described in a previous chapter. Its tunics are those of the other instestinal segments. It is as long as a palm, reaching along the hollow of the tail bone toward the kidneys. The lower part, the cul, or the fundament, is encased by two muscles which can open or close it. It is suppled by five veins called the hemorrhoidals. It is closely related with the bladder, and they share each others' ailments.

When you lift away the lowest intestinal segmant you can see the chief veins and arteries and nerves of the pelvis, and how they branch and go into the lower regions.

The most important parts that leave the pelvis are the didymes and the scrotum. The scrotum itself consists of the layers of the abdominal wall, The siphac passes from within the pelvis over the pubis and down into the scrotum. From that point it is called the didymus or 'twins'because it has two (ie'di-) layers of peritoneum. The scrotum contains three paired structures: First are the testicles, the principal organs of reproduction, which complete the digestion of the blood to form the semen. The flesh of the testicles is glandular and pale. The spermatic vessels are second; descending from above, the veins and the arteries that carry the spermatic matter to the testicles come from the vena cava and the aorta. The vessels that carry the semen to be ejected by the penis pass upward to near the bladder-neck (ie near the urethra). Contractile and sensory nerves travel with the others to the testes.

Therefore, within the scrotum and the didymes are four structures. It is clear that there is an opening in the abdominal wall and peritoneum that gives passage to three, the vein, artery and nerve. The fourth is the seminal duct, which carries the semen to the seminal canal of the penis. Also, you will note that when the opening in the siphac and mirach at the groin is enlarged, other intra-abdominal parts can descend into the scrotum via the didymus, such as omentum and intestines, and create a rupture. If they come down, they produce a hernia.[229]

---

[228] Trace the venous connections from the ovarian veins to the internal and common iliacs. Then to the external iliacs and into the inferior epigastircs, to the superior epigastric veins and the breasts. Another nexus connected the iliacs veins to the vena cava upward to the subclavuans and intercostals and breasts. (LDR).

[229] Joubert wrote that a hernia is an abnormal swelling: fleshy, watery, gaseous, or varicose, as Galen defined them in Treatise I, Doct. 2, Ch.7. He also defines a rupture, a so-called grevure, when omentum or intestine go into the scrotum. Therefore, hernia in ordinary parlance includes both. Rupture may have a double meaning: a simple dilatation or relaxation, or what we call a grevure. However, Galen meant that there was little difference between them except for size. See Treatise VI,Doct. 2, Ch.7 (J).

The penis is next. It is the instrument of human nature as well as a conduit for urine. It consists skin, muscles, tendons, veins, arteries, nerves, and thick ligaments. It is bound to the pubic bone, and it is tied by ligaments to the sacrum and near it. The veins, arteries, nerves, skin, and flesh come from above. Within it are two canals, or principal passages. The end of the penis is the balans, meaning glans, consisting of the meatus and the cowl (ie the prepuce). Its average length is eight or nine fingerbreadths, because it must match the length of the vagina. The perineum is called peritoneum in Arabic. It is the region between the fundament and the shameful organs. If you incise there, do so parallel with the scrotum and penis.

The groins (ie the lymphnodes) are the emunctories of the liver. They are glandular and lie in the fold of the thigh.

The buttocks are where all the muscles, cords, and ligaments that move the thighs are attached.

CHAPTER 8

The Lower Extremities

The entire leg, from the hip to the toes resembles the arm in many ways. (Galen, *Usum . . .* , Part 3.) and both have three divisions. In the leg one part is the thigh, the second is the lower leg, and the third is the foot. In Greek the thigh is Crus, and the lower leg is Tibia. In Arabic they are Coxa and Crux. Although, we do not treat names, we should know them and understand what we are talking about.

The entire limb contains skin, flesh, veins, arteries, nerves, muscles, tendons, ligaments, and bones. The skin and the flesh are the same as described before. The veins and the arteries travel together and we will describe them as such.

The descending vein (ie from the liver) divides at the last vertebra into right and left branches which are directed to the thighs. Once there each divides into internal and external branches, which ramify as they descend past the ankles into the feet. They appear at the surface as four veins for phlebotomy: the Saphenous over the medial malleolus at the ankle; the Sciatic over the lateral malleolus; the Popliteal behind the knee; the Renal between the fourth and fifth toes. Those four large veins can bleed copiously, and can be dangerous. There are many other smaller branches that the surgeon can use without fear.

The nerves to the leg differ from those in the arm (Avicenna). They come from the lumbar and sacral vertebrae and go to the muscles in front of the knee where they join muscles and cords that come down in back from the hips and attach to the femur. That combination of large muscles move the knee and the lower leg; others move the ankle, and others move the toes. All of that is similar to the arrangement in the arms, with only minor differences for the surgeon to be aware of. However, as warned before, he should be aware that the structure of muscles puts them at great risk when they are wounded or incised near the joints.

Large ligaments (ie tendons) come down through the lower leg and can be seen behind the knee, over the ankles, and around the joints of the toes. The soles of the feet are entirely ligamentous.

Now the bones: In the thigh there is only one, a marrow-containing tube. A ball at the upper end is called the vertebron; it bends inward and lodges in the acetabulum, which is in a lateral hump of the hip bone. At the knee there are two rounded eminences that sit in hollows in the upper end of the larger focile of the lower leg. Set over the knee joint is a large rounded bone called the Patella.

171

In the lower leg there are two bones, the fociles. The larger one is medial and anterior, and is the shin-bone all the way to the ankle where it becomes the medial malleolus. The smaller focile is lateral and goes from the side of the knee down to the ankle where it is the lateral maleolus, and is attached to the larger focile all the way. Both William of Saliceto and Lanfranchi disagreed with this description, and they were wrong.[230] You can see for yourselves: the shapes of the fociles are obvious. The larger has two hollows at the upper end to receive the two knobs of the femur and form the joint. The smaller focile does not enter the joint, and is fastened to its lateral surface and it is called the needle. After descending toward the foot both fociles join as a semilunar surface, joining the first bone of the foot at the ankle.

There are three rows of bones in the foot. The first rank has three bones that form a spherical mass. The first is called the Cahab in Arabic and Astragalus in Greek. It shaped like a nut with a groove with two rounded surfaces resembling the 'nut' of a crossbow. The upper 'round' sits into a hollow of the joined fociles and the foot is moved at that joint. The lower 'round' sets into a hollow in the Navicular bone which is the second bone of the foot. It resembles a ship with two hollowed out sides, one of which accepts the astragalus. The second rank of pedal bobes meets the navicular in front, but under the navicular and the astragalus sits the Calcaneus, shaped like a spur, attached in front to the rest of the foot. At its rear end the ligaments that move the foot are attached.

The second rank has four pedal bones placed against the navicular. One is the glandular on the small-toe side of the foot. All have rounded sides where they meet the navicular and hollows facing the third rank of five long bones. Each of them meet a toe. Each toe has three bones excepting the great toe which has two. The three ranks are the tarsus, the metatarsus and the comb (ie metatarsals and toes), as in the hand. Therefore, we count twenty-six bones in each lower leg, and thirty if we add the thigh (femur and patella.

All this information will help the surgeon when he sets fractures and dislocations. You will see that the most difficult reduction of a dislocation is at the ankle; the easiest is at the knee, and midway are those at the hip. Let God help us.

---

[230] My edition of William's *Surgery*, pp.199-200, describes the positions in the leg of the tibia and the fibula incorrectly, opposite to that given here by Guy. I assume that was a scribes error, detected by Guy. However, in most of his descriptions of the bones of the pelvis and the leg, Guy follows William almost verbatim.

My edition of Lanfranchi's *Surgery,* makes only a brief mention of the anatomy of the knee (p. 111). Perhaps Guy used Lanfranchi's *Minor Surgery,* not available to me (LDR).

HERE BEGINS THE TEXT OF TREATISE II
ON APOSTHEMS

RUBRICS FOR TREATISE II

DOCTRINE I: FIVE CHAPTERS

CHAPTER 1

Aposthems, Pustules, and Exitures In General

As defined by Galen in *Maladies and Symptoms*, Part 1, and Avicenna in *Canon* Part 1, an Aposthem is any of three types of lesions consisting of an accumulation.[231] That is a perfect definition, because, as shown by the Conciliator, and by Albert of Bologna, its very name defines it. Malady is a generic term, and an excess makes it different from other composite maladies as Galen described them. Aposthem was described along with complications in his *Tumors Contrary To Nature,* where he was wrong to declare that aposthem meant any lesion that could be labeled a Tumor, and not what that tumor should be, as to its size and the harm it may do in interfering with the body's actions. He repeated his definition in *Therapeutics* Part 3: "it is obvious that a tumor is any enlargement beyond the natural size. Perhaps a bad complexion is the initial and the principal impairment, followed by faulty healing, and third comes the swelling."[232] Therefore, a tumor does not only change the part's natural qualities, as is said in an Arabic translation of Galen's *Method*, Part 3, unless you wish to call all dispositions qualities, as did Galen. These definitions were completely explained by Halyabbas in his *Royal Book,* 8th Sermon, Part 1. He wrote, "An aposthem is an abnormal swelling in which enough, matter has accumulated to fill it and distend it. Call it a tumor if it is large; if it is small it is a complication (see his *Differences of Maladies)."* Yet what do we call a similar malady that is a composite and organic? And how do we speak of cause and effect, genus and species, complication or difference, all in terms of various considerations, and measure how much matter is present. All of this is a topic for Medical Theorists. *It is enough that we surgeons know that a tumor, an aposthem, an inflation, an enlargement, a bulge, an elevation, and an excrescence all are synonymous and mean almost the same thing.* That is what Henri de Mondeville said.

---

[231] Joubert asked." What is meant by an enlargement? It is not a tumor (ie swellimg) because that term is used to name a complication. Galen and Guy are not clear in this chapter. It would have been better to have said 'a diseased accumulation' or, even better, 'an accumulation that is an ailment per se (J).

[232] We will follow Guy's usage: Any aposthem is a swelling with abnormal contents (matter), and most tumors are swellings that may be aposthems. An aposthem with pus is an abscess (LDR).

When Halyabbas used the term 'against nature' he said that they are different from natural tumors of the head, abdomen and joints, "in which some humoral superfluidity, collects". They are different from swellings that appear after joints are dislocated or bones are fractured or bent where there are no humors (ie as primary causes) "causing repletion and distension", and he groups bad complexion with mal-union, and deformity.

So it is that we moderns, Bruno, Theodoric, Lanfranchi, and Henri simply define an apostem as a swelling (ie tumor), inflammation, or some unnatural enlargement in a region of the body. There are many kinds of apostems. Some derive from the substance of the affected part itself and others from accumulated extrinsic matter; some are complications which are alterations in the members themselves, and others from influent humors as efficient causes. As to the substance, Avicenna held that apostems were different in being larger than other swellings.[233] According to Galen (*Tumors Contrary to Nature*), large apostems are phlegmonous tumors that appear in fleshy regions. Small apostems (Avicenna) are small pustules, so-called bothors, or bulges, that appear in the skin.

As to their various contents, both Galen and Avicenna agree that apostems can either be warm or cool, using that term warm in a relative sense (ie more or less warm), not as to size or putrefaction. Warm apostems are either sanguine or choleric, whereas the cool are phlegmatic or melancholic. The gassy and aqueous ones can be classed with the latter.

As did the authors named above, we classify apostems according to their causes which are natural or non-natural humors, simple or composite, as we have defined those terms. We include suppuration, as the physicians sometimes do. They use From, meaning In, or Where, usually in general terms, applied when it seemed proper.

The true apostems derive from natural humors, well defined and uniform. Furthermore, the swelling (ie tumor) is an obvious condition and needs no other indicator.

Apostems derived from non-natural humors are not true: they are ill-defined, uncertain, and deformed. In them, the illness and the pain are more prominent symptoms than the mass; they are pustules, ulcerations, or exitures that are apostems.

Those that derive mainly from a single humor are simple and have simple names. Those that come from two or more humors have composite names, according to the Subtle Doctor's (ie Averroes) description of fevers in Part 3. of his *Colliget*. The names vary with the contents, especially if they are composite. They vary according to their qualities, quantities, and their situation: example: consider whether the pus drains from a breast or from the chest-wall. Those are the three principal items. (see *Different Fevers*).

What are the opinions at our School at Montpellier? We used other categories, and considered whether or not the contents were burnt or corrupt, et al. Some are sanguine and others are choleric, phlegmatic, melancholic, aqueous, or gaseous: simple or composite.

---

[233] Joubert could not make sense of the word substance, claimed by some to mean the essence of the malady. He thought it refered to the size, the amount, and the volume of the apostem, and he agreed with Guy when it came to treatments (J).

My Master Jean Jacques called the first (ie the sanguine) simply lesions. The others were bad lesions, worsened by fraudulence, or by erosive (ie biting) qualities.

And without a doubt Avicenna added a fourth category when he dealt with warm aposthems and others in the group derived from sanguine humor or choler: classed as good or bad.[234] It followed that sanguine humors may be from healthy blood, but if too thick, or too thin, it may be the sources for phlegmon, or for true erysipelas, called prickly (ie espine or dyn) (see Ch. 3, below),appearing at the surface (Galen, *Glaucon* Part 2.). Choleric aposthems do not derive from good bile, because all true aposthems derive from blood. Note that here we use the term 'blood' in two ways. Only the lesions called Formi, not erysipelas, come from bad and erosive (ie biting) bile,

As to treatments, we use the same orderly applications of topicals for both the warm and the cool. That topic is a topic for words and appearances (ie not surgical treatments.) Otherwise, too many choices, would make treatments unsafe.[235]

---

[234]   Joubert's Note: This sentence seems to me to be confused. We refer directly to Avicenna's text where warm aposthems are derived from good sanguine humor and bile, or from bad, both supplied by blood. Those terms refer to humors as well as to aposthems. One may say that the humor flows from one part of a body to another, or that it mixes with another, or that it sepatrates from the mixture and again becomes natural Just as (ie good) sanguine, choleric, pituitous and melancholic humors flow in blood, so do they when they are burnt and corrupt. (J.)

[235]   Joubert, again: Falco interpreted the passage thus: If we say that an aposthem derives from a good Natural humor, where will we find such? If it causes an aposthem it no longer is good or natural. Furthermore, the non-natural humor is an equivocal term; an aposthem is abnormal, per se. What derives from corrupt blood is not a sanguine aposthem, another equivocation. Finally, when blood is corrupt, a part is choleric, another is melancholy, etc. Therefore, there never is a good sanguine aposthem. Also the categorization of humors is not correct when we say that every humor, even the natural ones, can make a true aposthem. That explanation disagrees with Guy's contention that the four humors can make four different tumors: blood makes phlegmon, choler makes herpes, pituit makes edema, melancholy makes sclerosis. In his classification erysipelas is placed with phlegmon, but it is is more reasonable to put herpes with erysipelas, as some authors have done. Erysipelas is not derived from a single separate humor, as are true aposthems. The same is true for choleric blood or the thinnest part of blood, as Galen claimed. Therefore, one is wrong about herpes; it is on its own a completely different aposthem; unlike erysipelas which is not derived from a single humor, as are the four sovereigns: phlegmon, herpes, edema, and sclerosis. What derives from blood is obvious. When blood is of medium grade in temperature and consistency, it produces phlegmon, specifically and absolutely.

When blood is thick and very hot, it makes and blisters and anthrax. When it is somewhat thinner and hot it makes the kind of erysipelas called Holy or Persian Fire. All derive from the mass called blood, whatever its temperature or consistency. All of that will be dealt with in various pertinent chapters (J)

To sum up: Aposthems come from Natural and Non-natural humors; simple or composite; each with its own name. That is how it is, and that is how we will use the term.

There are difference in their complications as to pain and other symptoms that the patient can feel. And there are differences in where in the body they occur, as Galen wrote in Part 2 of his *Gaucon*. Some occur in the eyes as ophthalmias; some in the throat as quinsies; other in buboes[236] or emunctories; some are internal; others are on the surface; some affect the noble parts; others are in ordinary places; some occur in sensitive regions; others where there is less sensitivity; some come in replete parts; others not; etc.

The efficient causes vary according to Halyabbas as to their derivation or congestion; some are critical; some have internal causes. One should know the differences, because they are the principal indicators in our choice of treatments.

The causes for aposthems, pustules, and exitures are both general and specific. The General Causes are rheum[237] and congestion. The causes and its derivation are described in detail by Halyabbas in his *Maladies and Symptoms*. He said that there are six: the forceful entry of the humors; the inability of the affected part to reject it; the passages that transmit it; the proximity of the source; the quantity of the matter; the availability of a lower region (ie the rectum or bladder) to receive it (ie en route for evacuation). The causes for congestion are the state of nutrition (ie general health) of the region where the aposthem lies and whether the nutrition delivered to the part has been properly combusted.[238]

When the superfluidity increases bit-by-bit, the region becomes stuffed and distended, and the aposthem appears. Afterwards, the warm matter is more readily dispersed and the cool accumulates (Halyabbas).

When you recognize the sources of the aposthems, you can determine what to do and what not. Galen (*Therapeutics,* Part 3*)* used that information as the principal indication for his treatments: The source of the matter before it entered; simple or composite; fluid or static (ie congestion). When the aposthem derives from congestion, first cause the matter to flow.

Galen (*Irregular Fevers)* declared: warm rheum rapidly descends into muscles, and the larger veins and arteries are filled and distended[239] and the excess runs into the smaller branches, into even the tiniest, and then it leaks into spaces in the flesh and the membranes, and an aposthem is formed.

---

[236] Joubert said that bubo usually refered to the groins, and usually there is a phlegmon in that region (J).

[237] Joubert's Comment: Rheum in Greek connotes flux and distillation in France. Barbers thought it also meant a destructive and disordered movement of a humor. The movement could be upward, downward or lateral. Catarrh refered to a downweard flow, principally in the head. (J).

[238] Joubert commented that the term meant more than the quality of the nourishment and its assimilation. It included efficient evacuation of excrement to avoid accumulation (J.)

[239] Joubert added: Arterial blood (ie spiritual, not nutruitious) can excite a phlegmon. Furthermore, Erasistratus opined that only when the arteries are distended under pressure, making them pulsate, do they engender the phlegmon. See Galen's book *The Contents of Arteries Are Not Only Spirits.* (J).

So it is: We have described the antecedent matter in the veins, and the composite matter in the flesh, and we have explained General Causes. Now we proceed to the Special Causes. There are three: Primary (ie immediate), Antecedent (ie the materials), and Conjoint. The primary causes are such as falls and blows. Antecedent causes are the four humors, both the natural and the non-natural, each has its double.

The Naturals are the substances of nutritious blood. By that we mean the principal elements of nutrition, their essence not their quantity, or the ability to eliminate bad matter.[240] As such they are pure sanguine humor, and choler, phlegm, and melancholy. Whatever else they are they all are called blood in common. Galen (*Black Bile*), and Mass by Rabbi Moses, in his *Epidemics,* Book 2.[241] But Halyabbas (*Techni,* Part 2) called it a Bloody Mass, which is the sole source of nutrition, not different from other humors, as claimed by the teacher of S. Flour, not long ago, and expanded in the concordances of Jean de St. Amand.

The Non-Naturals are separate from blood, and because they are bad they are not useful nutriments, but they are delivered (with blood) to the places where they can do their evils or be evacuated. They produce aposthems, pustules exitures, excrescences, discolorations and sweats.[242] Sometimes they resolve without discomfort, sometimes they suppurate and cause fevers. They borrow the names of the four natural humors.

Galen (*Melancholy*) said that the natural humors in nutritious blood can form clots in various red colors, whereas non-natural humors do not clot and remain as liquids: pale pink, white, or black, as can be seen in the region where aposthems are beginning to form, as a sign that you should begin to treat with repercussives. This topic is taken more detail in Galen's *Elements* Bk. 2, and we shall do the same in the chapters which follow. on special aposthems, Now we shall deal with the Natural Humors that make the

---

[240] Joubert explains: The phrase meant "not to say it wrong". But Mingelousaulx understood it to mean "to eliminate all doubts and equivocation, and to declare once and for all that the meaning of Natural Humors, those which are principally, appropriately, and naturally, by a singular virtue, and by its very substance, can nourish, without regard to its amount or how it is used, which should be natural as is its proper destiny. (EN)

[241] Joubert: That is, blood is a certain mass of all four humors. But blood is more than just the four, because one can conceive of other humors in addition. Rabbi Moses' concept of the four-humor mass had misinterpreted Galen (J).

[242] Joubert explains: According to the barbers, Exitures were abscesses (in Latin), and apocima in Greek, when the matter in the tumor suppurated and drained. Therefore, exiture implied a tumor about to erupt. A pustule contained warm or cool matter within a sac. Thus a pustule was an exiture, but not the reverse. All of this is confusing. Nodes, scrofules, steatomes, atheromas, melicerides, etc. all may be called exitures. Pustules as such are small sanguine or choleric aposthems which are malignant or venomous. Evil discolorations are signs of choler or melancholy en route to the skin where they are trapped before 'exiting'. When they are dilute and liquid they can escape and drain. When the matter is is very dilute (ie subtle), it is absorbed without discomfort (J).

four true aposthems that usually are called phlegmons (see Galen, *Glaucon,* Part 2), but more properly are named Phlegmon, Edema, Sclerosis, and Sephyr.[243] And we shall see that the four kinds of non-true aposthems, pustules, and, exitures, sometimes take on the names of the true; and we will name two more, the aqueous and the gaseous.

In sum, there are six names for the simple and the composite aposthems, and the terms are used equivocally: correctly when they are true, and otherwise when they are non-true, as when they are pustules and exitures. Pustules are small, and exitures are single or clustered as glands or varioles or herpes.

Nevertheless, the most significant are the 'venomous. Those pustules involve flesh as well as skin, but the blistered variety involve skin alone. We will deal with them later.

The exitures, according to Galen (*Tumors Contrary to Nature,* and *Glaucon* Part 2) are either internal or exterior. The contents must contain some erosive substance and become pus, or other foreign matter resembling honey or wine lees, or grime, or stones, or fibers, as found in glands. That matter derives from non-natural humors which are liquid at first, or from purulent phlegm. Pus accumulates, as in the milk-abscesses in breasts.

Conjoint causes of aposthems and pustules are substances assembled and activated in an affected region. The surgeon easily can detect early signs of aposthems; some abnormal humor or solid matter accumulates in a region, and that is the aposthem.

True aposthems exhibit swelling, pain and warmth in various degrees. The non-true kind exhibit swelling and a collection of various amounts of erosive matter. The less warm surrounding region is said to be 'cool'. The temperature of the medications is graded by comparison (ie as cooler or warmer than the lesion), as taught in *Techni* and in Part 3 of *Simple Medicines.*

We will describe special aposthems and their contents in later chapters, and we should not jump ahead to them before we conclude our general concerns, as Galen taught in Bks. 7 and 9 of Therapeutics.

Avicenna wrote that a few particular aposthems are pure (a single humor), but many more are composites, also entirely of true humors. And there are many of the non-true kind that seem to be pure. We shall deal with the treatments of the simples first; from them you may be able to deduce the treatments of the composites.

The waxing and waning, the paroxysms, and the crises of aposthems are related to the nature and the amounts of their contents, as indicated by Galen (*Different Fevers,* Part).

An aposthem has four phases: onset, increase, stable state, and decline. The sign of onset is the swelling caused by the influx of matter.

The phase of increase is when the interior enlarges and begins to point. The stable state is when there is no further increase, and the phase of decline is when diminution or other changes occur. Those phasic phenomena are the essence of aposthems, and they can be distinguished by the changes in the matter within them. Recognizing the changes leads one to modify the treatments. Furthermore,the phasic evolution after the onset of

---

[243] 'Sephyr' is a corrupt term for a patch of painless sclerosis, perhaps a localized scleroderma or myxedema (LDR).

an aposthem is different from the changes due to a complication. Awareness of such, allows us to counter and to reverse them early on.

The final stage ends either by invisible resolution, by suppuration, by necrosis (ie putrefaction), or by induration. Galen (*Unequal Dyscrasia*) said that a cure by spontaneous resolution is better than the others, and that suppuration is better than hardening, and that putrefaction is the worst outcome.

When an aposthem resolves it softens and the throbbing ceases. When suppuration occurs, pain and throbbing and local heat increase. When there is a sudden cooling,[244] a sign of toxicity, serious fevers and other complications occur.

We will discuss the signs of pustules later on. The sign of corruption is dark discoloration and liquefaction.

The signs of exitures: Avicenna said that marked throbbing or persistence of the mass, or an increase in local heat, or persistence of the mass and tenderness will lead you to assume that suppuration is occurring and that drainage (ie exiture) impends. But when the signs ease, and the pain and heat subside, and when the aposthem comes to a head, and you can feel fluid, and the discoloration fades, you will know it has become a purulent abscess. Hippocrates (*Aphorisms,* Part 2) wrote that when pus is forming, the pain and the heat are worse than when the suppuration is complete, and you can sense it. Yet, he added (Bk. 6) that often a physician often is misled by his assessment of the tension (ie as it 'points') of the site and by the thickness of the pus.

When an exiture is in a noble organ or at a joint, or in a region rich with nerves or veins, it weakens the region and impoverishes its natural heat. When the matter is fatty (ie thick) and not liquid, and when the exiture is discoid and does not point, be suspicious that the process of maturation is defective, and needs help to suppurate and drain. But the outcome may not require such help; the maturation may proceed rapidly, and the abscess will open and drain on its own.

Sometimes the exiture will resolve, although more likely it will drain. The opening may come about naturally, and that is better than a surgical procedure; but when such is required, the knife is better than the use of caustics. Avicenna said that surgical drainage may exacerbate the virulence, contaminate the region and cause a fistula. However, when there is no other way, you have no choice but to act.

Hippocrates commented on pus in his *Prognostics*, that white pus is laudable, especially when it has no bad odor. Other kinds of pus are bad. We will discuss the varieties in another chapter.

Galen chose his treatments for aposthems based on all the above, as well as an assessment of the affected region. In general, he treated phlegmon as a true aposthem, derived from natural humors (*Therapeutics*, Part 3.) Part 4 deals with the no-true varieties derived from non-natural humors.

---

[244]   Cooling due to local necrosis (LDR).

Here we shall treat true aposthems, the usual uncorrupted kind, those which often resolve. When we say 'pustules' or 'exitures', we mean the non-true kinds that are purulent, corrupted, and converted (ie unusual). The treatment is determined by the disposition itself as well as the nature of the part, as stated above.

The term 'disposition' includes the quantity and the quality and the nature of the matter in the aposthem, its conjoint substance.

We deal with large masses as well as small; with those engendered directly from local substances, and with others by derivation; with those that result from congestion, or from heat, or cold, etc.

The nature of the affected member tells us that the aposthem is in fleshy or nervous tissues, or in an eye, or in the throat, or in emunctories, or in others parts that we will come to. That information must come first in diagnosing the nature of the aposthem (see *Glaucon*, Part 2) and its causes, before they can be eliminated, and to determine what is needed for treatments (see *Therapeutics,* Part 3). The usual source for phlegm is whole blood attracted abnormally to a part where it is congested: more than it needs, more than it can eliminate At times the affected part can send it off as a superfluidity, at other times as excrement, and at times by both routes. The faulty attraction may be due to local heat or pain, or the region lacks the vigor to reject what may not be an excess of inflow, or the inflowing veins may be too large, or the out-flowing passages may be too narrow. In view of all said, one must keep in mind all the three avenues for treatment.

First, prevent the excess of inflow. Second, ease the discomforts which may be the basis for continued abnormal attraction. Third, treat what already has occurred.

As to the first: Galen said that when the humors are in proper proportion (ie in the whole blood that is the source of the congestion) and the inflammation persists, and the heat thereof excites the inflow, the treatment must eliminate the overload by baths and exercizes, and by massages of the opposite limb[245]. These measures are to be used when there is no fever or a lot of pain. In such cases, add evaporatives and fasting, and other elements of a healthy daily regimen. But when the body has an excess of choler or melancholy or phlegm, or is too aqueous, and is enfeebled, the treatment must include purgation, directed at the humors at fault. We will discuss that in other chapters and in the Antidotary.

In regard to aversives (ie repercussion) and antispasmodicss in the affected region, they are to be used in all dispositions during the phases of onset and augmentation. During the stable and declining phases, it is harmful to use evacuations,[246] although the younger generations of physicians use them without much forethought, as Averroes complained in his Book 7. I will discuss phlebotomies below.

---

[245]   To induce hyperemia in the unaffected limb, attracting blood from the congested limb (LDR).

[246]   The repercussives et al. are topical medicines that drive the bad matter away from the lesion, and that may be inward. The evacuations by purgation attract the stuff into the bowel, whence to be eliminated. In an evolved aposthem the matter already is at the surface. Therefore avoid pushing it back in. Treat it by resolving it or attracting it to where it can be drained(LDR).

When an exiture is near a noble organ or at a joint, or in a region containing many nerves or veins, it weakens the region and impoverishes its natural heat. And when the matter is thick and not liquid, and when the aposthem is large and flat (ie discoid) and does not 'point', be suspicious that the maturation is defective and needs your help to suppurate and hasten to when it can drain. But the outcome may be better and not need a lot of help; the maturation may proceed rapidly and the abscess will open and drain on its own.

Sometimes the exiture resolves, but more often it comes to drainage. The opening may appear naturally and that is better than a surgical procedure; and a surgical operation with a knife is better than using caustics. However, Avicenna said that a surgical drainage may exacerbate the virulence and introduce contamination and lead to a fistula.

Hippocrates commented on pus in his *Prognostics* stating that white pus is laudable, especially when it emits no bad odors. Other kinds of pus are evil, and we will discuss them in another chapter.

Galen chose his treatments based on all the above in consideration of the nature of the affected part. Usually he treated phlegmons as true aposthems derived from natural humors (*Therapeutics*, Part 3). Part 4 deals with the non-true kinds derived from non-natural humors.

But when the body is full of yellow bile (choler) or black bile (melancholy) or phlegm (pituit) or if the humors are watery, all of which cause feebleness, the treatments should be purgation with potions designed for the humor at fault. We shall discuss them in later chapters and in the Antidotary. In regard to aversives (ie repercussives) and antispasmodics, the revulsion of undesirable matter by the affected part, their use is the same for all dispositions at the times of the onsets and during the phases of augmentation of the aposthems. But during the static phase and that of decline it is harmful to try to use evacuants for the affected part.[247]

The newer generations of Physicians use them without much common sense, as Averroes (Book 7) asserted, and we will discuss it when we deal with phlebotomy.

The second goal of treatment is to use sedatives to relieve pain, and to improve bad qualities of the humors, and to use restrictives (astringents) to block (ie constrict) the inflowing vessels, and to use relaxants to enlarge the normal pathways for discharging the unwanted matter.

The third goal is to get rid of the matter within the lesion. Use repercussives in the phase of onset, and evaporatives, in all excepting some special cases. Galen listed four of them; 1. When the aposthem is at an emunctory. 2. When the matter is venomous. 3. When the matter is thick. 4. When the matter is hidden. Avicenna made few exceptions except when the lesion was at an emunctory or when the matter could be driven into

---

[247] The theory of evacuation conceived of methods to withdraw the bad matter from a lesion to where it could be eliminated However, if the lesion was on its way to resolution, or suppuration and eruption, do not attract the matter back into the body. See fn. 16 (LDR).

nearby vital organs. Roger excluded only the venomous matter. The Four Masters added to Galen's list when the matter was congested, very cool, or at a critical phase, as when the humors suddenly deteriorate, and when the aposthems are near a principal organ. Bruno accepted Avicenna. Theodoric followed the advice of the Four Masters. Lanfranchi listed ten exceptions, and Henri listed nineteen. Master Dino of Florence, in his commentary on Avicenna's *Canon,* Bk. 4, accepted his third and fifth exceptions. As for me, I differentiate among the repercussives: the strictly defined ones are oxycrat, vinegar, plantain, morel, bol d'armenie, aloin, canelle, and others that penetrate deeply to reach and repercuss what is there. Others that are not truly defined but are commonly called repercussives are: egg-whites, malva, rosat-oil, camomille, mastic, white (ie clear) collyria, and similars which alter and prevent the member from receiving the superfluidities. Of the two I use the first type, the proper repercussive, at the outset of phlegmatic aposthems excepting in these ten cases: 1. When the matter is at an emunctory. 2. When the matter is venomous. 3. When the matter is very thick. 4. When it is hidden (ie not accessible to the topicals). 5. At critical times. 6. When the cause is immediate (ie after trauma). 7. When the body is replete. 8. When the body is debiliotated. 9. When the aposthem is near a principal organ. 10. When the pain is severe.[248]

I use the second types (the loosely named ones) in all phlegmatic aposthems excepting three cases: 1. When the aposthem is at an emunctory. 2. During a crisis. 3. When the matter is venomous.

In all cases, even when the inflow is complete (ie the third phase) and the aposthem is well established, and some of its matter has seeped into the neighboring tissues, try to resolve it and to evaporate it with familiar mild resolutives that warm and moisten (ie liquefy) the matter, especially in the last-named three cases, when we want to attract it into the aposthem and prevent it from getting back into the body. We use attractive plasters, and sometimes apply cups, as did Avicenna.

Let us accept this general policy: At the onset of all phlegmons—with the exceptions as noted—we apply repercussives. During the phase of augmentation we introduce resolutives, increasing them until the amounts equal the resolutives during the third phase, when we gradually reduce the repercussives and eliminate them during the phase of decline. In other words, we repercuss during the inflow until it ceases. Then we resolve along with the repercussives until the aposthem begins to resolve. We shall describe the various remedies in later Chapters and in the Antidotary.

But if the aposthem goes the route of the exitures, begin the treatment (*Glaucon,* Part 2.), with relaxants and painless evaporants, such as the soft and mitigative Triple Compound[249] to which you add some honey (see *Therapeutics* Part 3.) When drainage

---

[248] Consistent with his Theory of Humors, Guy avoids repercussing when he cannot displace the bad stuff, when he does not want to plug an emunctory or envenom the rest of the body, when the body already is replete, etc. (LDR).

[249] Galen's Triple Compound: Wheaten flour, water, and oil, cooked to a paste (LDR).

occurs, use suppuratives to complete the concoction. In should open others that fail to drain, as described (*ibid*. Part 4.) Choose the surgical method that you deem proper, considering three subjects: What will be the shortest method; what will cause the least suffering; what will achieve a complete cure, that is, what will satisfy three requirements?" 1. Complete emptying. 2. If that is not possible, at least you have given relief and not caused suffering. 3. The lesion will not recur by default (ie not having been drained).

Using these criteria you choose the method for surgical drainage: by incision with the knife or by erosion produced by caustic medications. Surgical incision works immediately when you cut into the contents, but you must not drain it all at once. The medical measures begin with evaporants and resolutives to affect the matter. If they fail, you apply suppuratives. and necrotizers. After drainage is establ.shed by either method, use detergents followed by incarnatives and consolidants to heal the ulcer that remains.

In Part 4 of his *Therapeutics*, Galen stated that suppuratives and plasters do not only treat the lesion; they also relieve the pain, and that is their first goal. Then, he said, the quickest treatment for a phlegmon is with dessicants and resolutives. Either they will resolve the entire lesion, or they will bring to a head a small point that has begun suppurate, and you can erode it with a caustic medicine. But when the skin in the region is very thin at the point and you wish to relieve the patient, use the knife. As Avicenna said, you cure the apposthem as such by removing the alien matter that caused it. We will discuss the effective maturatives in later chapters and in the Antidotary.

The result of the treatments will be suppuration, a change in the character of the matter, persistence of matter hidden within the cavity, failure to resolve the matter without drainage, spontaneous eruption after a while, the unexpected occurrence of some complication, or some other harmful outcome, or when you cannot trust the corrosives for some reason and you do not want the bad matter to be absorbed. Then you promptly should open it with an incision. As Avicenna advised, make the opening only large enough to get rid of the contents. The incision should be gently curved like a leaf of an olive or myrrh tree. Fulfill these seven condition: 1. Incise over the contents. 2. At the lower part. 3. Follow the skin-lines along the direction of the muscles. 4. Avoid nerves, veins and arteries. 5. Do not allow all the matter to escape at once, especially when the apposthem is a large exiture. 6. Be as gentle as possible at the tender places. 7. After the drainage apply detergents and consolidatives.

The detergents (mondificatives) are applied as packs, stoupes, plasters, and ointments, all to be described in special chapters, and in the Antidotary. During the first few days, egg-yolks or egg-whites thickened with alum will suffice, as used by William of Saliceto. Then use rosat-honey and the mondificative made from ache (ie apium), then to the Apostles Ointment and to the Egyptiac. Later apply the Basilicon, diachylon, diapalma, and other medications used for ulcers, because, after the drainage of exitures and the elimination of the excrescences and pustules, ulcers will remain.

When the patient will not accept the knife, you must establishe drainage with topical medicines. Avicenna favored linseeds, yeast, and dove-droppings, all combined with a soft soap or a mucilage of mustard. However, I advise that the eruptors made from quick-lime and soap should be used first.

Nicaise's Note On The Aposthems

In this Doctrine I, Guy dealt with Aposthems In General in simple members. In Doctrine II he dealt with composite aposthems in particular regions and organs. Here, I shall comment on Aposthems in General

Guy put them in two categories, the warm (sanguine and bilious) and the cool (phlegmatic and melancholic). Each group contained ten varieties, according to the humors that engendered them. Thus we have four classes of aposthems:

1. Sanguine Aposthems engendered by sanguine humor (blood).
2. Biliary Apsthems engendered by choler (yellow bile).
3. Phlegmatic Aposthems engendered by phlegm (pituit).
4. Melancholic Aposthems engendered by melancholy (black bile).

Finally, each of the four classes was divided into many or few species, and each species was naned by its humors, either Natural or Non-natural.

Aposthems from Natural Humors were True; those from Non-natural humors were Non-true. A natural humor became non-natural by mixing with another humor, by corruption on its own, by solidification or by liquefaction, in a way that resulted in many which follow.

GUY"S TABLE

	Made from Natural Blood	1. True Phlegmon. Could end as end as exiture
Sanguine		_____
	From Non-Natural Blood	2. Carbuncle, Furuncle or Braise, Persian Fire Or Malignant pustule
		3. Anthrax, Malignant Carbuncle

WARM—

		4. Esthiomene or Gangrene. St.Anthony's Fire. or St Martial
Choleric	From Natural Choler	5. Erysipelas
	From Non-Natural Choler	6. Choleric pustule—Herpes 7. and Formy
Phlegmatic	From Natural Phlegm	8. Edema.
	From Non-Natural Phlegm	9. Gaseous apothems Tissue emphysema 10. Aqueous Aposthems 11. Phlegmatic Excrescences Glands, Scrofules, loupes turtles, goiters., hernias, Melicerides, Steatomas. Lipomas

COOL—

Melancholic	From Natural Melancholy	12. Sclerosis, True Phlegmons
	From Non-Natural	13. Sclerosis, true Induration 14. Cancerous aposthems.

A Natural Humor became Non-Natural in several ways: By corruption of itself; by mixing with another humor; by solidification or liquefaction. In those ways there can be many different kinds of non-true aposthems, named according to the changes undergone by the natural humor, independent alterations that did not correspond with any others. The Table summarizes Guy's Classifications. His descriptions were adopted by his successors who simply copied or summarized, as did Tagault. Paré accepted a large part of the generalities in his *Tumors Contrary To Nature* as noted by Malgaigne, who stated that Paré's Doctrine in Book I was taken almost entirely from Tagault and Guy de Chauliac, and Tagault seems to have been Parés guide to Guy, who was carried into the analysis. Tagault's work was no more than a summary of Guy's. Paré accredited Guy with the large number of medications used for treating aposthems. In respect of his treatment of malignant ulcers (ie lupus), Malgaigne wrote, "The procedures are described with more detail by Guy and Tagault."

According to Paré, Guy's *Treatment of Aposthems*, was the source for generations of successors. Mingelousaulx's (edition of Guy (1683) gives full credit to Guy as well as a summary of the few differences adopted by Paré and Thévenin as to the number of exceptions.

Finally, the classic *Surgical Pathology and Treatrment* by Simon and Hévin (1780), both professors at the Royal College of Surgery, bears many marks of Guy. Their book is arranged on the same plan as Guy's, the classification of tumors is the same, while the number of types is only slightly larger.

CHAPTER 2

True Phlegmon And Other Sanguine Aposthems

Galen wrote that there are two types of phlegmon (*Maladies and Symptoms*, Part 1.). The first are inflammations anywhere in the body, commonly refered to as phlegmons. The second are aposthems due to pure true sanguine humor, and are more correctly named. But even that name had a double meaning, because the aposthem may be from true or non-true humors (*Glaucon*, Part 2). The true is engendered from benign and copious blood (ie sanguine humor), more than is required by the member. The non-true phlegmon derives from bad non-natural sanguine humors. Blood (ie sanguine) is warm and moist, derived from the most temperate part of the chyle. It, too, is double: natural and non-natural. The natural is warm, moist, and temperate in substance[250], is red in color, and has a pleasant odor and flavor. The non-natural is somehat like the natural, but has differences that make it another humor. That change happens in two ways: spontaneous or by admixture. The spontaneous occurs in two ways: 1. By thickening or dilution, and 2. By combustion: the thin part is converted to choler and the thick part becomes melancholy, without separation.[251] The Natural can become non-natural in another way, when another humor mixes with it, be it any other humor—phlegm, choler, or melancholy.

From all the foregoing you will see that blood (ie the sanguine humor) can engender four kinds of aposthems. 1. From natural and benign blood we get true phlegmons. 2. From blood debased by admixture we get three aposthems named for the added humors, as when we get erysipeloid phlegmons when choler is added, or edema when pituit is added, and sclerosis when melancholy is the supplement.

When sanguine humor becomes non-natural by combustion, and becomes thin and bitter, we get pustules, crusts, carbuncles, that deteriorate into esthiomenes, such as

---

[250] Temperate: Not as thin as bile and not as thick as pituit or melancholy. It tends to sweetness in flavor (J).

[251] It happens in whole blood. Coagulation yields a thick clot and a watery serum (LDR).

191

carbuncle, pruna, fires (Holy and Persian), esthiomene, and anthrax, but not formi.[252] All of the above is taken by me from the teacher of St. Flour.

The causes are immediate (ie primitive) such as falls or contusions, or by bad blood. The underlying (ie antecedent) causes are congestion by an excess of good blood that cannot be moved in the veins: it remains in situ exposed to native heat, or when it is hidden (ie not accessible to topical medications) as stated in the previous chapter. I refer you there, to apply the general material to particular aposthems. The conjoint cause is the sanguine humor already in the aposthem (ie causing it to enlarge).

The Signs: You can see the swelling at the surface. It is warm or hot. It is reddish and it throbs. It resists compression (ie whereas edema can be indented by a finger); and other signs that indicate an excess of blood.

A phlegmon has four phases: onset, increase, stable state, and decline. The onset exhibits the causes, as noted above. The phase of increase exhibits enlargement of the bulge and diameter. In the stable state the mass no longer increases. In decline everything begins to wane.

The phlegmon, terminates by resolution, suppuration, putrefaction, or as residual hardness. When it resolves, its size decreases as does the pain. When it suppurates the throbbing worsens, as does the heat, and the mass persists. When it putrefies (becomes necrotic), it darkens and and melts. When it petrifies it remains as a hard lump.

The phlegmon can suffer complications which delay and interfere with the treatments. The pain and tenderness may increase in severity. The contents may reaccumulate when it is at an emunctory. Esthiomene may spread when the lesion is too cool and the matter is squeezed.[253] A hard sclerosis may inhibit resolution. When you treat aposthems be alert to detect any of the complications that may happen along the way and attack them at once. (see *Glaucon* Part 2).

The treatments for phlegmons are general and particular. The General Measures were described in Chapter 1.

The Particular Treatments are directed at four elements[254]: 1. Restore a healthy life-style. 2. Restore the balance of the humors in blood[255] 3. Get rid of the hidden matter (ie the conjoint cause). 4. Treat the complications.

---

[252]   Formi: derived from non-natural choler, not from blood. They give rise to pustules or crusts (J). I think they probably were miliaria or a kind of herpes; an eruption with prickly discomfort, as if by crawling ants (LDR).

[253]   The undermining infection goes beyond the limits of the central necrosis (LDR).

[254]   This is a recap of the Hippocratic Doctrine: Diet-Regimen first; Medications second; Surgery third (LDR).

[255]   Joubert expleains "Equalize the antecedents": Tagault interpreted the term to mean the repercussion and resolution of the antecedent matter. But Guy meant to include congestion as well as inflow of humors, and to equalize meant to relieve the repletion and strengthen the feeble rejection that allowed the aposthem to occur. Usually that meant to replace the humors (ie hepatic combustion), and Guy sought to reduce the heavy humors and strengthen the thin ones by moderating them (J).

The first attends to the six natural factors in health and their annexes: the ambient air, food and drink, sleep and wakefulness, activity and rest, fasting and over-eating, and the emotions. Therefore, expose the patient to dry fresh air that is not rheumatic. Eat easily digested food that is not very rich, and avoid fats and sweets. Limit the soups to what is made from beans and dairy foods.[256] Strictly avoid spices, garlic, onions, and strong wines. Emphasize lettuce, spinach and borrach. Water the wine. When the patient is feverish, give him barley-broth (ptisane) and milk of almonds. Insist on sobriety and avoid late snacks. Keep the bowels open, and avoid lying on the side of the aposthem. Do not over-sleep, especially during the day and after meals. Live shamelessly (ie as to sexual activity).

The second Measure includes bleeding if the patient is replete, using the opposite-side limb during the phases of onset, but from the same side during the stable state and that of decline[257]. Also use purgations, as Galen did (*Therapeutics* Part 3). If you observe all the rules for phlebotomy, you will not only reduce repletion but also, when the lesion is large and painful, you will set the humors in balance. Pain and local heat (ie inflammation) themselves are attractive to inflow of humors, no matter what is the state of the body's repletion.

The third measure is accomplished with repercussives and restrictives at the onset, with the exceptions listed above in Chapter 1. Add resolutives during the second phase until they equal the repercussives during the stable phase 3. Reduce them and eliminate the repercussives during the entire phase of decline. If the aposthem goes the route of an exiture, use suppuratives, detergents and aperitives. Use desiccatives toward the end of the treatments in all cases.

There are four types of repercussive medicaments (the repellants and the restrictives). Galen's Type One (*Glaucon,* Part 1) included oxycrat with water and vinegar, weak enough to be potable, applied on a sponge.

Avivenna described the second type as a compound of joubarbes, heavy sour wine, barley-meal, pomegranate bark, and sumac, all ground to a powder before boiling to make a liniment.

Halyabbas compounded the third type from white and red sandalwood, memitte, cimolea, and bol d'armenie. All were ground to a fine powder and mixed with the juices of joubarbes and purslane or lettuce to make a liniment.

The fourth type is what we use for all wounds and contusions: egg-white and rose-water or rosat oil. Make a liniment and apply as stoupes or on cloths, renewed frequently.

---

[256]   The ordinary legumes: peas, beans, lentils, etc., all of which are cool and dry, consistent with the new life-style. They thicken some humors and slow their inflow. When the rich diet of legumes seems excessive, reduce the soups to broths which are less nourishing, yet are cool and dry (J).

[257]   'Opposite Side' diverts the inflow from the developing aposthem. 'Same Side' withdraws from and decongests the full-blown lesion (LDR).

During the phase of augmentation, use three topicals. First is rosat-oil; Galen (*Simple Medicines*, Bk 3) said that it serves both to restrict and to evaporate. Second, from Avicenna, is a compound of leaves of mauve, aloes, and roses, barley-meal and oil of camomille. Boil all in wine, grind it to a paste, and use it as a soft plaster. The third is boiled wine with rose-water, vinegar, and saffron. Apply it on stoupes or cloths; leave it in place longer than the repercussives.

These are three kinds of resolutives to use during the stable phase. Master Dino used parietary, leaves of mauve, fine bran or flour, aneth, fenugreek, and oil of camomille. Boil and grind to make a paste. Another is from Galen (ibid.): Bread-crumbs and wheaten flour soaked in boiling water for one hour. Filter and add honey to the crumbs to make a soft plaster. Avicenna made a third topical from diachylon and the basilicon. All of these recipes are repeated in other chapters and in the Antidotary. All of them should be left in place for long periods.

In the phase of decline we use lanolin on stoupes or a sponge. Or we apply felt pads wet with warm wine and squeezed almost dry.

Declination by suppuration is treated with three kinds of medications. First is Galen's tripharmaca: wheaten flour, water, and oil, boiled and applied as a plaster. You may add small amounts of saffron to color it. The second was used by both Galen and Halyabbas: a mucilage of figs, bark and roots of guimauve, and wheat-flour. The third has been used by most surgeons: leaves of mauve and senecio, roots of lilies, bark of guimauve, wheat-flour, powdered linseeds, and fresh pork-lard. Boil in water, grind in a mortar and make a plaster to be kept in place for long periods.

If the pus does not erupt spontaneously, open the apticosthem with a knife or corrosives. Then use detergents, incarnatives and consolidatives, as described before, as used for ulcers.

The fourth act is the treatment of complications according to their types. Relieve pain because it weakens the patient's resistance and interferes with the primary treatment. Use alteratives and dilatives such as rosat-oil with egg-yolks or bread crumbs in boiling water as a paste. Add the crumbs to the rosat and some boiled mauve and add oil until it is thin, and add oil of violets and saffron. If more relief is needed, use hyoscyamus with great care. Heat its leaves over warm embers and mix them with fresh lard, as used by Theodoric. Avicenna warned to take care to avoid too much moisture, especially dangerous at the onset phase,

Use evacuants and attractants and cups to prevent recurrence of suppuration. If you suspect that induration will ensue, boil the roots of cucumberts or use gelatin or asarum in water, usually all of them together, and sometimes add plump ripe figs. Finally mix in some wheat-water, fats of geese or hens and boil all. Make a plaster, as in *Glaucon,* Part 2.

If the apthosthem putrefies (necrosis) scarify the surface, wash it with brine, and apply a plaster of bean-flour or ers with oxymel. Treat it as we do for esthiomene.

An Explanatory Addendum About Carbuncle, Anthrax, Esthiomene, And The Bad Sanguine Pustules

Bad and corrupt sanguinous pustules leave deep scars after they have drained. Therefore, although a furuncle is a small phlegmon, it is not a corrupted pustule, because it does not leave a deep scar. After a phlebotomy, treat and alter it with a paste of wheat-flour and diachylon, and deterge with boiled honey and sarcocolla as did Rhazes who called it a carbuncle. Galen (Part 4 of *Therapeutics*) said that the pustules that scar are caused by thick, hot, and corrupt sanguine humor, and that they go on to become Persian or Holy Fire, etc. According to Avicenna, all of them are variants of Persian Fire[258], when the inflammation over-heats. As the combustion continues, the lesion becomes venomous and causes anthrax. That leads to putrefaction, gangrene, and esthiomene. Galen said (*Tumors Contrary to Nature*) that gangrene, esthiomene and carbuncle really are phlegmonous. Therefore, pustules are not caused by a mixture or an accumulation of natural humors. Lanfranchi and Henri agreed with Galen, although Henri expressed some doubts. As phlegmons, they are caused by thick blood that becomes thick sanguine (clot) as well as thin (serum) when combusted; that is, they are changed to choler and melancholy, and remain together. Avicenna said that the mixture will vary according to the amounts in it.

Therefore, a carbuncle or a burn or a Persian or a Holy fire (they are almost the same) is a phlegmonous pustule which can blister, cause burning pain, blacken the site, make it ashen grey, or make it dark red, raise blisters in the surround, or cause large scars as if it had burned the spot with a hot cautery.

The cause is a thick and moderately hot and corrupt blood before the clot and the serum have separated.

The Signs of a carbuncle at its onset are reddening, darkening, yellowing, induration, pain, heat, restlessness, and prickling, all in an area as small as a chick-pea. It enlarges rapidly and raises blisters all about. When it matures, the dead flesh is like an eschar except that it continues to discharge nasty matter as if it had roots feeding it from below. Sometimes it points in several places, but erupts at only one. One must not read the signs wrongly, because the lesions are partly venomous. Although at the onset they may not resemble ulcers, they all end as such and will be treated as such. In that sense carbuncles resemble the lesions of plague.

We treat carrbuncles in three ways: by establishing a healthy daily regimen; by restoring the normal balance of humors; by getting rid of the matter in the carbuncle.

---

[258]   Avicenna wrote that the cause was carbuncle and pruna; Serenus Sammoniccus called them Carbo and caharbon; Celsus said carbunculus, carbuncle, and charbon; Pliny said pruna, charbon, ardent, and braise (EN).

The first involves the six non-normal elements and their annexes, adapted to the type of phlegmon: tend to abstinence, coolness and moisture, because often there are fevers. Eliminate wine and meat and emphasize lettuce, purslane, pomegranates and sour foods. When more nourishment is required, you may allow chicken broth containing lettuce and sour wine. The second measure is phlebotomy. Early in the course use the opposite side, and latter on use the same side, when the site begins to darken and change. The third measure is used before ulceration occurs. Apply repercussives and evaporatives (resolutives) to prevent the reaccumulation of matter and to relieve irritation. This topical was praised by Galen (Therapeutics, Part 14) and by Avicenna as the Plaster of Plantains: plantain, lentils and bread[259], all boiled in water until reduced to the consistency of liniment to be applied over the entire region. Avicenna added oak-galls when he thought the blood was too thick or too thin. Then he applied a plaster of two pomegranates boiled in vinegar or in oxalis-water. After the pain has eased a bit, we apply Avicenna's plaster of figs, dry raisins, walnuts, and barley meal, all boiled in wine, all designed to mature and to erupt the lesion. Afterwards, mondify with celery, honey and flour, all boiled. Follow with consolidatives such as diachylon, as we do for ulcers. Always surround the lesion with an ointment of bol d'armenie, oil, and vinegar. If the lesion shows signs of corruption, scarify it all around it and wash with brine and dry the ulcer with calidicon (ie quick-lime) as troches moistened with wine. Loosen the eschar with a plaster of celery or with butter, and follow with the treatments for anthrax and necrotic ulcers.

William of Saliceto said that anthrax was a kind of malignant carbuncle. The matter in it derives from thick blood which has been heated long enough to become venomous. We call it a good bubo, a derisive name, because really it is an evil and threatening lesion. The word 'anthrax' comes from 'heart' because it threatens the heart.[260]

The signs, according to Henri, are those of a carbuncle when discolored by veins around it, like a rainbow. The limb feels heavy as if a tourniquet binds it. It hurts intensely. The patient is prostrate and loses his appetite and is nauseated. His heart races and he is feeble.

Anthrax is an acute dangerous disease; its cause is venomous, pestilential, and contagious. The worst lesions occur at emunctories or near principal organs, because the risk of recurrence is greater.[261] As is the case for any evil and venomous pustule, when it recurs it is lethal, especially when the symptoms reappear; its complications are violent. Therefore, look for the good signs of its is declination. The signs of improvement according to Avicenna are the reappearance of pink and yellow; black means all is lost. Anthrax lesions are more common during epidemics of plague.

---

[259] Bread: Galen at first insisted on bread, baked from flour from which the bran had been removed. Later he used home-made bread made with crude flour. He preferred it because it crumbled more easily. (J).

[260] Joubert chides Guy for his lack of Greek. Anthrax means black as coal! (J).

[261] The inflow of matter has an easy passage from below the surface (LDR).

We treat anthrax in four ways. First, attend to the general regimen (ie life-style). Second, comfort the heart. Third, void (repercuss) the bad humors (ie the antecedents). Fourth, attract and eliminate the conjoint matter.

The first deals with the six non-natural items. 1. Clean the patient's living quarters. 2. Keep him active and wakeful. 3. Limit his foods even more than for cases of carbuncle, more like the treatment of fevers. Barley-water and sweet rosat-water with milk of almonds in barley-broth should suffice as beverages until the fourth or fifth day. Then add pomegranates, oranges, lemons, and other sour foods. If needed, allow some chicken-broth cooked with lettuce and verjus.

The second measure uses a good theriac that has been well-tested: apply a paste of beans moistened with the juice of scabious; you may also dose it as a tea. If a fever rages, use rose-water (ie topical). Avoid between-meal snacks, because the time for the theriacs (ie oral) is not fixed, as Avenzoar noted at the beginning of his *Thesir*. Give it at any time, especially when it will be matter of first digestion (ie when the stomach is empty). Avenzoar dosed it every seven hours or oftener. Averroes gave it every nine. The timing is arranged to be midway between the total digestion of meals (ie from intake to stool), which is sixteen hours, according to the old saw that the body refills three times every two days. However, according to Avicenna, the time for emptying after a meal is twelve to twenty-two hours. Albert of Bolgna said that the emptying included the stomach and the intestine. The reason for all this dispute is the need to avoid mixing food with oral medicines, which makes the patient restless and increases the local pain (ie belly-ache), according to Avenzoar. Theriac is not only a medicine, it participates in all the disturbed bodily functions and mollifies the offensive actions of a fifth of the Simples, by cooperation rather than by its own action.

Now let us return to the subject in question. We apply a plaster over the heart, made from roses, violets, bugloss-flowers, both sandalwoods, citron peels, and if the fever allows, we add small amounts of melisse, marjolain, and saffron. Use a red cloth for the plaster, large enough to cover everything. Arnold of Villanova said that the betony and the tormentilla can replace the theriac.

The third treatment is by phlebotomy of the affected member[262]; apply ˋcups after scarification if the patient's age or vitality will not allow venesection.

The fourth measure uses coolants and defensives (repercussives) around the lesions, but not directly over it, simply to prevent the renewal of inflow of bad humors into the lesion. Use an ointment of bol d'armenie, rosat oil, and myrtle, all mixed with vinegar. On the lesion itself we apply attractives and other items, including cups and suction, used with great care. Be aware that we are dealing with a disease that leaves little room for error. Extreme maladies call for extreme treatments if we hope to cure. Sometimes we are forced to burn it away with an actual cautery, even when we the victoim says the treatrment is like a bite of a mad man. You may scarify around the lesion and irrigate it with warm brine to wash away the sanguine; it must not

---

[262] 'Of The Affected member': Every part of the body had designated veins and proper sides and correct half (ie upper or lower) of the body, as is best suited for phlebotomy (LDR).

be allowed to clot. Or you may apply corrosives based on arsenic. If, by chance, the progress of the disease is sluggish, allow the lesion to mature and erupt, and apply the paste of figs and yeast used by Avicenna. Then apply the plaster two or three times, as used for carbuncle, where the lesion is black and open. Then use detergents of apium. Finish with diachylon.

A topical of egg-yolk and salt will mature and erode the anthrax, as described by Theodoric. Janvier described the following for the same purpose: roots of apium, scabious, a geranium (called falcon's beak), marrubium, wheat-flour, linseeds, honey, oil, and aged lard. Boil in wine until it is thick enough for a pl;aster.

Another topical: consolida major and minor ground between stones. By some divine miracle the anthrax will undergo necrosis and slough away, all in one day. Then treat the defect as we treat ulcers, as did Roger and Theodoric. The Four masters mentioned it, but declared that scabious was more useful, because it can be made into a potion with wine, or eaten. It will convert an aposthem into an ulcer as it melts away insensibly. Henri said that the test for curability or lethality of an anthrax as attributed to Theodoric was a fable; it used an application of a pig's gall bladder. Henri himself said that the treatment of anthrax was like that for carbuncle and esthiomene, inasmuch as its severity is somewhere between them.

Esthiomene is not a pustule, although it may arise from pustules and its treatment in some ways is the same. It represents the death of a part of the body and its liquefaction. It is called esthiomene because it is hostile to humans.[263] The corrupted lesion is mushy, different from a lupus or chancre which end up as hard plaques. Therefore, contrary to what Theodoric, Lanfranchi, and Henri all claimed, they are not the same.[264]

The common name for esthiomene is St. Anthony's Fire, or St. Martial's, and by the Greeks it is called Gangrene. In Galen's book on *Tumors*, he called it Gangrene, and listed it among the major phlegmons, exhibiting necrosis. Avicenna classified all of them according to the degrees of necrosis.[265]

---

[263] Joubert again chides: Guy foolishly derives Latin terms from Greek words. Esthiomene means eating, from the verb to eat. Here it means a gangrene that follows the extinction of natural heat, although some life may persist in the tissues around it. Sphacelus (called asaschylos by the barbers, and syderation in Latin) is when the entire region is necrotic. (J).

[264] Joubert, again: Theodoric did not say that gangrene and esthiomene are the same as lupus or chancre; he did say that herpetic esthiomene was a lupus and an aposthem that eroded its tissues. Lanfranchi disagreed and said that gangrene was herpes esthiomene as well as lupus and chancre, in the sense that they destroyed themselves; and sometimes were interchangeable in name. In France they were called Our Lady's Disease, and in Lombardy were called St. Anthony's Fire, and elsewhere were called erosive erysipelas. (J).

[265] A modern Reader should not use the term 'Gangrene' as it is used today, simply connoting dead tissue. The mumbo-jumbo of terminology was dismissed by Henri Mondeville. He slyly suggested that we leave the nosological arguments to the academical physicians and leave the treatments to the surgeons(LDR).

The three causes of esthiomene and the necrosis of the part are: 1. When the part does not receive the sustenance sent to it from the heart (ie the vital spirit in arterial blood), and the complexion and the harmony of the deprived member grow cold, as in winter[266], or when one overcools an aposthem, or when it is overheated (ie burned) or when it receives poisoned blood, or when pustules are malignant. 2. When the necessary nutrition is blocked, as when a large aposthem obstructs the veins and pores, and its vital spirit cannot breathe. 3. When the passages that bring life to a part are shut off by a tight binding (tourniquet) or by a severe contusion (ie large hematoma).

The signs are these, according to Galen: The usual colors of ordinary aposthems disappear; pain and pulsations fade. All normal dispositions cease and sensibility dies. The affected part blackens and corrupts, and stinks of rot. Digital pressure indents it, and the dent persists when the pressure is released. The overlying skins floats loose from the underlying tissues.

The esthiomene is so ferocious that it will destroy an entire member unless it is attacked at once, and it will kill the patient as it spreads.

We treat in three ways: 1. Deal with the life-style. 2. Set the humors in correct balance. 3. Get rid of the corrupt matter and tissues.

First we deal with the six non-natural elements emphasizing the cool, dry, and abstinent. Foods should include a bread-crumb paste in water, barley-broth, oats with milk of almonds, chicken-broth containing lettuce as we use for treating fevers. Use purslane, pomegranates and other sour items. Protect the heart from the stench by dosing a theriac and cordials as we use them for pustules.

Second: use phlebotomy and purify the blood with catholicon, cassia fistula, tamarinds, hops, fumitory, polypode, and the like, because in these corruptions there is always heat, a commotion of choler, and an infection of the blood.

Third: we follow Avicenna's method. When we see the limb change color we must promptly apply an ointment of bol d'armenie, terra sigillata, and vinegar. If that fails, we cannot avoid making deep scarifications in different spots, and apply leaches with care to open small veins in the region. Then wash with brine until all clots are eliminated and unclotted blood appears. Place a resolutive topical on the scarifications to prevent corruption: a flour of beans with a vinegar syrup (oxymel), and irrigate the entire field with warm vinegar twice daily. After the area has been rewarmed and the accompanying inflammation subsides, apply the Egyptiac ointment made from verdigris, alum, honey, and vinegar, all boiled. It will prevent and resolve putrefaction and clean up what is already there and prevent recurrence. But if the disposition is already out of control and putrefaction and softening are apparent, cauterize with the hot iron or with caustics to separate the corrupt from the healthy. The corrosives include calidicon (ie quick-lime), aldaron, and asphodels. I agree with Theodoric and Henri that the best corrosive is powdered pure sublimated arsenic moistened with wine and smeared on charpy or cotton

---

[266]   The necrosis of frost-bite (LDR).

and laid directly on the lesion. When necessary, separate the slough from the healthy tissue and painlessly interrupt the spread into tissues that cannot resist it. I will describe the method in the chapters on glands, chancre and eruptors. Afterwards, attend to ridding the eschar by applying butter or other fats. Then clean the bare surface with vinegar washes and a plaster of William of Salicerto: honey, raw egg-yolks, and barley-meal. After two or three daily applications, dry the surfaces with a plaster of choice myrrh.

When you have to amputate to prevent the putrefied matter from it spreading, cut and saw as we will describe below. Then cauterize to insure destruction of anything left behind. We will describe what to about a dead limb that is not putrid in the chapter on embalming the dead. All of what we have described here, as taken from Avicenna, will be expanded in the chapter on putrid ulcers.

CHAPTER 3

Erysipelas and Other Choleric Aposthems

A lthough phlegm (ie pituit) is next to sanguine in the amounts of humors in whole blood, we will consider choleric aposthems now, because they have so much in common with the phlegmons that derive from sanguine.[267]

The choleric aposthems often are given the Greek name Erysipelas because they involve skin and hair. They are in the skin, just as phlegmonous aposthems are in the subcutaneous tissues, although the erysipelas ones may also extend below the skin, as noted by Galen in *Therapeutics*, Part 14. Erysipelas appears in two forms, the true and the non-true. True erysipelas cames from abundant natural choler, called "prickly' by Avicenna. Here, however, I will call it the thin (subtle) part of natural blood.[268] The non-true erysipelas derive from non-natural choler, and Avicenna called them Formis. Galen (ibid) said that erysipelas had two types: those with and those without ulceration. The first is uniform and is called Phlegmon, and the other is called Herpes or Formis.

The formis were called herpes. He described erysipelas in *Tumors Contrary to Nature:* You may be sure that the humors in the humors in erysipelas are pure and completely separated when the ulcer is in the skin only. But when the subtle (ie thin, watery) is mixed with some sanguine (ie not completely separated), there is more tumor than simply an ulcer. Therefore, we have either erysipelas or herpes. In his *Glaucon,* Part 2, Galen devoted separate chapters to each: a chapter on Formis and Herpes and then one on true erysipelas derived from thin choler. Therefore erysipelas are choleric aposthems of two types.

---

267 Another reminder to the Reader that Phlegm, the humor in blood also called pituit, is not the cause of phlegmonous aposthems or phlegmon as described on the previous chapter; they derive from the sanguine humor. In turn the sanguine often is called 'blood' by Guy, and we may confuse it with whole blood which consists of all four humors (LDR).

268 Joubert noted Guy's discrepancy. In Chapter 1 Guy called all aposthems uniform when they are derived from natural humors (J).

Choler is a warm humor engendered from the most liquid part of chyle, and it has two varieties: natural and non-natural. The natural choler is warm and dry, very liquid, orange-red in color, and slightly bitter in flavor.[269]

Non-natural choler derives from the other right from the beginning, and it is no longer pure choler, rather something else even as it leaves the liver. That change happens in one of two ways: a spontaneous change, or an admixture. The first occurs in either of two ways: the natural choler is overheated and corrupted, and is called choler-by-putrid-corruption, or when the yellow non-natural choler is is combusted in the stomach, or liver, or the veins, and becomes greenish (ie like leeks) or bronze as if its about to do harm. The second conversion occurs by mixing natural choler with another humor. When it is watery phlegm (pituit) the color changes to orange. When the pituit is thick the mixture remains yellow. When it is melancholy the choler is darkened.

Avicenna described six kinds of non-natural choler, whereas Halyabbas said there were only four,; he omtitted the two burned variaeties. Galen (*Natural Virtues*) said that the yellows were not non-natural and that greenish and bronze colors were the result of eating spoiled vegetables, or caused by unhealthy veins (*Prognostics*, Part 2). It is clear that choler can cause four kinds of aposthems. First are those from good choler (ie the thin serum of blood), and are called true erysipelas.

Three aposthems derive from the admixed non-natural choler: phlegmonous erysipelas; edematous erysipelas, and scirrhous erysipelas. Other aposrhems derive from the combusted non-natural choler, varying according to the degrees of liquidity of the humors from corrosive pustules to herpes and chancres. The herpes group includes serpigo and formy. See *Therapeutics*, Part 14.

True erysipelas, like true phlegmon, has three causes: immediate (primitive),[270], humoral (antecedent), and conjoint.

The signs and the diagnosis are similar to those of phlegmon, as described in Part 14 of *Therapeutics* and in *Glaucon*. Inasmuch as a true erysipelas is a kind of phlegmon, take the orange-red color as a first sign. Second, you can efface the color by pressing on it. Third, the swelling (ie the tumor) involves only the skin, Fourth, its intense local heat causes fever, more than an ordinary phlegmon. Fifth, the throbbing is less. Sixth, the pain is biting and prickly, and not as intense as in phlegmon. There are other signs that lead to a diagnosis of choler: Erysipelas often appears on the face, beginning at the base of the nose and spreading, varying with the mildness of the choler, and it is serious when bone is esposed and when there is slough.

---

[269] Joubert on flavors: Choler is slightly bitter at first but it becomes slightly sweet as it is admixed with others and forms the bile that we find in the gall bladder, according to those who have tasted it. They describe various flavors calling to mind the experts who identify the contents of various sauces that are not unpleasant. Some say it is like sweet blood, or watery tasteless pituit, or slightly bitter like peppery melancholy. Others compare it the foam floating atop the sea, or the foam atop the blood spilled into a phlebotomit's basin (J).

[270] Usually physical trauma (LDR).

As we described in other aposthems, true erysipelas has four phases. It does not often suppurate, and usually it subsides of itself.

Sometimes it has complications which are worse than the primary lesion, and we must reverse the order of the treatments and not leave the complicatuoins for last as we recommended for phlegmons. Furthermore, erysipelas may induce a third degree fever because its matter has more heat.

The treatment for true erysipelas is like that for true phlegmon except for the special cases. The general measures are the same: life-style[271], restoring the humors in balance, riddance of contents, and dealing with complications.

First we establish a healthy regimen, tending to cool and dry foods, such as we use in treating third degree fevers. Circulate fresh and cool air scented with willow leaves, vines and canes, marsh-reeds, roses, and violets. Abstain from warm and moist things and from fats, sweets and spices. Avoid wine and dairy foods, and favor lettuce, purslane, zucchini, barley, rice and other things that thicken and freshen blood. Live soberly; keep the bowels open; sleep long and quietly; live without shame (ie avoiding sexual activity).

Second, use evacuations and phlebotomy. The contained matter sometimes can be eliminated by medicines which reject choler, such as electuaries of rose-water, and tamarind-water: fresh tamarinds, damascus plums, and juices of violets; you may fortify it with more rose-water. Dose it as a potion in the morning. When you must bleed, follow the instructions given for phlegmon.

The third treatments begin with repercussives and coolants, excepting the list of cases in the General Chapter. Follow with evaporative topicals or by sudorifics. We need more cooling than for phlegmons because the violence of erysipelas is not only a matter of amount but in the degree of angry inflammation. Apply coolants until the color begins to fade. Avicenna favored irrigations with cold water, and Galen used the juices of morel, joubarbes, purslane, polycaria, hyoscyamus, and others as in the preceding chapter. Then evaporate (ie resolve) the retained matter with barley-meal. Using the signs of phlegmon as a standard for comparison, adjust your dosages for erysipelas according to the needs.

The fourth measures anticipate the recurrence (ie reaccumulation) of the contained matter, or the appearance of sclerosis or corruption as in phlegmon. Mitigate the pain and the inflammation with leaves of hyoscyamus wrapped in stoupes and heated over hot

---

[271] A note about 'life-style': I use the term to mean more than just 'diet' which was the Hippocratic term for the first-line treatment to be used by physicians before prescribing medicines and going to surgery as a last resort. In the centuries after Hippocrates, his 'diet' came to mean a general ordering of elements external to the patient's body (socalled un-natural things) which were extensions of the Hippocratic 'Airs and Waters'. The 'upper classes' (Emirs, Caliphs, Kings, Popes, etc) had personal physicians who prescribed their daily regimens; Maimonides' instructions for Saladin is an example. Soon, a wider distribution of such health books, appeared; The Regimen Sanitatis Salerni, or the Tacuinum, are examples whose influences persist today in the'Health-Food' emporiums throughout the world (LDR).

embers, mixed with populeum ointment or other greasy substances. Use more than what is needed for phlegmons. When it ulcerates, apply the white ointment or any ointment of litharge. If you have on hand some burnt lead shavings, use them to advantage.

An Explanatory Addendum To This Chapter

All the lesions that erupt, erode, and inflame with virulence are caused by bad choleric pustules, and include the entire group classed from herpes to chancre[272] and many that lacked special names. They are grouped in two categories. One is called Herpes in Greek and Formy in Arabic. They are derived from non-natural choleric humors that have different degrees if thinness. The thickest engenders what the Greeks called Herpes Ersthiomene, or Cancer in Latin, as defined by Galen in Part 1 of *Tumors Contrary to Nature,* and Part 14 of *Therapeutics,* and Part 2 of *Glaucon*[273]. Avicenna gave the name Formy to all choleric pustules derived from non-natural choler, the liquid part (ie separated from coagulum). Natural choler from natural nutritious liquid blood engenders the true erysipelas (epine, or miliaria).

---

[272] Joubert added: The term chancre here denotes herpes esthiomene, which Guy claimed derived from two kinds of pustules. One is called herpes (not the herpetic lesion more properly called serpigo or darte in French that is described in Guy's Sixth Treatise) which is called Cenchrias in Greek and Miliaria in Latin, and Formy by the barbers. The second type of Herpes Esthiomene was called Chancre by the Barbers and not by the Latin-speakers (ie the academics), because it derived from non-natural choiler produced by combustion until it is very dry and thick, a condition similar to melancholy, which really is the cause for chancre, as properly named. Guy's definitions in this chapter are similar when he explains his non-natural choler produced by combustion yielded the entire range of pustules, from herpes to chancre, each with its own degree of thinness or thickness, including herpes, serpigo and formy (J).

[273] In *Glaucon Galen* said that there were three kinds of herpes. One type gives its name to the overall group, and is the absolute herpes, derived from the most liquid choler. It invades only the epidermis. The second type creates clusters of blisters (pustules) that resemble grains of millet in shape and color; it is called cenchrias in Greek and we (ie in 1660) call it Miliatia, although some use the term Granular. Tagault said it was this blistered variety that Galen used in naming the entire group without secondary titles. This is not the important issue, since the same authors (Greek) also used the term Herpes for the undermining and erosive lesions (Herpes esthiomene).Even Tagault admitted his own confusion.

The third variety invades below the skin, and Hippocrates called it esthiomene because it erodes. It derives from the most pure non-natural choler that has not been diluted by pituit. The thin non-natural choler engenders miliaria (see above).

On the other hand, Avicenna called all forms of herpes Formis, and he used Galen's categories with different names: The first group were Ambulatives; the second were miliaria; the third were corrosives. Dino said that the ambulatives came from pure liquid (subtle) choler;

We combine the treatments of the Greeks and the Arabs to simplify our descriptions, of the two kinds of bad choleric pustules: In one category called Herpes are those derived from thin (ie subtle, liquid). The others are Formis from thicker choler. Let us deal with what they cause rather than worry about their names.

Therefore, let it be that herpes is one of the pustules derived from bad choler: erysipeloid, blistered, inflamed, pruritic and red (with a tint of orange). The name herpes is something other than erysipelas, blisters and ulcers. As Galen taught in part 14 of *Therapeutics*, the choleric humor that engenders it is very thin, and it can permeate more than fleshy structures, and even permeate the skin and its epidermis. It is the only humor than can erode the skin, although the skin can resist it and allow passage, like sweat, without ulcerating.

We have described its causes and its signs in the foregoing. As to differential diagnosis, we can see that herpes resolves more rapidly than formy. Also, it is clear that[274] erysipelas can degenerate into a formy, and the formy can become a chancre.[275]

---

the miliaria from choler mixed with pituit; and the corrosives came from the thickest and the most comvusted.

Guy lumped together the Greek and the Arabic names of Avicenna's classes and called them all Herpes: the most superficial, the tiny blisters, and the ulcerated erysipelas. The other two classes were formis derived from thick choler: 1. From the less thick came the erysipelas in the skin only, which spread quickly. 2. The thickest non-natural choler engendered what Hippocrates and Galen called esthiomene because it undermined. Even when an esthiomene (ie an undermining corrupt ulcer) was engendered from bitter and thin choler, it was classed as a Formy, not as a herpes.

Avicenna's second category was the miliaria engendered from moderately thick cholera contaminated with pituit. Here Tagault unjustly criticized Guy for confounding Formy with warts, and with erosive herpes. Guy had simply mentioned warts in passing as follows: "When a wart first forms it may resemble a formy, et al. However, it is not caused by very thin choler as are herpes esthiomene, ulcerated erysipelas, pustules, or other lesions in that catergory.". Formis are caused by thicker choler, as Guy said more than once. Tagault was correct in classifying as herpetic the roseolas and variolas that the Latins called eruptions and papules derived from pituit. Thse were called ecthymas and exanthems by the Greeks. (J).

[274] More Joubert: Guy classified Herpes in degree between Erysipelas and Formy. He characterized erysipelas as an ulcer and pustule, and formy as a bad erysipelas. Between formy and chancre was Phagedena., about which Guy wrote, "In that category we also put phagedena and ulcerated chancre, etc. The thinnest humors cause ulcerated herpes; the thickest makes a chancre. Next below that in thickness is the cause for phagedena (J).

[275] All this contentious, inconclusive and confusing nosology is based, at least in part, on the nature of blood, its thin and thick parts. An analysis of coagulated blood in the phlebotomist basin was so clear to the professionals that no further explanations were called for in their texts. I think that Nicaise was not aware of Henri de Mondeville's descriptions when he was working

The treatment of herpes is threefold, as used for erysipelas: improve the life style, set the humors in blood in proper proportions, and manage the conjoint material. The first and second are the same except that phlebotomy is used more often in treating herpes.

The third is similar, as described in *Glaucon*, Part 2. Cool the erysipelas and humidify it before it can become an ulcer. For herpes we use things that desiccate rather than moisten as do lettuce, purslane and cool water. Rather, we use topicals containing the tendrils of grape-vines and blackberries and plantain. Use lentils and barley-meal and other things that we use for phlegmons, adding honey when you need a detergent effect.

Treat the bad undermining corrosive ulcer with applications of the white metallic ointments which we will describe when we come to the chapters on virulent ulcers.[276].

Formis are bad choleric pustules, not large, yet inflamed and increasingly painful (prickly). They involve the skin and erupt and form ulcers. In brief, formis are two kinds of malignant herpes. One progresses more swiftly because it derives from thinner and more bitter choler. The other is slower because its choler is thicker, having been mixed with some phlegm. It is called miliaria, and its matter may be as thick as what we encounter in acne and sebaceous cysts.

Now we will discuss diagnosis and prognosis. Formy is slower to resolve than Herpes, and it does not form eschars, although it may corrupt and become virulent. Formy always come to a head (ie points), as claimed by Avicenna.

At the outset, a wart may resemble formy, but it does not contain matter, and it is rooted as if nailed down. Galen treated warts by coring them out and uprooting them with a hollow tube.[277]

Although formis do not begin as ulcers, usually they end that way, and we treat them locally as such. The treatments have three parts, as for the other lesions. The first two are the same as those for erysipelas and herpes. For the third measure Galen treated formis in Roman women by applying skim milk containing a small amount of scammony. On the other hand, Avicenna said that for miliaria we should add crushed turbith and epithyme.

Treat ulcerated formi with compounds of repercussives and resolvents. A good one is a plaster of plantains as used for carbuncle. Another, to be used before as well as after ulcers appear is the plaster of two pomegranates. Avicenna advised applications of the

---

on Guy's book. Three years later, when he produced his translation of Henri's *Surgery*, he came to a full chapter on the subject. I refer the Reader to my English Translation of Nicaise's edition: Vol. II, Ch. XI (see Bibliography)(LDR).

[276] The ointment was based on ceruse, and was used by Galen in treating Roman women (ie cosmetic). Guy's scribe foolishly translated the Greek word *phycos* (ie a red dye from sea weed) as *alga*. (J).

The White Ointment of Rhazes contained ceruse and litharge and many other items. The Greek word for sea-weed, and the Latin *alga* also referred to a red dye derived from it. I fail to understand Joubert's point here. Certainly, Guy referred to Rhazes' white ointment, not to the red (LDR).

[277] A tuyau resembled the cork-borers used in our laboratories (LDR).

sap that drips from burning grape vines for the treatment of miliaria, sebaceous cysts (ie acne), and small pustules. Also he praised the use of brine in which fish heads are boiled. Theodoric said that mille feuille and parietory ground with some salt are good for treating the same. For drying corrupt matter and miliaria Avicenna approved a liniment of bronze-flower and sulfur when added to the above.

Halyabbas favored an ointment of bol d'armenie and terra sigillata and rose water applied around the lesion, to be kept on throughout the entire course of treatments. When the lesion erupted and becames an ulcer, he applied a liniment based on troches (ie lumps) of quick-lime and aldaron mixed with vinegar and rose-water.If that failed, he resorted to the actual cautery or to caustic arsenic eruptors to get rid of corrosive conjoint matter (*Glaucon*, Part 2). In that way he stopped the undermining. Afterwards he separated the eschar with butter or other fatty substances. He also used the knife instead of caustics, or he tied the eschar as we do for the fig-like dangling lesions. We treat the residual ulcer as such.

CHAPTER 4

Edema and Other Phlegmatic Aposthems

In the same way that choler makes erysipelas, phlegm (pituit) makes Edema, as described in Part14 of *Therapeutics*. Edema includes two kinds of phlegmatic aposthems, the true and the non-true. True edema derives from intact natural pituit as is found in fresh nutritious whole blood and even when the humor is only slightly combusted.[278] All of the so-called raw humors in whole blood are lacking as good nutrients for the body as a whole or for the particular phlegmatic regions.[279]

The non-true aposthems derive from non-natural pituit, a cool and moist humor, engendered from the most cool chyle.

Natural phlegm is cool, moist, and raw. It is off-white in color and slightly sweet in flavor and odor. Non-natural phlegm differs in being more abundant than not only the natural phlegm, but more than all the other humors. It forms in two ways: The natural phlegm changes of itself, some of it remaining watery and serous, that is, more thin, and in turn it reacts on the rest of the phlegm which becomes thicker and more viscous, producing a mucilaginous or gypseous or glassy phlegm, over time. Then it can spoil and deteriorate and produce corrosives like salt-niter.

The other route that leads to non-natural pituit is the contamination (ie mixing with) by another humor. In that way phlegm becomes sweeter, choler becomes salty, and melancholy becomes astringent and bitter. According to Avicenna there are eight kinds of non-natural phlegm; Halyabbas said four: Galen said two, the salty and the bitter.

---

[278] Joubert explains: The combustion occurs at the outset of the formation of fresh blood, and the pituit is slightly affected. Second is when blood is well into the process but the pituit is not yet as complete as melancholy, and is still too serous. That kind of incomplete pituit gives rise to rheum (catarrh). Third is when the blood is completely combusted, as when choler forms. All of these so-called raw humors are not yet good nutrients, and that is why the physicians call them raw. The term comes from the aphorism of Hippocrates that state that one should treat with cooked substances and shun the raw. He meant that they should be matured to enable them to be more effectively evacuated by purgatives. In that sense he used the term 'cooked' in a different sense from cooking foods to eat (J).

[279] See above, Treatise I, Chapter 1, 'Simples', etc.(LDR).

However, in *Different Fevers* Part 2, in *Afflicted Places*, Part 2, and in many other places, he mentioned vitreous phlegm which he classed with the bitter, for convenience. From all this, we can infer that phlegm can engender at least eight kinds of aposthems.

1. True edema from Natural phlegm. 2. Three edemas derived from a non-natural phlegm that was formed by mixing: phlegmonous edema, erysipeloid edema, and sclerotic edema. 3. Four aposthems caused by a non-natural phlegm that was produced by ipso-alteration: gaseous and vaporous phlegm engenders gaseous aposthems; bitter phlegm engenders acute aposthems; raw mucilaginous phlegm engenders nodes, exitures, lupus, and nactes, which include glands, nodes, lipomas, turtles, etc. 4. from vitreous and gypseus phlegm come hard nodes and scrofules. In addition, corrupt and putrid phlegm causes scrofulous ulcers.

An Edema (in Greek) or Zimia in Arabic is a soft nontender swelling; it causes much less pain than erysipelas and phlegmon.

It has three causes, as do the other aposthems: immediate, usually a fall or a blow or a bad life style; antecedent, due to and excess of phlegm (pituit); and conjoint, the accumulated fluid.[280]

The signs and the diagnosis: the mass is soft and can be indented by digital pressure, and the dent persists when the pressure is released, and there is little discomfort. The color is off-white (the water); and other signs that we will describe for edemas.

Edema, as in other aposthems, has four stages: onset, augmentation, stable state, and decline. Edema usually subnsides and disappears on its own. It rarely suppurates, but quite often a lump persists, or another excrescence, as we described in the General Chapter. The phlegmonous swellings (edemas) are more common during the winter, and affect older and feeble persons.

The treatments for true edema, other than the general measures, have four elements: establishing a good regimen of diet etc.; balancing the humors in normal blood; getting rid of the contents; treating complications.

The first measures attend to the six non-natural things, favoring warmth and dryness, and keeping the matter liquid (ie for ease of resolution). Keep the patient where the air is dry and clear; offer well-baked bread; drink good clear wine (ie lacking sediment) and add some water to it; eat the meats of small meadow-birds and good mutton. Drink bouillons and soups containing herbs; abstain from unleavened breads, and sour foods. Eat no beans, cheese, or large fruits. Avoid fish except for small servings of rock-fish[281], and

---

[280]   The Reader will soon be aware that Guy refers to any abnormal collection of fluid, and/or gas, as an aposthem called edema! Occasionally, but infrequently, Guy distinguishes blisters, ascites, etc. The Reader often must decide when the text refers to local edema after a contusion, or residual edema after a dislocated shoulder, or a belly full of ascites—which does not resound when tapped, stasis edema in the lower legs, edema in the scrotum as part of ascites, hydrocoeles—which transmit light, et al. (LDR).

[281]   Other authors specify fish that are cought in streams with stony bottoms, in other words, from free-flowing currents that are not polluted. Guy's term 'saxatils' meant the same (LDR).

avoid other things cooked in wine. Roasted foods are preferred rather than boiled, pastry should be well-baked. Live soberly and eat moderate amounts and drink less. Exercise comfortably and avoid daytime naps. Avoid baths and other moisturizers.

The second therapies, the digestion of some of the matter, uses oxymel and other evacuants designed for phlegm such as cochia pills, herb Bennett, and diaturbith. If the patient is plethoric, bleed him.

The third measures are applications of repercussives at the onset and avoidance of coolants excepting those that resolve and desiccate. During the phase of augmentation add more potent resolutives, and in the stable phase use them alone. If you need to supplement spontaneous resolution during declination use consumptives. When that fails, promote suppuration and follow with our methods for draining exitures.

There are three types of desiccating repercussives that also are resolutives, to be used during the phase of onset: 1. Galen (*Therapuytics.*, Part 14, and *Glaucon,* Part 2) said that simple applications of water-soaked sponges, perhaps with some vinegar, will suffice.[282] That mixture also could be taken as a potion. Avicenna added that a double cloth could be used instead of a sponge. Renew the application frequently. A snug bandage should bring the edges together, beginning the wrap at the end of the member.[283]

The second type of of repercussive was Avicenna's borax water, ashes, and vinegar. The third type was Rhazes' and Avicenna's: Aloes, myrrh, lycium, acacia, a collyrium of memitte, souchet (ie sun-flowers), saffron, and bol d'armenie. Make a powder and add cabbage-juice and vinegar. Boil it to a paste for a plaster.

During the phase of increase, use any of the above fortified with vinegar. During the final two phases, when it seems favorable for resolution, use Avicenna's application: a sponge wet with a lye made with the ashes of grape-vines, fig-tree wood, and oak. Second use Bruno's and Theodoric's resolutive: alum. sulfur, myrrh, and salt. Grind them with oil of roses and vinegar and use it as a liniment. Another is the resolutive of Avicenna: cow-turds, frankincence, spica and aloine. Mash all with vinegar and cabbage-juice to make a plaster.

To mature phlegmatic aposthems that do not resolve, use Roger's diachylon as a cataplasm: malva, branca ursina, roots of guimauve and lilies, roasted onions, snails, yeast and linseeds. Boil all and grind; then add lard or butter. Another, from Theodoric: juices from hyeble, lapathum, lovage, and fennel. Boil and add dialthea ointment, honey, and oil, and make a plaster.

When the apostem has matured (ie suppurated), do not wait for it to erupt on its own; Henri said that the wait may be endless. Open it with the knife or by applications of caustics.

---

[282] Joubert explains: Guy probably used a new sponge, as was suggested by Galen, to ensure that it was not contaminated by prior use of unwanted substances, such as niter, aphroniter or the like. Guy's original Ms. said, "instead of a sponge, soak a folded cloth". Falces said that he meant a blue cloth because the blue was obtained by soaking the cloth in alum, favoring its desiccative effects (J).

[283] Galen and Avicenna insisted that it begin at the distal end (EN).

After the swelling has drained, use the apostles' ointment as a detergent or a celery-based mondificative, or the juice of aloin, or with Dino's attractant that deterges greasy pus: galbanum, ammoniac, pine-resin, terebinth, pix, cow-lard, and aged oil. Thin it with vinegar to simmer it and make an ointment. Finally, use what we apply on a sordid ulcer.

The fourth group of treatments attack complications, such as residual pain. Try such simple things as lanolin, wine, malva, branca ursine, oil of camomille, aloine, spica, and wax. For residual induration, apply bone-,arrow from cows, deer, and others. More of this in later chapters.

An Explanatory Addendum About Gaseous Aposthems

Now we shall discuss inflations. which Galen said should be treated as edemas (*Therapeutics,* Part 14.), because they derive from phlegm, and their surfaces dent when depressed. Yet this mass is air, sometimes beneath the skin or even deeper. Sometimes the collection is encysted, at other times it permeates the tissues. Sometimes there is pain, sometimes not.

Galen (*Maladies and Symptoms*) said the cause is the feebleness of the phlegmatic matter which is replaced by gas. The gas derives from phlegm and from unripe fruits, and it does not resolve beyond gas as it is heated.[284] Perfect cooking does not generate gas. It may attenuate (ie liquefy) the nutritive substances but cannot combust or resolve it. Strong heat will overcome whatever it faces and will liquefy the nutrients well beyond the gaseous state. In addition, the gaseous state can lead to its own troubles. It may becloud small collections when one or two belches could get it of it. When it is retained in a member, according to Avicenna, the gas will distend it.[285]

The signs and the diagnosis: When the inflation is tense it will resist compression and will resonate when tapped, as does a gas-filled leather bottle emptied of oil or wine. Often the gas spreads within the tissues and can be very painful. When the gas is trapped and is not dispersed, it causes much trouble in situ. When the gas inflates the body the suffering is intense, with much anxiety and fear, as if it was generated by venomous matter.

Treat gaseous aposthems in the three ways we emply for other kinds.

The General Regimen: Abstain from rich, starchy, raw, phlegmatic and gassy foods, such as sweets, peas, fruits, turnips, and chestnuts. Use warm dry, and soft items that will

---

[284] 'Heat': Exposure in the body's normal metabolism. For example, the liver 'cooks' the humors. Every organ's functions are classified in degrees as warm or cool, and every medication has its own degree. The Natural Spirit distributed via the arteries participates in the normal 'combustion'. A fever is a disturbance, not simply an abnormal combustion (LDR).

[285] Here we see that a gaseous aoposthem can be any kind of inflation. To the surgeon, gas was gas; be it in a dyspeptic stomach that seeks to belch; or a tympanitic ddistended belly after injudicious over-eating; or constipation; or intestinal obstruction; or an anaerobic infection; or ischemic gangrene; or a localized infection, etc. (LDR).

disperse the gas, such as barley-brerad with salt and cumin. Drink only white wine and claret. Serve purees of chick peas with onions, parsley, calamint, rue, and cumin. Serve meats of flying birds and other similars as in the preceding chapter.

Second: Comfort the digestion with good spices and recipes that contain cumin-seeds and calamint served as after-dinner dragées (ie digestives). Here is a good recipe: Anise, fennel, nut-meats, carrot, cumin, laurel-berries, sliced licorice,white ginger, galanga, cloves. cubebs, long peppers, seeds of rue, anis, and sugar-loaf (ie barley). Make dragées. After dosing one, massage the belly with oil of spica, costus, and rue.

The third treatments use topicals that have resolutive and evaporative actions, and are moderately astringent, yet cause no pain. Three kinds of such medicines were used by Galen (idem.) First was a new sponge wet with a mild lye-soap. He relieve pain with applications of resolutive oils or emollients until the pain eased. Then he used lanolin mixed with lye or soap that were boiled in wine before adding vinegar and sour wine, especially for use early in the course. The effective elements are the soap and the vinegar. Therefore, when you mostly want to relieve pain, use more wine and vinegar than soap. When you aim to repercuss, use more wine, and the wine should be a strong red. When you want more resolution, increase the soap. In respect of the vinegar, we use it for its double role, alone and as a vehicle. We make another third line medication, a liniment of cimolia and quick-lime, all boiled in wine and water.

When the gas stinks and is malignant and corrupt, excise the venomous matter and cut away the painful parts. When William of Saliceto encountered a localized collection, he elevated the limb and wrapped it before he incised the center of the mass with a razor or actual cautery and allowed it to drain. Then he packed the cavity with aloes and bol d'armenie moistened with oil of roses and vinegar. After three or four days, the place will show granulation tissues and begin to close. In such cases, he restricted the diet and purged the patient with laxatives. Then he added a good theriac.

An Explanatory Addendum About Aqueous Aposthems

Aqueous and serous phlegm engender aqueous aposthems. Galen (*Maladies and Symptoms,* Part 6) said that superfluous serous humors engender hydrops of the belly (ie ascites) and blisters (ie any watery cysts) in particular regions.

The cause is the damaging effect of cooling and the displacement (ie of the natural phlegm) brought on by harmful watery foods. In Part 14 of *Therapeutics* and Part 2 of *Glaucon* Galen compared the phlegm of these aposthems with the phlegm of hydrops (ascites) caused by bad dietary habits and said that they called for the same treatments, except that the latter needed more desiccation.

The signs and the diagnosis of hydrops are those of true edemas (the aposthems) except that the latter are softer, as tested by finger-pressure. The belly does not resound on percussion as does gassy distension, and the watery lesions can be transilluminated.

The watery lesions are less warm than the gaseous which are accompanied by cramps. They occur more frequently in the legs and the scrotum and in the head[286] and joints, because the watery matter tends to descend, and its warmth fades. (*Prognostics*, Part 2). The gaseous aposthems nearly always have some watery stuff, and the watery aposthems nearly always contain some gas.

There are five parts to our treatments. First, we deal the regimen; second, improve digestion; third we prescribe laxatives and we eliminate all moist and watery things; fourth we deal with diuretics and other evacuations; and fifth, we eliminate the conjoint matter with evaporatives. Galen (*On Diets*) put the first two treatments together (foods and digestives) and emphasized liquidity[287]. He discussed the other three in *Glaucon* and in *Commentary on The Seventh Aphorism* where he dealt with the liver, etc.

As to life-style, Galen repeats much of the topics in the preceding chapters, with more emphasis on dryness and heat. He eliminated all watery things: drinking-water, green vegetables, fruits, cheeses, milk, fish, beans, and pork, as well as bouillons and soups. He was careful about over-eating, and limited the beverages. We use barley-meal for baking bread and add some anise in the dough. Allow only small amounts of very good wines: Greek wines and red clarets are good. The juices expressed from chick-peas, sage, hyssop, calamint, boiled onions, and other spices are approved. Permit wild chives, oxalis, meats of hens and mutton, and all dry foods. Physical labor and periods of fasting are moderated, and the use of items that have laxative actions are recommended.

The second treatment uses purges designed to eliminate aqueous and serous matter, such as: bread made from barley-meal, milky juice of tithimal, powdered esula, tartar and small amounts of spica. Also use pills made from wild cucumber and spicea, as listed by Mesûe (ie Heben) in his *Simple Medicines*.

The fourth treatments use dragées as described above, with added celery seeds, parsley, and seed of baguenaudes and spica.

The fifth group supplements every thing else with desiccatives and three kinds of resolutives: 1. Galen's with added oxyrhodin and salt. 2. A sponge moistened with more of the preceding with added lye and other items listed for true edema, and with added aphroniter, alum and sulfur. 3. Avicenna's plaster designed for scrofules, and attributed to Galen's *Classes of Compound Medications*, which he claimed could resolve induration in less than one week, even within three days. Although I have not been able to find it in his book, I have used it to resolve the induration after watery

---

[286] I question the term "head (teste) used here. I believe Joubert meant 'testicle'. That seems more appropriate here, and refers to hydrocoeles in the scrotum, rather than to edema of the scalp (caput succedaneum) in the newborn infant (LDR).

[287] The essence of phlegm is watery, and the watery stuff can be resolved by evaporation, whereas thicker phlegm cannot so easily be ridded, and remains as the conjoint matter (LDR).

and gassy aposthems: seeds of mustard and nettles, sulfur, ecume de mer, aristolochia, bdellium, ammoniacm aged olive-oil, and wax. Make a plaster. If that fails, nothing else will succeed except for incisions with a razor or hot cautery, followed by our standard treatments for sordid ulcers.

An Explanatory Addendum For Nodes, Glands, Scrofules, And Other Phlegmatic Excrescences

Glands, scrofules, nodes, lupus, turtles, lipomas (nactes), hernias, goiters, and hard masses in the groins (buboes fugilis), wherever they appear in the body, all share some phlegm. Roger said so and I agree. Although they may evolve into hard melancholic tumors, they have phlegm at the onset. Avicenna said that sometimes it was a mixture of phlegm and other humors. However, for now, let us discuss the simples before they are compounded, and let us recall that there are many kinds of what we called exitures and excrescences.

A Gland[288] is a movable and partially soft mass usually found at emunctories.

A Scrofule usually is one of a firm collection of masses, usually encountered in the neck.[289]

A Lupus, like the strobile of the hops plant, is soft and is found at joints and other dry places.

---

[288] Joubert comments: Avicenna said that a gland is what the Greeks called a ganglion, as cited by Paul of Aegina and quoted by Tagault. But I disagree with Tagault who said that a lupus also was a ganglion and not a node or a stony lump in a nerve or tendon brought on by a blow or by hard labor, and is seen in hands and feet. Avicenna said that those were the signs of a gland or a node. But, according to Guy, a gland is something else: it is a single mass, separate from others and moveable, and has a soft part when it is at an emunctory where it may drain some of what becomes its matter. Its flesh resembles that of scrofules as noted by Leonardo and Aetius, and it swells as it accumulates matter. It differs from glands that always normally are situated at emunctories, where they persist as hard lumps after their contents have drained. Such a mass really is a scrofula, what the barbers call a bubo fugilin. Although that term should be reserved for a parotid gland, Avicenna claimed that 'fugilin' refers to glandular apopsthems, especially those that are beneath the ear. Guy, in Ch.5 of his Doctrine II, explains it all. I am amazed that Tagault questioned Guy's inclusive definition of phlegmatic aposthems (J).

[289] Joubert explains: Firm glands (ie lymph nodes) in the neck often become scrofules. Others are in the axillae and groins. Paul noted all such aposthems are encapsulated, as are also the steatomes, atheromas, and melicerides. What is common to all scrofules is the foreign matter, which distinguishes them from their prior status as glands. Where Guy wrote that a gland is soft like a houble, the hops fruit, which resembles an herb composed of layers of folded leaves forming a small tuber. That fairly describes a lupus as well. (J).

A Node is like a knot on a string: hard, rounded, and fixed where it lies, usually in nerve-bearing regions.

A Turtle is a large soft humoral mass, discoid like a turtle's shell. It is called a Mole when is on the head, a Goiter when in the neck, and a Hernia when in the groin. Sometimes it has crusts and fistulas

A Lipoma (nacte) is a growth: large, fleshy lke the tissues of the buttocks and thighs, varying in size and shape, sometimes like a melon or gourd. It may take its name from thew shape or the location.

An Excrescence has many names which one has no need for. Rather one should know what it is and how to treat it. Some excrescences are encapsulated (ie encysted), others are planted in surrounding tissues; some can resolve, others cannot; some are large, some small; some can suppurate, some are crusted; some drain like fistulas and chancres; and there are other differences.

All these aposthems have the usual three causes: Immediate (blows, falls, bad habits and diets); Antecedent phlegmatic humors, both natural and non-natural; Contents of conjoint matter—aqueous, gaseous, purulent, corrupted, sour matter, mucilaginous, honey-like, gruel-like, and fatty.[290] Sometimes they contained phlegmatic, spongy, or glandular tissues. Sometimes they contained stony (ie gritty) matter.

The Signs and Diagnoses include the foregoing descriptions. The encapsulated masses are movable under the skin. When they are immobilized and adherent you know that they are not within a cystic capsule.

The fresh lesions can be resolved; not the older and firmer ones. Redness, tenderness, and local heat denote suppuration, fistulas, and cancers.

The cluster of several of several scrofules and their bulging through the overlying skin, and their palpable heat indicate that they are beginning to suppurate (ie change from glands). Arnold of Villanove said that size of the bulge indicates how many there are under it. Avicenna said that they are increased by local trauma. Arnold added that draining them early is to no avail, whereas you should stress purgatives, diuretic beverages, and desiccative electuaries, and allow the young lesions to mature. Furthermore, young people with their injudicious gourmandizing and slim bodies more often fall victim to scrofules than do adults. Besides, they have smaller faces and narrow temples and larger jaws, and that disposes to scrofules forming in the neck by matter that is displaced downward, as Henri de Mondeville suggested.

---

[290] Joubert re 'fatty": These masses had Greek names: melicerides, atheromas, steatomas, according to their matter. Others, less common, had different names according to their shape or number. Some are named by their contents: honey-like, gruel-like, fatty. He called the larger tumors on the head Lipomas or Moles. Goiters were in the neck, Hernias in the groins, and some in the scrotum were called Sarcocoeles. Those of medium size were called lupus. Later on he called some of them glands, buboes fugilics and nodes (J).

Incisions and applications of caustics are very risky in the upper neck and pharynx (ie the entry to the stomach) and the commissures[291] because of the proximity of veins, arteries, and nerves, and body cavities. A large incision can injure the vein that carries nutrition from below, and it can cause a serious hemorrhage.

There are two kinds of treatment: general and local. The general is directed at the immediate causes and the antecedent humors, and the final phase attacks the conjoint matter in the aposthem.

The **General Measures** affecting the life-style regimen are described in the preceding chapter. Here we try to liquefy the matter to favor its evaporation or drainage by incision. As before, Avicenna avoided heavy meats and favored drinking cold water with as much salt as tolerated. The patient should endure hunger as best he can, thereby to have healthier digestion. Avoid damp living quarters and valleys supplied by unhealthy streams. Drink good wine and water containing alum and sulfur, because those minerals, especially those that have a bit of tartar for flavor can cause goiters to shrink, internally and as seen externally.

First among the **Particular Measures** are laxation, bleeding, diuresis, and evaporation brought out by consumptive and resolutive topicals. First try Avicenna's laxative powder of turbith and sugar; a dose of two drams may purge heavy phlegm without warming it; it will not irritate the intestines. Rhazes used a more potent laxative described in the chapter on abdominal pain in his *Almansor*. He increased the proportion of turbith, and he dosed three drams. I also suggest herb Bennett and hierapicra, and pills of agaric and hermodactyl.

Second, as recommended by many, I use this laxative potion: scrophularia, filipendula, pimpinel, tansy, red cabbage, garance, aristolochia, roots of spatula foetida, and raifort. Crush all and heat with white wine and honey to reduce it by half. Drink it three times a day for three days; and in the morning eat a handful of warm peas.

Third, Galen prescribed (*Therapeutic,* Part 14) especially for internal glands that enlarge toward the surface, a theriac with tansy and ambrosia, and medicines composed of Cretan nepiote (water calamint).

Our predecessors used many other potions and electuaries similar to the oils used for injection into the ears, and methods favored by lay practitioners and others, which I confess I have not employed in my own work (as a physician-surgeon). For example, I know that His Highness, The King of France, endowed with a divine virtue, has cured many scrofules simply by touching. With that, we end this section on General Measures.

There are many **Local Treatments**; they vary according to the contents, their size, and the regions of the body that are affected. The phase of the aposthem also affects the

---

[291]   Joubert explains 'Commissures' as the complex of vessels and nerves that supply the head. The proximity of the brain is the reason for their risks, as Guy noted in Ch. 1 of Dictrine II. All surgery in the neck is risky because of the jugular veins, carotid arteries, and the nerves for the voice, which Galen named the 'recurrents'. He described a patient who was rendered mute when a nerve was cut during an operation to remove scrofules (J).

treatment, and it always indicates the disposition (idem, Part 4). Although all the general measures can be used when indicated, we will concentrate on particulars, here on six lesions.

First are the small soft lupuses in firm regions of the body, before they have suppurated and become tender. Treat them with astringent and desiccative topicals. Second, when they are larger and not so solid and firm, soften them with resolutives and consumptives, just as we treat phlegmatic aposthems. Third, when they have crusts and soft centers (abscesses), use suppuratives that mature them, drain them (ie with eruptors), and deterge them. Fourth, when those measures fail, and the masses are still movable, incise and extract them. Fifth, when they are large and fixed, use corrosives and detergents. Sixth, when they are pedunculated, ligate their stems and cut them off.

Sometimes a primary maneuver can consist of smashing and squeezing the mass, and then binding over it a disc of lead that will resolve the lesion. First massage the lupus to warm and soften it. Then grasp the limb firmly and slap the lesion with a spatula or another heavy wooden object and cause it to disappear by splitting its capsule and dispersing its contents. Immediately lay on a disc of lead, larger than the lupus and use a two-headed bandage which you wrap tightly. Leave it in place for nine days. Roger always began with an ointment made from burnt lead with some ashes from elder trees, figs oil, and vinegar. Bruno (as copied by Theodoric) first applied a plaster of aloes, myrrh, frankincense, sarcocolla, vinegar, and egg-white. After removing the disc he applied bits of cloth moistened with egg-white and thickened with salt and sugar-alum.[292]

The Second Set of Treatments for Antecedent Humors augments what we described for phlegmatic aposthems. These also are useful for treating hernias., We use Galen's plaster (see); we augment what we described for phlegmatic aposthems. These also are useful for treating hernias. We use Galen's plaster (see *Types of Compounds,* Part 6.) to resolve large exitures, to treat scrofules, and other abscesses below the ears, for podagra, and for many other maladies. Take oil, chick-peas, labdanum, litharge, verdigris, and galbanum. Grind the litharge in oil and heat it until it thickens; then add the peas and vinegar. Last, add the labdanum and galbanum. Grind all and set it aside for use when desired.

Rhazes used another plaster based on diachylon and iris, as described in our Antidotary. Also, based on his own extensive experience, he favored deer droppings mixed with vinegar and honey while heating. Another is his plaster of fenugreek, linseeds, cabbage-seeds, and a mucilage of mauve.

Halyabbas recommended this: bean-flour, barley—meal, reglisse, roots of guimauve, peas, white wax, goose-fat, oil, and the urine of a young virgin.

Bruno used this plaster on all indurated aposthems and Theodoric copied it: ammoniac, galbanum, bdellium, vinegar, and bran.

---

[292]   Roger and Bruno used compresses only. No smashing. See Bruno in English, Ch. 6, p.67, and Theodoric Ch. 6, p 67) (LDR).

Roger used this plaster for scrofules: roots of fesire and asphodels, hyeble, wine, and sulfur.

My Master at Montpellier preferred this plaster: twelve snails cooked in wine, a lye made from the ashes of rotted wood or a capitel. To improve its actions he prescribed dry fruit or a jam daily (ie as a laxative.).

The Third Set of Treatments uses the following: all resolutives and softeners. When the matter resists resolution, use maturatives until it suppurates, especially when it is soft or contains blood. Halyabbas used this maturative plaster: barley-meal, farnkincense, peas mixed with a child's urine. Avicenna cooled an angry lesion with that recipe to which he added coriander-water. Add myrrh to the coriander-water to increase the potency. You will know by the usual signs of pus when the lesion has matured and can be drained (ie with a large opening) or a seton can be inserted. Then deterge with the apostles' ointment. That is familiar to Christians as a fine treatment for malignant scrofules. Or (ie as did the Arabs) use the Egyptiac ointment of Rhazes, as in our Antidotary, or with diachylon or diapalma. When the lesion has corrupted an adjacent bone, as we will discuss later, or if some evil corrsion occurs, go to our routine measures for ulcers.

The Fourth Set of Treatments follows the methods of Albucasis to drain an abscess. We use an introducer[293] to insert a seton. Over time, the matter will be drained and we can insert detergents. When the matter is not liquid, make a cruciate incision and turn up the corners use your fingers to manage the affair and assess the interior of the glands, scrofules, et al. Enlarge the incisons full length and evacuate the contents with a spoon-like spatula. Reflect the corners with hooks and cut away the corners to prevent them from falling back and blocking the drainage. When needed, use sutures and other maneuvers, as in treating ulcers.[294]

My technique: I cut away the flaps with scissors, shaping them like leaves of myrrh, taking away enough to suit the size of the mass. The rest of the procedure is as above. When bleeding is a problem, I use hemostatic remedies before continuing the operation. If the bleeding is scanty, I use sponges or stoupes moistened with water or vinegar before squeezing them almost dry. If by mischance I have cut a vein and the bleeding is brisk, I ligate the vessel and leave the threads long, to come away later on. If some of the capsule of the cyst remains, or some other foreign material, I consume it during the ensuing few days by filling the wound with cotton pledgets wet with brine. Afterwards I wet them with egg-white and alum. Later, I use the Egyptiac ointment or other corrosives. In all these procedures you should apply what is suited to the aposthem. However,a stoupe of egg-white and oil of roses is useful for all of them.

---

[293]   Joubert describes an instrument with a sharp point at the end of a small nut, about the size of a jujube. Set the point where the abscess is softest and gently spin it through the skin and into the liquid pus. Then insert a seton alongside it to full depth. Then remove the instrument and observe the discharged matter (J).

[294]   'Sutures': I assume that Guy meant to retract the corners rather than use hooks (LDR).

The Fifth Set of Treatments follows Bruno and his acolytes who used caustic substances to erode into the abscess in amounts determined by their sizes. Take care not to spill around or into adjacent tissues. He used an eruptor of quick-lime and soap that usually is effective within twelve hours. When more time is needed, no harm results. Then he incised the center of the eschar, full length, almost to reach living tissues, and he stuffed in a little more of the corrosive, or another strong medicine such as asphodels. Many practitioners have used others, and many of them are listed in the Antidotary. The sublimate of arsenic is the one used only by experts. We will describe it below.

Galen discussed at length how to prescribe and to administer these dangerous medications, in Part 3. of his *Therapeutics*. Arsenic is potent, and violent, and, even in small amounts it can cause fevers and other serious complications. The usual dose is the size of a half-grain of wheat. For stronger effects increase the dose, or reduce it for the opposite reason; a smaller dose to begin with is safer than too much. The procedure takes three days, during which we follow the regimen for fevers. Always protect the tissues around the lesion with populeum ointment containing morels vinegar and other coolants.

When everything has coagulated and the eschar is humped, we can infer that the matter in the gland has been corrupted (ie coagulated). We separate the eschar with applications of a paste of melted butter and wheaten flour, or with lard or other unsalted oily substances. After the eschar falls off we can see if any matter remains unaffected. We clear it out with the Egyptiac ointment. When all is clear, we apply consolidatives, as for treating ulcers.

The Sixth Set of Measures (ie for a pedicled lesion) ties a silk thread or a hair from a horse's tail at the base. Draw the knot tighter every day until the excrescence is completely desiccated. You may hasten the process by putting a small amount of the corrosive on the thread And you may mitigate any local discomforts with egg-whites and rosat with oil of roses, or by applying the populeum, or by prescribing some sedative potion. When the infarcted excrescence comes off, apply the standard medicines.

CHAPTER 5

Sclerosis And Other Melancholic Aposthems

Now we shall go to the two types of aposthems engendered from Melancholy, the true and the non-true, as are found for the other humors. Here the true aposthems derive from Natural Melancholy, which is that part of the blood that nourishes the melancholic structures. The Non-True derives from the non-natural melancholy. The humor itself is cool and dry, when first it is generated from the thicker parts of the chyle. The natural melancholy is the lees and the 'mud' of good blood.[295] It is dark red and its flavor is both sour and bitter. The non-natural melancholy exceeds the natural in amonts, thickness, etc. However, when it is too far in excess, it no longer is melancholy. It forms from the natural in four ways: First, the natural melancholy itself is overheated and deteriorates and becomes a darker and more bitter kind of choler[296], as one can see when it splashes into the pan (ie after a phlebotomy), and it is frothy like vinegar, and it repels flies.

Second, it is from combustion of other humors, as from choler. That kind is more malignant. It, too, is frothy and it repels flies. When sanguine and pituit are overheated they also become non-natural melancholy, sweeter than the others, according to Avicenna. However, Galen and Halyabbas do not mention the last two sources.

The third source is coagulation and hardening, as is found in the surfaces of phlegmons and other aposthems that form from natural humors when they have been cooled or partly resolved. In such cases, the more liquid matter is resolved and what remains is a solid mass coated with a melancholic membrane.[297]

Fourth, the conversion occurs when some other humor contaminates the pure melanchoily. Excepting the two knds of burnt choler, all the contaminats sweeten the melancholy.

Now one can see that melancholy can engender four aposthems: 1. Natural melancholy engenders a true Sclerosis, which is fixed in situ. When some pituit is added it will confer

---

[295] Again refer to the 'anatomy' of a healthy blood clot (LDR).

[296] Nota bene: 'melan- dark, choler-bile (LDR).

[297] Guy describes a clot in the phlebotomist's basin, or a hematoma in a body cavity. The clot organizes and contracts, the 'serum' exudes and the surface of the coagulum is the 'membrane of non-natural melancholy'. The froth on top was thought to be phlegm by many authors of the era. The body of the coagulum was mostly sanguine humor (LDR).

some sensibility without pain. 2. Non-natural melancholy (made such by mixing) can make three aposthems: phlegmonous, edematous, and erysipelous sclerosis. 3. Non-natural melancholy (made by coagulation and induration) can engender true sclerosis which is numb and nontemder. Fourth, non-natural melancholy produced by combustion can engender all types of cancers. Sclerosis is a hard, immobile tumor that is numb and painless.

As it is for other aposthems, those from melancholy have three causes. First is bad habits of diet, etc. which increase the melancholy and thicken the blood. Second are the excesses of the melancholic humors beyond the capacity of the spleen (ie the important melancholic organ) to attract all of it and eliminate it. Third is the melancholy itself trapped in the lesion.

The Signs and The Diagnosis: The tumor is firm and resists impression. Its color is very red; some physicians call it livid. When it comes from openings in large veins, the blood is thicker and darker, as is the blood in the hepatic veins.[298] See *Therapeutics*, Part 14.

Melancholic apostems usually begin as small nodes that slowly enlarge. Some are unilatereal and others become bilateral and are called Ferinos (ie fermos) by Avicenna. Often they resolve or persist as firm nodules. However, some go on to become cancers.

There are three targets for treatments: The life style, the antecedent humors and the conjoint matter in the mass.

The adjustment of the life-style involves the six non-natural elements[299]. It stresses sobriety, moderation, and emphasizes warmth and humidity. Provide good food and fluids that engender good blood, such as wheaten bread, good wine, good meats from grouses, kids, and piglets, thin bouillons, especially from grouses, will be restoratives. Also offer spinach, borrach, hops, lettuce, et al. All of them cleanse the blood. Avoid things that engender melancholy, as they are listed in Part 3 of Galen's *Affected Places:* such meats as beef, mutton, donkey, fawns, foxes, hares, dogs, and wild boars; large fish and shell-fish; and pulpy vegetables such as red cabbage.

Lentils are beneficial and so are breads made from bran and bean flour. Wines should be hearty reds, and cheeses should be well aged. Limit roasted, grilled, and fried foods, and such salty, sour, and bitter things as garlic, peppers, mustard, et al., all of which overheat blood. Avoid outburst of anger and moody states, and do not overdo physical exercize and keep late hours. Keep the bowels open and aim for relaxation and tranquility.

Heal the disturbed status of the antecednt humors with body-purges and phlebotomy. For purges use Mesûes laxative of senna, epithyme, polypodium, fumitory, hops, convolvulus, cassia fistula, myrobalans of India, and lapis lazuli. Other good laxatives are diasenna, catholicon, hierarufinum, et al.

---

298   The hemorrhoidal veins drained the hepatic veins (LDR).

299   A reminder to the Reader: Natural refers to things that are in and of the body. Non-natural does not mean abnormal nor does it have a pejorative connotation. Simply it means things that were not part of the body but are important in the affairs of the body: atmosphere, food, daily habits, climate, etc. (LDR).

In dealing with the conjoint matter, proceed wisely, because that stuff is malicious; it resists resolution and becomes very hard (ie sclerotic). If it softens too much, it may putrefy and become cancerous. Therefore, we use two methods: first we us softeners and then we apply resolutives. Because it may be difficult to precisely measure the two, we mix them at the beginning, adding a small amount of resolutives with time. Avicenna used many combinations. However, they were more useful against sclerosis engendered from a phlegm or an erysipelas, as we will show in the next chapter. He showed restraint early in the course by using Rhazes' medicine described as follows in his *Almansor,* Part 7: bdellium, ammoniac, and galbanum ground in a mortar with oil of ben or of lilies. Then he made a plaster by adding a mucilage of fenugreek, linseeds, and figs.

Galen (ibid) especially favored a plaster of the boiled bark of the roots of guimauve ground with the fat of hens.

When suppuration intervenes, be careful to avoid overheating and inducing inflammation and causing cancer. When it erupts and drains, use diachylon as a detergent to prevent putrefaction.

An Explanatory Addendum About Sclerosis And Aposthems Engendered By Non-Natural Melancholy Produced By Cooling And Induration Of A Phlegmon

As noted before, when a phlegmon has been over-cooled, its matter congeals, or if it has resolved (ie the liquid part), the residue is sticky and thick, and as it hardens it produces sclerosis, to be described.

Sometimes the lesion is hardened (sclerosed) by inflation with gas or fluid, as we described in their chapters. Sometimes the matter is hardened by dehydration; we will deal with that in the chapters on induration at joints.

The Diagnosis is made by the induration which is almost painless and insensitive. When it is completely numb to touch and pin-prick, it cannot be cured; but when the stimulus produces even a strange faint sensation, it may be curable, but not easily so (*Glaucon,* Part 2). When the induration is the result of dehydration and general feebleness, it is beyond treatment except to prevent further dryness, as we will discuss when dealing with weight gain and loss. When the sclerosis occurs in hair-bearing regions, all hope for cure is gone. Furthermore, when the induration has the same color as the body, it will be untreatable and will never change color.

The goal of Treatment of the conjoint matter (*Therapeutics,* Part 14) is to evacuate all of it. Find the correct way to dislodge what is adherent and difficult to evacuate. Some try to attack it with attractives and evaporatives before using soffteners, and they are encouraged by some early reduction in the mass. But that is to no avail, because all that is evacuated is the residual liquid, and what persists is even more incurable, a resistant stony mass.

In such cases no medications that heat and dry will do; only those that soften may have some resolutive potential such as bone marrows from deer, calves, and billy-goats, with added ammoniac, galbanum, bdellium, and liquid styrax, or the roots of tree-size guimauve or the leaves of mauve-weeds made into patties with the fats of geese or hens.

Avicenna preferred salted lard and added mastic, labdanum, the refuse of bath-water, donkey-turds, and the lees of the oils of lilies, alkanna, and kerva (ie castor beans). When the aposthem was large, Galen gingerly added vinegar, careful to prevent it from seeping into the surrounding region and injuring nearby nerves. When Galen treated sclerosis of the hips in boys he fomented with the oil of sambucus and then applied ammoniac and vinegar.[300] When treating sclerosis of tendons he used iron pyrites, marchesite, dust from mill-stones—all of them heated to redness before plunging the stuff into vinegar and fomenting the region. He said that he had cured many with those medications, as if by magic. First he applied oil atop which he laid a plaster of those materials. Later we shall explain what we use to treat residual sclerosis after fractures (see Chapter 16.) and Gouty joints.

An Explanatory Addendum About Cancerous Aposthems

The term 'cancer' has two meanings: either it is an aposthem (as here-in) or it is a kind of ulcer, dealt with in a later chapter. The aposthem is a hard, round, dark brown mass surrounded by many veins. It grows slowly and ineluctably, and it is warm and painful. Galen (*Unnatural Tumors*) said that when melancholy overtakes the tissues it devours it and erodes the overying skin and creates an ulcer. When the humor is less vicious it creates a cancer instead of an ulcer.

The cancer begins as a small lump, the size of a chick pea or a bean, hardly noticeable, as if it were a tiny bud on a flowering plant.[301] It continues to grow and even a child (a young girl) cannot ignore it.

The Signs leading to a Diagnosis are its hardness and its dark livid color; many veins surround the mass and appear as if they are the feet of the cancer; its tenderness: and its unusual warmth.

A cancer is the worst of the aposthems. Most often it occurs in breasts and in in glandular regions, especially in women who no longer are virgins. In men it is more common where it is blamed on hemorrhoids, where we consider the cancer to be a thief. The combusted melancholy appears in a region and disseminates through the body via the dilated veins and deprives it. (See *Glaucon*, Part 2). A cancer cannot be cured unless it is completed extirpated, roots and all. Although it is not an ulcer per se, it can ulcerate. Therefore it is a triple evil, as it was called by Hippocrates in his *Sixth Aphorism*. When the cancer is hidden, do not attempt to eradicate it and thereby hasten its progress.[302] The untreated patients survive longer.

---

[300] Stiff hips in boys: Here we see an extended use of the term sclerosis to mean stiffness or restricted mobility of a joint, as seen in congenital dislocations of the hip in children (LDR).

[301] I think that Guy had in mind a cancer in the breast (LDR).

[302] The medieval surgeon had little knowledge of and even less concern for treating lesions that began in the depths (ie hidden) and came into view as they appeared at the surface of the body. The anal carcinoma was thought to arise in hemorrhoidal veins because they were emunctories of the liver via the hepatic veins. Similar explanations are offered for cancers (ie metastases) in lymphnodes at other emunctories in the neck, axillae, and groins (LDR).

The Treatments are threefold: Life style and humoral Antecendents, and Local Treatments of special kinds.

The first measures copy the regimen prescribed for sclerosis, especially emphasizing foods which cool and moisten and engender healthy humors, such as barley-water stone-fish, soft-boiled egg-yolks, et al. Drink warm milk from which the butter has been churned away, and eat melon-like vegetables.

The second measures use the well known purgatives of which epithyme is the best. Galen dosed it three times a day for three days. Avicenna used hellebore. Galen insisted that phlebotomy be used when there are no contraindications. Halyabbas said not to exceed one or two doses in an effort to totally evacuate the matter.

The third measures against cancers include familiar medications that are not violent. They serve only to restrain the growth and resolve some of the matter. The substance of the matter is heavy and resists treatments, and even when we see that the some of the tumor remains, we do not intensify the local applications and cause the lesion to ulcerate (Avicenna). Galen preferred the juices of morels plus pompholyx. Theodoric used an ointment of pompholyx, oil of roses, white wax, washed ceruse and burnt lead.

Galen (*Miamir*) used an oinment of powdered litharge, ceruse, and oil of roses. All were ground with a lead pestle in a lead bowl in bright sunlight until the ointment took on the color of lead. Those surgeons who have used lead can testify as to its value in treating cancerous dispositions (*Simple Medications,* Part 9). Avicenna preferred a plaster of powdered shells of crayfish with added cadmia.

Avicenna recommended a total excision of small cancers with all their roots where it was feasible. Then mop out the blood in the defect before applying the red-hot cautery. Galen agreed, but hesitated for fear of hemorrhage and the resulting ulcer and the public outrage that would ensue when it fails. We shall discuss this topic in our section on cancerous ulcers.

With the Lord's help, here we end Doctrine I.

DOCTRINE II: APOSTHEMS, EXITURES, AND PUSTULES THAT ARE IN PARTICULAR PLACES; EIGHT CHAPTERS

A Note For The Reader: Many phrases, sentences and paragraphs in the Doctrines that follow repeat what was written in Doctrine I. Guy, as did many of the authors of the medieval treatises, claimed that his works were not only didactic, especially for practitioners who had not had formal schooling, but were to be referred to for assistance in treating a patient then in hand. Therefore, he would write materials easier to obtain from single designated chapters than by a search through books that had scanty indexes if any at all. Never mind that it repeated what had been written elsewhere in the same Treatise. Today's Reader must accept those repeated schemes of pathogenesis, diagnosis, nosology, therapeutics, diets, and recipes when he wades through Guy's precious book. And he must accept the numerous citations of authorities that represented the ancient writer's 'evidence' to prove his own statements as 'fact'. Indeed, such was the forerunner of a modern writer's list of references appended to his published articles.

The medicines mentioned by Guy and Joubert often are polypharmacies. In this translation I will repeat the names in the text and identify most of them in the Antidotary. The amounts specified in the recipes are omitted here; the ancient measurements only vaguely correspond with modern equivalents. However, some of the details of the formulations are included, perhaps of some interest to a curious student (LDR).

CHAPTER 1

Aposthems Of The Head

T he following deals with the so-called 'bad treatments' (ie surgical operations). They are diversified to suit the diseased parts. Four considerations determine our choice: the complexion, the composition, the virtue, and the location.[303]

As to Complexion: The warm parts of the body call for warmer medications, and the dry parts need drier medications. However, the fleshier structures should receive less dry, medicines than the parts that are less fleshy; in that their natural states are treated with opposites. As Galen taught, the malady also should be treated with its contrary, and the affected part is defined by its similars (*Therapeutics,* Part 5.)[304]

The Composition tells us to use medications differently in fat and thin bodies.[305]

The Virtue of a medicine defines its potency and its qualities as to its wild or cultivated form. For example: A sensitive structure such as the eye cannot tolerate an intense and heavy medication, whereas the bony cranium will accept them. Deep-seated pain requires different treatments than something at the surface, and that will determine what kind of evacuation to prescribe.

Therefore, we are aware that the treatments will be changed according to the organs that are affected as well as the nature of the malady. We choose differently for a turtle on the head, or an application for ophthalmia in an eye, or a quinsy in the throat. The

---

[303] Some authors called the last-named *Plasmacion*, here and elsewhere, to indicate more than just its place. Galen's list of the four things is more precise: the temperament and complexion; the contour; the composition or plasmation; the virtue or function, the position or seat. (J). The Montpellier Ms said *Plasmacion or formacio*n (EN).

[304] Canappe said that "the complexion is protected by its similars, according to Galen" (EN).

[305] There have been other Sets Of Four, and they are confusing and are not well explained by Guy. Galen was more explicit and elegant, in his *Glaucon,* Part 3. However, in selecting medicines to suit the affected organ, we also must consider their degrees and the dosages. (J).

The Montpellier Ms. stated, "The Composition dictates what kind of evacuation to use and other measures. And we should consider their own virtues, and whether they are wild or garden-grown (EN).

choices will vary with the diseases as they affect different regions, as we shall discuss them later. Here, however, we will deal with aposthems.

Aposthems on the Head may be warm, cool, purulent, non-purulent, nodular, glandular, or watery. We already have discussed their signs and their causes. The diagnoses are important especially if they are near the cranial commissures and have access to the underlying brain, as we described in the Anatomy. We agree with Roger's advice not to meddle with turtles, moles[306], and glands that adhere to the bone. When they are contaminated you may have to use the trephine. Lanfranchi describes a man (and I have seen such a case) who had an ulcerated mole on the frontal scalp that had rotted through the bone; he could see the pulsating underlying membranes. Knowing what that represented, he limited his treatments to palliative measures.

The treatments for warm and cool aposthems on the head are those which we use elsewhere, and involve three factors. First, deal with the daily habits as recommended for headaches, avoiding everything that engenders vapors and fumes. Second, evacuate the threatening humors. For the warm kinds prescribe electuaries of rose-water, and pills of myrobalans or fumitory. For cool antecedent humors use the hierapicra or coccia or dorée laxatives. Third, consider where it is and what kind of conjoint matter is in the lesion. If it is on the dome of the head, observe three precepts: 1. At the oustet do not apply potent repercussives and risk affecting the adjacent vital organ (ie brain). Use mild ones as oil of roses and other coolants. 2. When there is pus, do not delay opening the abscess for drainage, thereby preventing the pus from attacking the cranium. Third, when the abscess is large, make a triangular incision with the point directed upward. That will favor your use of detergents.

The treatments for nodes and glands are the usual.

'Water-on-the-head' of babies is treated with the desiccatives and resolutives used by William of Saliceto and his pupil, Lanfranchi: oil of camomille or aneth with some added sulfur if necessary. Lightly touch the surface with a hot cautery in two or three places, front and back. The water gradually will disappear. Then rub on some lanolin, or apply a stoupe of warm wine and oil, as used by Avicenna.

---

[306] A 'mole' is called such by its tendency to tunnel under the scalp (LDR).

CHAPTER 2

Aposthems Of The Face And Its Parts

The treatments here are no different from what are used elsewhere, be they warm or cool (ie their dispositions), nodes or glands, except that the incisions do not follow the skin lines, especially on the forehead where the muscles are not aligned with the wrinkles. Here the incisions should be in the ling axis of the body, except in the eylids where they should go from corner to corner to leave a smooth semilunar opening to accommodate vision. Below the ears and the jaws the same is true, as was claimed by Avicenna.

Ophthalmias, Exitures, Hypopyon, Painful Eyes: The Maladies and Their Signs

Ophthalmia comes first, as a phlegmon of the conjunctiva. It derives from itself, and can be a complication of other maladies of the eye. The exitures of the conjunctiva are pustules, blisters (ie clusters), single vesicles, and hypopyon.

The causes for all of them are the same as for aposthems elsewhere, but the discharge (rheum) here comes from within the head. The immediate causes are smoke, dust, wind and sunlight, all of which irritate the eyes, especially when the body is replete. Avicenna said that accounts for the rapid reaction to the irritants and for ephemeral fevers that may degenerate into other kinds.

One of the two types is mild, limited to redness and wetness. The second type is more severe, progressing to become white films that cover the iris. Jesus, son of Haly, said that it occurs in one of three ways, because the source came from within the eye. He said that the symptoms were those of complicated (ie inflamed) aposthems: swelling, pain, induration, heat, redness, dilated veins and a liquid discharge. The causative humors vary, and were defined by Alcoatin, Azaron, Galaf, and Albucasis[307]. When the cause is a sanguine humor, the signs are redness, heat, puffiness of the surrounding region including the temples, dilated venules, sticky rheum on the eyelids, abundant tears, listlessness, heaviness of the body and the head, and other indicators of the sanguine humor.

---

[307] Nicaise (see Sect, 2 of his Introduction) identifies Azaron, and Galaf as two of the many aliases of Albucasis (LDR).

When the cause is choler, the signs are pain, great heat coming suddenly, yellowish reddening of the region (ie bilious), abundant thin tears that are not very bitter or sticky. The pain is prickly and biting, as if sand is under the lids. The body suffers the usual symptoms of a choleric illness.

When phlegm is the cause, the swelling is pale rather than pink, but there is heat and pain. The patient is listless and exhibits other signs of phlegm.

When melancholy. is the cause, there is less redness, swelling, and discharge.

Ophthalmias have the usual four phases. At the onset the signs are minimal, and they are manifest during the phase of augmentation, with tears and particulate matter in the discharge from the nose as well as from the eyes. When fully developed, the signs are obvious. When the discharge begins to wane and become more mucoid, and the eyelids cohere, we can recognize the fourth phase of the illness. Jesus said that the sticky eyelids were the best sign of maturation (ie laudable pus). You will observe the changes in pus—more sticky, smooth, and white (ie laudable). It is easily washed away, whereas granular pus is evil.

A severe ophthalmia causes heaviness and aching of the head, redness and throbbing of the forehead and temples, dilated veins, and swelling, all of which comes from the outer membrane (ie the conjunctiva). When the discharge predominates, both from the eyes and the nose and the palate (ie postnasal drip), the source is within the head. If the patient vomits and is nauseous, the stomach is involved.

The patient is miserable with the keenly painful eyes, he can suffer nothing as bad as that (Galen); some victims want to die rather than suffer the unrelenting pain, night and day. We must try mightily to appease the headache that accompanies the pain in the eyes. That complex indicates that the pus is erosive and excessive, and that it emits fumes (Galen, *Therapeutics,* Part 13), and when there is fever, the outlook is grim.

When the topicals fail to abate the symptoms, you know that the rheum comes from within; the tunics prevent the applications from getting to it. The source continues to supply what is under the eyelids and sustains the ophthalmia.

Furthermore, ophthalmia has its ups and downs and paroxysms, much like the causes describe in Book 2 of *Different Fevers.* The acute spells may last as long as seven days, according to Jesus. Remember, too, that it is contagious and it easily involves the other eye. Know, too, that the disease may flow up from the stomach (*Aphorism* No. 6).

Gordon insisted that it should not be under-rated; it can be very difficult to cure, and it may lead to such serious complications as rupture of the cornea, or to permanent blemishes and corneal opacities.

The extent of the Treatments varies with the malady, complex or simple. In addition to the general measures for aposthems, we apply any or all of the four specifics.

First, the management of the patient's daily life. It includes limiting the intake of food and beverages, especially in the evening, avoiding the gassy items. Emphasize the simple and delicate foods. Avoid meat and wine, especially during the earlier stages, and omit hot and spicy things. Offer quince and coriander after meals to reduce the flatulence that can rise to the head. Keep the patient in dark rooms and favor dark clothing and bed linens—green, blue, or black. Avoid open air and sunlight with its brightness, smoke, dust,and wind. Abstain from coitus and fits of anger. He should not sleep prone, and should use pillows to elevate his head. Favor long periods of rest and sleep. Avoid rubbing or bruising the eyes, and staring at things. Keep the bowels open.

The second category of treatments is directed at the specific causes, as we have defined them, selecting the correct evacuations to divert and to dam back the flow of bad humors. Modify the rheum. Relieve the headaches with bleeding, laxatives, clysters, massages, applications of tourniquets on the limbs tight enough to cause local discomfort, apply cups, setons, and touch cauteries at the shoulders and neck. Apply plasters on the head to dry the discharges, using millet, salt, dove-droppings, camomille, anise, etc. Touch the dome of the head with cauteries. If necessary, follow Galen's advice, even when the offensive matter is hot, and open the veins at the temples and forehead to divert the inflow to the eyes (*Therapeutics,* Part 13). You may stuff the phlebotomy incisions with grains of wheat or of burnt frankincense, or a caustic medicine. You may have to doubly ligate and divide arteries, as we do for varicose veins, and cauterize the cut ends. You may wrap the head with a bandage to hold in place a plaster over the forehead and temples, using astringents: bol d'armenie, flour of lentils, oak galls, pomegranate peelings, acacia, aloes, frankincense, etc. and mix all with egg-whites. Replace it as frequently as needed for relief.

The third measures aim to modify and divert the offensive matter, as taught by Mesûe. Use the ordinary repercussives at the onset and add resolutives during the phase of augmentation. Eliminate the repercussives during the stable stage, and add desiccatives during the decline.

Therefore, at the onset when the disposition of the matter is warm, use such topicals as suppress irritation and inflammation: rose-water, egg-whites, purslane, endive, morel, and mucilage of psyllium. Also use the white collyrium without opium (a la Halyabbas) unless the pain is severe, because opium dims the vision and confuses the patient. Make it as did Galaf from Damascene's book[308] : washed ceruse, wheat-flour, gum Arabic, tragacanth, rose-water. You may make a paste of the collurium by adding oil. Dab them on with a cotton-tip stylus.

As the disease evolves, use mothers' milk, a mucilage of quince, fenugreek, and rose-water; add the white collyrium of Rhazes: wheat-flour, mashed sarcocolla, milk of anise, gum Arabic, tragacanth, and opium. Thin the collyrium with rain-water, or add egg-yolk to make a paste. I am amazed that Gordon was so emphatic in stating that all the sages agreed that sarcocolla should not be used except in the last stage, merely because he forbade its use as a powder at the onset.

During the stable phase, apply the mucilage of fenugreek with juice squeezed from melilot. Also use the white collyrium with climie as in the antidotary of Galaf: washed ceruse, flour, climie (ie probably silver), gum Arabic, frankincense, opium, and a decoction of fenugreek.

The yellow and the rosat collyria are used for the same purposes, and they can be made into pastes, using bread crumbs or apple-sauce made with rose-water. The rosat collyrium of Jesus was used to thin the pus and clear hypopyon: it contains freshly picked red rose-buds, verdigris, copper flakes, spikenard, gum Arabic, burnt and washed cadmie, saffron, and opium. Grind

---

[308] Damascenus (Janus) was Serapion the Elder (LDR).

all and add rain-water. The yellow collyrium of Alexander: wheat-flour, memitte, sarcocolla, tragacanth, gum Arabic, myrrh, and opium, or you can make a confit by adding rain-water.

During declination, use baths (ie eye-washes), fomentations, and evaporatives containing rose-water, camomille, melilot, and fenugreek. The collyria of tuthy and of the three powders are good.

The powders: 1. Ground tuthy, calamine, cloves, honey, white wine, rose-water, and camphor. 2. The standard powder-base for the ordinary collyrium: Tuthy, aloes, camphor, rose-water, and pomegranate juice. 3. Master Arnold's powder was used to relieve Pope John's red and teary eyes: Tuthy, antimony, pearls, powdered red coral, raw silk from the cocoon or the eggs of the silk-worm. Grind to a fine powder and add a pinch of verdarain. Apply it with a cotton-tip stylus.

The yellow powder of Rhazes to use at the conclusion of the treatments: sarcocolla, memitte, lycium, aloes, saffron, and myrrh.

When the matter is cool, use Mesûe's collyrium of spikenard at the outset: sarcocolla, spikenard, roses, saffron, flour, aloes, gum Arabic, tragacanth, opium, and rain-water. Apply this plaster over the eyes: leaves of mauve and aneth boiled in wine.

During maturation (suppuration) foment the eyes with decoctions of fenugreek and melilot, and apply the yellow powder with soft bread that is moistened with wine. The patient can drink the wine squeezed from the bread. That accedes to the *Aphorism* of Hippocrates that states that painful eyes are relieved by pure wine, washes, fomentations, phlebotomy and medicines.

The fourth elements treat and prevent complications. We will deal with them later.

Painful Eyes

Relief is provided by medicines that stop the prickles and biting, and are somewhat narcotic and sedative, depending on the state of repletion and flatulence of the patient. Mesûe prescribed beaten egg-whites with a decoction of opium; or a watery mucilage of psyllium; or the juices of the mandragore fruit or lettuce with added opium,applied as needed. Do not use too much of the narcotics; they prolong the stable state and delay maturation, and the recovery of vision. You may add some of the sedatives to the white collyrium and you can control the dosage of the opium.

Azaron used this fomentation: opium, plantain, saffron, memitte, aloes, gum Arabic, and acacia. He also used this plaster (see his antidotary, Ch. 21): saffron, memitte, lycium, aloes, acacia, and mouse ear, breast-milk from a woman who is nursing a girl.[309]

---

[309] Joubert: Was the sex of the infant important? Jean de Vigo missed the pont when he used the term 'pregnant' (ie rather than nursing), that is, during pregnancy, and that later she nursed a girl-child. When did the question of the sex of the infant become important? We cannot be very precise in the answer. Was the source of the milk 'cool' simply because the mother nursed a girl, who, being female is inherently cool, and was the milk different when she nursed a boy? Wa

The harmful effects of narcotics can be ameliorated by fomentations (ie of the eye) with decoctions of camomille, melilot, and fenugreek.

Azaram's powder made of shells of hen's eggs was used by Jesus to induce sleep. But I do not approve of the powder of tuthy, sarcocoilla, and sugar which Bienvenue used during all the stages of ophthalmia. All powders used early in the course are irritants and harm the eyes. I agree with Jesus on that score. Sleep (ie sedation of pain) may be induced simply by adjusting the dose of pavot (poppies), violets, water-lilies, sawdust of sandal-wood, hyoscyamus-tea, applied on the forehead.

Treat the sticky discharges by moistening the lids with warm water daubed on with a cotton-tipped stylus. And, in all cases of ophthalmias and other ocular maladies, observe the dicta that will follow in the Treatise on The Eye.

Hypopyon: Pus Behind The Cornea

After a prolonged ophthalmia pus may appear behind the cornea. Treat it with a collyrium of frankincense used for chronic ulcers and for profuse suppuration as described by Jesus in the chapter of Dubellat: ceruse, opium, sarcocolla, tragacanth, gum Arabic, frankincense, and (according to Mesûe) and sel ammoniac. Add rain-water, or use a mucilage of fenugreek instead of the rain-water to make the collyrium. If that resolutive fails, use diaphoretics and consumptives, especially as infusions with steam, and use the collyrium applied in the fourth stages of cool aposthems.

Galen recommended (*Therapeutics*, Bk. 14) used the collyrium of myrrh that was recommended for treating cataracts; it contained opoponax, euphorbia, etc. The soothing effects are wonderful. But if the resolution fails, Jesus and Alcoitin advised incising (ie a nick) at the juncture of cornea and sclera to allow the bad matter to escape. Galen (idem.) said that an oculist in his time had cured a patient's hypopyon by seating him and shaking his head vigorously with his hands placed on the sides. The pus was displaced down and away.

Blisters and Vesicles (Phlyctanea)

What you cannot eliminate with resolutives you should cut away. And what you cannot eliminate with that trimming, you may ligate with a thread. Treat the residual defects as ulcers. If a scar or a blemish remain, deal with them as we will describe in Ch. 6 on Ulcers (God Willing), where we deal with lachrymal ulcers.

---

it true that the doctrine of similars held that during pregnancy the female fetus withdrew cool blood and the male attracted the warm? Then, after the birth, what remains in the mother's blood was warmer after the birth of the girl, and is cooler after the birth of the boy. The common folk believed that the milk delivered of a girl is better than that after delivering a boy. But ignorant of the cause, it is said that the newborn boy needs cool things, and otherwise for the girl (J).

Aposthems of the Ears

They may be warm, cool, purulent, deeply embedded, or superficial. Some are at the base of the ear. Their causes are those described for other aposthems, with some unique manifestations, such as more severe pain, especially when they are warm and situated deep within the ear-canal.

Severe earache is a menacing sign, leading to fevers, fantasies, and syncope, even to death. That pain is more dangerous in the young, and they may die within a week before suppuration is complete (ie and can drain). Adults usually can endure it until that occurs. The degrees of pain are discussed in Part 3 of *Prognostics*.

Treat these aposthems according to their dispositions, just as we treat others like them, with differences according to their locations and the pain. At the onset treat both the deep and the superficial lesions with our usual repercussives designed to repel the faulty matter, but avoiding repercussives when the lesion is in the emunctories of the brain, at the base of the ears, which Galen called the parotids (ie the upper cervical lymph-nodes).[310] Instead, we use attractants to bring the matter to the surface, much as we described for treating aposthems in the armpits and the groins, contrary to what Henri de Mondeville had questioned. If the pain is worse than the cause, treat it first; you may have to use narcotics.

The treatments involve the usual four elements: life-style, antecedent humors, conjoint matter, and complications/

The first and second categories are the same as what we described in the general chapter and in those dealing with the head and with ophthalmias.

The third treatments vary with the type of matter in the lesions. For warm matter use cool applications such as Galen described in *Miamar*, Part 3.: A cataplasm of oil of roses and vinegar; troches of memitte with one twelfth part of gum Arabic, thyme, and rain-water and wine. Halyabbas used the white collyrium with milk. Avicenna recommended mothers' milk introduced repeatedly for three days before he applied a mucilage of linseeds, fenugreek, and the juice of convolvulus. During the third phase of the aposthem he prescribed vapors of various sorts: decoctions of roses, camomille, and melilot dripped through a funnel; and injections of foils, slightly warm, such as oil of almonds.

Afterwards, Galen (*Miamar*), Part 3) used the basilicon ointment with oil of roses when the abscess remained warm, and he used spikenard if it had cooled. In that case, Avicenna repeatedly introduced the following on a cotton-tipped stylus: warm fox-fat (or lizard, or duck) and butter, or the marrow of a calf's femur. Mesûe approved.

When the aposthem was not warm, he recommended application of bull-fat, honey, pork-lard, oils of laurel, nard, rue, and balm (ie mint balsam). When the malady was mild, some authors made a perfume from hyssop, marjoram, betony, fennel, and rue.

When suppuration occurs, apply Mesûe's plaster of bean-flour barley-meal, camomille, melilot, violets, and guimauve.

---

[310] In other words, do not drive the bad humors back into the brain (LDR).

When the abscess is at the surface and there is little risk of delay, Avicenna's plaster (v.s) may be successful in treating a vesicle in the ear: a decoction of figs and wheat-flour. Treat the pus with wine and honey and other things used for ulcers in the ear.

The fourth treatments befit the complications. Treat pain according to the causative matter (ie humor). If it is warm, use troches of opium, castoreum, and boiled wine when it retains some warmth. Instill it into the ear on a bit of soft wool. After a fomentation, let the patient rest with a pad of warm wool over the orifice and the entire lobe. When necessary, repeat the fomentation, but take care not to press on or molest the opening. Avicenna used this analgesic: oil of rosat, warm oil of violets, egg-whites, camphor, (the oil of violets is a better sedative than the rosat, by virtue of its mollifying action), mothers' milk, a decoction of morel, rosat-oil or oil of almonds in which you have boiled some earth-worms or centipedes—which are round like bean pods, as found under rocks or slab—melon seeds, and oils of nenuphar, poppies and willow. Instill it into the ear as above. The glorious Avenzoar prescribed and applied an oil of egg-yolks. It relieves the pain and hastens the suppuration and eruption.

If the pain is caused by gassy matter and the abscess does not point, mix the above with astringents: aphroniter (which Galen called nitre), the two hellebores, the two aristolochias, rue, centaury, wild cucumber roots, roots of bryony, arum, dragantum, costus, cannelle, and cubebs.

Avicenna explained why he described so many medicines (ie the numerous contents in his prescriptions) because many substances may not be available in various places.

He treated the heat by applying cups filled with warm water, millet seeds, and salt which he spread on warm cloths. Mesûe evaporated through a funnel containing a decoction of camonmille, mellilot, aneth, fenugreek, red cabbage, marjoram, wind-herb, and parietory. Avicenna and Averroes insisted that we restrict the use of narcotics to avoid fainting and strange behaviour that occurs when the offending humors are cool; the result is bad. In that event, use castoreum as the antidote.

Treat scrofules in the region as above.

Treat aposthems in the regional emunctories as we do those for the heart (ie at the axillae). Be aware of the nearby large veins and arteries when you incise for drainage, and do not injure the nearby recurrent nerve and render the patient voiceless.

We shall deal with the other lesions of the face later. We avoid treating nasal polyps and oral aphthous ulcers (ie alcola). Although they begin as simple swellings and pustules, they may become ulcers as the result of treatment. Aposthems in the mouth are treated as we do quinsy; see below.

CHAPTER 3

Apposthems of the Neck and The Back

These apposthems are either external or internal. The first are called simple protrusions, as glands or pustules, and are treated as such, as we have described. All of the second group are called *Quinsy,* and each has a specific treratment.

Quinsy[311]

A quinsy is a tumor in the throat that can obstruct breathing and swallowing. Avicenna described four types according to where they are situated, based on what Galen wrote in Part 4 of *Affected Places.*

One type involves the muscles of the throat and it may show itself at their surfaces. The second lies in the posterior muscles over the vertebrae or can be seen in the throat past the tonsils, or when it involves the tongue, The third is in the esophagus (ie pharynx) and cannot be seen, and it is most serious when it affects respiration. The fourth affects the muscles of the pharynx and epiglottis (ie larynx) and interferes with gargling. Galen added (ie a fifth type) scrofules that appear after dislocations of the first or second vertebrae, and we will deal with them later.[312]

The causes are those for apposthems everywhere, according to the sources and dispositions. (ie warm or cool).

The signs and the diagnosies vary with the sites, but all of them affect swallowing and breathing. The victim cannot rest supine and his tongue is pushed forward, and when

---

[311] Guy made few distinctions among them and called all quinsies what we recognize as tonsillitis, paratonsillar abscess, septic throats, trench-mouth, and what must have been diphtheria, although he makes no mentioin that the last was contagious. His quinsies were classed only as internal and external apposthems, and according to where they appeared in the neck and throat. That confusion of entities was not new in the medieval epoch, and it continued well into the 18th C. An interested Reader may consult R.H. Major's *Classic Descriptions of Disease,* Charles Thomas, Baltimore, 1939, p. 149 ff. He will find descriptions of what was called Diphtheria from the times of Aretaeus (50 A), Baillon (1576), Fothergill (1754, and others (LDR).

[312] After torture or hanging ? (LDR).

he swallows some of the fluid comes back through his nostrils. Avicenna said that the eyes bulge and the impaired tongue causes nasal speech (ie twang). When the symptoms worsen the situation is threatening: breathing becomes stertorous;, the patient sighs (wheezes); swallowing is blocked in the throat; the extruded tongue is covered with foam, as seen in a tired horse. The tongue, the lips, and the eyes are discolored; the limbs are cold; and the heart fails.

When improvement occurs, the patient is less restless, he sleeps comfortably, pain is dulled, and breathing and swallowing are easier.

Quinsy is very dangerous because it develops rapidly. Hippocrates (*Prognostics,* Part 3) said it was most serious when you cannot see it in the throat and it does not appear at the surface of the neck, and when the pain is severe, and when breathing is obstructed. That type of quinsy can choke the patient by the second, third, or fourth day. Another bad case is when you can see it in the throat but not on the neck; but when it is slower to mature, it less threatening. A third kind that appears both within the throat and on the neck is even slower than the others. A fourth kind appears on the surface; it is the least threatening.

A quinsy is not contagious. When it drains, the patient rests more easily. But if suddenly he feels faint (ie syncope), the symptoms may recur, and death is imminent. In *Aphorism* No. 5, Hippocrates wrote that suppuration occurs within a week when quinsy involves a lung, and that empyema ensues (*Aphorisms* 6 and 37). When the pus drains externally, the prognosis improves, because the internal organs are rid of it. Avicenna said that all quinsies that obstruct breathing that do not resolve or drain will enlarge and kill.

The good signs indicate resolution. You observe the changes: a rapid decrease of inflammation and discomforts. The mass softens and contines beyond the critical fourth day. The signs of lethality, on the other hand, will be obvious if you are alert. Suffocation is the worst sign.

The treatments are the usual four groups for aposthems in general: lifestyle, reduction of the contents of the mass with early use of repercussives, resolutives, and maturatives, and altering the matter already in place. But do not use repellants for any type of quinsy. Begin at once with evacuants and attractives. The details follow:

First (the regimen): The usual diets should contain hydromel and sugar. Then add a decoction of lentil pods and barley which assuage the thirst, the bitterness, and the inflammation. Then cook chick peas with barley broth and add bean and wheaten flours and some honey. Afterwards, add yolks, chicken soup, etc. to the diet. Do not overfeed. Keep the bowels open. Sleep in short naps, because the patient may suffocate during a long sleep.[313]

The second tier of treatments consists of evacuations and the usual diversions: bleeding, properly selected laxatives, clysters, and vigorous massages to divert the humors from the quinsy. Immediately afterward bleed from a saphenous vein, and then from an arm-vein when the patient can tolerate it, and then from a vein under the tongue. If the patient can swallow it, on the following morning, give him a potion of diaprunus or of the catholicon fortified with an electuary of the juice of roses. When the matter is cool, give these laxatives: hierapicra, cochia pills, and agaric. To arrest the inflow of rheum (ie

---

[313] Seven hundred years before the cautionary treatment for sleep apnea!! (LDR)

from the head) use millet and toasted salt applied on the head; or use dove-droppings, or have the patient suck diapapaver pills, or storacines under the tongue.[314]

The third set of treatments follows Galen (*Miamar,* Ch. 6.) who used repercussives only at the onset of internal quinsies. He used lenitives for the externals to block the accumulations from the throat; and he used resolutives at the end. He mixed the two during the middle stages. He did not use repercussives alone lest the bad matter drift into the internal organs. Select the medications to suit the disposition of the region. Therefore, at the onset, use potions made from nuts: the dianucium and the diacaryon, with added traces of roses, balaustium, lentils, blackberries, galls, sumac, memitte and other astringents. Or use diamoron (ie from mulberries) with myrrh, saffron, et al. Finally, use diahirundinum from arondelles[315], and add a pinch of each of dried figs, calamint, oregano, pouliot, hyssop, satureia; add sulfur and niter if needed. You may use feces of wolves and of common chicks that have been fed lupins, or use the heads of herrings and other salted fish, and the herbs called devil's bite (ie scabious).

Recipes from Galen: Diamoron: Juice of two kinds of mulberries and honey, five parts to one. Dianucum: Juices of the cortex (ie not the nut-shells) of walnuts and honey, five-to-one. Those are simple recipes, suitable for women, children, and feeble men. Galen said that he had found that dianucum was better than other astringents at the onset of phlegmons, when he added saffron and myrrh. In the 'steady phase' he added diaphoretics, as above.

The Diahirundinum[316]: Incinerated arundelles (the ashes), saffron, nard, myrrh, and honey to make a paste.

Galen prescribed gargles, sucrets, inunctions, and insufflations within the throat. For internal quinsies Avicenna prescribed a light neck-wrap of raw fleece moistened with olive oil

---

[314] The Greeks and Arabs had different recipes. Rhazes (in Almansor Ch. 55) and Mesûe (Ch. on Coughs) differed from the text above, which was taken from Galen, Aetius, Paul of Aegina, et al: styrax, myrrh, galbanum, and opium, all cooked in wine and reduced to pills (J).

[315] The feces of arondelles young swallows that have been fed only bread and lupin with a little wine for three days. Galen insisted that the fecal matter was a better resolutive if it was not malodorous. Canine feces was used on other lesions as well against quinsy. Galen highly recommended it for dysentery when it was mixed with iron-milk. What Galen wrote about a commonly held opinion is even more incredulous, that everyone should eat an arondelle chick once a year to fend off quinsies. It is hard to believe:what Galen wrote about human feces; mix them with honey and apply as an ointment after phlebotomy, and use it when a quinsy resisted other treatments. Also, we can doubt what he wrote that the cloth in which a viper had been strangled and thereby colored purple was a remedy for quinsy, because its virtue is obvious, as potent as that of a squid against all swellings in the neck (J).

[316] Galen took his recipe from Asclepiades' *Another Oral Medicine Made From Arondelles Without Wild Rue*, Ch. 6. Whereas Avicenna described another Diahirundinum that contained many more simples. Also, when Guy wrote 'nard, myrrh, etc' we should understand that he meant Galen's Indian Nard. (J).

and camomille. Toward the end of the treatments he added attractants such as borax, sulfur, costus, mustard, castoreum, and other medications that cause reddening and blisters.

When the foregoing are successful, the quinsy may begin to resolve. On the other hand it may suppurate. In that event Mesûe prescribed this plaster: barley-meal, linseeds, date-meats, dried plump figs, and soft bread, all cooked in wine and boiled down to make a paste. Butter made it better.

Roger used a plaster made from the roots of sambucus, senecio, aloin, barley-meal, linseeds, honey and pork-lard.

Lanfranchi's recipe: Boiled swallow's nest (in water) and sieved. Add lily-roots to the filtrate, and guimauve, briony, leaves of guimauve, mauve, violets, and parietory. Add a strong yeast and flour of linseeds, and add oil to make a paste which he declared was a wonderful resolutive and maturative for quinsies. He added water, wine, and honey, and a decoction of guimauve, figs, linseeds, fenugreek, etc. to make a gargle.

When the quinsy is full of pus, let it erupt or lance it. Then deterge with ache or the like. If you cannot see it back in the throat, rupture it with your finger-nail and wash out the drainage with gargles and aperitives such as decoctions of figs, dates, and yeast. The strongest gargle recommended by Avicenna has borax, niter, myrrh, peppers, feces of arondelles, and of wolves, mustard and wild rue.

Roger tied a long string around a chunk of half-cooked meat and had the patient swallow it beyond the quinsy. Then he jerked it out and ruptured the abscess. One may use a piece of sponge for the same purpose.

Then deterge the drained abscess with gargles of wine, honey, etc.

When the matter in the quinsy is cool, begin the treatment with gargles of oxymel, and later add cannelle, spikenard, pyrethrum, and asafetida. Lay on plasters and oils. If the mass is slow to soften, use diachylon with chicken-fat, etc., or use softeners described in our chapter on sclerosis.

The fourth tier of treatments will deal with treatments suitable for each complication. For intense pain gargle with warm-milk, which is the best for the purpose. Or use a syrup of violets and poppies, or penidium with a mucilage of linseeds, psyllium, and quince dissolved in a decoction of astringents, as roses, plantains, and morel. Halyabbas used cassia fistula soaked in a decoction of reglisse. When swallowing was badly impaired, he applied cups around the neck to enlarge the passage. Avicenna said that Halyabbas sometimes introduced a gold or silver cannula to assist in respiration.[317] Sometimes to relieve suffocation or choking he forcefully pressed the shoulders.[318]

---

[317] The cannula was curved and could admit air or liquids to bypass the obstruction behind the tongue. But when the obstruction was in the pharynx itself, he had to pass the tube into the trachea (ie larynx) or into the esophagus, as did the peasants successfully intubate the strangled Cathars with wood tubes (ie tubes of bark of elder trees) (J).

[318] An ancient anticipation of Heimlich's maneuver ! (LDR).

When the quinsies are more intense and the medicines are not effective and one can expect a fatal outcome, there is only possible cure, and that is to open the trachea or larynx but not the pharynx, between two rings and allow the victim to breathe.[319] Leave open the wound no longer than three days, until the obstruction was relieved. Close the wound with sutures.

Albucasis reported a case that he had witnessed; a chambermaid had been stabbed in the chest and the knife had cut a duct (ie bronchus) in the lung. Avenzoar repeated the injury in an operation on a goat.

Goiter

A goiter is an aposthem or exiture or excrescence in the neck; it is filled with altered humors.

The signs are those of other aposthems with some special diagnostic criteria. The natural goiter (ie inborn or inherited) is not curable because it is interlaced with arteries and veins, defeating any safe operation. Also, leave it alone if it is large and bilateral. Arnold warned us to be fearful of the consequences of using the knife or corrosives. It is a malady of certain regions or countries, and many cases are hereditary, appearing in many persons in isolated districts.

The treatments differ little from what we use for other protrusions and glands except that they do not yield to resolutives, maturatives, desiccatives, and consumptives that are designated for other lesions in the neck. Roger treated them with a hot cautery and two setons, placed up-and-down and transversely. Every morning and evening he sawed them back and forth until all the matter was consumed. If some matter remained, he destroyed it with asphodels or the like. Afterwards he treated a simple wound.

If after excising just the overlying skin, you can see that the goiter is free of the network of vessels, you may enucleate it intact with its capsule, and cure it. Or you may choose to get rid of it with corrosives, as we do scrofules.

---

[319] Some authors have 'read' this as 'the trachea or larynx' as equivalents, as did Albucasis and many others. Joubert was more correct in translating the tube (ie trachea) but not the larynx (epiglottis), which the Greeks would call laryngotomy, which could mean to resuscitate the suffocating patient who is near death. The barbers call it subscannation, derived from the common word 'escanna' which means suffocation or strangulation. Paul of Aegina (Book 6) described the maneuver as taken from the great surgeon, Antyllus. Aurelian wrote that it was the invention, a terrifying fable, created by Asclepiades, Aretaeus even before them had condemned it, that the new wound made matters worse and increased the suffocation. Furthermore, the wound never healed (EN).

The desiccatives for that purpose were approved by the old masters. Roger used an electuary for internal goiters: roots of white bryony and wild melons, cyclamen, polypode, wild myrrh, asparagus, round aristolochia, roots of wild cucumber, arum, mullein, sea sponge and palm-tree moss. He added honey to make an electuary to be held under the tongue at bedtime. In the morning the patient drank a full cup of wine mixed with a decoction of the roots of aristolochia, mullein, polypode, betony and branca ursinus. He repeated the measures for ten days, and then he added the milk of a sow after she bore her first litter. Of course all of that was empirical (ie not methodistic). However, the treatment was warm and occasionally there was improvement because the recipes are diuretic, and purgation by diuresis is recommended for the treatment of all glands.

Some resolutive plasters have been applied to the neck by some masters, made of roots of lapathum, radishes, wild cucumbers and saxifrage, all cooked with some lard. But we use maturative and resolutive plasters as we described them for scrofules. Lanfranchi offered another beverage of wine with dcoctions of the small nut plant with its roots and some peppers. Master Dino made a powder of combusted sponge, squid-bone, juice of cyclamen, rock-salt and common salt, roots of chelidoine, sunflower, ginger, poppies, bedegar, cypress-nuts, and marine-palm.[320]

Aposthems of the Back

In our Anatomy we placed the back in the section on the neck, and we do the same for aposthems of the back because they have much in common as to causes, dryness, and treatments. They are more at risk because of their proximity with the neck, which as we have stated, is the lieutenant of the brain.

Galen (*Usum . . .* Part 12) said that the complications in the neck resemble those of the brain. That is why we insist that at their earliest state you should treat with our usual alteratives and repellants, such as oil of rosat. When suppuration occurs, do not wait for complete maturation before you open the abscess. When you cut deeply, avoid injury to the nerves.

Gibbosity is a special malady of the back, but it is not an aposthem, although it may be the late result of one, or it may result from a dislocation, to be discussed later in thie book.

---

[320] Joubert's 'balle marine': Guy loved to give examples. When he wrote 'balle' he meant a palm that is rooted in the sea. It is a mossy growth found on the banks, resembling goat-leaf (caprifolium). He thought it was the fuzz torn loose from algae and floating on the surface, and often called 'paille marine' by such as Arnold, Roger, Dino, et al. Latin authors (Constantine, Dioscorides) called it 'ulva' (ie sea-lettuce, a sedge). In France it commonly is called 'sagne' or, in slang, 'sea hay' (J).

CHAPTER 4

Aposthems Of The Shoulders And Arms

T he treatments for these aposthems is the same as for others anywhere in the body, excepting when it comes to open drainage; here no delay is permitted, lest the pus damage the nerves and ligaments, and enter the joints. That could engender fistulas, especially near the elbow joint, a region that is a web of ligaments and bones. In that region be sure that you enter the pointing abscess at one side where the repeated movements will not delay healing.

Aposthems After Phlebotomies

Sometimes an incision for phlebotomy in the arm becomes the site of an aposthem. That is why Avicenna advised us to bleed from the opposite arm very early, and to apply a plaster of ceruse covered with an epithyme of coolants.[321] I use a plaster of bol d'armenie and egg-whites, and I use Janvier's fomentation of softeners and resolutives as a stoupe bound in place.

Aneurysms

An aneurysm is a thin-walled aposthem filled with blood and gas, as defined by Avicenna (*Canon,* Ch.4*)* and Galen *(Abnormal Tumors)*. The mass that appears when an artery has been cut or injured is called an aneurysm. Usually it happens when artery has been divided underneath an incision through an overlying scar[322], more often at creases in the neck or groins. Or it can result from a spontaneous rupture (ie an internal cause). The post-phlebotomy aneurysm occurs more often in the arm. The lesion may appear after a blow strikes an artery and ruptures it.

---

[321] Guy means, "When you are treating an infected phlebotomy wound, avoid that arm for further phlebotomies" (LDR).

[322] The artery is adherent or the scar is thinner than anticipated. A surgeon's error (LDR).

The treatment has two elements. First apply pressure over an astringent plaster that has been bound in place. Second: Locate and ligate the artery on both sides of the aneurysm, and divide it between the ties. Then treat the wound. Albucasis closed the wound by transfixing the wound with needles and lacing the ends, as we will describe for treating umbilical hernias.

Chiragra

This is a fleshy and phlegmatic swelling in the hands. Its causes and symptoms are the same as those of all phlegmatic aposthems.[323] The spontaneous (ie Natural) and chronic lesions should not be treated, and the others are treated as phlegmatics. In addition there are certain special measures, such as cauterization along the length of the forearm with a cultelary (ie see Instruments).[324]

The designated topicals are: First apply the plaster made by my Master at Montpellier: red cabbages, sambucus, and water-thistles. Cook them in a lye made from wine-lees and vinegar; then add salt. Second use the plaster made by my Master at Toulouse. He filled a leather glove or sack with a lye made from the ashes of fig-tree wood, serment, and bracken. Add salt and vinegar to reach the consistency of a paste that adheres to a hand (ie with the chiragra) dipped in it. You can save and re-melt it repeatedly with warm water in a storage jar, and continue the coating treatments until the hand has healed.

Other authors say to first bathe the hand in a decoction of squills or other aromatic flowers, and then cover the mass with resin or wax. After it dries, snatch it off roughly and leave behind the intact softened skin. Then apply perfumes, fumigants, vinegar, and sprinkle the area with marcasite or mill-stone dust that has quick-lime actions. Then apply a plaster of galbanum, ammoniac, etc.

I prefer this at first: I apply sponges moistened with a strong lye in which I have added sulfur, aloin, and salt. I bandage it snugly over the mass, as I do for phlegmatic aposthems elsewhere. It will work better when treating elephantiasis.

---

[323] Chiagra is the counterpart of podagra in the feet that is treated with wrappings. The barbers call it chiragra when it is painless. When it is large and unsightly, it may not resolve (ie under the topicals). Then some authors label it as elephantiasis, and when it involves the legs and the feet and is discomforting. The bulges resembles an elephant's legs, and that may be the source of the name (J).

Guy includes several tumors as chiragra, including caries and suppuration of the carpal and metacarpal bones (EN).

[324] The cautery was used not only to drain the mass, but also to 'cook' the undrained residue to prevent recurrence (J).

In all such case the matter can be attracted by massages on the opposite side. Then if needed, you may go to the cautery.

Fistulous Aposthems Of The Finger

Sometimes hard phlegmatic masses appear at the joints of the fingers. They are dark in color and are surrounded by dilated veins. They may ulcerate and erode the bones, and exposed tendons may be seen to slither as described by William of Saliceto. The lesions may appear in several fingers as scirrhuses, called Fermos by Avicenna.

The causes, symptoms, diagnosis are the same as those of flaky scrofules. Treat the early stage with oils of spikenard and lilies, and apply the diachylon ointment over that inunction. If they ulcerate, deterge with the apostles' ointment and a powder of asphodels. Use arsenic when needed, as we do for some sctrofules. If the bone is eroded, you must use the cautery to eliminate the corrupt matter, because the cautery is best for that purpose. Then treat the ulcer as such.

Paronychia[325]

It is a warm exitural aposthem, with causes, symptoms common to all.[326] The special features are the severe pain, bad enough to cause a fever, which is difficult to treat because the suffering is so intense, and which may cause hallucinations, syncope, and death. Sometimes the lesion will erupt its corrupt matter, flesh and bone, and ulcerate and undermine as an esthiomene that can destroy an entire finger. The drainage is thin and foul.

Treat the lesion with the usual measures that stress coolants and repellants and analgesics at the onset.

Use phlebotomy, evacuations, and diversives. Instigate a warm diet. The repellants at first should be vinegar mixed with a mucilage of psyllium seed, galls and pomegranate-peels. Halyabbas said that the wise Hippocrates (*Epidemics*, Part 4) treated Dobohim[327] with green galls and vinegar; Avicenna used camphor. During the phases of augmentation and steady-state, he applied vinegar and barley-meal or bran. Later he used oil with frankincense and nigella as resolutives.

---

[325] In Bk. 4, Ch. 49 of Dioscorides, and in Galen's *Simples*, Bk.8, we find an herb named 'Paronychia', which was the chief treatment of the lesion named such by the Greeks (ie 'alongside the nail). Paul of Aegina (Bk 5, final Ch.) described it as an aposthem of the base of the finger-nail. The Latins call it Redivie; the barbers call it pannarice, which is a corruption of the Greek (J).

[326] Guy makes no distinction between paronychias and felons, although it is clear that his emphasis is on the latter (LDR).

[327] 'Dobohim': Some read this as dolor. Joubert read it as dahasen after Halyabbas. Others called it paranychia. His treatment was green galls for pain as well a cure for the lesion (J and EN).

When the tumor enlarges, you can hasten the maturation with a mucilage of psyllium mixed with lard, or a plaster of William of Saliceto: cooked egg-yolks, flours of fenugreek and linseeds, a mucilage of guimaulve, and fresh butter.

When it suppurates, open it through a small incision and empty the matter. Then mondify the ulcer with honey and the flours of lentils and lupin. Use aloes as the best incarnative.

Cut away the overlying nail and dry the underlying ulcer with asphodels or chalcidon or arsenic. If the bone is necrotic, expose it and cauterize it, as recommended by Albucasis. When the lesion undermines (an esthiomene) scarify and treat it as such. If the rest of the finger is involved and is threatening to spread, cut it off and apply a cautery. In all cases, cut proximal to the pus and apply medications at the stump and on the hand, using rosat oil and bol d'armenie. Avicenna used opium as an analgesic; he added as much as needed to a mucilage of psyllium. It works perfectly.

William used an ointment which he took from Halyabbas: rosat oil, opium, hyoscyamus, celery seeds, and vinegar.

I insist that you to apply all the plasters on soft cloths, and cause no additional—suffering.

CHAPTER 5

Aposthems of the Chest

And a description of the Plague Years

The surface of the thorax can give rise to a large number and many types of aposthems, including buboes in the armpits, abscesses in the breasts, and in the ribs. First we shall discuss buboes.

There are three kinds of them. The true buboes are in the axillae. They take their name from the animal that hides in walls[328] Second are the abscesses in the three emunctories: that of the brain appear in the neck below the ears; those of the heart in the axillae; and those of the liver are in the groins. In those regions are glandular tissues also found in the breasts and the testicles. Although all of the glands are noble and principal members (ie as emunctories) none of them are necessary for the existence of an individual or of the species. Only a few of them yield to treatments with repercussives. All of them are discussed at length in Part 4 of Avicenna's *Canon*.

Although this chapter deals with buboes in general, we here will take on two aspects of the apposthems at the emunctories, principally those of the heart which is the most important organ.

Some of those apposthems are warm, some are cool, and others are hard. The latter were called fugilics by Avicenna.[329] All are classed according to the matter that is in them. Most of them arose by derivation, and they came to drainage by natural eruption., and the sources usually were ulcers or other apposthems of the extremities.

---

[328] The French call the owl a Chathuant and Hybron. It lives in deserted, inaccessible, and unexplored places, and most of them live in caves. The Greeks call them buas—not buboes—because in Greek the word bubo is the term for the groins and an inflammation there. We have come to use the term for phlegmons of glands in other places, apposthems derived from sanguine humors. Those from choler are called phugethlas (in Greek), and if they suppurate are phumaia (Galen, *Glaucon*) (J).

[329] A fugilic was a very hard mass, set deeply and not movable. It did not discolor the overlying skin. It caused little discomfort and rarely did it suppurate. It occurred most often in the axillae, and contained pituitous matter (J)),

We have discussed the special causes and symptoms in our general chapter. But there are some particular signs: A fever may ensue with little provocation; and, as Hippocrates said (*Aphorisms*, Part 4), excepting the ephemerals, all fevers in these cases are bad, because the buboes are engendered from within; that was Galen's opinion, too. And since the internal causes are near the principal organs, the buboes themselves are dangerous.

The Great Plagues of 1348 and 1360

We ourselves saw that internally caused buboes were lethal during the plagues, and we will describe them. The first arrived at Avignon in 1348, during the 16th year of the Papacy of Clement VI, whom I served.[330]

Nicaise's Note On The Epidemics In The 14th C.

We noted in our biography of Guy that it seems that he had not written about the Plague of 1348 other than what appears here in his Major Surgery. Therefore, this remains as the best description by authors of his generation.

Most of his contemporaries called it The Great Mortality, as translated from the Latin. The Italians called it Montelega Grande, or The Florentine Plague, because it was there that it first attacked Europe. It is still called the disease of the groins, anguinalgia (ie inguinalgia) by Simon de Covins, the mortality of the boxes by G. de Machaut, and the bubonic plague, the dark plague, from which came the black plague—although Michon said that the Latin 'atra' did not mean black even in the 14th C, and that it really meant 'terrible', which referred to the disease and its victims, and the terror of the peoples.

All the authors dated it as the plague of 1348: by Villani, the 14th C historian, by Boccacio in the Decameron which was written during the plague in Florence, and by Petrarch, and others.

As to medical reports, other than Guy's we have those of Chalin of Vinarion, his contemporary, who described the epidemics of 1348, 1360, 1373, and 1383, all in a book published at Lyon in 1542 by William Lothier, a surgeon at Montpellier, and other accounts by several more recent physicians.

In addition we name Simon de Covins who wrote a short poem, Les Con published by Littré as A Small Work About The Plaque of 1348 Written By A Contemporary, in the Bibliography of the School of Maps, Vol. II, p. 201, 1840-41. In the poem Coven attributed the plague to the stars.

Another 14th C poet, William of Machaut, wrote, in French verse, a description of the plague first published in 1860, by Michon.

During the plague, King Philip VI of Valois ordered the physicians of the School of Paris to meet in consultation to discuss ways to combat the plague. The findings were

---

[330] This confirms the fact that Guy was already in the Pope's service when he was recalled from Lyon to Avignon to fight the plague (EN).

written in Latin in 1348, and it is the oldest document from that school. It was not published until 1860. The published document is joined with another consultation from Montpellier, in Latin, dated 1349. Those are very important documents from the 14th C.

Another consultation from the School of Paris was translated in French verse in 1426 by Olivier de la Haye. The original Ms at the Palace of St. Peter in Lyon. It was published by Georges Guigue in1888 with an interesting essay and a valuable glossary of strange and obscure terms as used in 1426.

When seeking the origins of the plague, Michon pointed to the Orient. The epidemic of 1348 was renewed three times during the same century (1360, 1373, and 1382), truly a plague of buboes. The treatises of Hecher, Haeser, Ozanam, and Littré have confirmed that hypothesis, and Europe has been revisited as recently as 1720 at Marseilles, and in Egypt, Palestine, and Syria.[331]

In 1348 it was recognized at once by Guy and others that the plague was contagious, thought to be spread by particles of powders and dust in the air, all caused by the conjunction of three planets, Saturn, Jupiter and Mars (as in Machaut's poem).

Goiffon (1658-1730), a physician at Lyon, who wrote about the plagues in 1721, considered the particles to be earthly rather than celestial, and that they were from living animals such as worms and insects. "Some invisible poisonous insects which we cannot detect, carried here from some foreign country, perhaps with trade-goods, from which they spread in the air of the town and cause all the lethal effects that characterize the plague." Humbert Molliére, a physician at the Hotel Dieu brought Goiffon's ignored concepts to our attention in 1885, and said that Goiffon had anticipated the theories of microbiology.

The plague of 1348 claimed many lives at Avignon. Achard in the Ms of his History of Avignon) the notes of which I have been privileged to examine, agreed with Guy that the plague began in January and lasted for seven months. In a single three-day period after the fourth Sunday of Lent, fourteen hundred persons died, including seven cardinals (one was Giovanni Colonna, the sponsor of Petrarch, who died on July 3). The papal city and the county of the Venaisson buried one hundred-twenty thousand victims during that spell, including Laura, Petrarch's beloved ideal. She took ill on April 3, had a fever and a bloody cough and died on April 6, while Petrarch was in Parma; he did not learn of her death until May 19. She was Laura de Noves, wife of Hugh de Sade; she left behind nine children, six boys and three girls. Her will named her husband as sole beneficiary.

During the epidemic of 1361, between March 29 and July 25, seventeen thousand persons died, including ten bishops and five cardinals. A large number of priests and officers of the papal court died, and nine members of the faculties of the Holy See. The plague of 1348 killed many more of the nobility than in 1361.

Polycarpe de la Riviére, in his History of Avignon, described the epidemic of 1368, during which Guy died (in July). I cannot determine that he died of the plague, neither from that source nor from any other that I have been able to find.

---

[331]   See Duhammel, *The Great Epidemics At Avignon and in The County of Venaisson*, Chaissaing, Avignon, 1885 (EN).

The epidemics were lethal at Montpellier, as described in a chronicle of that town, The Petit Thalamus, that Michot brought to my attention. "This curious work contains some interesting details about various aspects of the plague at Montpellier. In 1348 six city councilors died and were replaced. During the same spell, two of the six also succumbed.

"In 1383 the plague recurred with new violence, especially affecting children under age-twenty. In praying for the Lord's mercy, the citizens measured the perimeter of the city with a cord, which they burned on the altar of their Notre Dame."(Part 4 of the Petit Thalamus, in Mem.de la Soc. Archeologique de Montpellier, 1840). (EN)

Return to Guy's Text

I trust that you will not be annoyed by my recital of its phenomena. Let me alert all of us to what may happen again.

The Plague began in January and lasted seven months. There were two types. The first occurred during the initial two months; its victims died within three days, with continuous fevers and coughing of blood. The second type occurred throughout the course. The victims also had fevers, and they developed carbuncles on the surface, especially in the axillae and groins, and they died within five days.[332] They were very contagious, especially those who coughed blood, spreading the disease to those who lived near them as well to those encountered by chance. The dying patients were left alone, without servants or priests. A father did not attend his son, and the reverse. Charity was nonexixtent, and Hope was struck down.

The word 'Great' meant that the entire world suffered. Few questioned the fact of its origin in the Orient, from which it flung its darts against the world, going beyond us into the West, leaving only a fourth of the population alive.

It was said that there never had been anything so fierce. We have read of the city Cranon (in Thrace) and of Palestine and of others in the *Epidemics* of Hippocrates and of others in later Roman times as described by Galen in *Euchymia*, and of Rome when Gregory IX was Pope. None of them was as large as this. Those plagues were local, whereas this one involved the entire world. The victims in those plagues were treatable in some ways; here there was no cure.

Because it was to no avail, the physicians were put to shame; they did not attend the sick because they feared for themselves. And when they did attend, they obtained no cures, and they earned no fees. All of their patients died excepting a few near the end of the scourge who were left with draining buboes.

Many were at a loss as to the cause. Some towns blamed the Jews for poisoning the world, and they massacred them (ie the upper class), and mutilated the paupers whom they hunted down. Some who were of the nobility, in fear, shuttered themselves. Others who where guardians of the city gates, blocked the entry of strangers. If they encountered

---

[332] Guy defines pneumonic plague with its fulminating course, different from the more common bubonic type (LDR).

anyone who bore traces of topicals—powders or ointments—they shunned him for fear of venoms, and made him swallow them.

The popular opinion held that the plague had two causes: general and particular.

The General Cause was derived from the conjunction of the three superior planets, Saturn, Jupiter and Mars that had been preceded by another conjunction on March 24 1345, in the 14th degree of Aquarius. Those great conjunctions, as I explained in my book on astrology[333], signified amazing, potent, and terrible things, such as changes in kings, appearances of prophets, and of great plagues. Those things vary with the Zodiacal signs in which the conjunctions occur. Do not be astounded that such a great conjunction can indicate a terrifying great plague, not simply great, but very great. Because it was human (ie Aquarius), it was directed against humans, and because the zodiacal sign was fixed (ie not at the cusps), it meant a long-lasting thing. Because the plague began in the Orient, it came soon after the conjunction, and it lasted through the following year in the West. It envenomed the air and other elements that favor iron, and they changed them (ie by rust), and they altered the humors by combustion and poison. Those humors collected and became aposthems. Fevers and bloody coughs appear while the tumors formed in strange ways. When the disease occured in a person, Nature tried to expel it through the emunctories, especially at the axillae and the groins and caused the buboes and other aposthems, which really are external manifestations of internal aposthems.

The Particular Causes affect the patient according to his own complexion (ie disposition), be it feeble, debilitated, obstructed (ie constipated), and if he is a laborer (ie not sedentary) or has unhealthy habits.

The treatments can be prophylactic to avoid the attack or be applied after it happens.

The best preventive is to flee the regions where the plague is flourishing, and when there, to use purgative pills of aloes, and undergo some phlebotomies, to keep the air pure with fires in the hearth. Comfort the heart with a theriac; and apply sweet-smelling items includimg bol d'armenie that rectify the humors; and resist suppuration with astringents.

Treatment of active cases consists of phlebotomy, laxation, electuaries, and sweet cordials. Mature the abscesses at the surface with ground figs and boiled onions mixed with yeast and butter. When they drain, treat the ulcers. Gassy carbuncles should be scarified and cauterized.

Rather than be accused and vilified, I never kept away. And although I was fearful for my own welfare, I perservered with the treatrments. Nevertheless, toward the end of the plague, I fell victim to a continuous fever and had an aposthem in my groin. I was sick for six weeks and all my associates feared for my life. The bubo suppurated and I treated it as above, and I survived, as God willed it.

Afterwards, in 1360, during the eighth year of the Papacy of Innocent VI, the plague returned, spreading south from Germany. It began just before St. Michael's Day (ie September 19) with fevers, bosses, carbuncles, and anthrax. It spread slowly and sometimes remitted, but it lasted until 1361. Then it burst forth furiously for three months

---

[333] A small pamphlet, no longer extant (EN).

and laid waste to half the populations in many places. It differed from the first plague in which the common folk were more often afflicted. In the attack of 1360-61 the rich and the nobility were affected more than the commoners, including their children and their women.

During that plague, I put together this theriac as an electuary, combining the recipes of Arnold of Villanova with those of the Masters at Montpellier and Paris: Juniper seeds, cloves, mace, nutmeg, ginger, turmeric, two aristolochias, roots of gentian, tormentilla, betony, oregoano, enula, sage, red balsamita, mint, pouliot, chelidoine, laurel-buds, doronic, saffron, seeds of oxalis and citron, basil, mastic, frankincense, bol d'armenie, terra sigillata, spode, the heart-bone of deer, bits of ivory, pearls, sapphire, emeralds and red coral, aloe-wood, red sandalwood, muscatel-grape vines, conserves of rose, bugloss and water-liles, and a tested theriac; sugar-loaf (ie penidium). Add the juice of scabious, rose-water and a little camphor. I took that as a preventive theriac, and I was free of the disease, as God willed. Bless his name through all the centuries.

The Treatment of Buboes[334]

Avicenna said that we treat them differently from other aposthems where we use evacuations and repercussives, because these are products of a crisis or of expulsion from vital organs[335] and you must use evacuations at the onset and apply attractive topicals instead of repercussives, and even use cups, as in our general chapter.

If the source of the bubo is an ulcer of the leg or arm as the primary cause, or is in a replete patient, or is spontaneous in origin and hurts (see *Therapeutics* Part 3), evacuations are the basic treatments, to thin and diminish the impetus of the inflow. In those cases. If you wish to repel the incoming humors at the onset use alteratives and comfortives such as oil of roses and camomille, and perhaps softeners. But do not follow Henri who indifferently used repercussives at the onset, after evacuations and did not use our usual softeners except when his were inadequate or not tolerated. Avicenna also said that repellants are avoided because they reverse the inflow back towards the noble organs. And as to the use of softeners, be careful lest they attract toward the mass. What we fear in these two situations is reversed by the evacuations.

Galen (*Miamir*, Part 13) said that even though the parotids are phlegmonous, we should not at first try to repel the matter to where it arose; on the contrary we use attractants, especially cups. Furthermore, if the inflow is vehement and rapid, we should do nothing and simply await what Nature will accomplish. Too successful attraction will increase the suffering. In time, the fevers will appear and the force of the augmentation will be dissipated.[336]

---

[334] Here the buboes are not of the plague. The term also applied to all abscesses in the axillae and groins (LDR).

[335] Buboes in the axillae are at the emunctories of the heart, and those at the groin are from emunctories of the liver (LDR)

[336] Parotids were inflamed upper cervical lymphnodes at the emunctories of the brain. Therefore, do not repel the bad matter to its source. Let them suppurate and erupt (LDR).

251

You may be able to mitigate the pain with special mitigative plasters and yet not intensify the violent inflow of the bad humors. They are warm and moist and in addition to relieving pain, they treat the abscess and hasten suppuration. Galen (idem., Part 13) said to use them to warm and moderately moisten the tissues overlying the glands in addition to relieving pain. On day one apply wool fleece moistened with warm oil. Don't add salt, as do some others. Rather, later on, use warm brine as a fomentation, and apply as a cataplasm; you may resolve some of what has accumulated.

After the pain has eased, digest what remains and get rid of it by suppuration. A good cataplasm for that purpose is the tripharmac of boiled wheat-flower, water, and oil. There are other good remedies.

Fugilic Aposthems At The Emunctories

Galen called them scleroses; Avicenna called them fugilics. They are not easy to cure in situ. Galen (idem, Part4) treated them as scrofules, but Avicenna used a plaster of the ashes of sea-shells and lard. Galen, followed Archigenes, and used honey instead. Many surgeons have incised the hard surface and shelled out the normal glands beneath the plaque. I do not do that, because the ensuing scar is abnormally hard and the glands at the emunctories no longer can accept the superfluidities from the important organs.

However,if you must open the axillae or the groins, Albucasis insisted that you make semilunar incisions. Avicenna said the gently curved incision in the upper neck should be in the long axis.

Aposthems of the Breast

As are other aposthems, these are abnormal collections of matter. Here it is the obstructed flow of milk (ie cloudy fluid) that swells the breast.

The warm and cool aposthems are like others as to causes and symptoms. However, the tumors often are due to retention of the menses. That kind is treated with phlebotomy of the saphenous veins.[337] There are signs and diagnostic clues that are special here; the aposthems always are warm and tender when they are due to curdled milk. There are other signs: enlargement of the breast, often on one side, especially at the onset and when the woman is big (ie pregnant) or after childbirth.

The swollen breast may lead to mania, as noted in *Aphorism* No. 5. That was suspect by Galen who changed the sense of it in his *Commentaries*. "When the (ie menstrual)

---

[337] We have mentioned the ancient concept of the connection by veins between the uterus and the breasts. Here, I remind the Reader that Guy, Joubert, and even Nicaise in 1890 had no concept of endocrine glands and hormones that could explain the phenomenma of lumpy breasts. A phlebotomy in the leg 'attracted' the humors from the uterus, derived from menstrual blood which was the source of the milk (LDR).

blood resists its conversion (ie to milk) because of its virulence and excessive amount, it offends the brain."

Although Lanfranchi accepted Hippocrates' opinion, I never have; no more than did Galen.

Avicenna thought that Birsen [338] was the result of the matter in the aposthem of the breast or that it had become pleurisy; not the opposite.

The treatment of mammary aposthems differs from others, where we use repercussives, because the breast is near the heart. Early on, treat with warm applications of rosat oil and vinegar, or with warm sea-water and vinegar. Later, apply a plaster of bean-flour, and leaves of morel and melilot; or oxymel and oil of sesame seeds or almonds.

When the tumor enlarges, use suppurative plasters, especially this one preferred by Avicenna: soft bread, bean-flour, roots of guimaive, flour of fenugreek, cooked egg-yolks, saffron, myrrh, and asafetida.

After it is mature, open it at its lowest point through a curved incision as did Albucasis. Do not insert a bulky drain and cause more pain. Deterge as usual.

If the aposthem is cool, apply the oil of spikenard and of lilies, et al. When it hardened, Avicenna applied a plaster of rice cooked in a sweet wine with oil of violets, and egg-yolks.

If the hard aposthem becomes a cancer, there is no other way to cure it except by amputation of the entire breast, a fearsome enterprise. It is better simply to palliate than to attempt what will bring down infamy on yourself.

---

[338] 'Birsen': the term may be a corruption of 'Sirsen' in Avicenna's text. He explained that Birsen was a Persian word derived from 'bir' meaning chest, and 'sen' meaning aposthem. Sirsen, also in Persian: 'Sir' means head. Sirsen denoted the frenzy accompanying inflammation of the meninges. He also explained that 'Karabite' meant an aposthem of the brain itself. Therefore, pleurisy of the diaphragm led to the dream-like state and sometimes to Sirsen. Avicenna explained when discussing the symptoms of pleurisy, that the difference between Sirsen and Birsen had to do with the state of 'revery'. In sirsen revery occurs early, but in Birsen it comes late, often just before death.

When Guy wrote 'or because pleurisy' he was badly translated. One should read Avicenna's text. Birsen often resolves into a mammary aposthem, and one must suspect the existence of pleurisy (ie in such a case). The Reader should interpret Guy as follows. "Aposthems of the diaphragm often are converted into mammary abscesses, with success (ie the pleuritic abscess has necessited). Also, at times, we should be concerned that a breast abscess may be converted to a pleurisy (ie a Birsen) by the unwise use of repellants. When Guy wrote 'not the opposite' he did not mean that Avicenna contradicted Hippocrates, that a breast abscess never could become a sirsen, because it seems that Guy also was as wrong as others, but really he meant only what Avicenna did not say (J).

This is one of the places where Nicaise and this translator are left to wonder, and, in turn, leave the matter to the Reader (LDR).

Curdled Milk

This tumor begins cool and can be resolved by many medications used against cool aposthems, and by bathing the breast with warm water and decoctions of white chard, celery, mint, and calamint. Lanfranchi used this plaster: soft bread, barley-meal, fenugreek, linseed, roots of guimauve, and leaves of eruca and oil.

Aposthems of the Chest Wall[339]

There is no special treatment for these abscesses except for a more judicious use of repercussives. When you open them for drainage, incise parallel to the ribs; do not wait for complete maturation lest the pus enter the chest and engender a fistula, as described by Hippocrates (*Prognostics,* Part 1.), when he said that abscesses that communicate inside and out are very bad. Galen said that Nature would cause the abscess to drain downward and would heal with new tissue and scar. Early treatment of a fistula is not good, as we shall explain in our chapter on fistulas, where we will deal with a fistula that drains from within the body. The abscesses that suppurate early should be opened with a knife or cautery. If the pus is laudable, drain it. If the pus is foul it is lethal (Hippocrates, *Aphorism* No. 7.). Also we will discuss later where to incise.

---

[339] Guy's description lacks a clear distinction between chronic empyema, Pott's disease, osteomyelitis of ribs, infected hematomas, etc. (LDR).

CHAPTER 6

Aposthems Of The Abdomen

A posthems of the abdomen differ little from those on the chest. Here, too, we use resolutives and maturatives mixed with aromatic astringents and comfortives, but avoid repercussives lest we harm the internal organs of nutrition, and deprive the entire body. We use the oils of nard, quince, myrtle, mastic, aloine, and quince, which Galen used to comfort the intestine and the liver (*Therapeutics,* Bk.3).

Therefore, we begin with oils of rosat, quince and myrrh, and add camomille, aloine and nard. When suppuration occurs, apply artomel (bread and old honey) which was used by Attalus the pupil of Soranus[340] to which we add oils of mastic and aloine.

Aposthems of the Stomach[341]

Be alert when it begins to harden during the stage of decline as happens easily in those tumors. They are difficult to treat, when facing the risks of hydrops (Lanfranchi). William of Saliceto applied the diachylon when the aposthem affected the orifice of the stomach, and made it a thick paste by mixing it with a powder of hierapicra (ie a laxative!) and oil of absinthe.

Induration of the Liver

Apply a paste of bdellium, opoponax, flours of fenugreek and linseeds, terebinthe, and oil of camomille. Heat the gums before adding the flours. Improve the plaster for the sake of the liver by adding oils of rosat, quince and aloine.

---

[340] Here is a bit of amusing history. In his *Methods,* Bk. 13 Galen deals with an inflammation of the liver that had affected Thegunas the Cynic. The treatment of Attalus dosed him with relaxants, and he died within three days, as a result of the arrogant physicians ignorance (J).

[341] 'The Aposthem of the stomach' is within the abdomen and is not seen by Guy. He describes a hard mass, palpable through the abdominal wall. In time, not affected by his treatments consisting of applications on the abdomen, the patient's cancer spreads and malignant ascites is manifest. However, if the patient's tumor at the pylorus is a benign ulcer, it may subside during the course of treatments, which then are given full credit for the cure (LDR).

Induration of the Spleen

Apply a plaster of bdellium, opoponax, oil of spikenard, terebinth, flours of fenugreek and linseeds and lupin.

Abate the coolness and humidity (ie ascites) that accompany these tumors and others like them as did Albucasis by touching, skin-deep only, with a pointed cautery in three or four places, befitting the size of the tumor.

Hydrops (Ascites, Anasarca, and Tympanites)

Bruno said that the term derives from 'hydros meaning water, and from pisus, meaning a lesion. Hence we have watery maladies, especially in the belly. The only kinds of hydrops that are treated by the surgeon are swollen bellies distended by a watery matter and gas, caused by faulty digestion in the liver. In *Affected Places* Galen wrote that it occurs only when the liver is abnormally hard. Sometimes the malady is caused within the liver, and sometimes it is related to other organs. Hippocrates (*Prognostics,* Part 2) said that the kind of hydrops that was caused by the liver begins in the flanks: the liver being too cool, having lost heat as a result of an intra-hepatic defect. Or, the lost heat may be a complication of over-treatment of a hot malady of the liver. Gordon expressed wonder that if hydrops has a warm cause, that its temperature (warm or cool) could come from the same liver. In one sense the matter was conjoint, in another it was antecedent; in one sense it was Natural, in another it was contrary to nature; in one sense it was formal, in another it was material; in one was sense it was apparent, in another its was occult. From the cooling, according to Galen in *Maladies and Symptoms,* Part 6, and *Unnatural Tumors,* comes the accumulation of most of the watery matter of ascites, gas and tympanites, and all of the phlegmatic humors which we call anasarca and white phlegm. Therefore, there are three kinds of Hydrops.

We surgeons need not seek its causes beyond what we have written about phlegmatic aposthems. Let the physicians delve into the minutiae.

The signs of all the three kinds are bloat, a discolored face and extremities (ie puffy pallor), and diminished urination. The special signs for ascites are thin arms and enlarged legs, and when the belly is shaken one can hear sounds like those coming from a leather sac half filled with water, much as we described as signs of aqueous aposthems.

The special signs of tympanites are similar, except when we tap the belly wall, it resounds like a sack distended with air, in addition to other signs of gaseous aposthems.

The special signs of anasarca are puffiness of the entire body. When you indent anywhere with a finger the hollow persists, in addition to symptoms we have described for phlegm. In all three types, the belly is distended.

The physicians add their own diagnostic signs based in feeling the pulse, examining the urine, the feces, and other excretions.

In *Prognostics* Part 2, Hippocrates said that any hydrops is worse if it is accompanied by an acute fever; and when the fever persists it becomes difficult to treat, if at all curable. It is especially bad if the patient's complexion is warm and dry, because the functions of the entire body are impaired. A hydrops caused by the liver is worse than that from the spleen or others. Ascites is the worst kind; hyposarca (anasarca, swollen legs) is the most curable; tympanites falls between. Why do we treat hydrops when it is difficult to cure and when we cannot promise a good outcome?

Ascites is the only kind that we can treat surgically, repeating what we use for phlegmatic aposthems, with some special elements. First emphasize strict daily habits and dry things. Second, try to improve the liver and comfort it. Third, frequently dose laxatives and diuretics. Fourth, try to evaporate rather than suppurate. Follow Galen's instructions about aqueous aposthems in his *Glaucon* Part 2, and *Commentaries on the Aphorisms,* No. 7., as to the liver, because in cases of ascites, which is our subject here, the matter should be voided with diuretic pills of rhubarb, as stressed also by Rhazes: Rhubarb, juice of eupatorium, seeds of scariola, agaric, and daphne.

A diuretic potion was used by Rhazes: Bark and roots of celery and fennel, celery and fennel seeds, ammi, schoenanthum, red roses, and spikenard, all boiled in water and reduced to one-third before drinking the dose.

Galen (*Temperaments,* Part 3) used combusted cantharides mixed with cherry-juice and wine. Dose only one grain; it is a potent diuretic, useful against hydrops.

To treat the liver internally, use troches of the juice of barberry seeds and scariola, citrouille, poplar; roses, rhubarb, spikenard, a bit of tar from a shield. Dose one, swallowed with vinegar-surup.

Another troche for the liver: epithyme, sandalwood, cannelle, roses. Moisten all with a sour wine.

When you use any of the foregoing to treat a warm liver, add endives, scariola, chicory, and hepatica. And when you want to evaporate the matter, apply this plaster over the abdomen: barley-meal, feces from a ewe, souchet, niter, bol d'armenie, cimolia, snails with their heads and shells, some lye and vinegar. Shake well, and expose it to sunlight, between the liver and the head, because warmth from sunlight is beneficial.

When the above is not effective or when you cannot obtain some ingredients, dose the juice of iris-roots, and bring on a purge as well as a vomit. Another like it is water distilled from the pulp of elder-tree roots and its flowers. Gordon favored it. When mixed with the other remedies, the diuresis will be more effective.

When all these treatments fail, yous should do as Albucasis did, and apply claval cauteries to the skin surface only: four times around the navel, three over both the liver and the spleen, and one over the stomach, and two applications of a cultelary between vertebrae in the back. Leave them on just long enough to produce blisters and attract the hydropic fluid.

A seton in the scrotum is a good treatment.[342]

But, if no treatment has succeeded, you may have to undertake a dangerous treatment, too dangerous for feeble and debilitated persons, and children, and the very old who

---

[342]   If the processes vaginalis communicates with the scrotum, the seton may tap off some of the ascites (LDR).

cough, have diarrhea or other complications that contraindicate any operation with a knife, as agreed to by Albucasis, Halyabbas, and Avicenna. Make your small incision to drain ascites due to the intestine, about three fingerbreadths below the navel. If the liver is at fault, incise on the left; if the spleen, incise on the right, so the patient can lie where there is less discomfort and not be over the incision, and yet where you can control the outflow.

The Procedure: The patient should sit or stand facing you. Assistants standing behind him will support him and press the ascites forward in the belly. Then pinch up about a fingerbreadth of skin (ie and panniculus) where you plan to incise., and pull it up or down and expose where you will cut through with a razor or spatula-knife until you are through the siphac and some ascites spills.[343] After discharging a suitable amount release the skin over the inner opening and let it cover the slit in the siphac and prevent unwanted out-flow. Now, provide a boost for the patient and give him some toasted bread dipped in wine. Again pull up the skin and slip in a bronze cannula and allow the ascitic fluid to drain through it, as much as you judge to the patient will tolerate, and not too much. As Avicenna said, it is better to empty small amounts several times than to let all of it out at one session, and cause the patient to collapse or to die. Remember the risk every time you drain, and after every episode, withdraw the cannula and let the skin fall over the opening which you bandage firmly.

If you repeat the procedure on succeeding days, take time to comfort the patient and feed him well with liquid foods and beverages which are aromatic and pleasant. Repeat the drainage for several days and empty most of the ascites. Then let the wound heal.

Some surgeons make the initial incision (ie below the pinch) only through to the siphac. They retract the inner abdominal layers and nick the siphac beyond that second retraction. That means that the channel is doubly angulated.

Tympanites is reduced by drainage as above and by diuretics. Use clysters and suppositories of oils of rue, cumin, and borax. Comfort the liver as described above, and use internal (ie oral) medicines to expel the gas: diacumin and an electuary of laurel berries. Rub the abdomen with garlic and evaporate with a plaster of millet and sulfur. You may have to use cups.

Anasarca is treated with purges of agaric pills: agaric, eupatorium, rhubarb, and round aristolochium. Or use pills of oxymel and squills.

The liver is comforted with troches of lacca: lacca, rhubarb, celery seeds, ammi, ginger, spikenard, bitter almonds, mastic, squinanthus, costus, cabaret, garance, aristolochium, gentian, and eupatorium. Dose the troches with Rhazes' apozeme of roots.

Also, apply epithymes of spikenard, mastic, souchet, squinanthus, cannelle, aromatic sedge, saffron, myrrh, and an astringent wine.

Keep the patient in a warm room, either directly exposed to sunlight or near a bake-oven. That (ie sweating) will resolve the matter.

---

[343] The fold of skin in the 'pinch' is pulled toward the head or feet and the stab wound is made through all layers above or below the pinch. Later, when the pinch is released, the intact skin and panniculus falls over the inner opening. The mneuver is repeated as necessary to readmit the cannula (LDR).

CHAPTER 7

Aposthems of the Pelvis and Its Parts[344]

Some of the aposthems of the pelvis are on the surrounding parts, and they are no different in origins, causes, signs, and treatments from similarly situated aposthems. The internal aposthems are not the concern for surgeons, and are not part of this book.

The aposthems we deal with are those of the groins, the testicles, the penis, and the anus, and we already have discussed those of the groins and the axillae. As to the others, hernias at the groin and scrotum will come first.

Some hernias of the testicles really are aposthems: there are five types: humoral, watery, gassy, fleshy, and varicose. In addition are the herniations that are bulges or ruptures containing omentum and/or intestines. We will defer consideration of the latter group until Treatise VI.

Humoral Hernias[345]

These are either cool or warm, suppurated or not, and are derived from natural humors and develop in the scrotum. Other than the usual causes, signs, and treatments, the exceptions for humoral hernias relate to the sensitivity of the structures, and their dependent situation that almost makes them emunctories. Although they are covered (ie hidden), they are easily damaged. They are sources of embarrassment, and are difficult to treat, requiring special agents. These are purges by certain suppositories designed by Avicenna, which attract the matter from the scrotum to the anus in a wonderful way. Also, use repercussives of cimolia and vinegar. Although the testicles are principal organs for the

---

[344] Joubert's Note: The Pelvis continues the lower abdomen as far as the hips and the genitals as explained in Ch. 7, Doctrine I, Treatise I. The 'containers' are the layers of the abdominal wall, the omentum, and the bones. The 'contents' are the rectum, the matrix, the urinary bladder, the spermatic vessels, nerves, veins, and the arteries that pass through. The external parts are the testicles, scrotum, the penis, the vulva, the groins, the buttiocks, and the muscels that attach below to the thighs (J)

[345] I assume these are inflammations: orchitis (perhaps mumps), epididymitis, hematomas, not connected with the abdominal contents (LDR).

species, they are not vital for the individual. Use resolvants of bean-flour and decoctions of cabbage. Use fresh hyoscyamus leaves taken from the new growth as analgesics when the matter is warm. When the matter is cool, apply fenugreek and cumin added to the bean-flour resolvants.

If the contents suppurate, drain them completely through a vertical (in the body's axis) incision at the bottom of the scrotum to forestall a fistula.

If the mass hardens, use our usual softeners and Avicenna's special emollient for indurated testicle: finely ground bran in a paste with oxymel and ammoniac. Apply it warm, and often. In all cases, suspend the scrotum with a truss and belt.

Aqueous And Gassy Hernias

These are apposthems in which the scrotum is filled with watery fluid or gas, or both. The wind-hernia distends the entire scrotum and the watery type is contained in a membrane (like an egg in a shell, as noted by Avicenna)[346] that descends from the didymus as does the testicle.[347] Sometimes it passes through a gap in the peritoneum. That leads to the misconception that it is a rupture.

The cause is faulty digestion, especially in the liver, and bad habits, as we noted for hydrops and other watery and gassy apposthems such as anasarca and tympanites, and is a guide to our treatments.

The signs of an aqueous hernia are the weight of the swollen scrotum, and the water transmits light. A gassy hernia is less heavy, it resists indentation and it is translucent. The watery hernia develops slowly, whereas the gassy hernia develops rapidly, as noted by William of Saliceto. Athough the apposthems rarely occur alone and derive from more than one faulty humor, they are named according to the predominant one.

All hernial apposthems affect the testicles which have had prolonged contact with foreign matter that can alter or corrupt them, as many authorities have stated.[348]

The treatments—refer to the section on treating hydrops—according to Galen (*Therapeutics,* Book 14), include topicals and surgical operations. The medications for aqueous hernias are those for hydrops: Avicenna's plaster: niter, wax, oil, peppers, and laurel-berries. For gassy hernias use cumin, rue, calamint, and oils of costus and nard.

---

[346] Most of the medieval surgeons realized that the wind-hernia was a trapped loop of bowel, frequently a sliding hernia of the cecum or sigmoid colon, and they were aware of the risks of attempting to drain it, and of the difficulties of reducing it. Note that aside from a brief comment, Guy does not explain why he does not operate to drain wind hernias. The anatomic basis for the 'slide' was not clearly defined until the 20th C (LDR).

[347] Joubert, again: Really the didymus is testicular, but Guy describes it as the processus and tunicca vaginals extending from the peritoneum, along which the pass the spermatic vessels. It is called didymus because it is doubled (twinned) (J).

[348] Another justification for orchiectomy as part of herniorrhapy (LDR).

Manual procedures are reserved for the aqueous type. Albucasis said that we should use the cautery (for touch and not for incision) alone, although Halyabbas said that there are several techniques, as he used for aneurysms.

Galen's first maneuver was to aspirate the water with a syringe or with a seton. Insert the seton as follows: Compress the lower scrotum and apply a flat-blade, perforated forceps where the scrotum is empty[349]. Pass a needle mounted with a thread through the holes. Pass the needle through and leave the seton in situ and remove the forceps. During the first few days apply oil and egg-white on the cord. Later rub it with a cabbage leaf. Work it back and forth until all the fluid is gone. Avicenna, and William of Saliceto, pushed the testis away from the openings for the seton and used the large phlebotomy blade to enlarge them for drainage. When the field is dry, the wound will close. To prevent a recurrence, they placed a plaster over the groin region and held itn place with a truss. If, as Albucasis reported, the fluid returned within six months, he repeated the insertion. Meanwhile, the patient had enjoyed his comfort. Both Albucasis and Halyabbas placed their setons through the center of the scrotum, up near the groin. After emptying the fluid, they sealed the didymus[350] to prevent flow into the emptied sack. After the eschar came away, the wound healed.

Others perform the entire operation with a cautery. They are successful if the water will not return and if they have not damaged the testicle (ie by injuring its vessels). You recognize that when the scrotum begins to stink and blackens. Then, as Albucasis did, deliver the didymus and cut away the damaged structure after burning the end of the stump. Treat what is left as an ulcer.

Fleshy and Varicose Hernias

The fleshy tumor, according to Galen (*Tumors Against Nature*) is a sclerosity of the testicle and the surrounding tissues[351] that is not congenital.

The varicose hernia—a recent term—is named for the abnormal veins in it.

They are caused by an inflow of humors into tissues that cannot resist it.

The signs of the fleshy hernia is its firm bulk, slow to accumulate and rarely to diminish. You may be able to feel the enclosed testicle in the mass. The signs of the varicose hernia are the distended tortuous veins, somewhat like a grape vine wrapped around a soft testicle.

Your decisions for treatment should be based on the difficulties and the risks of operations; often it is better to leave them alone. Nevertheless, I will describe the operations that the Masters have performed.

Let us assume that we have attended to improving the patient's daily habits. Then we have applied softeners and resolutives as used in our chapter on sclerosis. After offering

---

[349] Although not stated, it is clear that he displaced the water in the hydrocoele and included some of its wall in the grasp of the forceps (LDR).

[350] With a cautery (LDR).

[351] The 'surrounding tissues' are the tunica albuginea of the testicle. The term 'sclerosis' was read differently by other authors, but it refers to a fleshy hernia in which the tumor is stony-hard (J).

a sincere prayer (ie a sine qua non), and having provided a prognosis of the serious risks, We make an incision, a la Albucasis, in the skin of the scrotum. Then as best we can, we dissect the flesh (ie the mass) and cut it all away and discard it. If we cannot separate the diseased flesh and the sclerotic tisticle, we apply corrosives, ligate the didymus, and cut away the distal mass and cauterize the stump. We deterge the wound.

For varicose hernias, we doubly tie the base of the hernia in the groin and divide it and remove the scrotal contents. We then deterge the region and treat the wound wound. We may choose to do the operation with caustics and corrosives, as did Master Peter de Orlhac, or as we did for glands.

Aposthems of the Penis and the Matrix[352]

The treatments are nearly the same as those for aposthems of the testicles, although these structures are warmer and more readily inflamed than the testicles, and they can better tolerate applications of astringents at the onset, even though they are emunctories. Although they are important organs for the survival of the species, they are not vital for the individual. The treatments must be supplemented with potent analgesics because the organs are very sensitive, and they readily lose fluid and wilt. To treat the warm structures, Avicenna chose lentils, pomegranate-peels, a decoction of roses, and rosat oil. The populeum ointment and morels also are good. For relief of pain, apply fomentations of mauves and hyoscyamus, and epithymes of violet-oil and egg-whites. Another: the pulp of white bread soaked in milk mixed with soft-boiled egg-yolks, some opium, saffron and oil of poppies.

The discharged fluid should be mopped away frequently with mondificants, and the opening in glans should be stented with a wax or cloth strip. The dressings should be held in place with a supporter, as used for the scrotum. That will help relieve the discomforts. The penis that is inflated with gas, called priapism, will be discussed later.[353]

Aposthems at the Anus

In general, these are similar to all others, excepting some particulars. in the use of maturative and sedative topicals. Here you add tassus barbatus. When the abscess suppurates do not delay draining it through a semilunar (ie perianal) incision that will allow free and complete emptying. That will prevent the occurrence of a fistula.

We will discuss hemorrhoids later.

---

[352] Although 'matrix' meant the entire female generative organs and tissues, here, as stated early in the chapter, we deal with external parts of the pelvis. Therefore, in this chapter 'matrix' means the vulva ((EN)

[353] Galen and others believed that erection and detumescnce of the penis occurred rapidly because gas was the inflating agent (LDR).

CHAPTER 8

Aposthems of the Thighs, Legs and Feet

These aposthems are like all others in general, except that you must not incise for drainage near the kneecap. That could cause serious complications which often are fatal, as described by Avicenna and confirmed by Henry.

Elephantiasis, Varices,and the Meden Vein

Some abnormal enlargements and bulges are called Varices, the Meden Vein, and Elephantiasis. Varices are tortuous dilated veins that give off branches from the thighs down to the feet and.resemble a vine that wanders over a wall. The Meden vein (of Avicenna) was called crural by Albucasis, and famous by Halyabbas. It is tortuous like other varices but breaks down readily.[354] It begins with painful inflammation and blisters. Elephantiasis is an unnatural enlargement of the tissues of the limb, and the appearance is that of an elephant' leg.[355]

The causes are the same for all three: thick melancholic blood and combusted phlegm which descends into the leg and does its harm. It happens more often in melancholic people, in those who perform hard labor, and after acute illnesses when the matter was delivered in the lower limbs. The signs are visible for each type, by pain, and by the life-styles of its victims.

Early diagnosis is not easy, and one hesitates to treat before they are chronic and well established. When the bad matter has taken its unhealthy place and is retained and expands upward, it harms the patient and his disposition, as stated by Avicenna, and as Hippocrates wrote in his 12th *Aphorism* about treating chronic hemorrhoids. He advised to leave at least one untreated lest you induce hydrops or phthisis[356] One should not treat it and shorten one's

---

[354] Stasis ulcer (LDR).

[355] Did Guy ever see an elephant? The fantastic drawings by medieval artists probably were the only images available to him. The elephants that Hannibal led over the Alps left behind in France little more than the strange pictures. (LDR).

[356] The ancients believed that hemorrhoidal veins were emunctories of the abdominal organs, and were convenient for the discharge of certain superfluidities via small bleeds. They served much as did the prophylactic phlebotomies which were offered as regular 'doses' to maintain good health. (LDR).

life. I have seen that happen, and Lanfranchi has reported it. Furthermore, other lesions appear, usually ulcers that are not easy to cure, which may lead to cancer (ie necrosis). The new lesions vary with the geographic region, and may run in families, especially the meden vein which may occur more often in certain places, as do other maladies. Albucasis treated it by binding over the crural vein a strip of lead as long as fifteen or twenty hand-spans. Neither I, nor Galen, as Avivenna reported, have seen it in use.

The treatments have three purposes: 1. Establish a life-style that opposes the things that engender thick melancholic humors. 2. Prevent the entry of more of the antecedent humors. 3. Get rid of the conjoint matter by voiding it or drying it.

Rhazes dictated abstinence from heavy melancholic foods, such as beef and venison, and from sweet dishes of honey; lentils, cabbage, boiled wheat, and unleavened bread. The patient should conduct himself in the manner recommended in the chapter on melancholic aposthems. He should limit ambulatory activity and things that require prolonged standing. The food should be light and not rich. Choose the kind that produces thin blood, such as the meats of pigeons and kids; delicate vegetables; soft-boiled eggs. A delicate white wine and well-baked bread are good.

The second tier includes phlebotomies, pills of hermodactyl, hierarufinum, epithyme, polypode, lapis lazuli, and a single dose once a week of tryphere as a vomit.

The third tier consists of astringent desiccatives and potent resolutives applied on the leg and foot. Begin the wrap below, as we do for podagra and cheiragra. Repeat it as needed, re-applying it at least three or four times (ie interspersed with embrocations). First wash the leg with rusty water containing cimolia and vinegar. The second time wash with the epitheme of Rhazes made from ashes, a decoction of baggage-seeds, stoichas, lupins, fenugreek, niter, and goat turds. The third wash also is from Rhazes: add myrrh, aloes, acacia, hypocistus, alum, and vinegar. The fourth taken, from Theodoric, is for matter that is more phlegmatic and gassy: roots of asphodels, sambucus and royal fern (ie feuchere), elder-tree leaves, parietory, and red cabbage.

All the above recipes are improved by grinding them with the lees of wine (tartar) while they are heatedt on a stove. When the region is inflamed and blistered, as at the onset of a meden vein, Avicenna suggested that we apply a plaster of the juice of cool quinces and the two sandalwoods, psyllium aloes, myrrh, and camphor.

If yet not improved, you must incise behind the knee or on the thigh where the varices arise. Or, if no incision, catch a vein with 2 hooks through the skin, separated by about two finger breadths. Ligate it with a sturdy silk thread. Then pull up and divide the skin and vein between them. Then loosen the tie on the lower segment and milk out the contents of the varix, as much as you can, and cauterize the cut end as well as the surfaces of the incision, using a hot iron or some arsenic. Later mondify and consolidate the wound. If any of the bloody contents remains, consume it and dry it as above.

Some surgeons perform the entire procedure with a cautery. Some open the varix and empty it; others divide it.

Make your incisions as follows. Wrap the leg snugly, beginning below, up to where the tissues are healthy. Incise the skin over the varix in several places. Hook the unopened vein beginning below (ie just above the wrapping) and pull it out. Follow along from incision to incision until you have stripped out the entire segment of vein. Compress the wounds (ie for hemostais) with wool-fleece wet with oil and vinegar.

The first method seems safer to me. That is the way that Galen treated rheumy eyes (*Therapeutics,* Bk. 13.*)* He wrote," That is how I treat varices at the temples, tying them in place and dissecting them out between the ties. Halyabbas agreed (*Royal Dispositions*, Sermon 9.

We will deal with podagra and painful joints, and related aposthems, wounds and maladies of the bones, in Treatise VI.

HERE ENDS TREATISE II.

Figure 4. A Wounded Patient
From The Treatise on Wounds In Ms 396. Bibl.iothéque Nationale, Paris. 14th C

HERE BEGINS TREATISE III: ON WOUNDS

TWO DOCTRINES

RUBRICS FOR TREATISE III

A NOTE: COMPARISON OF TRANSLATIONS OF

TREATISE III AND V

I have at hand W.A. Brennan's translation of Treatises III and V, which I believe is the only version in English other than this, of Guy de Chauliac's Major Surgery.[357] Although I had read Brennan's book many years ago, for reasons of personal pride-of-performance I did not consult it here until after I had completed my own translation. Now I am pleased to note how few differences there are in the substances of the two versions. They differ in many places in the choice of words, in the arrangement of sentences and paragraphs, and only slightly in the arrangement of topics, and I have made a few changes where I think Brennan's translation is more accurate, Brennan's formal British English of 1923 used by an Irish professor of anatomy transplanted to Chicago a century ago 'reads' differently than my American English of the present era. Furthermore, Brennan's version is more literal, that is word-for-word, based on a Latin manuscript edited for print at Venice in 1546. Mine is more a translation of the context of the French edition of Joubert printed around 1582-1583) which he based on a collation of several Mss and editions in Latin and French. Nicaise used Joubert's 16th C French in his edition of 1890, which is the source for this book.

I am pleased to note that Guy's orderly presentation, as seen in the extended table of contents in Brennan's Treatise, was not much changed through the centuries between the 14th C of origin and Nicaise of 1890, and now this of 2006. LDR.

---

[357] See Bibliography (LDR).

DOCTRINE I WOUNDS IN GENERAL, THE TISSUES: FIVE CHAPTERS

CHAPTER 1

Wounds In General

A wound is a recent disruption of continuity[358] in soft tissues, a bloody defect that is not yet corrupted. The general term, disruption of continuity was used by Averroes (*Maladies and Symptoms*) primarily for a malady involving simple tissues, but not excluding the composites (ie organs and regions), which I believe is a better use for the term.

Nicaise's Note

Here are a few comments to supplement what I wrote about wounds in my Introduction.
Some of the remarkable features of Guy's Book are his classification of wounds, and his detailed descriptions of the finer points in suturing, dressings, etc. All of it testifies to his own experiences as a careful, observant practitioner. He accepted the teachings of Hippocrates and Galen in their recommendations to use dressings that contained desiccatives (antiseptics) to avoid wetness (suppuration and corruption) in wounds.
In 1683 Mingelousaulx praised this General Chapter as follows: "Every surgeon should study Guy's General Chapter and observe how forthright was his summary of everything that the ancients and moderns had written about the topic. Courtin, (in his Ninth Treatise on Wounds In General) added nothing to what Guy wrote. He simply pieced the material and expanded it into several chapters to make the topics easier to find and to remember by the Reader. Paré, according to Tagault, compressed the material a bit, but eliminated nothing of importance. He wanted to lessen a student's dismay when confronted by the vast collection. Fabricius of Aquapendente (Treatment of Wounds) expanded on the materials in Guy's Chapter and presented them in various places in his own book. He did not deny that he did little more than embellish and use a sharper modern language in expressing his agreement with what was not new. He expected a

---

[358] In the ancient treatises the commonly used term was 'solution of continuity' which certainly meant a disruption or a separation within previously intact tissues. This translation will use 'disruption' which in English connotes a more violent cause for a wound than simply 'solution' or 'separation' (LDR).

Reader who to apply himself to study what others had written about wounds, could find everything in what Guy had placed in this chapter. Tagault deemed the chapter to be so beautiful that he took it verbatim in his own Surgery, asserting that all surgical authors agreed unanimously in adopting what Guy taught. He added that it was accepted by all the most famous physicians. That was a potent support for Hippocrates and Galen who continued to be vibrant and fulfilling sources for the well-being of Man, which no one should divert or alter without incurring infamy from those who need the benefits of surgery."

Guy sometimes used 'ulcer' when referring to a wound, as did Galen. Hippocrates always used 'ulcers' to distinguish true ulcers from wounds.

Continue Guy's Text

My own definitions of a wound, that it is recent, bloody, and not yet corrupted, will define the difference from an ulcer, which is a corrupted lesion. The restriction to soft tissues sets wounds apart from fractures in bones, the hard parts. While accepting this general definition, Joannitius, in his own *Surgery*, stated that Surgery was two-fold, that for flesh and that for bone. Flesh included the muscles, veins, and nerves, which Galen (*The Medical Art)* had placed in a category called soft and moderately hard.

Avicenna said (*Canon,*) Fen 2[359]) that there are many kinds of disruptions of continuity: wounds, ulcers, openings, punctures, incisions, rips, fractures, et al. We shall deal with each of them, but we shall not meddle with terms beyond what they denote. Furthermore, their identities often have been changed, as in Greek, wound and ulcer mean the same[360], whereas in Arabic they are different. The Greek version is better, because Galen (*Therapeutics*, Part 4) insisted that wounds and ulcers are very different; one has no preexisting cause, while the other has. Halyabbas (*Royal Dispositions,* Sermon 3*)* said that wounds and ulcers were the same.

The various disruptions of continuity differ principally in three ways. First are differences in the affected parts. As explained in *Techni,* Part 3), some involve simple tissues and others are in organs (ie composites) Some wounds in Simples involve soft tissues as in flesh or fat; those in hard tissues may be in bone or in joints (ie cartilage); others in the medium-hard (ie or medium-soft) are in nerves, ligaments, arteries, and veins.

Disruptions in organs may involve the principal ones, as are the heart, liver, and the brain. Others may involve the structures which serve the chief organs, as the trachea, the esophagus, the bladder, et al. Some of these disruptions involve autonomous organs

---

[359]   A Fen was part of a Book of the *Canon* (LDR).

[360]   This refers to the Latin translation from Greek by Nicolas of Reggio. Until then, Galen was available only in Arabic translations into Latin. Guy had both versions, and, as he wrote in the Singular Chapter, the translation by Nicolas was more accurate (J).

such as the eyes and the ears. Albucasis said that wounds vary as to where they occur: the head, neck, chest, abdomen, etc. They also vary according to the offending instrument, as we shall note.

The second category of variables includes the essence of the disruptions (see *Therapeutics,* Part 3). One is simple, the other is composite. A simple disruption is uncomplicated; the composite is complicated by two or more dispositions. They are not the principal causes of a wound, but they must be considered in the treatments (ibid. Part 3). We will deal with them in the next chapter, and in the chapter on ulcers.

In the third category are the physical characteristics of the dispruptions: the size, the entire wound or just a part of it, its directions—straight or angular, and similars. All of these elements are important in diagnosis and in plans for treatments, medications, etc. Galen (ibid. Part 3) said that in addition we must determine if the problem is surgical and what we need to know in order to assess the particulars of the injury: the substance, its functions and uses, and where it is in the body. Furthermore, the attending physician should know what you intend to treat, so he can prognosticate its curability, and that help in the selection of the medications.

The Causes

Galen said that there are two kinds of causes: external and internal. However, we state here that the causes of fresh wounds always are external: instruments that can penetrate, or can break a part of the body.

Halyabbas (idem, Sermon 4) described them as inanimate objects: swords, spears, stones, etc.; or animate things such as stings by venomous animals, or bites by savage beasts. He said that slashes are different from concussions and bites, the complexions of which will affect the treatments. In *Therapeutics*, Part 4 Galen wrote that the dispositions are the primary bases and the weather also is important.

The Signs and Symptoms

The signs are obvious. The assessment beyond what we observe must include knowledge of the substance, the function and the purpose of the affected part, and its dispositions, as we described above. Galen wrote (ibid. Part 4) that all wounds and disruptions, both large and small, are dangerous in degree, when considered in three ways: The affected part, the severity of disfunction, and the amount of damage. All injuries to the head, the chest, and the abdomen are dangerous,especially as to the damage incurred by the structures within them. We know that all injuries that enter joints soon deteriorate, because the adjacent tendons, nerves, and exposed bone cause pain, disturbed sleep, convulsions, and delusions., Furthermore, large wounds that require sutures, such as those that transect muscles, those that cut large veins, arteries, veins, and release bone marrow, all of them bode ill.

275

Judgments—Diagnoses and Prognoses

Galen (*Aphorism* 6) claimed that all wounds which enter the bladder, brain, heart, diaphragm, small intestine, liver, or stomach are lethal. And in *Aphorism* 5 he said that wherever ulcers cause local swelling, convulsions will not occur.[361] Where there is no edema around large wounds, the prognosis is poor. Hippocrates, and Galen, said that wounds in the head or at the insertions of large muscles are malignant, especially when the latter involve nervous tissues. Be alerted to the dangers and to the threats to life when any part of the body dies, or to the threat of permanent disability, paralysis, and loss of sensibility and normal functions—an eye becomes as functionless as a stone or a bronze ball, as Aristotle described it (*The Spirit,* Book 2, and in *Meteors,* Part 4.

Surgeons must offer prognoses during the course of treatments. He should know which wounds are lethal and which are very serious. Galen explained it in Aphorism 6, where he mentioned wounds of the urinary bladder. He said that large cuts are always mortal, but smaller ones are not necessarily so. The same holds for all wounds that usually are lethal.

Wounds Which are Lethal

Wounds of the heart are such. Its natural function, to receive all the blood that is delivered to it, including all of what appears elsewhere as hemorrhage, or in tumors, or as resolved matter from aposthems, or in congested vessels; all of what must be sent forth to animate the rest of the body is lost. The heart cannot withstand an open wound or a warm aposthem[362] and yet live.

Other lethal wounds are large openings in the brain, liver, diaphragm, stomach, small intestines, kidneys, trachea, esophagus, lungs, spleen, gall-bladder, and the other vital organs and the attachments to them that are necessary for life. They are fatal whenever there is no available treatment, as we now shall discuss.

Wounds (ie in the vital organs) that may not be lethal are small and superficial and do not penetrate deeply into the organs or into the muscles at the joints. If they are treated properly, they can heal, even when the wounds threaten life. However, if not treated properly, even the salvageable wounds will be lethal. I have seen a wound in the back of the head that extruded brain. The patient survived, and he recovered some of his ability to remember. I cannot claim that all such victims are so fortunate. Theodoric reported the survival of a wine-dealer who lost a small amount of brain.[363] And did not

---

[361] When there is edema there is circulation, hence no gangrene and tetanus (LDR).

[362] Pericarditis (LDR).

[363] Guy's report here is incorrect in its details. Theodoric reported (1265), "I have known a man whose wound cavity was wholly emptied of brain tissue, and finally refilled with flesh in place of the brain substance, and he was cured by Master Hugo. And since it was the cavity of memory,

Galen report the survival of two wounded men that he saw at Smyrna, treated by his Master Pelops? Brain exuded in one, and in the other it was injured within the cranium. And he had seen recover at Smyrna (see *Usum . . .* Part 8) another man who had lost no brain substance but with penetration into one of the lateral ventricles. You may well believe that those survivals were due to the Grace of God; otherwise they would not have survived the treatments. Galen also credited the use of several special instruments and his knowledge of anatomy. So much now for the rare successes of rarely applied treatments, more of this later!

One of the reasons for lethality of wounds of the brain and its membranes is the subsequent lesion in the lungs and the respiratory apparatus. That disorder allows the heart to overheat and suffer its own damage, and the entire body perishes. (Galen, *Afflicted Places*, Part 5.)

Also, I have witnessed recovery after small wounds of the margins of the lobes of the liver when they have not cut deeply into the organ and have not sliced off parts. Galen described the same (ibid). When the damaged liver cannot provide nutrients for the heart, that organ will die, and the animal spirit is lost, especially when the malady lingers.

Injuries of the diaphragm usually involve the nervous (ie contractile) parts and are incurable. They lack blood, and are in continuous motion.[364] Wounds can heal only when they are at rest and the edges can join. Nevertheless, some wounds will heal when they are filled with proud flesh. Galen agreed.

Because they are inflamed, wounds of the lungs often are incurable. Furthermore, our topical medicines cannot reach within and be applied to the ulcer (ie the unhealed wound). And the organ is disturbed by breathing and coughing.

Large wounds of the trachea, especially those in the cartilaginous rings, are seldom cured, because the tissues are hard and their blood supply is deficient, and the wound is disturbed by respiration.

The act of swallowing food and beverages causes wounds of the esophagus to be incurable.

Large wounds of the jugular veins are incurable because hemorrhage overwhelms efforts to staunch them, and bindings around the neck will suffocate the patient.

For obvious reasons, wounds of the stomach, small intestine (especially in the duodenum), the matrix (ie the internal parts), the kidneys, and the urinary and the gall bladders, all of them are lethal. They are nervous and bloody, and are the normal means to evacuate bad humors, and as such they are vital. And they, too, are not accessible for applications of medications.

I saw that Master Hugo was greatly amazed over this; for the man had memory just as before; for he was chair-maker and lost only his skillfulness." See *The Surgery of Theodoric,* translated by E. Campbell and J. Colton. Vol I, p 109. Appleton-Century-Crofts, New York, 1955 (LDR).

[364] Hippocrates insisted that motion disturbed a wound and prevented the cohesion of the surfaces (LDR).

Wounds of the spleen, a vital organ like those in the liver, are threatening, but less so than the others listed above.

Stab wounds that enter the internal parts are said usually to be fatal because they permit entry of air that harms the organs. Also, they are vents for the escape of the vital spirit, leading to a weakened state that may be beyond rescue.[365] And if the victim survives the wound, a fistula and empyema will follow, both of which can be lethal.

The Signs of Particular Wounds

Much of this material and others will be dealt with later in the text. Now I shall be brief. Furthermore, wounds and punctures in muscles and nerves are open for direct inspection, and do not need more description Wounds in tendons and ligaments near (within three fingerbreadths) joints usually are lethal, as Galen said (*Medical Arts, Part 3.*).

In respect of puncture wounds of nerves and tendons, know that they can cause convulsions, because their innate sensibility transmits the pain to the brain, and that can be the cause of death. The damage suffered by the brain is transmitted to the respiratory apparatus, and that is the direct cause. Hippocrates (*Aphorism 5*) wrote that a wound that is attended by convulsions is mortal but Galen commented that such was not always fatal.

Particular regions may be considered at high risk when the principal veins, arteries, and bones which nourish and support them are cut or destroyed. The early signs of necrosis are the dark color and the local softening, and the cadaveric stench resembles that of esthiomene when it leads to amputations of arms or legs. When the limb is paralyzed or weak, you will know that its nerves have been cut or destroyed, and the loss of function is followed by shriveling (ie atrophy).

On the other hand, you can judge that a wound is curable when the victim is robust and the injury is in fleshy tissues that are not rich with blood vessels and nerves, if the treatment is suitable, and the wound is not deep. In such cases, fevers and other bad complications are avoided by good care. You may offer a good prognosis, barring unanticipated complications which are not yet apparent.

Although those wounds sometimes are not cured. They are like the occasional curable wounds at the insertion of muscles (ie near joints), at the head, the chest, and the abdomen. The successful cure follows good surgical measures diligently appled on a cooperative patient, and when the equipment and other materials are at hand. Lacking such, the outcome may be mortal.

A wise prognosis should include all of these matters. You should add that you, the surgeon, are subject to the rules of the Papal Court: its permission to attend to the wound obtained

---

[365] I assume that Guy here implies stab wounds of the thorax and that it was hemorrhage of arterial blood, the vehicle for the vital spirit (LDR.

by your descriptions of the wounds[366], the injured part, and your plan. I beg you not to be hasty in your assessments and prognoses. Be deliberate and use foresight. Our forefather Hippocrates said that judgment is a difficult matter. As I shall explain in the section on head-wounds, a wound per se can last for a maximum of forty days,[367] and be prognosticated as early as the seventh day, and more commonly by the fourteenth, as the intervals during which most complications will appear, and you can recognize the good and bad signs, such as fevers, heart-failure, delusions, convulsions, etc. In your assessments you should observe the pulse, the urine, the vomit, and other excretions. Note the appetite and how the patient tolerates the treatments, his appearance, etc. Then, following the precepts of Hippocrates and Galen, assess the patient's vitality and offer a wise prognosis to the patient's entourage as to morbidity and mortality. It is not proper to tell a wounded patient that he would not die if he had not suffered the wound, not matter how defective were the treatments. And that he may die despite your best treatments. Tell him that he will recover if he follows your recommendations, and that both the patient and the surgeon must act together, and that there are no preventive measures that can deal with whatever may happen, by chance, or unanticipated, as described in *The Art of Medicine.* The human spirit may be spared in various ways even when it has been exposed to risks, as we have described them.[368] We listen to Avicenna (*The Removal of Arrows,* Part 4) who wrote that even if a wound is lethal, one must not back away from doing a surgeon's special duty.

But an honest prognosis must be given in language that is understandable by illiterate folks, because unexpected miraculous cures occur. And if we abandon the dying patient we forego the possibility of success, and we can deem ourselves insensitive and pitiless.

The Different Kinds Of Healing: The Intentions

You should know that injuries in vital organs will not heal (see *The Lesser Arts*, Part 3*)*. when their internal connections which carry nourishment and vitality are cut, and they suddenly lose are their subtle humors and functions, properties not found in lower animals and plants. Their failure to heal is not the result of resistance to medications, as

---

[366] Here Guy describes the precautions taken by the surgeon before he undertakes to treat wounds that were deemed at high risk, where failure could lead to censure of the surgeon, or punishment, or criticism in the community that deameaned his reputation. Theodoric described how Hugh of Lucca obtained permission from the ecclesiastical authorities before he undertook to treat a patient whose lung had herniated through a chest-wound. And Henri de Mondeville consulted with the patient's family and entourage as well as the priests in similar cases. He took pains to excl;ude witnesses at operations lest they faint and take his attention from what he was doing (LDR).

[367] After forty days an unhealed wound is an ulcer or fistula (LDR)

[368] In other words, spare the patient a hopeless prognosis. In this section about dealing with the usually lethally wounded patient, Guy exhibits a gentle humaneness that marked him throughout his career (LDR).

explained by Halyabbas (*Commentaries*). On the other hand, simple flesh will heal by First Intention and bones will heal by Second Intention.

First Intention means the union of the disrupted wound's surfaces without intermediate non-local tissues, but by a conversion of the sealing matter to tissues normal for that region.

Second Intention means that the union is bridged by different tissues, as pieces of copper are joined by lead (ie solder). For example, in the healing in bones, that material is called porous sarcoid (ie callus); it derives from a humor that is thicker than flesh but less dense than bone. The hardness of bones is not suited for first intention which is only for moist tissues (Galen). The supply of nutritive blood is feeble, to explain why bone is partly cool (Halyabbas). It is only partly a spermatic substance, and he insisted that spermatic nutrition that reached bone was converted (ie to callus) and became bone.[369]

Avicenna cited Galen (*Therapeutics*, Part 5) that the hardness of nerves and veins is between that of flesh and bone. They heal by incarnation when the wounds are small and the body is moist. That is not the case for other wounds. Galen said, "My own experience tells me why. I have seen arteries heal in children and women because their bodies are soft. And I saw a partly divided artery heal in a young man. Hippocrates confirms me in *Aphorism 6*[370] as to the following: When a bone is broken or a cartilage or a nerve are injured, or the skin of a cheek or the prepuce, it will not grow (ie lost tissue) or will the wound unite." The exceptions are the bones of babies that can unite directly by bone. Galen said it was due to their soft bones soon after birth. The same holds for teeth which are engendered not only in babies but at other ages[371], and their matter is unique, and unusually profuse as a result of good nutrition rather than in their initial formation following conception—so said

---

[369] Spermatic tissues derive in utero from the union of the male and the female sperms, and cannot be replaced after birth. The other tissues derive from the retained menstrual blood, and can be renewed after birth (LDR).

[370] I find this item almost identically stated in my edition of Hippocrates (Adams). That causes one again to reflect on the roles of translators! Nicaise's Edition (which I translate here) used Joubert's translation which was based on several Mss, all of which were variant versions by scribes of Guy's own Ms. Guy cited Galen from a Latin translation of an Arabic version, which was a translation of Galen's Greek, probably by Paul of Aegina at Alexandria who used Mss transmitted by generations of scribes between the 2nd and the 7th Cs. And Galen wrote his *Commentaries on Hippocrates Aphorisms* based on what had come to him after five centuries. And who wrote the Hippocratic Opus ? And when? And who was Hippocrates?. *Aphorism # 6* Indeed! (LDR).

[371] Joubert described a gentlewoman in the Languedoc who had no teeth, and, according to the testimony of reliable witnesses, suddenly sprouted five or six teeth at age 70. Another witness said that he had observed the reappearance of teeth removed before age 60. The new teeth were smaller and weaker than the first set (J).

Albert of Bologna in his lectures on the *Aphorisms*. And Hippocrates (ibid.) wrote that ulcers in edematous persons do not heal easily.

Avicenna added that disruptions of continuity and ulcers were alike in rapid rates of healing in persons with healthy complexions, whereas the opposite is true in persons with bad complexions. When wounds occur in patients with hydrops, the union is insecure and defective. And he cited *Aphorism* 5 that cold irritates ulcers and produces a tough and tender dark membrane associated with chills and spasms.

We can infer that large wounds near joints may heal poorly and lead to atrophy of the limbs distal to them. That is due to the interference with the passage of a weakened vital spirit (ie the large vessels and nerves were divided).[372] We shall discuss these matters in the special chapter in The Treatise On Ulcers, where ulcers and wounds share the text.

The Treatment of Wounds

Our goal is to bring together all disruptions of continuity (see *Techni,* Part 3). In every case we first assess the injury so that we may apply a contrary to oppose it. That is accomplished in two ways: Use what Nature provides with nourishment and special powers, and use our Medical Arts: five measures applied in order, one after the other:

I. Clear the wound of foreign matter. II. Bring together the separated surfaces of the wound. III. Restore the normal conformation. IV. Preserve normal tissues V. Treat the complications as they occur.

I. Eliminate The Foreign Matter

When we cannot see it in the wound, objects such as a fragment of bone, or an arrowhead or thorn, enlarge the wound. When the wound is large enough, retract the edges, gently and painlessly, with your fingers, tenacula or forceps, etc.

Ia. The Instruments for removing embedded arrowheads or other objects. Select those that are designed for special tasks, suited to various items and the various parts of the body, and know how to use them. It will be impossible to describe every situation because the variety is infinite. I will advise you to first examine the path of the offending object. Avicenna discussed them in eight categories which I simply will give examples: Metallic items, thorns, bones, et al.: Some are barbed, some arrowheads are still attached to the wooden shafts, and some of them are pinned to the shafts, some are envenomed.

You should know the anatomy of the wounded region: some parts are 'principals', others are fleshy or bony, and all may embed the objects, more or less fixed. Some missiles

---

[372]   See Chapter 3. (LDR).

are trapped in perforating wound and can only be detected near the opposite surface where they may almost come through. Others will be hidden in the depths.

Ib. I keep in hand the eight instruments designed for these cases, which suffice for most of the countless varieties of wounds.[373] First are the tenacula of Avicenna, with teeth set in curved blades. Second are the tenacula with blades resembling a bird's beak. Third are tenacula with tubes, for removing barbed arrowheads. Fourth are threaded gimlets with which to impale and extract a barbed arrowhead after screwing it into a socket no longer occupied by the shaft.

Fifth is a straight gimlet to dilate the tract. Sixth are impulsors, hollow and solid, to enter the empty socket of the arrowhead and push it forward and out through the oppoite surface of a limb. Seventh are scissors for dilating flesh to free the trapped object. Eight is the arbalest (cross-bow) for sudden extraction of a penetrating arrow.

Ic. How to remove an arrowhead. When the impaction is so tight that it resists an easy extraction soon after the injury, let it stay for awhile until the tissues around it soften a bit and suppurate. Then you may be able to get around it and rock it back and forth to loosen it and gently remove it.[374] That policy contradicts Henri who recommended that it should be forcibly removed, whereas I agree with Avicenna, Albucasis, and Bruno.[375] Then treat the wound as usual, after evacuating the contaminated blood which will cause the wound to putrefy. Wash it out with warm oil that will also ease the pain. If the wound has been envenomed, treat it as you do venomous bites. When these nice measures fail, you must go further and remove the arrowhead, after offering your careful prognosis. Grasp the object with a common tenaculum and twist it to loosen it for removal. And if

---

[373] I have added brief descriptions to Guy's list of instruments. More detail as well as drawings can be found in Nicaise's Glossary II, at the end of the book (LDR).

[374] The wood shaft or part of it, is attached to the arrowhead and protrudes from the wound (LDR).

[375] The policies re removal of embedded arrowheads were discussed at length by Henri de Mondeville, in Chapter 1 of his Treatise on Wounds. I shall repeat some of it: "They (the old-timers) believed that a firmly embedded object deep in a wound which does not yield to gentle extraction should be left in situ. They had three reasons for that practice. 1. To avoid further loss of blood. 2 To allow expected suppuration to lubricate the wound . . . 3. Nature on its own can expel the article if given time. We moderns . . . . extract the objects as soon as possible . . . . Everything that is lodged in a vital region produces a wound . . . . and causes swelling . . . . pain, infection and fever, simply by being there. Its removal will reduce complications, and it must be done immediately when it is in a vital organ or a threatened site . . . . Even while arguing against the rationale offered by the old-fashioned surgeons, we must admit that although we know how to prevent and to quench hemorrhage, occasionally we will recognize an obvious case when it may be preferable to leave the object in place. With that exception we do not leave foreign objects in wounds. The Ancients did not know how to control hemorrhage and they renounced prompt extractions . . . ." see *The* Surgery of Henri de Mondeville, Vol. I, p.348. English Edition by LD Rosenman (LDR).

that fails, you must go to more violent maneuvers. When the arrowhead is barbed and can be reached through a tubed tenaculum, insert the curved (hooked) gimlet into the socket. If some of the wood shaft remains in the socket, screw in the straight gimlet. If you need more exposure, enlarge the wound with a razor. If the impaction is in bone, drill around it with the straight gimlet or with a trephine, and then remove the loosened object. Failing that, bind the arbalest to the tenaculum that grasps the arrow, hold the patient securely, and fire the bow, That will tear it out. When the arrow has penetrated deeply and resists withdrawal through the wound of entry, use the impulsors to push it to the opposite side where you can easily see or feel it. Incise over it and remove it. If none of the maneuvers are successful, leave the object in situ for Nature to deal with. Albucasis described his experiences with arrowheads that remained for long periods, and the patients suffered little harm. Some were recognized by Nature and were rejected and cured. An incantation of Nicodemus [376] was favored by Theodoric and Gilbert. I am not influenced by such drivel.

The medications that I have used in removing embedded thorns, fish-bones, stones, bits of glass and bone are these as recommended by Avicenna: yeast, honey, bee-hive ordure, oak-tree sap, ammoniac and oil. Roger used mashed roots of reeds mixed with honey; he applied them directly on the object and removed it without pain. Many other medications are in our Antidotary. Now we leave Intention I.

II: Approximating the Wound-Edges

We use surgical (ie manual) actions to bring the surfaces together and restore the normal contours in ways that cause as little pain as possible.

III. To Hold in Place What We Have Restored

This is accomplished by bandaging.[377] When necessary we use sutures. Avicenna (*Canon,* Part 4) said that there are three types.

1. The Incarnative Bandage is useful for fresh wounds (ie ulcers) and fractures). We use a double-ended bandage and wrap from the center over the lesion toward both ends of the limb, and cover as much adjacent normal tissues as is expedient. The wrap over the wound, will be drawn most tightly, while taking care not to be excessive. The patient's tolerance is the guide. Suture the plies to prevent slippage. If you need to

---

[376] Joubert quotes Theodoric's about Nicodemus' incantation: "While kneeling, the victim utters a paternoster prayer three times a day, while grasping an arrow in both hands, and say that Nicodemus had removed iron nails from both the hands and the feet of our Lord (ie a crucifix) by pulling on the arrow, and the arrows come out" (J).

[377] Joubert-Nicaise's text uses the term 'Ligature'; to describe wrappings that bind the wound. Our term is 'bandage' (LDR).

repeat the dressings, use the same method. It will prevent swelling (see *Therapeutics,* Part 6). Some surgeons use a two bandages and end the wrap over the wound.

2. The Expulsive Bandage is useful in wounds (ie ulcers) to prevent puddling of matter in the hollows, and to dam back further accumulations. Use a single-ended wrap and begin below the wound, increasing the tightness up to the wound, and lessening it as you pass proximally., over normal tissue. Galen (idem. Part 5) said that this bandage directs what comes from the heart and the liver (ie to the ulcer). My technique fits the wrap to the contours of a taperted limb. I turn over the bandage as it crosses in front and let it lie flat as it passes behind the limb. The Lord knows how many have benefited by that method.[378]

3. The Bandage To Hold in Place the topical medications. This is suited only for the region that can be wrapped, and not for others as at the neck or the belly. or over aposthems and tender lesions (ie cannot tolerate the pressure). In those cases we use a many-tailed bandage, centered over the wound and tied on the reverses. For a gentle application, the center of the bandage should be softened by moistening it with wine.

Galen used a pure linen strip as wide as needed by the part. For example a bandage for the shoulder should be six finger-breadths wide; five for the thighs; four for the leg; three for an arm; and one for a finger. They should be long enough to easily encircle the part. The wrapping should demonstrate the surgeon's ingenuity, an obvious sign of his art and gentle anner,as noted by Damascenus; and his knowledge of the contours of the body. Galen (idem. Part 4) said that a person who wants to bandage a wound well, will suit it to the shape of the wounded region.

IV. Suturing a Wound

This process used in the treatments of disruption of continuity.has three features: incarnation, hemostasis, and maintaining what has been restored

IVa. An Incarnative Suture is used to close wide-open wounds that will not come together with bandages alone, which are fresh, and may have retained foreign material, and older wounds that are prevented from closure by loss of skin or by a membrane over the deeper surfaces. There are five kinds of incarnative sutures.

1. In uncomplicated wounds use a long strong unbraided thread of silk. Place the first stitch at the center. Place the second half way between the center and one end of the wound. Continue by halving the distances between the sutures on both sides of the center, until they are about one finger-breadth apart

The needles should be long and smooth, triangular in cross-section (ie cutting edges), with a hollow near the eye in which the thread can lie parallel with the needle

---

[378] The 'chevron-design' pattern (LDR).

and lessen the diameter of the puncture for needle-plus thread Place a perforated tube against the opposite side to guide the exiting needle so to match the insertion. Pull the needle and thread through while you support the second side with a flat blade as a counter force. Then tie the threads. The first layer should have a double turn, and the second should have one.[379] Cut the ends long.

2. The second incarnative suture uses needles and a drain, either a cannula or a strip if cloth transfixed by the needle to treat a deep and widely open wound. Or you may use fine threaded needles and wrap the threads around the exposed ends of the needles that are left in place until the wound is solid. Women use this method to mount their needles and colored threads on their sleeves when they are sewing

3. A third type of incarnative suture uses tubes of equal length, made of tightly rolled cloth, or reeds or a central tubes of a feather, each about the length of a finger nail or longer. Pass the needle through both sides of the wound but leave a loop on the second side when you pass the needle and thread back through the same punctures. Thread a tube through the loop and pull it snug as you draw the wound edges together. Then tie the ends of the thread over the another tube on the first side to maintain the closure. Cut the ends long. Leave the entire arrangement in place until the wound is solidly united.[380]

4. The fourth type of incarnative suture is Galen's. He used hooklets of the correct size, each with a hook at both ends, to catch both sides of the wound after the edges are brought together. They will maintain the closure, in the same way as used by wool-cutters.

5. The fifth method uses cloth, and leaves no suture scars after the wound has healed, an important cosmetic consideration for facial wounds. Cut triangles of a fine thin cloth to size. Smear them with a glue made from sangdragon, frankincense, mastic, sarcocolla, pitch, fine mill dust wheat-flour, and egg-whites. Apply the triangles on opposite sides of the wound, about a thumb-width back from the edges. When they are dry and adherent, suture them carefully and that will close the wound.

IVb. The Hemostatic Suture: This employs the same continuous over-and-over stitch as used by leather-workers, when the simple interrupted stitches cannot control hemorrhage. Also, it is useful for sewing intestines and membranes, and in places that lack underlying flesh. Be aware that when one of the bites of a continuous suture breaks, the entire row will loosen.

IVc. The Conservative Suture is like the previous one but is not so tightly drawn. It is placed only in the edges of the wound to hold until the tissues below it are united. It is useful for wounds that were torn open instead of cut, and for wounds where tissue has been lost. Suturing the skin edges may close the opening and hold. It is useful when you re-close a wound that you have had to reopen to remove something.

---

[379] A simple description of what always has been called 'the surgeon's knot' (LDR).

[380] The tubes are bolsters that prevent the sutures from cutting through the skin (LDR).

How to Remove Sutures

Leave the sutures in place until they have accomplished their purpose. When you wish to remove one, slip a narrow flat blade under the suture near the needle-puncture through which it will be withdrawn when you pull it out on the other side, and cut it over the blade. That will avoid tearing open the wound.

The Use Of And The Quality Of The Pads

The pads compress the wound and serve to retain the Natural Heat, and they relieve the uncomfortable pressure of a firmly wrapped bandage.

They have been called 'feather pillows' since ancient times when they were made by stuffing feathers between two layers of cloth and sewing the opening. That required a lot of work, especially since they had to be renewed frequently. That led to the substitution of stupes made from hemp-cloth, fluffed and laundered. Sometimes wool or cotton are used, and often simply a double or triple ply of folded soft cloth, or a sponge. One may use two or more or as many as needed. Sometimes they are dry, sometimes softened and moistened with egg-whites, or wine, or oxycrat, or oil—to suit the case. They may be round, triangular, or square. Avicenna preferred the triangles, one on each side of the wound to conserve the heat and to absorb the discharges. The square ones are used to protect the wound from the pressure of a too-tight bandage.

The Use And The Quality Of Tents (Drains) And Packing

There are only eight uses for them. First: when we want to dilate or cleanse or remove matter from deep in the wound, as when fluid puddles and has to be counter-drained. Second: In excavated wounds when we must await the formation of proud flesh. Third: In wounds that have been polluted by exposure to air and must be deterged. Fourth: In contused wounds (ie hematomas). Fifth: When draining an abscess. Sixth: After bites. Seventh: When wounded bones must be exposed for treatment. Eighth: When treating ulcers.

We treat all other wounds without drains and packs. But you should know which to use when indicated. Some are made for deterging; use charpie or soft old cloth. Others simply to keep open the wound need only a clean stupe made from strips of of cotton (a la Rhazes) or a perforated cannula, as inserted into nostrils to permit the entry of air, or in deep wounds to drain foul matter and to keep it open. To enlarge a perforated wound they are made from strips of brine-free sponge, or a root of gentian.

A long strip of packing should have parallel edges, whereas a tent should be shaped like a chevron. They may be inserted dry or covered with ointments designed for particular cases.

Other matters involving the Third Items are left to the ingenuity and the Natural Spirit of the surgeon.

IV: Preserve The Normal Tissues

IVa. The Use of Plasters. This matter is concerned with the preservation of the normal tissues of a part of the body, to avoid suffering, abscesses, and other complications. All are treated with plasters and with applications of egg-whites, and cool things during their early stages (Halyabbas). Then we apply astringent wines or corrosives, and contraries; or we open the wound for drainage, or apply splints and bleed and purge; and we establish healthy habits.

Although many writers have claimed that wounds per se do not require such things, and I confess that may be so in rare cases of wounded patients who are robust and have healthy complexions. But only God knows where to find them. In most cases and in all debilitated people, we must order the correct topicals to preserve their bodies. Galen (*Therapeutics,* Part 6) said, "Let us suppose that we encounter someone who has been stabbed and wounded, and that the wound is clean and has a good complexion. That fellow may need no special medications, and he will not come to a bad end. But if the wound is contaminated, and the patient suffers, inflammation will soon appear. In all tissues of the body there are nerves and veins that are involved in bad injuries. So be it that Hippocrates (*Prognostics*, No. 1) said it always is better to anticipate what may happen.

IVb. On Phlebotomy: Wise Rhazes and Albucases said that surgical phlebotomy is unnecessary if the wound itself had bled sufficiently, but if indicated, bleed from the opposite side. And Galen (ibid., Part 4.) said that Hippocrates favored purgations (including phlebotomy) as diversions.

IVc. On the Bowels. Avoid constipation by prescribing suppositories, clysters, and gentle laxatives that contain cassia fistula and manne.

IVd. On Wound Potions: I do not use them routinely in treating fresh wounds; they are warm astringents that are not hemostatic and they favor the formation of abscesses, and also are laxatives. But in cases of chronic ulcers, fistulas, and when there are retained blood-clots or hemothorax, and for internal glands, and for hernias, I do occasionally prescribe them. As you will see in later chapters, all the old masters, including Roger and the Four Salernitan Masters prescribed them indiscriminately for all wounds: beverages containing garance, red cabbage, Herb Robert, columbine, Herb Bennett, dog-tongue, pimpinelle, pilosell, etc. They squeezed out their juices or boiled them in wine and honey. They dosed them every morning and evening and placed red cabbage leaves over the wounds, and bandaged them in place. Those empirics claimed it was a bad sign if the patient vomited the dose; but it was good if he could tolerate it. God help them! Both Theodoric and Henri encouraged the use of potent potions for head wounds and chest wounds. I will not frankly decry that practice, but I will say that Galen did not use them.

The Diet For Wounded Patients

During the first week after the injury, the period when abscesses appear,[381] the life-style of all patients should include a diet that is cool and dry, especially if the victim had been well-nourished and is young, and the weather is warm. He should avoid wine, especially undiluted wine, hearty meats and fish, unleavened and incompletely baked bread, cheese, fruits, garlic, onions, mustard, other strong spices, and salted and sour foods. Let him eat chickens, partridges, small birds, seasoned with rose-water. He may eat oats, barley, almonds, wheat, spinach, bourrache, plantains, and bouillon with eggs. Let him drink water boiled with soft bread or drink barley-water with a bit of astringent red wine. He should not be depressed, and disturb his sleep (ie with worry), and he should avoid women. When past the time for abscesses, he may gradually return to his prior more rich diet.

That will include good wine, fresh meats of grouse, capon, and lambs, and every thing that engenders blood and repairs Nature. The good habits that I list have been time-tested. They favor a virtuous life and do not incite fevers and abscesses and excite the flow of blood. Cooked foods should be nutritious and seasonable. They were favored by Galen, Rhazes, Halyabbas, Avicenna, Bruno, William, and Lanfranchi. On the other hand, Theodoric permitted warm wine from the first day, and Henri who was bred at Paris amid the Philosophers (Aristotelians), followed Theodoric's practices.[382] As to the English (ie Gaddesden) I am not surprised that they said nothing except what they took from Henri; their reasoning is worthless, limited to saying that we should comfort the suffering. Galen said the opposite (*Aphorisms*), "We should not hasten to intensify the force of the malady; rather let it wane moderately, and conserve what is there, at least in chronic problems. Only when the patient is very feeble do we set aside the rules that we apply in treating common wounds. And even then, let us not blindly forgo our use of contraries, but keep them in mind and seek a middle course (*Therapeutics*. Part 7). And near the end of Part 8, he added "The physician must be experienced and well-trained in assessing every indication and comparing one and another, until he has learned the habits of the patient." The earlier authors had been misled by the Arabic translation of the fourth part of Galen's *Therapeutics* that said (penultimate chapter), "one should abstain from wine when the aposthem is warm, but not in other cases." A correct translation is from the Greek Ms which says, "It is a well-known rule to avoid wine in all cases of phlegmons but not at other times." That agrees with the policy of abstention during the first week, when inflammations occur. After that there is no need to forbid it.

This concludes Section IV about conserving the substance of the part.

---

[381] "Abscess":If any appears let it be 'good pus' (LDR).

[382] Was this a criticism of Henry or of the Parisians? (LDR).

V: Remedies For Complications

The treatments are specific for each of the various complications. Most common in cases of wounds are pain, abscesses, dyscrasias, fevers, itching, convulsions, paralysis, faintings and delirium. I emphasize the fact that we should not deal with the wound until the complications have been corrected; they may be more threatening than the wound, and they will change the order of the treatments (*Glaucon,* Part 1).

Va. Pain. Pain attracts humors to the wound and in turn they engender abscesses which above all should be avoided. Usually we mitigate pain by fomenting with warm oil, especially that of rosat. Add egg-whites or similars when there is much local heat; they will not muddy the wound. If the whole region suffers, add poppies, or, when more is needed, add some opium and mandragore, as did William. The Four Masters also used the roots of morels mashed in pork-lard. Theodoric favored a potent plaster of leaves of mauve boiled and mashed before adding powdered bran. That was improved by adding rosat oil. Also, he used the pulp of wheat-bread wet with boiled water.

Intense pain signifies an injured nerve; we will deal with that in the chapter on nerves.

Vb. Aposthems (abscesses). We manage them as we instructed in Treatise II. I repeat: Avicenna applied his plaster to treat all abscesses, from head to feet. That plaster is made from sweet pomegranates heated in sour wine, then mashed and applied. When the abscess could not be aborted or resolved, he applied suppurative plasters. Roger bathed (embrocated) the wound with malva, aloin, artemisia, and wheat-flour. All were heated before adding wine, honey and lard. If the abscess does not erupt on its own, open it for drainage in the correct place.

Vc. Dyscrasias[383] When they are hot, they will be red and blistered, and they should be treated with coolants; do not use hyoscyamus or mandragore because they are too cool (Galen). Instead, we use roses, plantains, and the white ointment which dries as well as cools. When the dyscrasia is cool, it may be soft and will not be red. Warm the place with resin, pitch, bitumen, wine, the black ointment (fuscum) and the basilicon. Untreated dyscrasias lead to ulcers and that will be dealt with in the section on ulcers. No matter if the dyscrasias are dry or moist, apply contraries. More about that later, too.

Vd. Fevers. If they occur, use coolants as we described for warm aposthems. Also, in such cases, consult the physician.

Ve. Convulsions. Averroes used the term to describe sudden contractions of the limbs as well as rigidity that prevented flexion or extension. Unlike paralysis. they cannot be moved (ie straightened, etc. by attendants). In convulsions (see Ch. 5 of *Maladies and Symptoms*) the movements are violent and beyond the voluntary control by the patient. Although it is a disorder of the nerves (*Canon,* Part 3), the muscles contract involuntarily and prevent relaxation.

---

[383]   Local inflammations, etc. (LDR).

The Causes of Convulsions

The two underlying causes are repletion and evacuation[384]. (Galen and Hippocrates in *Aphorisms.*). In phlegmonous dispositions repletion gives rise to convulsions. In cases of violent and very dry fevers, it is evacuation that causes convulsions. In other words, the nervous tissues that are affected are either too full or too empty, more or less like the strings of musical instruments, they can snap according to how tightly they are strung in a humid versus a dry house. That is why the players loosen them before setting them aside. The same sort of phenomenon involves straps placed near an open hearth or when the air is very moist. Avicenna said that also there is a third factor: how much the harmful matter affects the brain, constricting it because the injured nerves that discharge what the brain usually delivers into them cannot unburden themselves.

Therefore, there are three kinds of spasms: of inanition, of repletion, and from the suffering brain.[385] The first is caused by such as profuse diarrhea, or escessive heat, or by profuse flow of thin pus (*Aphorism 5*). Item 26 states that is better when a fever follows a convulsiuon than when it precedes it.

The second (ie repletion) is caused by abscesses, edema, and swellings that fade away (*Aphorism,* 5, item 6)[386] Excessive cold that blocks the nerves and causes them to bulge (ie and contract) is a factor in the convulsion.

The third cause is pain due to direct injury (ie puncture) of nerves and tendons[387] or by erosions of nerves by bad humors or by venoms. In Galen's *Commentaries on Aphorism 5* he said that Hellebore can be a cause.

When the convulsion affects the entire body, the source is the brain which cannot reject the injurious matter coming back from the nerves and nervous structures; therefore the body convulses. When the spasm involves only a part of the body, the brain is not the cause. The part is immobilized, as described by Avenzoar. He said that a convulsion is universal when most of the body is affected and it is different from epilepsy which involves the whole body and is caused by excessive humidity. However, in epilepsy the contractions are intermittent and not continuous, and the victim is insensate, whereas a spasm is rigid while the patient may remain lucid (see *Affected Places,* Part 3).

The entire subject of convulsions, its differential diagnosis, its causes, etc. is amatter for serious consideration (ie by Physicians). The surgeon needs to know what is given here as to its signs and its treatment.

---

[384]   "Evacuation": Leading to depletion, as diarrhea, bleeding, etc. (LDR)

[385]   Elsewhere the favored term was 'spasms of sympathy' (LDR).

[386]   The Numerical citations here do not correspond with numbers in Adam's translation of Hippocrates *Aphorisms* (LDR).

[387]   'tendons': A reminder that the terms 'cords' 'nerves' 'tendons', 'ligaments' referred to objects near a muscle, entering it or leaving, with little concern for what was what (LDR).

The Signs of Convulsions

All convulsives exhibit difficulty in voluntary movement, have stiff necks, grimaces as if to laugh, pursed cheeks, clenched jaws and teeth, and a tight throat. The eyes bulge out of the face. The special signs of convulsions of inanition appear bit-by-bit before and after diseases of consumption. Those of repletion appear rapidly after aposthems and repletion, and chills. The signs of convulsion by compassion accompany external causes, with pain, cramps and a sense of being bitten. If the respiratory apparatus is the site of origin the patient will die suddenly. Once in full force, even that of inanition, cannot be rescued; the complete attendant dryness remains beyond cure (*Therapeutics,* Part 7). But before it is in full force, the patient may be saved, especially if it is a convulsion of repletion.

The Treatments For Convulsions

Avicenna said that the result of treatments for a dry convulsion are bad. The most appropriate treatments are baths. Afterward repeatedly apply wet oily ointment. If possible, make it a milk-bath, and irrigate the nostrils, use gargles and clysters, and finally provide potions. These measures work best if the fever has no local origin. For others use only water and oils in which you have boiled leaves of water-lilies, applied over the joints and over the attachments of muscles. Feed thick soups and light solids made from almonds, barley and good sugar, and the gravies of meats of sheep and goats. You may provide a small amount of a good wine to help the absorption of the food, but do not stray far from the strict regimen.

When the convulsion is moist, use potent evacuants to eliminate thick humors, such as pills of hierapicra and agaric; irritating clysters; gargles and sternutories also are good. Anoint the neck, armpits, and groins with warm oils of lilies, costus, spikenard, and pouliot, fortified with castoreum and euphorbia. Then cover with large amounts of wool fleece. In these cases Roger used this ointment: oils of musk, petroleum and olives, butter, wax, styrax (both the calamite and the red), mastic, frankincense, and gum of hedera. Mix what can melt with the others that are mashed. Keep the patient warm and anoint the whole exposed body. Theodoric added the herbs of Aragon and Agrippa to the ointment, and used an oil of castoreum with the grease from red snails. It is effective against spasms of nerves. The fever that appears is a good sign, even if only for one day.

Steam vapors and dry perfumes that induce sweating are good treatments for convulsions of compassion, Medicate for pain in delicate regions and with mitigants, as used for spasms due to wounded nerves. If a puncture is caused by the bite of a wild beast, dose a theriac and apply cups over the bite. If it is due to irritation of the stomach, provide emetics to obtain relief.

Comfort the brain in all cases of spasm: anoint the head, the neck, the back, the axillae, and the groins with the oils of lilies. Avenzoar claimed that was a good medicine for humid spasms.

Insert a rod between the teeth to prevent the jaws from closing. As a last resort completely divide the affected nerve in the wound. Rhazes said that it is better to lose the function of a limb than to lose an entire body.

Paralysis

Paralysis is a complications of wounds and contusions, especially those on the head and the back (ie and neck), as were described in *Afflicted Places,* Part 3. Of note is the fact that it may follow a wound on the same side or on the opposite side of the head. It represents a softening of nerves and the loss of sensibility and movement, whereas a convulsion is due to hardening of nerves and faulty alternating movements (see *Colliget,* Part 3.) As stated in *Afflicted Places,* Part 3, Apoplexy is a collapse of the entire body, whereas paralysis affects only one side, either right or left, and sometimes only a part, a leg or an arm. We name it a Universal Paralysis,when an entire side is affected, and a Particular Paralysis of a part.

Causes

There are extrinsic and intrinsic causes. The externals include falls, blows, incisions, dilatations, cold, aposthems, and other things from without the body which block the transmission of the vital spirit (ie arterial blood). The internal causes are thick and viscous humors that plug the nerves at the brain or in the neck, and the so-called Lieutenant of The Brain.[388] How that happens and why it eliminates sensibility and movement in many different ways are matters for fine guess-work. We surgeons need know only what we write here, and to be able to identify the harmful source in terms of the correct anatomy. A Universal Paralysis affects nerves coming from the brain, the harm is universal. A Particular Paralysis of the upper limbs comes from the neck. If it is in the legs it comes from the lower vertebrae. Paralysis in-between arises from mid-level vertebrae. The evil humor is identified by its well-known characteristics, whereas the patient himself can tell you what was the external cause.

As you are aware, paralysis, as are all maladies of nerves, is difficult to treat, and is worsened by the fact that they (the maladies) are warm in nature, which is the basis for our treatments. Older patients suffer more. Furthermore, you will note that chills and fevers are good signs in paralyses, and a paralysis after a contusion that does no serious injury to nerves may be curable, as Galen said when describing the case of Pausanius, (ibid., Book 6). However, if the nerve is badly injured (ie partly severed), there is no hope, because nerves do not heal perfectly. Avenzoar wrote that the patient will suffocate and die if the nerves for respiration are injured. When the wounded part does not lose color, there is hope for recovery. The opposite means that the paralysis will be permanent (Gordon).

---

[388] The medulla oblongata and the spinal cord (LDR).

Treatment

Treating paralyses due to internal causes is a physician's responsibility. Heben Mesûe used a two-fold system: common and specific. The common deals with the part, the illness, and sometimes with the diet. First, he carefully examined the region of the nape of the neck which is a common site of origin for five disorders: paralyses, spasms, tremors, stupor, and torticollis.[389] Then he applied topicals designed to sedate the nerves, such as acoris, ive, and castoreum. Third he limited the intake of food and beverages as in our chapter on phlegmons.

The treatments for Particular Paralyses had four elements. First restore the balance of humors. Second, reduce the amounts. Third, repercuss the residue to the opposite side. Fourth, treat the complications. Let the attending physician explain how he does it.

Avicenna improved the results of his treatments of paralyses of external origin, such as contusions that became abscesses that drained pus, using phlebotomy and applying warm and evaporative medications. They included ointments and plasters applied on the affected places, and occasionally he applied cups. My own customary treatment uses Mesûe's favorite liniment for relief of painful hearts; I apply it on the neck, back, and the wounded region.

That Angelic Doctor (Heben Mesûe), described it in *Inunction Of The Spine*. I believe that expert physicians and philosophers have buried the traditions, the memories, and the praises for one of Nature's great gifts, and they have erased it from their treatises. It is this elegant preservative of the substance of life. You know that bones and nerves begin at the neck, taking origin from the brain, and the spine is where all the arteries, nerves, and spirits and virtues are exposed and where the spiritual organs areconnected, and which contains a very moist marrow. Therefore, that is where you should direct your efforts. By comforting the substance within, that of the vital spirit, the nerves, the beating heart and the tremors, you will bestir the palsy. It works more rapidly as a sedative than any other medicine: Myrrh, hepatic aloes, spikenard, sangdragon, frankincense, mummy, opoponax, bdellium, carpobalsamum, mastic, gum Arabic, liquid and red styrax, musk, and terebinth. Grind all before adding the terebinth. Distill it in an alembic and recover the distillate in a glass jar, as it becomes a balm. Sometimes, I add herbs of paralysis to make it even better.

Syncope

Galen defined Syncope as a sudden weakening of the Vital Spirit (*Therapeutics,* Part 12) which usually follows profuse evacuations or severe pain. You will observe a weak pulse, pallor, difficulty on opening the eyelids and performing other movements, and cold sweats, especially at the neck. Syncope should not be under-rated because it may portend death, and it has been called the 'little death'.

After it has occurred, there is little you can do about except to protect the patient by shooing away the crowd of curious people who overheat the room and frighten the patient.

---

[389] Joubert's term is 'torsement'. The Montpellier Ms used 'torture', which meant 'twist' (EN).

However, if you can see it coming, give the patient some wine soaked white bread with a touch of rose-water; Galen used warm wine (idem) to hasten its spread in the body. Bestir the patient and cense his face with rose water or cold water. Slap his limbs and tug at his hair, his nose, and his ears. Shout his name and slap him, and other actions recommended by the physicians.

Delirium

In *Maladies and Symptoms,* Part 5, Galen wrote that these are defects of self control, that is to say, by the ruling spirit. Avicenna *(Canon,* Part 3) called them aberrations. Although they are real complications per se, usually they are tagged on to another, by communication after open wounds or contusions near joints (*Therapeutics*, Part 4). They resemble the responses to things that cause flushing, or after an upset stomach, or dream-like visions. And as it is in similar situations, the delirium will come sooner when the malady is nervous. Sometimes it is caused by heat alone, a wave rising to the head, or by vapoprs and fumes. Finally, they may be caused by desiccation (Avicenna). Galen also said (idem, Part 13), "As we have shown in other places, listlessness may be caused by cold, heat, restlessness, bad humors, and by madness."

The treatment that is surgical—the physician always should be called upon—is to divert the humors from the head with massage, tourniquets on the limbs and unpleasant vapors. At times, Avicenna favored clysters as part of the treatment, and mitigation of pain. Also, he said to slap the patient to bring him to reality. Galen (ibid) prescribed oxyrrhodon at the onset to repercuss the humors and vapors from the head. He also favored somniferous infusions made from poppy-seeds, and he allowed the patient to sniff aromatics, and to anoint the axillae, the nose, and the forehead with them. Avicenna said that he annointed the head with an oil made from the head and the feet of sheep, and he used bryony roots as a plaster, or had the patient eat them with foods that disguised their bad flavors.

We will discuss *Itching* in Treatises V and VI, and *Persisting Induration* after wounds and in Treatise VI on painful joints, and in the Antidotary.

CHAPTER 2

Flesh-Wounds

As Galen wrote in *Therapeutics,* Part 3, a flesh-wound occurs in a fleshy region of the body; it bleeds, and it is not putrid. A simple wound suffers no loss of tissue, but it may be deep or shallow, and large or small. The lost tissue of a compound wound may be skin alone or contain some underlying flesh. Both types of wounds may be complicated, leading to ulceration and thereby increase the problems for treatment, as we will discuss in the Treatise on Ulcers. Why that happens is described in Ibid. Part 3, as also are dyscrasias, pain, aposthems, and undermining (ie esthiomene). All of them have been discussed in the previous chapters. Although Galen made a great fuss about differential diagnoses and dispositions, he said that such niceties have little to do with cures. Although naming is not curing, we must be correct in diagnosing what we encounter.

Causes

As stated earlier, the causes may be external: anything that can perforate, fracture, or erode. We plan our treatments—the remedies and the applications—based on those elements, and on the complexion of the affected part, and on the various complications we may encounter.

Signs

These are described in the General Chapter.

The Treatments

The treatment of fresh flesh-wounds has the five elements common to all wounds. Here, special attention is given to the control of excessive blood-loss (Avicenna). A moderate 'bleed' may prevent the formation of an abscess, or the obstruction of a nerve[390], or a fever, all of which can seriously interfere with the treatment of the wound

---

[390]   See below: Convulsion and Paralysis (LDR).

itself. Hippocrates and Galen agreed, noting that a dry (ie bloodless) wound more readily suppurates. But when there is too much bleeding without natural control, we must use the measures we described for hemostasis in the chapter on wounds of veins.

In addition to our general measures there are special techniques for flesh-wounds.

A Simple Small Incised Wound Without Loss Of Tissue

Galen (idem. Part 3.) said that we should close a wound with a single suture. If we are careful to approximate the surfaces, the wound will heal nicely without need for other measures. Rhazes agreed. However, today's surgeons follow Lanfranchi, who inserted a thin cloth wet with beaten egg-white into the wound. Galen (*Simple Medicaments,* Part 11) said that can control bleeding (ie surface ooze), relieve pain, and prevent infection, not only in wounds of the eyes. Pain from any cause provokes the inflow of humors that cause suppuration. Therefore, we should mitigate it. Galen noted that his nicely closed small wounds were agglutinated in just one or two days (sometimes longer). If that was not secure, the wounds could reopen.

A Large Superficial Wound

Here a single suture will not suffice. Therefore, Galen used the method of suturing by equal spacing. Afterwards we usually apply the red suture powder made of frankincense and sangdragon. Albucasis added quick-lime, as did Lanfranchi. I use bol d'armenie instead of quick-lime, and Halyabbas used sandalwood. Take care that the powder and other foreign material such as hair and oils do not get into the wound. You may mix the powder with egg-whites and saturate one or two strips of cloth as a stupe and apply it over the closed wound and over the punctures for the sutures, lest they be loosened. You may apply oil of rosat around them for relief of local pain, hoping to prevent the influx that causes abscesses. After the suturing and the applications, do not dress the wound until the fourth day unless there is severe pain or other signs of serious complications. After the fourth day, and you see that the wound has sealed, wash it with a warm sour (ie astringent) wine, replace the wine-damp stupes. Repeat the dressing daily until the wound is solidly healed. Galen (*Against Thessalus,* Part 4) said that he needlessly prolonged the process for a month, whereas he (ie Galen) needed only six or seven days.

You may use egg-whites and powder at the first change of dressings, trying to prevent hemorrhage, to lessen pain, and the appearance of an abscess. Go to the wine dressings at the time of the second change. Galen insisted (ibid. Part 3), that wine was a good medicine for all wounds, unless there is a special counterindication. A wound, no matter how large must have dry surfaces (ie free of pus) to cohere. Wine serves two purposes: The major benefit is to wash away any wet (ie pus) residue and to repress further influx into unsealed gaps which also must be dry.For that purpose the consolidatives medicines should be more desiccative than incarnative, up to the second degree. The second benefit derives from the new wine's warmth in the first degree, and older wine's in third degree, and middle-aged wine is in between in the second degree of wamth; the degrees of dryness are parallel with those of warmth (*Simple Medicaments,* Part 8). As Theodoric noted, the effects on drying and consolidation vary from wound to wound, but they do not humidify or cool. That is what Galen cited from Hippocrates (*Therapeutics,* Part 4), who used wine to irrigate ulcers, saying that dryness is good and wetness (ie pus) is unhealthy.

Master Arnold (ie of Villanova) said that fresh wounds washed with hot eau de vie (ie brandy) being dry, get an early start toward healing. After the initial dressing, some surgeons apply ointments and plasters, and lay on stupes. Galen liked the black plaster (fuscum); Avicenna preferred a linseed ointment. I prefer an ointment made from the red powder and washed terebinth. Other formulations are described in the Antidotary.

A Deep Wound with Recesses

Such wounds can be treated with sutures and bandages. Galen advised us (*Techni,* Part 3) to use properly placed counter-incisions to drain whatever fluids that may collect. Avicenna added that the counter-incision should be placed at the low point of the wound to favor the escape. When discussing deep wounds in the thigh, Galen (*Glaucon,* Part 2) described a deep wound into the thigh extending to the knee that he drained through the primary wound rather than make a counter-incision near the knee. He obtained dependent drainage by elevating the limb. The same scheme can be used for wounds in the arm. However, if drainage through the wound is not feasible, make the counter-incision, as did Galen. If the matter to be eliminated is liquid, do as described above. But, if not, try to evacuate the matter by wrappings to express the contents of the wound. When more is needed, devise measures to befit the case. Sometimes you may have to lay open the entire cavity, or you will simply open its lowest part. Your choices will depend on the anatomy and the size of the wound. When you are uncertain where to incise, and the wound is large, make the counter-incision or simply enlarge the wound and use compression bandages. Bruno placed drains in the counter-incision and in the primary wound. I use a long strip of cotton as a seton through both openings. That favors a better cleanout with less pain than with pack tents. I use an introducer, like a large threaded needle and I insert a flat wooden stylus into the depths to identify (ie by palpation) where to makethe counter-incision. I apply detergents on the cloth strips, twice daily.

Excavated Wounds With Loss Of Tissues

There are two elements in the treatments because there are two defects: the wound itself and the cavity (Galen, *Techni,* Part 2) that must be filled with new tissue. The idiots call the treatment a repletion of flesh, because it is a flesh-wound. The problem for the true artificer is how to bring it about, using his intelligence, his recognition of local cues, and a good plan of action. There are four local indications that lead to the plan: First is the essence of the wound. Second is the nature of the body. Third are the conjoint matters. Fourth is determining the contraries.

The first concerns the generation of proud flesh from congealed blood, a natural process involving two superfluidities: one is thick (ie the clot) and the other is liquid (ie the serum). Therefore, we need two medications to get rid of the excesses. One will dry the wet surfaces and clean away unwanted debris, not a potent detergent, but a mild one in the first degree. Too strong a detergent will consume and congeal the matter that will become the new flesh. Use frankincense with the flours of barley, beans, ers and iris, aristolochia, cadmia, panax, and terra sigillata. Those medications differ only in degree[391]; aristolochia and panax are stronger desiccants and are warmer than the others; the flours of barley and beans are much weaker desiccants and less warm. Frankincense is moderately warm and a weaker desiccant. The flours of ers and iris are midway in degree between aristolochium and panax.

The second element assesses the qualities of the body and its parts: are they more or less warm or dry or humid. What is Natural should be preserved and what are contrary to nature and should be rejected. That which we regard as Natural should not be destroyed by contraries. Natural warmth should be supported by warm remedies, and cool ones by cool. That is, to restore the loss, we must provide a surplus of nutrition so the new tissue will resemble the original as closely as possible. Try to restore what formerly had been very dry flesh with very dry new flesh. For humid matter the opposite is true. In such a case, frankincense will engender dry tissue in naturally moist tissues, and will dry the pus amid natural moisture.

The third intention deals with annexes—the conjoint matter. First determine the complexion of non-natural matter. For example: If by chance or simply by delay, the wound has become an ulcer, and the wounded flesh has become too warm or too cool, you will need more than a desiccant, something that cools or warms to a degree exactly enough to restore normalcy. In so doing, you will have to consider the temperature of the remedy as it affects the abnormal excess. That is why Hippocrates used cooler medicines in warm seasons and warmer ones in winter, always conserving the natural degrees.

The fourth intention deals with the simultaneous use of contraries in groups, not sequentially. Consider a patient with too humid a complexion, but the wounded part is too dry and the wound is too wet, and the annex is too dry. You must choose between a desiccant in the second degree and another in the third or you may need to apply one in the first degree. And you should be aware that the malady may have a very different disposition (ie complexion) from that of the affected region, or they may be nearly alike. All of those assessments represent conjectures, and so too will be your choices of

---

[391]   Degree of warmth: Again we remind the reader that this term does not indicate the actual temperature of a substance, but is a measure of its potency. All medications were classified as to warmth, coolness, astringency, corrosiveness, etc. The scale was established in early times, and came to be accepted through later centuries. The degree of a medication determined its property as a 'contrary'. Galen insisted that a balance of contraries must be obtained if the treatments are to be effective (LDR).

medications. That is the essence of good medical practice as it affects complexions and medications. It is apparent that a person who undertakes to treat a wound (ie ulcer) must strive for correct evaluations of many things. Although every one knows that you should use a desiccant to treat a dry wound, be it warmer or cooler. Which one will serve better in the treatment of the many varieties of ulcers and affect the nature of a certain patient or an annex. Thessalus failed in those matters, as do all the 'Thessalizers' who believe they can treat everybody in the same way. They are like the cobbler who makes the same size shoe to fit everyone, as described in *Therapeutics,* Part 13., and in *Preservation of Health,* Ch. 11.

The usual methods for treating these wounds—after arresting hemorrhage and mitigating pain and its attraction of the matter that can form aposthems—is to irrigate the wound with a warm mildly astringent wine. Then apply a powder or an ointment known for its regenerative effects. Then fill the defect with pads or charpie overlaid with a plaster of regeneratives, all of which are described at greater length in the Antidotary. Cover all with stupes, dry or moistened with wine, and wrap with a bandage. Dress twice daily during the summer, and once a day during the winter.

Wounds With Loss Of Skin

Galen (*Techni,*Part 3) said to add this treatment after an excavated wound or ulcer have been filled with proud flesh. When the new tissue overfills the defect, the skin must grow over the top from the edges and that may not be possible if the edges are too hard (ie scar); something like skin may suffice.[392] That film will form a scar when treated with desiccants and astringents that do not erode, yet are potent in the third degree.

The desiccants must be in the third degree, but the incarnatives and agglutinatives should have only first degree dryness. The cicatrizers must be most potent of all (third degree), because they must dry the natural moisture of the proud flesh as well as the discharges (pus). The result will be a callus-like surface, as if it is skin. That process is abetted by cool and dry astringents, such as the juices squeezed from green galls, pomegranate-peels, and the fruits of Egyptian spines (ie spica), and others to be described. Real scar will not be affected by applications of chalcathum, alum, copper, copper-flower, vitriol, et al. if used only in small amount of combusted and well-washed substances.[393]

Scar

Bad scars may be improved with bits of diachylon or litharge. Excessive scarring can be abated with oil of balsam, or can be shaved off with a razor, or burned off with a cautery, followed with applications of fats from grouses or ducks, with added mastic.

---

[392]   The dry film atop the exposed granulation tissue (LDR).

[393]   The surgeon wants to skim off the film to allow growth of true scar to cover the wound (LDR)

Superfluous Granulation Tissue

Galen (*Therapeutics,* Part 3) told us how to get rid of excessive overgrowth of proud flesh. It will not go away on its own; medications are necessary. On the other hand, the agglutination and the condensation (ie from blood clot) of proud flesh are natural processes, perhaps abetted by medications. We use potent desiccatives to get rid of the excess: ink, couperose, vitriol, sponge, roots of asphodel, hermodactyl (all on stupes cut to size), alum, green ointment, etc.

Contused Wounds That Are Altered By Air, And Are Painful

and Engender Aposthems

Treat all contusions (ie hematomas) with phlebotomy, purges, and a healthy regimen. You know that the above complications attract humors to the site of the injury, and lead to ulceration (ie delayed healing).

The local measures in addition to the general should avoid consolidatives and desiccatives. Apply repercussives on the region around the contusion itself. Use oils of roses and myrtle, or an ointment of bol d'armenie, oil, and vinegar. Apply mitigative oils, softeners, and maturatives over the bruise itself.[394] If also there is a wound (an arrow or a knife), immediately apply suppuratives and hope to avoid inflammation. The damaged tissue must be cleaned away before healthy granulation tissue can form. So it is, that our treatment has two elements.

First we use such warm and moist maturatives as boiled mauve, roots of guimauve, the tetrapharmacon, wheat bread, and other applications for treating aposthems, as we describe them in our Antidotary. Place drains in such wounds, moistened with rosat-honey or the apostles' ointment. Atop that lay a detergent plaster, then dry cloths, then a bandage that wraps all. Repeat until the cleanout is complete.

Second we engender flesh as it accumulates behind the gradually withdrawn drains, allowing it to proliferate and consolidate. If the initial wound gaped and you used sutures, be sure to tie them loosely, barely to bring the skin edges together. When a complication occurs or the wound stinks, reopen the wound. Later on, resuture it.

A Contusion (Hematoma) With No External Opening

Galen called it an Ecchymosis (*Therapeutics,* Part 4). Avicenna (*Canon,* Part 1, Fen 4) said that it was a kind of disruption of continuity, and dealt with it as such. Therefore, a contusion is a disruption of continuity, a tearing apart within muscular tissues, often

---

[394]   The 'contusion' is manifest as a hematoma and edema. The surgeon want s to keep the matter in a liquid state, the better to disperse it or drain it—including causing it to suppurate so he can drain it (LDR).

a rupture. It is painful and it distends with blood and often becomes an abscess. On the surface it may appear as a discoloration, and sometimes as excoriations. A large hematoma puts a limb at risk for corruption, and the rest of the body shares the risk. You may note that the skin alone is elevated and an indentation is slow to refill. That will lead you to open it with a knife or with corrosives. Leave open the drainage wound, because exposure to air will help the recovery (Avicenna).

The Treatment of Contusions (Hematomas)

There are as many methods as there are the multitude of injuries.

First: Repress the inflow of blood and evacuate what is there. Avicenna and others said that there is no reason not to bleed, even when the body is feeble.

Second: Relieve pain by applying cool things and astringents. Rhazes and Lanfranchi preferred an ointment of rosat-oil, and they dusted the region with a powder of myrtle. Then they wrapped the limb with a moderately tight bandage. At the outset in all cases we apply egg-whites and rosat-oil (to relieve pain).

Third: After above, try to resolve the collection while it still is liquid and is close to the surface. If the resolutives are ineffectual, use cups and scarification, and treat as we do abscesses. The familiar resolutives are wine, honey, and salt. Go on to barley-meal, calamint and wine. If more is needed, use cumin and wax, or go to camomille flowers, melilot, stoechas, and cumin boiled in wine. Or even stronger use mauves, bran, absinth, and cumin or aneth boiled in wine. A sixth resolutive is barley-meal, fenugreek, saffron, and a bit of orpiment, all boiled in a water with calamint and salt.

Certain beverages can help resolve suggilated blood: bdellium, costus, centaurea, syrup of vinegar, et al.

We will discuss falls and blows when we also describe lividity and other complications.

Bites, Venomous And Other

We will devote little time to these uncommon surgical problems. Furthermore, when they occur, a surgeon rarely is called; the common folk use folk remedies: garlic, onions, and oil. Nevertheless, we should learn the details by reading the treatises of Avicenna, Rhazes, Rabbi Moses (Maimonides), and Henri. The poisons themselves are topics for the physicians. The surgeons deal with the wounds.

There are two kinds of bites: non-venomous and venomous. The non-venomous kinds usually are bites by dogs, hogs, horses, fleas, bees, lice et al. The venomous kind may be caused by a mad dog, a lizard, a serpent, a scorpion, honey-bees,etal. All of them as wounds, are treated alike, with exceptions for particular kinds. None should receive desiccants, or repercussives, because they require attractives, softeners, detergents and incarnifiers. The non-venomous kind need attractives and maturatives, such as onions and garlic (wild and common) heated and mashed before mixing with yeast, oil, and salt. But the venomous bites, as indicated by their aching and biting pain, changed color, agitation, local heat, and numbness, all of which should warn you of the danger before the poison reaches the heart. However, a rapid evolution is not typical of mad-dog bites. As Gordon reported, the onset of hydrophia (fear of water) may be delayed a month or more, or even a year. He described one case that was delayed for seven years. Yet once hydrophobia appears, it is incurable.

Galen,(ibid. Part 3), wrote this, concerning the treatment of venomous bites, "When a bite by a wild beast causes penetrating and biting pain, apply two kinds of treatments. First, mitigate the pain; second, get rid of the venom, and modify its harmful character. Evacuate with things that are warm, and by others are potent attractives, as cups and cones." Some surgeons suck at the wound. The cautery and corrosives may help by coagulating. Use every means at hand for evacuation.

A different sort of topical acts as a contrary as well as an attractive and antidote. First apply a plaster of galbanum, opoponax, asa foetida, myrrh, peppers, sulfur, calamint, menthastrum, droppings of doves and ducks, wine, honey, and stale oil. Some create suction by pulling a feather from the anal region of a grouse or another bird, and set its anus over the bite. If the bird dies, you know that the venom has been attracted.

Second: use an ointment of Master Dino: wax, black tar, resin, lanolin, stale oil, and galbanum. That ointment also relieves pain.

CHAPTER 3

Hemorrhage From Wounded Veins And Arteries

Having dealt with flesh-wounds, now we will go to Galen's treatments for wounds in veins and arteries in *Therapeutics,* Part 5.

A wound in a large vessel causes a major hemorrhage. Needless to say again, bleeding is our first concern when treating any wound; the wound itself is a secondary matter. The blood comes from an opening in the wall of a vessel or it comes from the ends after it has been transected. Much blood also may seep through openings in many small vessels. The latter usually is treated by physicians, whereas the former requires a surgeon.

The vessel may be an artery or a vein, one or several, large or small, from a simple wound or when some tissue has been lost. Sometimes the vessel has been corroded. Hemorrhage is a potential risk when a buried arrow must be removed. Sometimes the cause is apparent, as from corrosion, or from an embedded arrow. Those variants determine our methods of treatment.

Causes

The causes here are the same as those for any wound: those that slice or pierce (ie swords or arrows) or are corrosive. A hard blow with a heavy object may injure vessels. The corrosives (ie intrinsic) derive from incisive humors: from choler or combusted melancholy[395]; they can damage vessels. Therefore, a vessel can be injured only when the surrounding tissues or the overlying skin have been wounded, and that injury (the damaged tissue) must be treated after the hemorrhage has been arrested.

Signs

Arterial wounds bleed in spurts and produce thin pink blood. Venous bleeding is sluggish and the blood is dark red. However, all hemorrhage threatens life, and can cause fainting, convulsions, delirium, and hiccoughs, all of which are bad signs (Avicenna).

---

[395] To explain a rupture or erosion of a vessel when no external cause is noted (LDR).

Master Anold said that a transected artery can be controlled more easily than one wounded in its long axis.

Treatments

Galen said that there are two elements, and Avicenna added a third, stopping the bleeding by diverting the flow, along with the general and local measures.

The general diversions are more effective for controlling venous bleeding. They include applications of dry cups with flames, massage, bandaging tightly over the bleeding region, loosening as you wrap over adjacent normal regions. Other diversions are by evacuation: phlebotomy on the opposite side, as from an opposite arm, or from the same diameter[396], as from a right foot to divert from the right side of the head. Those were the prescriptions of Hippocrates as transmitted by Galen, and were his rules for phlebotomy in general.

For systemics, we use such blood-thickeners as lentils, rice, jujubes, quince and all astringent fruits, as we use them for treating diarrhea. Others are stupefacients: cold water to drink and to dowse the region, avoiding the wound itself (*Aphorism 5*). All very cold things are useful. When fainting occurs, the outflow of blood will decrease as the body chills, and its blood is retained.

Avicenna described eight local measures; I have condensed them to five: by suture, by packs, by excision of the vein, by ligation, and by drying.

1.   Suturing is useful where there has been no loss of tissue. Empty the wound of clots, etc, bring the wound-surfaces together with your hands and sew them with the continuous furriers' stitch. When the bleeding is vigorous, include enough soft tissues to control it. Apply an astringent and cooling powder, and stupes wet with the same powder mixed with egg-whites. Then bandage. In that way we both approximate the wound-surfaces and cool the region, as was Galen's goal (idem). Avicenna (Part4) said that often you may have to control vessels within the wound with deeply placed sutures which bring together the surfaces and supplement the bandages and pads that compress the veins. Theodoric and Henri agreed, despite Galen's admonition against suturing the openings in vessels and intestines, because they are bloodless membranes and cannot unite cut edges. However, although we all revere him, I claim that Galen did not forbid such. We can read his text: he said that sutures in those tissues alone (arteries and veins) will not heal, but if they are united by suturing the wound, as we do with the peritoneum (ie when suturing the abdominal wall), they will heal by second intention.
2.   Packs are used to control bleeding in wounds that were deprived of tissues. Apply astringent powder in the wound and on the pads and stupes and bandage them snugly in place. Galen testified to his successes, "The opening in the vein will be plugged with clot and compressed by the medicated pads."

---

[396]   The body is divided in two sides and in upper and lower halves (LDR).

3. The complete division of a vein is a means of control of bleeding deep in a wound. The two ends can retract and be compressed by the overlying flesh and skin. Apply your powder and medicated strips on top.

4. Ligation of vessels deep in wounds is a good way to control bleeding arteries. Avicenna exposed the vessel, caught it with a hook, encircled it with a silk thread and tied it. Then he applied incarnatives and bandaged them in place. Galen said that he tied the base of the bleeding vessel, where it entered the wound, on the side where it came from the heart and liver; that is, on the lower side of a wound in the neck, and the upper side of a wound in an arm or thigh. He applied incarnatives and let the threads come away later, on their own. If the flesh does not promptly grow over the opening in the artery, an aneurysm will form.

5. Cauterization with a hot iron is good way to close veins that have been eroded. Also, you may use caustic medications that are hot and astringent. Do not use quick-lime; it is not a good astringent, and its eschars often detach too soon. Use astringents that bind to the tissues and remain as lids until the vein is sealed from within; do not hasten the separation of the eschar and risk the recurrence of the hemorrhage when it will be even more difficult to arrest. Avicenna insisted on applying a hot iron until the char was thick and extended deeply and was not easily separated. Theodoric favored sublimated arsenic as his caustic because it stopped the flow with a thick and firm crust.

Another special technique is useful when you anticipate bleeding when you extract an object (ie an arrow) fixed in the tissues. Perforate the center of several pads and stupes. Wet them with your hemostatic medicines and thread them down the shaft that extends from the wound. When you pull out the arrow push the pads onto the wound and lay others, not perforated. atop them for compression.

Supplementary Treatments

The following are instructions for the surgeon who treats damaged vessels.

1. In *Therapeutics*, Part 5, Galen described how to quickly stopper a bleeding vein by placing the ball of a finger on the opening, an effective and painless maneuver. While you keep it in place, the blood within can coagulate and plug the hole, and other clots within the wound will lend support.

2. When blood continues to ooze after your applications of astringent powders, lay on stupes moistened with vinegar and water. Then add more powder and wrap a snug bandage.

3. Avicenna discussed bandages. Use linen. Take the initial five or six turns over the bleeding vessel. Then loosen the succeeding turns as you cover normal tissues. That allows the matter to be repercussed.

4. The limb should be properly positioned: an important supplement for all the other remedies. Elevating it will abate the pain. When the limb is lower than the rest of the body the humors cannot flow out, and they remain as a source of inflammation.

5. Galen said not to disturb the initial dressing for three or four days. Then carefully remove all of it: moisten them for several hours, using the original medicines or with beaten egg-whites and oil, or with wine. That will ease the removal of the pads and stupes.

6. Cover the patient's eyes or work in a dimly lit room so he will not see his blood or other red things, while you continue to tell him that there is little or no further bleeding. That will support him and boost his spirits, and relieve his fears.

Finally, I will describe the astringent medicines that Galen mentioned (ibid.): Mix pulverized frankincense and aloes with enough egg-whites to reach the consistency of honey. Finely minced rabbits'-hair (pili leporis) and apply on the bleeding vessel, or in the wound (ulcer). I use it in several ways: mixed with aloes and more frankincense for soft bodies, in equal amounts for hard.

A second medication comes from Avicenna: bol d'armenie, sangdragon, frankincense, and socotrin aloes.

A third is from Bruno, who took it from Rhazes (*Book of Divisions*) and Albucasis, as a hemostatic: quick-lime, sangdragon, plaster, aloes, frankincense, vitriol, spider webs (ie airagne), and egg-whites. Halyabbas highly recommended burnt oak-galls mixed with vinegar. Roger use powders of combusted consolida.

Treatments

Now we will treat the wound, after the hemorrhage has been controlled. Galen said that a wounded artery is more difficult to treat than a vein, and a vein more difficult than a flesh-wound. The general treatments for all three are similar, differing only in the amounts. All other things being equal, wounded arteries need more desiccants than veins, and veins more than flesh. When there has been a loss of substance, you should use more of the consolidative medicines than in the treatment of simple wounds. The excavation may be result of the wounding, or it may appear when a large eschar comes away after a cauterization, or when sutures slough away from a ligated vessel. In all cases use the medications listed for treating excavated ulcers.

CHAPTER 4

Wounds Of Nerves, Cords, Ligaments[397]

According to Avicenna (Part 4), wounds of nervous structures can be punctures, cuts, breaks, or concussion. Some punctures are closed (ie into the nerve) or open (ie through it). Incisions may in the long axis or across it; Some are exposed in excavated wounds, in others there has been no loss of tissue. Some are very painful, others develop abscesses and cause spasms. Some are bland. All of this influences our treatments.

Causes

All the instruments that can cause wounds can injure nerves, and all nerve-wounds are part of flesh and skin-wounds, and they share the problems of injured veins and the attendant hemorrhage and other complications.

The Signs

The severe pain and the anatomy, and the loss of sensibility and motion all indicate that a nerve has been wounded.

Galen (*Techni,* Part 3) said that wounds of large nerves and tendons are painful because they are so sensitive and because they are connected to the brain. They cause aposthems, convulsions, and delirium. In such cases (*Aphorism 5)* a transient soft swelling is unfavorable and a soft swelling is good, whereas hardness is bad. Galen (*Therapeutics,* Part 6) said that a partial division of a nerve was more serious than a complete transection, because the former still can transmit bad matter to the brain. On the other hand, when a nerve is transected, there is complete loss of sensibitiy and motion, as we described in the section on spasms. Also, he reminded us that nervous tissues in wounds are more sensitive to cold than is the flesh.

He wrote (ibid., Part 6), ligaments and joint capsules are similar to nerves and tendons and they are to be treated as such, differing only in the required amounts. Ligaments need larger

---

[397] Nerves, tendons, ligaments and thickenings in joint-capsules all were considered to be 'nervous' by Galen and all subsequent writers. They did not discriminate in their names. A Reader must determine the structure in the context, as to its type (EN).

amounts and more potent medications, almost as much as bone. Those that enter muscles, less dangerous than the tendons and nerves, are more at risk if they are not treated properly.

Treatments

The general measures are the same as those used in treating all wounds, except that here the matter of immediate importance is the pain. Yet we should not interrupt the other remedies while we relieve the pain. Therefore, first remove all foreign matter. Second, close the wound. Third Apply contraries. Fourth, prevent further damage (ie loss of tissue).

Let us begin with a puncture wound in a nerve; there is no large wound to close and to preserve tissue. Remove objects like an a arrow fixed in the wound, as described in previous chapters. Apply remedies other than sedatives to prevent the formation of an aposthem (ie abscess) that could lead to spasms, and deal with items three and four, above.

First: Deal with general health. Second eliminate the antecedent matter that is the cause for the pain. Third prevent spasm. Those three items are the elements of treating all nerve-wounds. A fourth measure is particular to puncture wounds. Get rid of whatever can allow air into the puncture in the nerve, and thereby mitigate the pain.

1. The regimen for nerve-wounds includes a completely liquid diet (as used for other wounds). A soft bed in a humid environment will favor comfortable rest. (Galen)
2. Prevent any superfluidity from accumulating in the body. Use phlebotomy from the opposite side when the patient is robust, and medications when the patient is feeble.
3. Treat spasms as we described earlier when dealing with wounds of the head, neck, and back. Use such comforters as warm oils of lilies or olives (Galen, Ibid. Part 6). Halyabbas, and Avicenna Avicenna treated incipient aposthems with plasters of minerals and vinegar, but not suppuratives, or water as used in phlegmons. Suppuratives are bad for nerves. These are discussed in Part 3 of *Types of Medication,* some of them are not in our Antidotary: The Plaster: Chalcathum, tragacanth (Joubert used vitriol), frankincense, galbanum, wax, oil, and strong vinegar. Mix, grind, and let stand for four days. Then melt and cool, and place it in an earthenware jug. Repeat until the mixture is uniformly smooth I have not used it, but I respect its source. I prefer Galen's and Avicenna's plaster of barley-meal, bean-flour, and ers cooked in a lye-water. Avicenna added honey and vinegar.
4. Galen (ibid. Part 6.) used many medications to dilate the puncture wound to gain access to foul collections. First he enlarged the opening in the skin with a knife or a cautery (which Henri preferred). Then he used desiccants to dry the pus (*Techni,* Part 3), which mostly are liquid and can reach the perforated nerve at the bottom. Then he introduced warm oil of sambucus (not rosat or myrtle, as many do) because it does not plug the track and prevent resolution as do the other medications. I repeat, avoid cool medications that are the enemies of nerves. Another, favored by Avicenna, is made of resin and terebinth. for use in children and women and other adults with soft flesh. He added small amounts of euphorbia when the flesh was more firm.

After the above, we use an ointment prepared in advance: wax, terebinth, pitch, and euphorbia. Also, in more muscular (harder) patients I use propolis (ie bee-wax) with euphorbia, serapina, and opoponax, after I soften the region a bit with oil and terebinth. I have seen melted sulfur—alone it is too liquid to be useful for nerve-wounds—be effective when mixed with warm oil to make it sticky. I have used it, as did Avenzoar. Washed lime can be used on denuded nerves. You can make stupes with all these medicaments. Hold them in place with bandages.

For incised nerves, in addition to the initial three general measures, we use three or four others for special purposes. First, when there is no loss of flesh, we support the nerve by suturing the flesh-wound. Second, we insert a drain at the lower end of the sutured wound. Third, we apply sedatives and incarnatives directly over the sutured wound, medicines suitable for nerves. Fourth, we bandage over all after laying on a pad of soft lamb's wool.

Here are a few words about the most useful method of suturing. A full thickness careful approximatiuon of the skin and the underlying soft tissues will cover the nerves and protect them from dangerous exposure to cool air. Avicenna added this recommenation: Suture a transected nerve, end-to-end before closing the wound., claiming that the nerve will not heal unless that is done. William of Saliceto and Lanfranchi agreed, opposing the opinions of many authors who cited Galen's advice not to sew the nerve, claiming that he believed that puncturing the nerve with the suturing needle can induce a convulsion. In spite of that common view, I say that Galen did not forbid it, and indeed he did it. In *Therapeutics,* Part 13, he wrote, "There is no danger in sewing a transected nerve, but you may damage some of its substance. But it will heal, as do other sutured wounds. etc. Certainly other wounds are sutured to bring together the separated parts." Also, he said in *Techni,* Part 3, not to treat nerve-wounds differently from others, excepting puncture wounds. In Part 6 *Therapeutics* he discussed denuded nerves, transected or not, and their atrophy, and he discussed their healing in injuries to the abdominal wall, when the divided nerves are brought together while the other tissues are united by sutures. He discussed the folly of the argument that suturing cause pain and damage by puncturing a nerve that already has been cut through. Furthermore, they argue that a sutured nerve cannot function because it cannot heal by first intention. If one argues that it is to no avail to repair a transected nerve if it will heal only by second intention,because the union produced by foreign (ie scar) tissue which blocks the pores through which flow the spirits, and the numbness and the paralysis are permanent, I will reply. There is double benefit from the repair. In infants the healing may be complete and the losses will only be partial. The same is true in young people when the nerve-ends are brought together, and the amount of scar is small. The sensory loss will recover to a degree and the appearance of the limb is more normal. I can affirm my own experience with several cases of transected nerves that were repaired with sutures and other remedies; people doubted their own eyes in disbelief that the nerves had been cut.

Do not fail to insert a soft drain. If omitted, the pus that forms will collect near the nerve and will corrupt it. Halyabbas added' "when nerves are part of a wound, do not hasten to approximate the tissues around them until they are completely incarnified. (ie leave no pockets in which pus can form), during the several days during which abscesses and spasms are more likely to appear." But, Halyabbas said that the precaution pertained only to deeply incised (ie narrow) wounds.

The green ointment is useful here: cynoglossus, plantains, piloselle, two consolidas (major and minor), earthworm-oil, white wine. Grind and let stand to macerate for seven days before adding ram-fat, black tar, resin, ammoniac, galbanum and vinegar. Boil off the wine and vinegar, and when it cools, add terebinth, mastic, sarcocolla, and saffron. Apply this precious ointment with a spatula.

Roger added some mille-fuille and so did Lanfranchi, after suturing the wound, fomenting the sutured wound for two days with rosat oil in which he had cooked earthworms. Then he applied the suture powder enriched with dried earthworms. Galen had written (*Simple Medications,* Part 11). "Chopped long slim earthworms are wonderful for treating cut nerves." In Part 7 he said that they were effective in agglutinating large wounds when other consolidatives failed. Avicenna had seen it work, and he added centaurea heated with chopped meat. The Germans at Prague encase the entire limb in a glossocome[398] as we use in treating fractures to discourage movement while the nerve heals (See Ch. 5).

Denuded Nerves

When a nerve is fully exposed, do not use any of the medicines that contain euphorbia or other acerbic substances. The exposed nerve cannot tolerate them it unless it is protected by the overlying skin. Instead use washed lime diluted with oil. Pompholyx (ie tuthie) also is given after it is washed and dissolved in rosat oil. Use fresh water for washing your medicines during the summer. All metallic substances should be washed to eliminate their erosive qualities when you use them as desiccants. If the patient is robust and the wound is very purulent, but the body as a whole is not edematous, you may use more potent medications. Sometimes I use dilute troches of polypode in syre, which is called hepsenna in Asia. Here we call it mulled wine, diluted with warm water.[399] I moisten pads with it. Also, you should irrigate the wound to wash away any pollutants. Use lanolin in warm wine, but avoid water and oil. Water corrupts nerves and the oil leaves smudges. Oil should not be used directly on a nerve except when it is covered by skin.[400]

---

[398] A glossocome was a bulky wrapping that sometimes included flat and cylindrical splints to support the hands and feet (LDR).

[399] Sirio is sapa, a syrup of raisins (EN).

[400] "Covered by skin": This remark will remind the Reader that the surgeon was confident that his 'topicals' applied on the surface of the body affected what lay beneath the surface, including the major internal organs. The benefits of oil were obtained whereas the ill effects of oil on nerves were blocked (LDR).

When you need detergemts, use the ointment of earthworms with honey, terebinth, barley-meal, and bean flour. Or use ointments of resin, and others which are described in the Antidotary. Roland and Roger said that touching a hot cautery on the cut ends of a severed nerve, with care to avoid the surrounding tissues, will improve their healing. That is what the French surgeons now do.

Compression Or Concussion Of Nerves

When nerves have been concussed and the overlying skin has been mashed, special medicines should be used, desiccants and astringents. Begin with rosat oil and egg-whites to mitigate the pain. Then apply an astringent wine. If the blow did not damage the skin, foment the region repeatedly with warm oil that will resolve the bruise (hematoma). In Galen's experience with gladiators he observed two cases successfully treated with cataplasms of oxymel with bean-flour alone. When a concussion was very painful (ie due to the nerve-injury), he applied a liquid tar. That was Lanfranchi's medication for treating bruised feet and twisted hands.[401] I use the same. When you need a more potent desiccative, add the flour of ers, or, even stronger, illyrican iris. As to the life-style and the use of evacuants in these patients, the usual measures are taken.

---

[401] Certainly they were victims of torture (LDR).

CHAPTER 5

Wounds Of Bone And Cartilage

G alen wrote that all disruptions of continuity of bone were called καταγμι in Greek *(Therapeutics,* Part 6). In Latin they are Fractures. A fracture may occur with or without an open wound. A slice with a sword may cut into or all the way through a bone. It is obvious that a bone cannot be injured without a wound in the surrounding tissues, and that may cause hemorrhage and other complications.[402]

Signs

The diagnosis is obvious. Galen (**Techni,** Part 3) said that a fracture cannot heal by first intention but almost always by second intention, because the union is held by callus (ie porus sarcoidus) that binds the opening. The exceptions are in infants in whom healing by first intention may occur, as is well known. Besides, Hippocrates (*Aphorism 7*) said that a bone that is stripped bare may be the victim of a bad erysipelas, fortunately a rare event. (Galen, *Commentary).*

A fracture (see fn.) hurts more during a cold night, and that may lead you to suspect the diagnosis. According to Roger and Lanfranchi, when a sword slices completely through a humerus, femur, or the two bones in the arm or leg, and releases marrow, the risk is grave, perhaps leading to gangrene of the entire limb, because such a wound will divide the veins, arteries, and nerves that give life to the part. On the other hand, William of Saliceto, who rarely questioned Avicenna (Part 4), said. "He (Avicenna) wrote that entry into a marrow cavity will be lethal, and treatment will be to no avail, because the marrow is soft and viscous and cannot be cut away. Avicenna meant fractures without open wounds when the marrow can ooze because it is viscous. But who can doubt that marrow cannot be cut when a bone is completely transected by a sword? And who can doubt that loss of some marrow is of itself not fatal?" I agree with him. Albucasis reported the survival of a thirty-year-old man who suffered loss of marrow from a wounded femur, and he is not the only one. A limb will die (ie in such injuries) because its sources of vitality are divided, as we stated in our General

---

[402]   Guy defines a fracture as a wound into the shaft of a bone (LDR).

Chapter. It is as if a vital organ is cut off. So much of the limb is cut that survival is impossible (*Techni,* Part 3).

Be careful not to manhandle the limb and tear fragments of bone. That can lead to a fistula (ie osteomyelitis), spasms, fever, et al. (Avicenna, Part 4). It is better to wait before removing them, and help Nature with applications of medicines. Exercise patience, as we described in procedures for removing arrows; avoid violence.

Treatments[403]

The usual four elements of the treatment of wounds are modified as they apply to bones. First remove foreign matter, such as arrows, bits of bone, etc. Then gather the larger fragments attached to soft tissues (ie shape them) and suture the wound. in depth. Second, apply suitable topicals. Third wrap bandages in such a way that you can expose the wound for applications and yet not unwrap the entire limb. Fourth, after the early period (ie about one week) when the wound is at risk for abscesses, apply the medications that favor the development of callus.

1. Closing the Wound: Suturing the soft tissues will bring together the bony elements and protect them from further exposure to air. Galen (*Therapeutics,* Part 6) said what is written above. The sutures supplement bandages and splints in holding the fractured fragments together. Galen cited Hippocrates who used medications suited to bleeding wounds, but he (Galen himself) thought that sutures and bandages were the best treatment. Avicenna, Halyabbas, Albucasis, William of Saliceto, and Henri all agreed and followed that treatrment for compund fractures. Only Lanfranchi varied; his general rule was not to close unless the fracture was properly aligned, to hold until the union by callus was complete. But can the repair of the bones occur without the support from the surrounding tissues? I can say only that callus is supported by nutrition that comes from the stomach and the liver, carried by veins into the flesh, and from the flesh into the bones, every part serving its normal functions. You should favor the formation of new flesh to fill the wound and to cause it become firm callus, assisted by applications of desiccatives. Without that tough tissue, the gaps between the bony fragments will never heal (ibid. Part 2.).

2. Fulfill the second goal by inserting a drain through the lower end of the sutured wound. That will allow you to clean out bits of bone and other matter, and pus, if any will form. At first apply our powders on the sutures and add egg-white, every day. Later apply incarnatives, adding terebinth to the powder, and others as you see the need. Keep the drain soft with rosat and honey, and add myrrh to the powder. Avicenna said that will protect denuded bone. If the bone lies exposed, cover it

---

[403] Here Guy treats wounds. Special treatments of fractures are deferred until Treatise V (LDR).

with charpie pads with the powder. Then lay on a plaster (see below), and stupes moistened with wine.

3. When the bone had been transected, and after the wound has been closed, wrap the entire limb in such a way to leave the wound exposed. Use a long two-ply cloth previously saturated with egg-whites and astringent wine. Begin at the lower part of the limb, void the space over the wound, and continue upward toward the root of the limb. Then reverse the wrap and go back to the other end. Then apply two or three very smooth splints along the limb, but not over the wound. Bandage all. Leave all of it in place until the bone heals, unless severe pain or undermining infection or an abscess intervene. In most cases, dress the wound daily (through the window) with medicated pads and stupes. Clean it, dry it, etc. as for all wounds. In cases of wounded nerves, apply a glossocome with strips of wood to splint a drooping hand or foot.

4. We will discuss fractures in the designated chapters.

The induration that appears after fractures will be dealt with in the treatments for gout, and painful joints, and in the antidotary.

Here Ends Doctrine I of Treatise III

DOCTRINE II WOUNDS IN PARTICULAR PLACES AND ORGANS: EIGHT CHAPTERS

CHAPTER 1

Wounds Of The Head

A lthough Galen and Avicenna had provided fine descriptions of the treatments of wounds of the head and the abdominal wall, they left much to be said about other wounds, as to bandaging, suturing, the selection of instruments and the roles of medications. Galen (*Therapeutics,* Part 5) said that those matters depend on the nature of the structures, just as our choice of desiccatives depends on 'similars'.

Therefore, there are four elements to consider in the nature of the parts of the body, as we noted in our chapter on aposthems. Now, we will deal with wounds in particular parts of the body, having dealt with the generalities in the preceding Doctrine I. Here we will begin with wounds of the head.

Head-wounds may be open cuts or closed contusions; both types may have cranial fractures. When that occurs, the injury may or may not penetrate the bone. The wounds may be large or small; some of them are simple, whereas others may be complicated by severe pain, or abscesses, or with injured membranes (ie dura or pia mater).

Some details: With open-wound fractures some bone may be lost. If the bone is simply fractured, a part may be depressed, or the bone may be comminuted. Fractures of the dome of the skull are different from those of the sides. A contusion of the head may not depress bone or drive spicules into the brain, but a major fracture may do both.

We follow Paul of Aegina (Book VI) in planning our treatments of the various wounds noted here. Galen limited his descriptions to large simple fractures of the dome and the sides, penetrating or not. He paid little attention to small (ie linear) fractures underlying the injuried soft tissues. Larger injuries were managed as we shall describe below.

Albucasis added his descriptions of depressed fractures of the cranium as resembling a copper pot that had had been dented,[404] as occurs more commonly in children. We cite Avicenna's (Book 4) description of a fracture of the cranium on the side opposite that

---

[404]   The 'copper pot' ascription to Albucasis really belongs to Paul of Aegina (Book VI, p 429 in Adam's transl.) (LDR).

of the blow. That sort of injury was denied by Paul (ibid.).[405] Avicenna was concerned with incisions as well as contusions as well as fractures, and defined the various kinds of simple cranial fractures.

Causes

They are common to all wounds: swords, arrows, etc.

Signs and Diagnoses

We shall deal with fractures, injuries to the brain and its membranes, and collections of fluid that compress and impair the brain and its membranes.

Signs of Fractures

First we seek the cause for a fractured cranium: a fall, a blow? Second: the severity of the injury: is there a large contusion or an open wound? Third: where is the injury in the head and what hurts? Examine the site by palpation. Are there fragments of bone attached to the soft tissues, or are there free bits ? Does watery fluid leak when the patient sighs or holds his breath? Fourth: do complications occur soon after the injury, such as apoplexy, scotomas, vertigo, loss of speech? Fifth : what about the voice? A blow on the head of a young girl may make her hoarse. Sixth: does the patient grimace when you tug at a cord which he holds between clenched jaws or when he tries to crack a nut or bites through a reed?[406] Seventh: you observe ink or some mastic placed on the suture line of a linear fracture to see if the black remains or the ointment remains dry.[407]

There are many signs of injury to the cerebral membranes. First: the pain immediately is severe and the patient faints, or is giddy. The face and the eyes redden and pustules appear. The eyes bulge and vision is clouded. Third: the nose, the eyes, and the palate bleed. Fourth: the vital spirit is impaired, body movements are halting, the speech is mumbled, the patient is distraught and terrified. There are chills and fevers. Sleep is disturbed. The appetite for food wanes, and the patient may vomit. He is constipated and he may pass only a small amount of urine.

---

[405]   Guy seems to have overlooked the fact that Paul *preceded* Avicenna by more than three centuries (LDR).

[406]   This was an ancient and trusted method of determining the severity of a head injury. The patient bites one end of a cord while the examiner jerks it (and the head). If the patient suffers, the injury is proved (LDR).

[407]   To prove that the fracture is in only the outer table of the cranium (LDR).

The Signs of A Wounded Brain

What comes out of the wound? Is it greasy and lumpy, like bone marrow, not like pus. Second: the vital functions are impaired. The patient becomes irrational when the front part of the brain is damaged, and his memory is impaired when the posterior part is at fault. The senses are disturbed and delirium is common.

Signs of Abscesses

Here, too, there are many indicators. First: note the swelling that bulges the membranes: they are red and no longer throb (pulsate). Second: the eyes redden and are inflamed, and they bulge, and their movements are blunted. Third: note fevers and restlessness. Fourth: the vital functions are suppressed, and the patient is frenzied.

Signs that indicate the Nature of the Discharged Matter

Cerebral matter coming through the wound may irritate the membranes and produce the symptoms described above, coming on somewhat latter.

Diagnosis And Prognosis Of Fractures

Prognosticate the risks. First: when the brain has been entered, and when the injured membranes have been corrupted, Hippocrates said that the injury usually is lethal unless the wound is very small, as cited by Galen (*Prognosis and Diagnosis*). Serious wounds often are complicated by fevers, tremors, convulsions, delirium, fainting, loss of speech, bulging eyes, impaired vision, flashing, and blindness. All are life-threatening complications, especially when they continue and show no improvement.

Avicenna (Part 3) described the results of open wounds that reached the membranes: weakness of the same side of the body and spasms of the opposite side. William of Saliceto said that a wound of the right side of the brain caused paralysis of the left side of the body, because the nerves to the left side of the body are rooted in the right side of the brain. Let the Reader accept Galen (*Maladies and Symptoms, Book 2*).

Furthermore, you must accept the risks for serious complications that may be blamed on head-wounds for up to one hundred days, according to Roger. However the laws that pertain and judges, have accepted them up to forty six days; The Four Masters said fifteen. Furthermore, a blackend dura mater is always fatal if honey fails as a detergent (Paul of Aegina). Also, a cranial fracture suffered on the day of a new moon is specially perilous (Roger). If proud flesh appears in a wound with a healing fracture, and a small abscess undergoes a good resolution (digestion), those are good signs. However, a large tumor, that disappears of a sudden and is not inflamed is bad sign.[408]

Another note: Callus will appears late and it may not become thick and tough for thirty-five days.

Treatments

Choosing the right treatments for so many varieties of wounds is difficult. Galen, Paul, Halyabbas, Avicenna, Albucasis, Roger, Jamier, Bruno, and William of Salice treated all wounds of the head, with or without damage to the cranium, with the same methods: diagnose, scrape the bone (ie ruginate)[409] before trephination, remove fragments of bone after freeing them with a hot cautery, detect depressed bone, cut away bone to drain subcranial accumulations,then deterge, and dry.

Other authors, including Anselm of Genoa, some Paduans, nearly all the French and English (who simply copied other surgeons) treated with incarnatives, consolidatives, plasters, beverages of wine, and bandages. They based their methods on the belief that they could release what was beneath without removing any bone. I agree that we can succeed with good medicines if we do not engender pus, and inhibit it. But if pus appears, we use proper detergents to get rid of it.

Some authors, as Theodoric, Henry, and Lanfranchi (whose exposition is the clearest), followed their own middle-of-the-road routines, all of them different. Theodoric treated the wound with stupes of wine and gave potions and wine generously as beverages. Henry used his own plasters and abstained from potions. Both of them used trephination four or five days after the injury (as did the ancients) to elevate depressed fractures. Lanfranchi, with two exceptions, used incarnative pads covered on both sides with honey, and with rosat oil on one side. On top he applied a detergent of barley-meal, honey and wax and the capital powder. In both exceptions he scraped the bone before trephination. In one case, the bone had been pressing on the brain, and in the second case a spicule of bone had been driven through the membrane or into the brain itself. He described the risks of exposing the tissues to air, the pain caused by the procedures, the risks of abscesses, etc.

As for me: In light of the disputes of the Masters, and considering their work as compared with my own experiences, I set aside what I thought was improper or suspect, and I was left with two or three to choose from. I remembered Galen's advice (*Aliments*, Part 1.)not to pass judgment without testing by personal experience, and Halyabbas (*Royal Dispositions,*Part 2) had said that direct experience is more reliable than what is read on a page. However, testing medicines on human patients puts their lives at risk, Therefore, I accept what has been the common practice and has been time-tested. With God's help I chose to follow Galen (*Therapeutics*, Part 6*)* to the letter; he had taken

---

[408]   Probably an abscess that ruptures into the subdural fluid (LDR).

[409]   The Rugine was the surgeon's periosteal elevator (LDR).

Hippocrates' methods for treating head-wounds (ie fractures) from a book he had written on the subject, that indicated what he did in such cases. I shall comment on that at the end of this chapter.

I do not dissent with everything taught by Halyabbas, Paul, and Avicenna, and I try to interpret what our Physicians use, as I explained in Chapter I of Doctrine I. I realize that you may accuse me of prolixity, but I deem it necessary and wise to deal at length with the perils. It is not a waste of time to repeat what things are common in treatments, and what are specially different. I shall divide my exposition into Nine Sections.

The Nine Matters To Assess When Treating Head-Wounds

1. The most important differences between these wounds and wounds elsewhere in the body are the nearby brain with its noble substance, the round shape of the head, and the fact that a wound in the brain cannot unite even when properly bandaged.
2. The treatment of head-wounds, as in other wounds, must attend to bleeding, laxation, (a daily bowel movement) with garlic et al., or with clysters, suppositories, or other softeners. The diet should be liquid. The removal of fixed objects should be less traumatic than elsewhere. Hemorrhage must be controlled. Complications should be prevented or treated.
3. Before treating a fractured cranium, shave the scalp after anointing the hair with oil and water, as taught by William of Saliceto, and take care to keep hair, oil, and water out of the wound; they will prevent union. Do not delay to treat pain and to protect the exposed surfaces with egg-whites, and to use incarnatives. Apply an ointment of bol d'armenie and rosat oil on the scalp around the wound; it will relieve pain and prevent dyscrasias and abscesses.
4. Protect against cold, which Hippocrates said is the enemy of nerves, bones, and brain. Minimize the exposure to air, a cause for complications in the principal organs. William advised us to keep a kettle of water boiling on a stove in the room during dressings and to use candles for light when the windows are shuttered. Cover the bandaged head with a sheepskin cap.
5. If pus appears, dress the wound daily during the winter and twice daily during the summer. Gently clean the wound with cotton fluff, charpie, or soft cloths, and avoid painful actions.
6. Lay a soft sponge over the pads to soak up what comes through.
7. The bandages should conform to the head. To apply the incarnatives, use a two-armed bandage. The rolls each should be as long as the length of the oustretched arm, and four fingerbreadths wide. The center should be flat, about two handbreadths wide. Begin to unwind over the forehead toward the ear opposite the wound, and unroll the other toward the wound, but do not cover the ears. Continue the first to meet the second near the opposite ear, flip it over before you again pass over the head, carefully maintaining the medicated center in place. Unroll the second arm over the head to

pass over the flat part. Take as many turns as necessary. The Bolognese tie the ends below the chin; the Parisians sew them over the forehead.[410]

But if you simply want to apply an incarnative medicine, use a many-tailed bandage. I use a large piece of cloth, about three hand breadths long and two wide. Cut into the long ends toward the center to make strips about three fingerbreadths wide, leaving an uncut center about one hand breadth wide. Tie two of them behind the head, two over the head and behind the head, and two tied in front under the chin.

8. When spicules of bone lie in the wound, irrigate them briskly with wine, and if the patient is not feverish, apply a Capital powder of pimpinelle, betony, caryophylla, valerian, and royal fern, and add enough piloselle to equal the others.

9. Immediately put the patient to bed, to lie on the less damaged side. If pus appears, he should lie with the wound down, to favor the drainage.

Now, after the Nine Generalities, we will go to the Particulars.

An Incised Head-Wound Without A Fracture.

When there is no loss of tissue, suture the wound and bandage it. Then use incarnatives, as in other simple wounds. If tissue has been lost, incarnify and cicatrize, using suitable pads, powders, ointments, and plasters.

Suturing simple wounds is a time-tested treatment. It unites wounds of all sizes limits the harmful exposure to air. Suturing is described in *Techni,* Part 3, and in many places in *Therapeutics.* Avicenna said (Part 4), "Whether or not the wound is incised, if it its large, suture it". But he is even more emphatic when there is a fracture of the skull, as we discussed in our chapter on the subject. He added, "Only when it is indicated. Because a wound on the dome of the cranium should not be sewed, whereas yes for wounds on the sides." That policy was followed by William of Saliceto, Lanfranchi, and Henry, although Henry went only so far as Theodoric permitted, denying Avicenna's and Paul's use of rosat oil in the wounds. They used it as a softener and as a sedative when the incision went only so deep as the nervous layer that covered the bone (ie the galea and the periosteum). They used it also to soften the adherence to fragments of bone to make it easier to remove them with less pain; I use honey for that purpose. The arguments against suturing are not worth much because bandaging incarnatives on the head is not easy to do properly, as we shall show. Nor are the arguments about the oil valid, because it works well in simple wounds; it relieves pain and shortens the recovery from complicated wounds. Indeed, Galen said to use whatever are the proper contraries.

---

[410] The description fits what we now call the 'head-melon' (LDR).

Incised Head-Wounds With Fractures That Do Not Penetrate[411]

When the wound is small (commonly called a Rimule or Little Ditch), treat it as if there is no fracture, because such wounds seldom suppurate, and the pus is too thick to drain through a small hole.

A large wound on the side of the skull is treated as a simple wound, except that we insert a small drain at the lower end of the suture line, to favor the discharge of whatever may accumulate in the wound. But if the larger wound is in the dome of the head, do not suture it. Follow Galen's instructions in *Therapeutics,* Part 6. Simple fractures that enter but do not penetrate the diploe call for the use of sharp narrow rugines (ie scrapers); the surgeon should have available a supply of various sizes. Use a large rugine to scrape away the overlying membrane (ie at the fracture line), and follow with a series of narrower ones in the fracture until he exposes the diploe. Then (unless the attendant pain calls for mitigants) apply drying agents of increasing potency until you arrive at what we call the capital (to be described) applied on pads, et al.

Why do we not suture wounds on top of the skull? Because in that position they cannot drain on their own. If we inhibit suppuration with medicated packs, we can avoid trapping pus between the tables of the cranium that can infect the bone itself.

A Wound With A Fracture That Penetrates The Inner Table

Without Loss Of Bone

In some of these cases there are loose spicules, and some bone may be depressed. In others, the tables are not displaced and the surface is smooth.

Sharp spicules may penetrate the dura mater. They should be removed at once with a lenticular chisel that elevates the depression. When the fracture does not perforate the dura mater, Galen said to use the scrapers especially when the fracture is at the dome (ie the bregma).

However, depressed fractures at the sides of the head are not easily elevated with instruments; They are better sutured and treated with drains and detergents. Galen cited two cases in the chapter. In one the parietal bone was fractured along with a smashed temporal bone. He treated (ie with rugine) only the parietal bone. The man recovered and survived for many years. If he had not opened the parietal bone, the underlying membrane would have suppurated and the bones would not have made callus. Galen explained his two methods. When the wound does not drain fluid and there is no accumulation within, there is not reason to cut the bone. And when the fracture is only in the temporal bone and there is no leakage of fluid, there is no need to use your instruments. But in the case of a parietal bone, and there is a discharge of fluid, you may have to enlarge the opening

---

[411]   Fractures only of the outer table of the cranium (LDR).

for drainage and to insert packs to encourage the outflow as well as to allow drainage of what had already collected.

In his second case, he also left the temporal bone alone, fearing to injure the brain and to make a hole through which brain could herniate, and put at risk the nerves that emerged from the sides.

A Similar Wound When Some Bone Is Lost

As in the above, remove sharp spicules and elevate the depression with the lenticular chisel and other instruments. Then lay a piece of thin cloth saturated with honey and rosat oil in the defect and with a flat probe tuck the cloth under the surrounding bone to protect the pulsating dura mater from their overhanging sharp edges. Then lay on pads wet with the same medication, and cover them and the surrounding bone with a similarly wet cloth, all of which to prevent pus from spreading under the bone. The Ancients, at the initial dressing, placed a disc of metal (ie probably silver). It fooled the assistants who expected it to remain and to replace the lost bone; they were absent at the second dressing (ie when it was removed). Nowadays we fill the defect in the soft tissues of the excavated wound with dry pads or sponge to absorb the pus. Over all we apply a perforated plaster, the openings in which allow pus to escape. Finally we apply wine-wet warm stupes and wrap the head gently with a dry bandage. Later, when we are ready to use detergents, we remove the entire dressing and apply a capital powder. Later, we use incarnatives until the wound is ready for cicatrixative powder. I give credit to Henry de Mondeville for this procedure.[412]

A Contused Wound Without Fracture

Avicenna told us to treat with repercussives mixed with egg-whites and with mitigatives for relief of pain. The residue of the hematoma was resolved by applications of salt and honey and other medicaments described in the chapter on contusions. If the hematoma suppurates, use maturatives until it is ripe for open drainage.

A Contusion With A Small (ie Linear) Fracture

Galen treated such as if they were simple contusions and ignored the fractures except to note the diagnostic signs. The treatment was directed at resolution before suppuration could occur. Avicenna (*Canon,* Part 3) treated the headache that follows a concussion of the brain, and he tried to resolve the matter before it became an abscess. He used evacuations to attract the matter (ie hematoma and edema) with phlebotomy on the opposite side, irritating clysters, and coccia pills. At the beginning he applied comfortive plasters that contained

---

[412]   Indeed, Guy took most of his treatments for head-fractures from Henry (LDR).

the saps of myrtle, salix, and shepherd's purse; oils of myrtle, lilacs and rosat; powders of roses and pomegranate flowers, and cypress; aromatic roseau (reeds), lentils, camomille, melilot, bol d'armenie, alum, myrrh, frankincense, quince, and grape jam. He prescribed beverages containing stoechas, water, and hydromel, which will prevent delirium.

If the blow causes bleeding in the brain, you must treat it with roasted brains of grouses and pomegranate water. Theodoric applied plasters of laurel berries, cumin, anise, salt, mastic, frankincense, and powdered bran, all heated in wine. The last item is not to my taste, although I agree that harmless potions may be used if the injury is a minor one; the wound will heal of itself anyhow, without help from treatments.

A Contusion With A Large Fracture

Here we must follow Galen (ibid. Part 6), and Avicenna (Part 4), and operate. We have three good reasons:

1. This kind of fracture cannot be reduced with bandages, nor can we prevent aposthems and collections of matter (hematomas) which are serious concerns in a all depressed fractures. Furthermore, the shape of the head does not lend itself to bandaging (ie and splints). Therefore:
2. Lesser treatments, as we use for fractured arms with hematomas, are inadequate in the head. Here the brain is our reason.
3. Although an operation may be avoided (ie as in an arm) in favor of treatment with medications, that method always uses bandages (and splints). Here we always must expose the fracture and open (spread) it to expose and clean the meninges. We must ignore Theodoric and his followers, and Anselm of Genoa; they treated all head-fractures with pigments[413] as potent potions, hoping to avoid operations to elevate the bones. Although that may succeed when the depression is shallow, I consider all depressed fractures too serious for it. The reason offered by the Conciliator[414] is worthless. He said that potent topicals can elevate depressed bone. All such concepts are suspect, and the application of too strong medications may induce aposthems, especially in those who are sensitive and often are weak, as Dino wrote in his critique of Avicenna's Book 4. My denial includes Henry who tried to cure all wounds without suppuration. The violent force that can cause large contusions often leads to suppuration (*Therapeutics,* Part 4), and Nature is impotent, except in resolving small hematomas. The surgeon must open and drain large collections, just as he must drain empyemas through the fourth rib (as we shall describe). I am even

---

[413] 'Pigment': a special powder mixed with honey, wine, and spices, for use in treating wounds of the head and the chest (LDR).

[414] The Conciliator, Pater of Abano, 1250-1316, author of *The Conciliator.* See Sarton, Vol. III, p 439 (LDR).

more dismayed by the claim that potions do not work after the fourth day to suppress the inflow of matter and to relieve pain and to prevent abscesses. The dissenters do not say what to do if their potions fail. They reminded Galen (ibid. Part 5) of a bad captain whose carelessness loses his ship, who then gives a board (ie torn from the deck) to every sailor with which to save themselves if they can.

Now we shall summarize Galen's treatments for fractured skulls, (ibid. Part 6). When there has been a great concussion you must incise and elevate what has been depressed or encircle it, beginning with the screw of the trephine followed by scissors or gouges as soon as possible. Here I shall comment on two parts of Galen's Epilogue, rather than extending my own brief summary.

I will list the eight elements of most usefulness in operations as performed by Galen, Halyabbas, Paul, and Avicenna.

First: Do not operate on enfeebled victims while they lack vitality.

Second: Before doing anything, clearly state the risks. In that way you will avoid the recriminations (*Canon,* Part 4) that may follow.

Third: Whenever possible make your incisions to avoid the commissures. That is where the dura mater is attached to the bones and it may fall away or be injured.[415]

Fourth: Be aware of the risks of operating during the time of a full moon, when the brain is puffed up against the dura mater, as noted in *Critical Day,* Part 3.[416]

Fifth: Perforate the cranium at a low place in the fracture line, to favor dependent drainage (*Therapeutics,* Part 3).

Sixth: When ruginating, scarify only as much of the bone as needed to make an opening through which pus can drain.

Seventh: When the bone to be removed is still attached to membranes, soften those attachments with rosat oil to allow you to remove it painlessly.

Eighth: Operate as early as you safely can, especially when the membranes are depressed or perforated, when they are least resistant to abscesses and other complications. When the matter comes from above, do not await beyond the seventh day during the summer or the fourteenth during the winter, because then it may be too late to do any good.

Consequently, I follow the precepts of Avicenna (Part 4) in performing the operations, as described in Chapter 1. First shave enough scalp to allow you to make a cruciform incision (ie two cuts), or a v-shaped one a la Lanfranchi.[417] One limb of the incision should lie over the fracture. Elevate the corners to expose the damaged

---

[415] The medieval anatomist said that the dura mater is suspended by the vascular and nervous connections that supply the epicranium, which pass through the commissures (LDR).

[416] The lunar effects on the tides were thought also to affect the brain which is afloat in its own 'sea' (LDR).

[417] Guy wrote, "shaped like the number 7." We note that the letter V was written as U in Guy's times, and a U-shaped flap was not suitable (LDR).

bone, where you will open it. If your incisions bleed briskly, pack the wound with a cloth soaked in water, vinegar, and egg-white. If there is no hemorrhage, use a dry cloth pack, and overlay it with pads wet with warm oil and wine, and bandage snugly. The following morning undertake to ruginate through the bone while the patient sits with his ears plugged with cotton to lessen the fearsome noises of the scraping. First unwind the wrapping and remove the packs from the wound and gain exposure by trimming away the tips of the flaps, or insert threads to allow you to retract them. When the bone is soft, you may cut it with a scissors instead of using a rugine or a lenticular chisel. When the bone is very hard, pierce it with the screw of the trephine in several places to encircle a disc. Scrape from one hole to the next until the disc is loose and can be elevated with your fingers or instruments. Gently smooth the rough edges of the opening with a hammer and chisels. Treat your incision as any wound, as described in the section on incised wounds with fractures.

Treatments For Complications

An abscess usually is caused by a depressed fracture or by a spicule of bone driven into the dura mater or brain, or by improper use of drains or bandages, or by exposure to cold, or by a bad diet. When an abscess appears, hasten to elevate the depressed bone, and use evacuations: bleeding and other methods. Use rosat oil or water boiled with guimauve, fenugreek, linseeds, camomille, etc.; a plaster of mauves also works well. If the dura mater darkens apply abstergents and detergents containing rosat oil and honey. If the blackening is not do to external contamination but is of itself and is deeper than what you see at the surface, there is no chance for recovery, because it is due to loss of vitality (ie gangrene); so said Paul.

Excessive granulation tissue and other complications are to be treated in the usual ways.

The Capital Medicines

Treat head wounds from the time of injury until past the usual time when abscesses appear; use mitigatives containing rosat oil and honey, or use the 'old-timers' ointment of Galen (*Therapeutics,* Part 6) to deterge and act as a contrary.[418] When the pain is severe, Galen used a stronger detergent as a contrary: honey and rosat oil. After the risk-times of abscesses are past, we use a desiccative that is not corrosive, such as the capital powder of Galen that comntained illyric iris, flour of ers, manne (which here is minced frankincense),

---

[418]   This was a recipe of Galen's named after Eudeme, a respected contemporary, who used it in head-wounds. Galen applied it on the exposed dura mater as a potent desiccative, and covered it with oxymel. He claimed successes in many cases, better results than with the less vigorous medicines (J).

aristolochium, and the bark of the roots of panax. Bruno added myrrh, sarcocolla, and sangdragon. Lanfranchi and William added myrtle-seeds and cypress nuts.

A Capital Plaster of betony is placed on top of the powder. Henry's recipe is juices of betony, plantains, and celery; melted resins and wax; terebinthe. Cook until enough liquid evaporates to make a paste.

A plaster of Centaury is good for head-wounds: lesser centaury soaked overnight in white wine. Boil the fluid to half-volume. Pour off the liquid and boil it to the consistency of honey. Add mother's milk, terebinth, melted wax, resin, frankincense, mastic, and gum Arabic.

The Instruments For The Operations

There are six kinds, and three sizes for each: large, medium, and small.

The Trephine for elevating depressed bone is used in several ways. Galen used the drill with a chaperone (ie a guard) above the sharp point to limit the penetration short of the dura-mater. The Parisians had drills with several holes for pegs at different distances from the tip of the drill suited to the thickness of the bones, to eliminate the need for a chaperone. The points entered the bone (ie the outer table, up to the lowest protruding peg) and the rest of the penetration was accomplished with a hook-chisel (ie a lenticular) which they inserted through the puncture. The Bolognese use a drill shaped as a triangle; only the point can perforate and the widened part prevents an undesirable perforation.[419]

Two kinds of *Separators* to enter the slit in the bone: the French variety and the Bolognese, which is curved, and can be used as an elevator.

Elevators to lift the depressed bone after trephination.

Rugines (scraping chisels or rasps) to enlarge the slit.[420]

Lenticular chisels which were Galen's favorites. They can elevate the depressed bone and restore the normal flat surface after separating the irregular edges of the fracture-line. It is shaped like a scythe with a button at the tip of the blade.

Hammer used to tap the thick rear edge of the lenticular chisel. It is made of lead: the weight helps, and its soft metal dampened the less frightening sound.

A Note By Nicaise

Nicaise summarized Malgaigne's History of the Trephine in his Paré, Vol. II, pp 55 ff.

---

[419] Trephines: In Guy's era the crowned trephine in use today (ie 1890) was not known. Joubert's drawing of Guy's instrument is wrong—anachronistic. Guy used Galen's drill with a disc (bourrelet) above the point to prevent too deep a perforation. He called it the abaptist (ie non-immersible!). The drill used by the Parisians replaced the protective disc and used a chisel in the drill-hole. The Bolognese drill had a knife-edge at the tip (EN).

[420] The Montpellier Ms. described this as resembling the barrel-makers shaving tool, a chisel-like tip on a hooked blade (ie scythe-like). Other rugines also are described in the Ms (EN).

Trephines

The shape of trephines has varied. Hippopcrates knew two: the crowned and the perforating drill. The second alone remained in practice and was used until Galen's era. The concern for penetrating too deep and injuring the dura mater led some surgeons to arm the drill with a circular bourrelet that would limit the depth. They called it 'unsinkable', the abaptista. Albucasis and Guy for our benefit described it and the different types devised by surgeons of his era to prevent the deep penetration. But even then the crown was not remembered, and Malgaigne did not know why Sprengel could write to the contrary.

The first indication of its rediscovery is in Jean de Vigo's work of 1517, *The Brief Surgery,* where he called it the divine mespila, the 'divine medlar-like instrument'. Nicolas Godin, in his translation of Vigo's small book, wrote that his copy lacked drawings of Vigo's instruments.

Marianus Sanctus (*Compendium of Head-Injuries*) mentioned three instruments: a rugine, a terebella (drill) and a trephine. He did not describe them or comment on Vigo's (Marianus was his teacher) invention which was printed after Marianus had departed Rome. It seems that again the abaptist fell out of favor. The crowned trephine and the perforating drill were found in Celsus' treatise after he was 'rediscovered'. Were Vigo or Marianus aware of it? That question is not easy to answer. However, we know that until Guy's epoch the abaptist trephine was the only one in use, and that Marianus—two centuries after Guy—did not use it. Indeed, we have his statement that he preferred the drill, but it was inconvenient unless used by experienced operators. Vigo said that was what led him to his invention, to help neophytes who easily could injure the brain and the meninges without it. Finally, we know from letters written by Langius that German surgeons did not known about the ababtist trephines because they had not read Celsus. Malgaigne commented that Langius had no proof for his mockery. Indeed, it is probable that the German barber surgeons continued to use the instruments of their progenitors (ie as far back as Guy's era) which were abaptists, which they knew by other names. In fact, that term was not used by the Arabs or the Arabists who used local rather than Greek terms. Even in Vigo's epoch, Berengario da Carpi's book (before 1550) had an engraving of an abaptist trephine which he labeled as a terebrum non-profundus.

Now, let us return to Vigo's instruments. We will go to Andre de la Croix for a more precise description.[421]

Andre had drawings of a large number of trephines; we regret that he did not label them with the names of their inventors, to make his book more useful for historians of surgery. What he called a 'safe instrument' probably was Vigo's; it is a crowned trephine armed with a bourrelet placed above the drill point, truly an abaptist. Above that drawing are two others labeled as modiolis mespilati, which brings to mind Vigo's term mespula.

---

[421]   Giovanni Andrea de la Croix; Venetian surgeon, ca 1560 (LDR).

Andre explained the term as referring to the cone-shape of a medlar, which is mespula in Latin. In the drawings, the drill is encircled by discs placed, one in every turn of the screw ascending with increase diameters, creating a pyramidal profile, largest at the top. We really can see the profile of a medlar, as Vigo called his 'crown'. One stack of bourrelets has a drill point, and the drawing is labeled 'male instrument). The other drawng has a stack with out a drill, labeled 'female instrument.' In the index of Andre's book, it is called the abaptist, or a safe instrument.

Berengario da Carpi's illustrations contain at least eight trephines without crowns in addition to one with it. All of them are in Andre de la Croix's book, which lacks attribution to Berengario. All of them are labeled as trephines with two or more wings, a trephine with a limiter, and a trephine 'a image'. The last is a trident. More to the point: we can see that Berengario had a crowned trephine to prevent unwanted penetrations, and here, perhaps less nicely engraved than that of Vigo, we see an instrument with the same purpose. And it is in Berengario's book that Malgaigne found the term 'brace and bit' (arbre du vilebrequin) for the first time as applied to the trephine.[422]

Neither Vidus Vidius (before 1569) in his *Commentary on Hippocrates' Head-Wounds*, nor Andre seem to have known that technique of turning the screw in the trephine, which we use even today.

In his book, Andre described the trephine used in the 16th C, without attributions. Others in the rich collection go back to Berengario and before him. He wrote that the trephines were too numerous for him to describe, and he added, "Indeed, more than once I decided to make one myself, a new instrument for treating cranial fractures, unlike any that I had known or had been used".

We note that Paré used an abaptist based on Vigo's.

Malgaigne noted that Andre did not picture Paré's instrument, probably because he did not know Paré's book. We may conclude that all of the instruments in Andre's book which also are found in Paré's were not invented by Paré, rather they were his modifications of what were in the surgeon's arsenal during his era.

Therefore, his elevator with three feet (page 13 of Vol.II of Malgaigne's edition) seems to be a copy of leaf 30 in Andre's book, and the second forceps on page 16 is on the verso of Andre's leaf. The deep drill on p.12 is an almost exact copy of Berengario's terebrum non profundum. The same claim is made for Paré's lenticular chisel, hammer, forceps, and rugines.

---

[422] Henry described his drill-trephine with instructions to turn it with one's hands, implying that a mechanism for turning the drill was available and was not as easily and safely controlled. See Henry de Mondeville, Vol I. (English Transl) p. 457 (LDR).

CHAPTER 2

Wounds Of The Face And Its Parts

The special concerns for wounds of the face are the elements of beauty and respect (ie honor), requiring particular attention to healing with good scars. Whenever possible, suture cloth attachments (ie rather than skin). If that is not possible try to transfix firm deeper tissues (ie back from the skin-edges) and wind the threads around the exposed ends of the needles which are left in place until the edges unite. However, if the wound is deep, close it with a continuous suture as used by the furriers (leather workers). Whenever possible, close the wound with incarnative bandages instead of sutures.

Albucasis insisted on closing recent wounds of the nose, ears, and lips only if their surfaces were moist (ie bloody). Lacking that, he gently scraped the surfaces to refresh them before sewing as we do in the abdomen. However, the face shares the problem of roundness with the head, and that defeats most bandage-closures, excepting where the wound falls together on its own, We, as do many surgeons, fit the patient with skull-caps, with or without brims, made of heavy linen to serve as an exterior membrane to which we can sew the bandages. When we dress any of the wounded many parts of the face, we may use small pads and strips of medicated cloth rather than large over-all stupes. That kind of dressing is easier to apply and to remove.

Treatments For Wounds Of Parts Of The Face

The Eyes

These wounds threaten loss of vision and injury to the nearby brain. I, like Bienvenu, have seen wounds around the orbits lead to damaged optic nerves (ie blindness) and cataracts. When the wound enters the eyeball itself, the ocular humors escape and the organ mortifies, having lost its function. Galen (*Maladies and Symptoms,* Part 4) described a child with a puncture wound in his eye with loss of the aqueous humor albeit a rare event, the wound healed. Rabbi Moses challenged Galen saying that the case was not exceptional, because we all have seen spermatic structures heal in infants.

Jesus[423] said that the goal of treatment is the prevention of loss of ocular fluids. If the eye is not bloody, use a collyrium of tuthy with a bit of camphor. If the fluid is bloody, treat the wound with the blood-stone,[424] a potent medicine when applied with egg-whites and applied as a plaster over the orbit.

Bienvenu[425] favored fresh eggs, beaten and mashed in a mortar to make a paste. He called it "God's Gift".

All direct injuries to the surface (conjunctiva or cornea), a wound, or smoke, or dust, or pebbles, or straw, or bran, were treated by Jesus with irrigations of mother's milk [426] or with soft water. That will wash away most irritants. If necessary, fold back the upper lid and remove particles with a probe or a finger covered with a bit of soft cloth. If that fails, use a tweezers, followed by an irrigation with the milk.

When hypopyon is caused by a blow or a nearby wound, Jesus used the mother's milk with egg-whites with added blood taken from the undersurface of dove's wing. Also, he favored a plaster of soft bread and wine. When more was needed, he instilled an oil of ammi and rock salt, diluted with barley-broth and hyssop. If more was called for, he soaked powdered arsenic in water, and used the water as an irrigant.

Other medications for hypopyon: A collyrium of blood-stone, burnt bronze, coral, pearls, gum Arabic, tragacanth, pepper, ceruse, red arsenic, sangdragon, yellow amber: all in a liquid of grouse's blood and mother's milk.

The Eyelids

Sew them with round (ie in cross section) needles[427]. Use small bolsters to sustain the closure even while the lids are frequently blinked, especially when the wound crosses the gently curved hair-bearing edge. The lid contains cartilage and is very slow to heal. Dress the repair with ingenious use of powders and bandages to hold them over the wound.

The Nose

The nose may be incised, bruised, or smashed. Here we shall deal with special problems of incised wounds; they are the most frequently encountered. Most of the other kinds are managed as other wounds of the head.

---

[423]   Jesu Ali: Issa ben Ali: famous Persian Oculist, 10th and 11th C (LDR).

[424]   'Blood-stone': One source claimed it as an herb. Joubert's Ms of 1559 said it was malabrathum, or the leaves of laurel. Canappe said it was hematite (EN).

[425]   Bienvenu, better known as Benevenutus Grassus of Jerusalem, a Salernitan ophthalmologist, 12th C, author of the most popular medieval treatise on diseases of the eye. Sarton (Vol. II, p. 243), says that Guy was the last person to quote him! (LDR).

[426]   'Mother's Milk': Jesus insisted that the lactating mother must be nursing a girl (J).

[427]   Square or triangular needles had sharp edges (LDR).

The Sliced Nose: At times it is completely cut off and at times it is left dangling from the upper lip. You cannot restore a completely detached structure because it is organic. We have no disagreement with what Galen wrote in *Techni*, Part 3, although there have been many old-wive's tales to the contrary. But if the nose is only partly detached and it continues to bleed, it can make a scar. We can sew it nicely and skillfully with techniques described for wounds of the eye, using straight needles that we heat and soften so we can bend them to suit the contours of what we are sewing. Insert tubes of cloth or use the hollow shafts of goose-feathers in the nostrils. Cover them with a wound-powder and put the same on the pads that are of sizes and shapes to fit the irregular sides of the sutured nose. Cover the entire nose with another pad; all the pads are wet with egg-white at the initial dressing. Later we use white wine and incarnative and consolidative ointments. Bandage with skill.

How to bandage a nose: This has been a much-debated topic. Albucasis and Avicenna favored their own methods without bandages; Lanfranchi and Theodoric, insisted on two bandages, one under the nose to support it, and the other to hold the medications in place (Henry disagreed). Roger and William used one bandage with a slit in the center to expose the nose. Henry did not approve, saying that no bandage over the nose is better; avoid the risks of a compressive dressing that can displace the structure, or of a too-loose wrap that weakened the effects of the topicals. In both such cases, the nose-flap will darken and die. That will not happen if the bandage passes only beneath and not over the nose.

As for me, I remain aloof from the disputations, because, as I have said, one cannot be dogmatic about incarnative bandages on the face. The surgeon must decide what seems better in each case, insisting only on the use of the skull-cap or bonnet to which you can affix (ie sew) the bandages that hold the stupes and pads in place and support the sutures that close the wound while the medications are in direct contact. To prevent the complications of exposure to cold, Henry used the natural heat of a (ie split) chicken laid over the healing wound. If the reconstruction fails, remove the structure, but delay that as long as possible to avoid public censure. Then dress the wound until it scars.

The initial dressing should remain for three or four days. Then change it twice daily

Fractured Noses

Restore the contours with a finger or rods within the nostrils while you mold the outer part with the other hand. Then insert drains as packs in the nostrils and place pads moistened with egg-whites alongside and over it. Later, use diachylon creamed with rosat-oil and wheaten flour and red powder held with a bandage.

The restored fractured nose should be secure within eighteen days.

Contused Noses

Treat them as if they were fractured. If the cartilages have been dislodged, they will reattach.

331

Wounds of the Ears and Lips

There are no special measures. Sew the wounds as neatly as possible and apply bandages, and treat as we do for other flesh-wounds.

CHAPTER 3

Wounds Of The Neck And The Back

B ecause wounds in the neck may involve more than simple flesh and bones, and involve the ligaments[428] on the sides, the organic veins and the conduits for swallowed food and for respiration, there are particular concerns for treatments other than the general elements that affect all wounds, as to bandages, prognoses (and diagnoses), and medications.

Bandages

When we wrap incarnatives on neck-wounds (perhaps to replace sutures) we use a long two-headed bandage, centered on the opposite surface of the neck. That allows us to bring the ends around the neck to cross behind the wounded side and over the wound and under the axillae before returning to the neck where you fasten them. If you use the skull cap, suture the bandage to it to prevent slippage. Roger divided the ends of his double bandage in three and passed one pair over the ears on both sides and fastened them at the forehead; the other pair passed under the armpits and around the chest where they were fastened. The third pair went around the neck and were tied in the middle.

Prognosis and Diagnosis

As to neck-wounds, Roger noted that a full range of motion of part of a body will be lost after a cord is cut. Furthermore, if a wound penetrates the neck from back to front (ie dividing the spinal cord), it will be fatal, because the functions of the spinal cord are like those of the brain, as explained by Galen in *de Usum Partium*. But if the wound does not involve the spinal cord, it may be curable, although the parts served by nerves in the neck may be deprived of sensibility and movement, similar to what happens after fractures of the lower vertebrae, to be described. Furthermore, wounds in the neck that damage the recurrent nerves (of Galen) will cause permanent hoarseness. And, so it is said, if certain

---

[428]   Another reminder: 'Ligaments' are tendons, nerves, cords, joint capsules, et al.(LDR).

veins near the ears are cut the patient will be rendered infertile. The same is wrongly said, by similar fools, about injuries to the spermatic veins, as mentioned in our Anatomy. All injuries to major veins and arteries may cause life-threatening hemorrhage that may kill the patient rapidly. Wounds of the esophagus and trachea are perilous because their functions are vital for the entire body, and they heal poorly, because they move continuously while swallowing and breathing.

Treatments

As to flesh wounds, we add nothing more than the general measures: sutures, powders, wine, et al. As to nerves and cords, they must be approximated by deeply placed mass sutures and with topical oils and plasters.

As to the larger veins and arteries, we use mass sutures, plasters containing Galen's powders, hare's beard and egg-whites. Failing that we must expose the bleeding vessel and ligate it. As to the esophagus and trachea, we may try suturing and topical applications of powder, et al. We sedate the patient with draganthum and diasymphyton (ie consolida).

As to the nape, we irrigate with warm rosat-oil and apply egg-yolk until the pain is eased.

If pus appears, we use detergents and the plasters of William and Lanfranchi: rosat oil, barley-meal, terebinth, wax, resin, myrrh, sarcocolla, mummy, and oil of mastic.

CHAPTER 4

Wounds Of The Shoulders And The Arms

W ounds here that are limited to flesh and bones require special attention only to prognoses, bandaging, etc.

When a wound in the shoulder damages nerves to the arm it causes unusual pain, and a loss of sensibility and motion; those are diagnostic signs. Wounds at the elbow and near other joints are painful and are at particular risk for hemorrhage from large veins, abscesses, induration, and spasms. Deep cuts may injure bones and ligaments, and they are difficult to deterge and pus may be trapped and solidified, and the joints will become stiff. We already have discussed the bad prognoses of wounds within two or three fingerbreadths of joints.

Treatment

Except for one precaution, the treatments are the same as those we use for most wounds. Here we insist that the sutures used in shoulder wounds be sturdy enough to withstand the size and the weight of the dangling arm.

Bandages

After suturing, use a two-tailed bandage. Place a soft ball in the axilla and set the center of the bandage over the wound and unroll the tails under the axillae and back over the wound, criss-crossing it finally to reach the opposite axilla, where they are tied or sewed. A simple dressing (ie only for applying medications, not incarnative per se) can be wrapped over the pads and tied under the opposite axilla, using two tails, creating a kind of sleeve.

After bandaging an arm-wound, suspend the arm against the chest with sling around the neck. except for wounds behind the elbow. In those cases, keep the arm extended and avoid tension on the sutured wound.

The surgeon should devise his own bandages for wounds of the fingers and hands.

Soften the residual induration (ie and stiff joints) after the wounds have healed; use the methods that we described in the chapter on sclerosis.

CHAPTER 5

Wounds Of The Chest

I am amazed by the paucity of materials about this topic in the works of Galen, Halyabbas, and Avicenna, especially about injuries to the structures within the chest, whereas they are full of information about the anatomy and functions of those organs. Other authors offer opinions which often are discordant.

Roger, Roland, Jamier, Bruno, William, and Lanfranchi all said that we should not close penetrating wounds and allow blood to collect within the cavities. To that end they inserted packs and drains coated with detergent ointments. They applied attractant plasters on and in the wounds. They reasoned that the retained blood could envelop the heart and other organs and kill the patient.

On the other hand, Theodoric and Henry closed the wounds with sutures and without drains. They prescribed their pigments and used white wine as sedatives, just as they did for head-wounds. They reasoned they would prevent loss of natural heat and the entry of cold air.

As for me, lacking precedents from Galen et al., and avoiding the later disputations, I use what seems best ad hoc.

The Different Wounds

Some wounds are in the chest wall alone, front or back; others penetrate. The latter may or may not involve structures within. Some, however, damage the heart, or lungs, or the diaphragm, and the attendant bleeding may be much more than minor. Therefore, the treatments will vary.

Causes

As for all wounds, the causes are any things that stab, cut, and penetrate the surface, including darts, arrows, and swords.

The Signs

A penetrating wound produces sounds of breathing through the opening, even when the nostrils and mouth are shut. The air flow can be detected by the flicker of a candle's flame or of a wisp of cotton placed near the wound. A probe that passes through the wound into the chest is less reliable.

When the heart has been wounded, the emitted blood will be dark, and the victim's limbs will be cold, and he will sweat profusely, and he will faint. The site of the wound is under the left breast.

When the lung is wounded, the blood is pink and frothy, and the victim is feverish and pale and he coughs. The site of the wound is between the ribs. A lot of blood spat through the mouth means that a vein has been cut. (Galen).

A wounded diaphragm causes the patient to breathe rapidly, and the deep coughs are painful and bloody. He may be delirious; he is thirsty and loses his appetite for food. He belches (ie hiccups) and suffers chills. The site of the wound is at the false ribs.

When the blood is retained within the chest, it will suppurate and become foul. Galen (*Aphorism*) said, "When the blood settles near the abdomen it must suppurate." It's own weight causes it to sink to the level of the false ribs. The victim's sputum is purulent and he is feverish. Jamier called attention to the victim's foul halitosis, and to the discharges in the dressings which will contain foul-odored clots.

Prognoses

Wounds that enter through the back are more dangerous because the veins, arteries, nerves, esophagus, trachea, and the pericardial membranes are placed closer. A wound in the spinous ligament also is serious.[429] Wounds that do not enter the cavities are less threatening.

Other prognostic matters, as to causes, symptoms, et al. are shared with other wounds.

Treatments—Bandages

Not much is special about treating wounds that do not penetrate, except a method of closing them without sutures. Incarnative wraps use long and wide two-tailed bandages that begin on the side opposite the wound. Unroll both so to cross over the wound, around and around as often as needed. Sew the ends in front, away from the wound. As did Henry, we sew strips to the wrap-around, front and back, that pass over the shoulder and under the crotch.

Roger used a wide bandage simply to apply medicines. He cut a slit in one end and divided the other in two. One went over a shoulder and through the slit and then back where it met the other strip. If he could succeed in closing the wound without sutures, he wrapped the bandages less tightly if the wound was near the breast.

A bandage for treating an abscess in the axilla has two strips cut into both ends. One pair goes around the chest to meet in the opposite axilla. The other pair crosses over the same shoulder. one in front and the other in back, to meet and be tied under the opposite axilla.

---

[429]   Guy called it the Nape (LDR)

When a penetrating wound causes none of the signs of internal injury, and if there is no accumulation of blood or other fluids, treat it as if had not penetrated, and do not use drains. Apply incarnative plasters and ointments, and cover with wine-wet stupes that are not frequently changed. Galen and his followers paid little attention to such wounds, calling attention only to their practice of not inserting drains to allow the escape of whatever may accumulate within. I do not use drains. After all, we have assumed that whatever small amount of matter may accumulate will be resolved by nature. I do not question the principle that nothing can resist the actions of a healthy nature. It can resolve matter through membranes and even through bone, as Galen wrote in *Aphorism* 7, "You can help Nature with your potions, whereas the unnecessary insertion of drains can lead to other problems, such as feebleness and the complications that follow exposure to air." Avenzoar (*Treatments, 4, or 10*) described them in cases in whom quinsy follows an incision in the uvula.

When the penetration enters an internal organ and there is no suppuration, treat the wound in the chest wall, as above. But internal damage requires the use of viscous and adherent potions. Galen (*Therapeutics,* Part 5) tried all kinds of desiccants applied on the outside and he made potions of them by diluting them with water and wine. He called the best of them 'diaspermatons', and I use one of them made from cassia as a comfortive for all cases of chest wounds. Avicenna added diacodyon (a kind of diapapaver) and diatragacanth or diasympyton containing cassia. We usually apply it as a topical. Also, we irrigate with wine and use desiccative plasters containing rosat-oil and quince during the summer. We use nard in the winter, and a plaster containing chalcanthum. It was first described in the *Catageni (the Contents According To Types).* I believe that really it was diapalma, which is used for ulcers. In the penultimante part of *Therapeutics* Galen wrote, "The common goal in treating all injured internal organs is to provide nutriments and medicines that are most useful for the patiuent, and to avoid such contraries as verdigris, cadmia, pompholyx, litharge, ceruse, et al.[430] Galen also said (*Temperaments* and *Simple Medicines*) that medicines used as consolidative topicals may irritate and open internal organs.

Therefore, we should offer foods that promote agglutination and scar-formation, that are bland and viscous, and that do not irritate, and provide potions of hypocystus, balaustium, galles, pomegranate peels, terra sigillata, sumach, and the juices of roses, acacias, et al., and decoctions of quince, the tips pf blackberry and grape vines or myrtle, and sour wine. But do not administer the decoctions during the initial period of risk for inflammation (ie the first week). Instead we should add gum-tragacanth and gum Arabic.

When indicated, use purges of mild abstersives, the best of which is warm honey. In fact, all such medicines work better when mixed with honey. Honey is the best means for spreading medicines through the body; it is the vehicle, the chariot that opens gates, and it is harmless in ulcers.

---

[430] This list of metallic substances attributed to Galen is an interesting "contary" to the common claim that Paracelsus was the first to advocate metallic medications (LDR).

No applications on the surface can improve the situation within as well as potions of wine. Roland and Theodoric approved of it. They agreed in spite of Theodoric's disclaimer that Roland had cured a patient with applications of powders on an incised lung.

However, when the signs indicate an accumulatiuon of matter within the chest, do not wait for spontaneous drainage[431] Follow William's advice to enlarge the wound to improve the escape of blood and pus. Insert a properly made cloth drain, long enough to dangle and be fastened outside to a long thread to prevent its falling in and to make it easier to remove. Wet it with warm oil. Roger then placed the patient on a flat table and rolled him side-to-side to spill out the contained fluid along the drain. Jamier used a syringe, as for clysters, and irrigated the cavity for three or four days, rolling the patient each time. He compared the measured input with the output. When the latter was less, and it was clean and clear, as compared with what he had injected, he allowed the wound to close around the drain which remained until all drainage ended. He urged the patient to sleep on the wounded side to favor the drainage. Albucasis wrote that after three days without spasms and irregular heart beat or exchange of air through the wound, and when the patient's complexion was good, the surgeon could be assured that the wound was healthy and that the patient's natural forces were strong enough to eliminate what it could not manage before. Then, bit-by-bit he removed the drain.

When that treatment fails, or the patient cannot tolerate a prolonged course, simply treat the wound and shorten the drain; apply this detergent plaster over it: rosat with honey, myrrh, frankincense, sarcocolla, barley-meal, and fenugreek-flour. You may add some terebinth. I suggest that you also prescribe a potion containing costus, nepita, garyophyla, pimpinelle, piloselle, cannabis, tendrils of red cabbage, tansy, garance, and reglisse. Boil all in wine and honey, and dose the patient a small cupful every morning.

When you are not certain that the wound had entered the chest, demonstrate the fact after applying drains softened with rosat and alum.

When the patient cannot endure the frequent irrigations and rolling, and the stench is awful, and if a lower intercostals space begins to bulge, and there are other signs that matter is trapped in the 'valley' of the diaphragm, and a robust patient agrees to it, follow William's advice to make a counter-incision at the lowest part of the bulge, just beneath a rib, slanted to the vertebrae, along the rib and the skin lines, between the fourth and fifth rib or one interspace above. If you encounter the diaphragm adherent to the under surface of the ribs where it will obstruct drainage, even at the higher level, inform the attending physician that it will be better to use the lower (ie 4-5 interspace). After making your entry, insert a wick wet with rosat, only deep enough to reach the abscess. Determine at once that there is no exchange of air along the drain, and that you have not entered a vital structure. After establishing drainage of the matter, use detergents in the new wound, a large amount of camomille wine (a decoction) with lupin-flour, frankincense, myrrh and rosat-honey. Inject about a pound of the warmed mixture, at each treatment, using a clyster syringe.

---

[431]   'Necessitated' empyema (LDR).

Twist and turn the patient, and have him lie on the new wound. Then insert a drain wet with rosat-honey and lay a detergent plaster over it. While dealing with the new incision allow the initial wound to close after you remove the drain. This surgical (ie artificial) treatment was attested to by Galen (ibid.,Part 5) when he described an arthritic patient with a thoracic empyema which he drained after removing a segment of infected rib. He washed the wound frequently with honey and water, encouraged productive coughing, lying on the affected side, and sucking at the wound with a pyulco (?). When he was confident that all the pus and his medications were eliminated, and that the wound (ie ulcer) was clean, he discontinued his ministrations.

Avicenna agreed (*Canon,*Part 3) saying in re empyema, "When you think there is more than a simple pleurisy and that there is much matter, deterge it with topicals for about forty days, or until the sick person shows signs of phthisis. Pierce the chest wall with a hot wire—cautery where the pus is pointing, Keep it there until all the matter is dry. Then remove it gradually while irrigating with honey and water. When the returns are clear, treat the wound."

Halyabbas (Sermon 9 of Part 2 of *Royal Dispositions)* withheld the knife and the cautery, because the diagnosis and the localizations were uncertain, when the patient is neither delirious or moribund, or has a draining fistula which resisted local treatments. He said that he always provided an honest prognosis and obtained a full permit. He used a potential cautery, the roots of the long aristolochium with oil, a powerful caustic. I have not used it, but I believe what he wrote. Albucasis agreed that there is good reason to treat any unhealed wound (ie ulcer) when you are sure that it contains a fistula. We will discuss that problem later.

CHAPTER 6

Wounds Of The Abdomen

We shall deal both with the abdominal wall and with the abdominal contents.[432] Wounds of the abdominal wall may or may not penetrate. A penetrating wound may or may not damage an internal structure, such as the omentum (ie coëffe), the stomach or the intestines. The prognoses and the treatments will vary accordingly.

The Causes

Here they are no different than elsewhere: swords, lances, arrows, et al.

The Signs

One can determine if the wound has penetrated by inspection of the wound and by probing it. The probes will show the opening if it has not manifested itself by the extrusion of omentum or intestine. When the herniated omentum is altered (ie by exposure to air), its surface is greasy and its veins are filled with purple or black blood. When intestine has been entered, it will leak fecal matter. When the wound is above the navel, the leakage will come from small intestine; a leaking lower wound indicates a wounded colon. A wounded stomach leaks chyle from the uppermost abdomen. A wounded liver bleeds from the right side; the spleen bleeds dark red blood, like wine lees, from the left. Blood from a wounded kidney is watery, and comes out through the flank.

Prognosis

Galen (*Therapeutics,* Part 6) said the most dangerous wounds are the most difficult to repair with sutures, and are in the center of the abdomen rather than in the flanks, under the ribs where there are masses of muscle which are easier to close and the intestines are not as close to the surface. Furthermore, if you do not close the wound promptly and

---

[432] Guy ignores the organs that defied the surgeon, and limits his concerns to certain structures: the lower stomach, the small intestine and the colon (LDR).

reduce whatever intestines have herniated, they react to exposure to cold air by inflating with gas and creating difficulties for replacement. Hippocrates (*Aphorism 6*) said that unless we promptly replace herniated omentum, it will corrupt beyond recovery, and the surgeon must cut away what was exposed. However, although it is a common occurrence it is not always the case, as Galen noted in his *Commentary.* He also said (*Therapeutics,* Part 6) that the large intestine is easier to repair than the small, and that a wounded jejunum cannot be cured, because is so rich in blood vessels, and has thin nervous (ie membranous) walls. Furthermore it is close to the liver and receives all the pure bile.

Whereas the wounded lower end of the stomach can be sutured and heal because it is fleshy, and will retain medications to act locally, the upper stomach lets them slip by and that impairs efforts to cure wounds by continuous contact.

Other causes and prognoses are not special.

Treatments

Wounds that do not penetrate can be treated as we do wounds in the chest wall; we use incarnative[433] bandages unless we need sutures. Afterwards, apply the usual incarnatives. Treat penetrating wounds in the same way if no internal structure has been injured, except we use sutures instead of bandages alone.

Suturing the abdominal wall

We follow the recommendcations to suture all layers of the abdominal wall (ie mirach and siphac) as one, because the siphac is a nervous membrane without its own blood supply, and it cannot hold sutures on its own, and it will come apart when the distensible mirach is stretched.

The Methods

1. Pass the needle through the skin and the mirach (about a fingerbreadth back from the edge) but not through the siphac, then come back up through all the layers including the siphac. Use a long thread. Then place the next bite of a continuous suture about a fingerbreadth away from the first point and, bring it back through all layers except the siphac. Continue the same pattern, spacing yhe sutures down the line with as many bites as needed to close the wound.
2. Galen used this method, and Albucasis followed. It is in common use, but it is not as secure as the first. Sew through and through all four layers and tie a knot. Repeat with separate stitches instead of a continuous one.

---

[433] "Incarnative' means to unite the wound by approximating the surfaces, by first or second intention. Both bandaging and suturing are incarnative. Incarnative medicines induce healing with granulation tissue and scars (LDR).

3. Albucasis used this method as well as Galen's. He placed long needles through both sides of the wound, and wound long threads around the exposed ends criss-crossing the wound tightly enough to close it. This is much like what our fine ladies do when closing slits in their sleeves (ie winding yarn around buttons on both sides).

4. Lanfranchi and Henry did it this way: They passed the threaded needle through all layers on one side, and came back up through all layers on the second side. Then they returned the needle through a point about a small fingerbreadth away on the second side, up and out on the first side, at the same distance away from the initial puncture, and tie the knot. They repeated the double sutures to the end of the wound.[434] They put medications over the sutures, and bandaged as in the chest.

Bandaging a Penetrating Wound

When an internal organ has been wounded, you may have to enlarge the wound in the abdominal wall to bring the organ to the surface where you can repair it with sutures. That may be successful in wounds of the fleshy lower end of the stomach and in the colon. Use a continuous leather-worker's stitch rather than the foolish ant-head method of Albucasis, which fails because they fail to hold.

Roger, Jamier, and Theodoric inserted a hollow stent (a tube of elder-tree bark) into the damaged segment of intestine to prevent leakage of feces as it passes through the damaged segment. William and others inserted a segment of an animal's intestine or trachea as the stent, as described by The Four Masters; I think that is no better than the bark-tube. Nature soon rejects the foreign body, passing it beyond the sutures of the repair, and it no longer acts as a stent. Therefore, I think it wise to suture the bowel as is, after emptying its feces and afterwards applying the suture-powder, before replacing the repaired segment within the belly.

When the omentum (ie coëffe) has been extruded and it is dark and necrotic, tie off the bad part and cut it away distal to the tie, and replace the healthy omentum within the belly. Bring the long ends of the ligature as well as the ends of the suture in the intestine out the wound. The dangling threads will be cast off when the wound suppurates.

Do not keep open a gap in the sutured abdominal wall while the repaired internal organs are healing, as advised by Jamier, Roger, and Lanfranchi and his acolytes. They reason that further exposure to air will add to the ill effects already suffered.

The pain that is the complication of the exposed wound of the intestine may lead to convulsions and death. The enlarged wound invites herniation later on, and that is a dire and threatening complication which must be attended to promptly.

---

[434] In other words, a series of mattress sutures (LDR).

A large ant's pincers were held against the closed incision. After they grasped the edges, the bodies were cut away. Curious but not very successful (LDR).

As a topical for treating internal injuries Avicenna applied centaury and terra sigillata, and other medications that he used for wounds in the chest and were effective here. Galen (*Simple Medicines,*Part 6 and *Therapeutics,* Part 6) used equisetum when treating wounds in the intestine and the bladder, and administered clysters of warm sour red wine, whenever the wound had entered the belly.

The diet and the elements of life-style, especially during the first week, should be very delicate and not laxative or suppurative; rather they should encourage consolidation. The Four Masters recommended this bouillon: after soaking wheaten bran in warm water for one hour (rain-water is best), decant the water and add flour, tragacantrh, gum Arabic, sangdragon, consolida and hare's beard. Provide one small cupful three or four times every day.

When the patient is debilitated, feed him chicken-soup with some tragacanth and gum Arabic, enough to tease his appetite. William favored a decoction of frankincense and mastic.

When the intestine or other organs were extruded—whether or not they had been penetrated or sutured or ligated—Galen and Avicenna treated the patient in four ways: Replace the herniated structures; Suture the wound; Apply topical medications; Try to prevent distension and to relieve pain.

1. Replace the herniated organs with your hands. You may have to suspend the patient by his limbs while you gentle shake him. If the intestine is too distended with gas and the wound is too small, you must eliminate the distension or enlarge the wound. I prefer the former and emphasize heat, because it is the exposure to cold that causes the inflation. I foment with applications of a soft sponge wet with water or wine; wine is better because it stimulates the intestine to contract. Roger and Theodoric and other surgeons, slice open live piglets or other small animals and lay their open abdomens on the distended loops. Halyabbas suspended his patients by their limbs and immersed them in a tub of warm water and shook them. Before the dip he lubricated the intestine with oil of violets, whereas Jamier used melted lard. If all those measures fail to reduce the extruded distended gut, Galen enlarged the wound with his 'syringotome', an instrument for dilating fistulas. Avicenna described it; it has two curved blades slightly (not acutely) bent outward at the tips. Cut upwards when the wound is low in the abdomen, and downward when the wound is above the navel. Take care not to cause more injury within the abdomen.
2. Close the wound with the continuous suture while an assistant with palms down keeps the organs inside the abdomen.
3. Use the medicines that Galen called 'bloody', as we described them earlier, such as suture-preserving powders, stupes with wine, and incarnative plasters and bandages.
4. Although we are not very different here from treatments of wounds elsewhere, we wrap the patient with lamb's fleece, moistened with oil, from his axillae to his groins, and we administer clysters as prescribed by Avicenna (*Canon*, Part 3) when he treated edema and ascites with small incisions and punctures (with the cautery).

He wrote, "When the incisions are painful or prickly, apply a wash of oil of aneth or cammonille followed by a plaster of fenugreek, linseeds, guimauve, et al." Henry and others covered the wounded region with a large sac that was filled with a hot mixture of salt and wine, boiled until it was thick. He wrapped bandages over it and replaced it frequently, as hot as tolerated, when it cooled until the pain eased.

If pus accumulated low in the belly after the wound had been sutured, do not hasten to treat it, because it is in a place that is not rich with blood. William said that Nature will resolve it or bring it to a head in the groins where it can be treated as any abscess in that region (ie the deep pelvis).

Wounds of the back side of the abdomen are treated as we described for the back of the chest and the nape.

CHAPTER 7

Wounds Of The Pelvis

Some of the wounds are limited to the tissues surrounding the pelvic cavity, some involve the organs within, and some structures are attached on the outside. The wounds of the containing tissues are treated like those in the abdominal wall. The contents, that is the urinary bladder and the organs of reproduction (ie matrix, also called the amarry) are treated as if they were intra-abdominal organs, differing in their signs and prognoses.

The sign of a wounded bladder is the escape of watery blood through a suprapubic wound. A wounded matrix drains blood from a wound below the navel.

Treat by suturing, especially at the bladder neck and in the lower uterus where the substance is fleshy and will heal. We irrigate by injecting the medicines that we used in the chest.

Wounds of the penis, testicles, and buttocks are treated as flesh wounds. Wounds of the hips are difficult to close with incarnative bandages, but they do well with applications of the topicals used for treating ulcers.

CHAPTER 8

Wounds Of The Thighs, Legs, And Feet

These wounds are no different than those in the upper extremities, and elsewhere, and the prognoses are almost the same. Wounds near the knees and ankles, as are those near the elbow, are most dangerous because of the proximity of bones, ligaments, cords and nerves. In the knee region, matters are even worse because the dependent location attracts humors. Avicenna said that wounds which involve the knee at the level of the knee-cap are very bad, and they can lead to many complications, including delirium.

But all wounds of the limbs lend themselves to closure by incarnative bandaging, even in the feet. Use a long bandage and begin at the ankle, then wrap it around the foot down to the toe nails before proceeding up the leg as far as you need. Some surgeons apply a bandage just to apply topicals on the heel. They apply it like a spur and tie it on the opposite surface. The site of the wound commands the patient to bed. That confirms the old Lombardic adage. "Hold an arm to the chest, and keep the foot in bed."

Here, With God's Help, We Come To The End Of Treatise III

TREATISE IV ON ULCERS

RUBRICS FOR TREATISE IV

DOCTRINE I: SIX CHAPTERS ON ULCERS IN SIMPLE TISSUES

CHAPTER 1

Ulcers in General

According to Galen (*Therapeutics,* Part 4), ulcers are disruptions in tissues that for any of many reasons have not healed. Avicenna added that they may lead to suppuration and corruption. Henry added that we need not wait seven days for the corruption to occur to name a dead fish. 'Dead Fish.'[435] Therefore, the term is based on what happens in the disrupted tissues on whatever day it happens. Galen argued against the Thessalan claim that seven days must pass before an unhealed wound becomes an ulcer, and I think that Henry's claim is worthless, that an uncured apposthem becomes an ulcer after forty days. His statement refers to fistulas, not ulcers. Strictly speaking, an erupted abscess can be called a fistula as soon as it begins to drain, but (ie the drainage tract) of a real fistula is indurated, and that takes time to form, at least several days.

---

[435]   The term in Guy's time was 'rhumb', a turbot, and the proverb implied that one should not make a foolish metaphor. Here Guy implies that time is not a factor in determining when a wound becomes an ulcer (J).

   Migelousaulx agreed.

   However, I (Nicaise) have a different explanation of' "time does not make a turbot". The term refers to the Mistral rather than to the fish, and the Mistral blows in all seasons. Both words, fish and mistral, derive from the Greek ρομβοσ, which denotes roundness in shape. The Abbe de Sade, in his *Life of Petrarch,* of 1764, p. 25, wrote that the wind called 'rhumb' cames from the west, north or south, and is most fierce at Narbonne, Beziers, and Agde; then it tapers off as it enters the sea, and it misses Montpellier.

   I thank G. Bayle of Avignon, for his memorandum to me in which he agrees. "The ancient Provençal phrase is 'leu tem fai ren au raú'. It refers to the northwest wind that most people call the Mistral—in Provence it still is called raú. The power of the wind is overwhelming; it is called the king of the winds. The word 'raú' is derived from the Greek ρομβοσ, which describes a dust-ball produced by the wind. The Latins called the wind 'cistius' because of its turbulence. The proverb means that the Mistral can blow in any seaspon, very true for Avignon." (EN)

   I carry the question one step further. The French word 'temps' can mean both 'time' and 'weather'. Consider here seven days (time) and rhumb (weather) (LDR).

Therefore, the definition above (ie Galen's) is good enough. The term 'disruption' is generic, and many kinds were discussed in our Treatise III on Wounds. Wounds of flesh and other soft tissues are different from corruptions of bone, which really are ulcers. We consider fractures as disruptions, as did Avicenna (Part 4).

There are differences between ulcers and wounds. A wound is a disruption per se, unrelated to any preceding or subsequent condition, including anything that prevents its consolidation and leads to it becoming an ulcer. I do not deny that there may be factors in the wounds themselves that cause them to become ulcers, such as size (large or small), loss of substance, one edge of the wound may be longer than the other, etc. None of them contradict the general purposes of treating wounds as we discussed them in Treatise III. It is obvious, however, that wounds usually are simple, whereas ulcers always are compound, and include their dispositions per se. Avicenna also said that ulcers involve the pus, the corruption, the putrid matter, and the foul scabs and crusts which make them ulcers rather than wounds.

Halyabbas classified ulcers (*Royal Dispositions,*Part 1, Sermon 7) as to three elements: causes, maladies, and complications. We will be even more brief, and avoid confusion with Galen's dispositions (ibid. Part 1), and aim for a clearer understanding of Avicenna and others who classified ulcers according to two elements: causes and complications. The nature of those elements were thoroughly discussed in our Treatises II on Aposthems and on Wounds (III). We shall supplement that material in the discussion that follows here.

As to **Varieties** we include 5 types of ulcers, all well known: virulent and corrosive; sordid and putrid; excavated and deep; fistulous; and cancerous.

As to **Complications:** we include the common types, which to a lesser degree included wounds. Here we specify dyscrasias (ie fevers), pain, abscesses, contusions (ie hematomas), proud flesh, both too soft and too much, hard edges, corrupted bone, varices, chronicity (ie slow to heal), and some other obscure matters. An ulcer is said to be virulent, corrosive, or invasive when its bad elements go beyond simple virulence and mortify and despoil the affected part.

A sordid and putrid ulcer ulcer putrefies the region, or it melts the soft tissues and produces foul crusts,and the stench of dead bodies.

A cavernous ulcer has a small external opening that enters an unexpectedly capacious defect below the surface.

A fistula is like an ulcer with an indurated and sclerotic lining.

A cancer is a large, horrid, sordid ulcer with hard rolled edges.

A dyscrasia is an abnormal bad quality that predominates in an ulcer.

A painful ulcer represents the contrary nature of an ulcer.

An aposthematous ulcer is a swelling which contains an abnormal fluid (pus).

An ulcer with superfluous granulation tissue exceeds that which complicates a wound.

An indurated ulcer is firm and is surrounded by a dark zone, discolored but not malodorous.

An ulcer with corrupted bone is one in which granulation tissue appears on the bone and emits irritating matter that seeps through the dressing pads.

A varicose ulcer appears over clusters of dilated veins which supply its matter.

An ulcer that defeats treatments has no obvious reasons for slow healing.

Causes

These are either antecedent or conjoint. Dino wrote (*Commentary on Canon,* Part 4) that the cause is either Primitive,[436] the result of a wound, or conjoint. Suppuration is not the immediate consequence of a wound. It comes later.

Antecedent Causes

These are the harmful humors, in amounts that can erode and corrupt anywhere in the body. They are engendered by bad habits (ie diet, etc.) and abuses of all or parts of the body, especially the liver and the spleen.

Conjoint Causes

These are the dispositions that were produced by the antecedent causes in the presence of wounds or drainage from abscesses or open pustules. So it is that formy and herpes engender corrosive ulcers as well as carbuncles, anthrax, sordid abscesses and excavated deep ulcers. Galen (*Therapeutics,* Part 4) wrote, "There seem to be three types of ulcers that resist treatment. In one the flesh itself is intemperate (ie lacks resistance); in the second the blood that flows to a region is defective; in the third it is its quantity." He added a fourth cause where the intemperate life-style is the basis for the bad disposition, or the association of an abscess. And he added a fifth in his *Commentary* on the sixth *Aphorism* which I have taken from the Latin translation of the Greek Text (ie not the Arabic)[437]: an ulcer that drains for a year is caused by diseased bone. In his *Therapeutics,* Part 4, he stated that some or all of the dispositions, alone or all together, are the bases for all the various kinds of ulcers, which we treat as simples or as compounds.

Pus

The causes for suppuration and the humoral origins of pus in ulcers are the same as are found for abscesses. Pus is an altered and putrefied derivative of blood (ie its four humors) or the residue of crushed tissues affected by impaired natural heat, just as ashes are the remains of incinerated wood. My term, 'altered', uses what Galen wrote in

---

[436]   The Primitive Cause is a wound or a contusion or a burn. It is followed by suppuration or corruption which is not the actual external source. However, a wound that suffers a loss of tissue cannot heal without suppuration (J).

[437]   See Nicaise's Introduction. The first translations of Galen in Latin were taken from Arabic translations from Greek. Late in the 13thC competent translators used the Greek originals to provide superior versions. Guy was careful in identifying his citations (LDR).

Simple Medications, Part 5. He said that there is a triple alteration. One occurs as the Natural heat acts on the normal nutrition. Another is combustion by foreign heat acting on corruptible matter. Third is the effect of natural heat on a mixture of moderately abnormal matter. The first alteration yields normal nourishment (ie normal humors); the other two yield pus. We say that pus derives from blood (ie sanguine humors) or from burnt flesh. When pus from blood appears in an ulcer it is corrupted when the region of the wound is too weak to defend itself, or it attracts a superfluidity from nearby regions. Or it may be caused by ointments that are too emollient and soften the tissues too much. Dino said that such was both a primary and a conjoint cause, and he added that the superfluidity cannot be perfectly controlled by the body's natural resistance after the flux has begun to deteriorate into pus, and he called it the material cause. In that conflict, the body's natural heat continues and becomes abnormal. The result is the conversion of the humors into pus. Galen (*Commentary 2*) said that pus is made from an inflamed humor, just as ashes are the residue of combusted wood. Hippocrates (*Aphorisms*) said the same about the generation of pus, and the causes for the pain and the fevers that accompany it.

There are two kinds of pus: one is white, smooth, and free of stench. It is the pus in most abscesses. Some pus is thinner and is called virulent; some thickens and is called sordid or excrement. What persists as part liquid and part thick simply is pus. Small amounts may appear in saucerized wounds or in ordinary incised wounds, and large amounts appear in ulcers. Henry said it well, that pus in ulcers exceeds that of wounds and you can prognosticate that moderate amounts of pus come from a moderate superfluidity of humors, and it is off-white in color. Galen (*Techni*, Part 3) said that organs which engender sperm (white) also produce white pus, as we observe when tissues are washed for a long time.[438]

Virulence is a thin superfluidity derived from aqueous humors, warm, cool, or serous, or pink Pus in a sordid ulcer is thicker and is engendered from three thickened humors. There are three types: One is fatty, lumpy, curdled and white. One is dark. The third is like cinders. The third has hard flakes, resembling fish-scales, produced by the body (ie not foreign matter) and deposited at the margins.

A crusted ulcer is like the scaly one but the crusts are much thicker and larger, and they lie on the surface of the ulcer.

The Signs

We have defined them in earlier chapters. Whenever you encounter a draining wound with unlaudable pus, it is an ulcer; no other diagnostic sign is needed. Hippocrates (*Aphorism 6*) said that all ulcers that last for a year or longer contain rotten bone

---

[438] Joubert noted that persistent irrigation of flesh will wash away the red color of blood. One notes that well cooked meat has lost its color, a normal occurrence. Meat that is half-cooked (ie rare) retains its color (J).

that must be expelled (ie sequestrum) and the ulcer then will heal with an indented scar. And Galen (as taken from the Arabic translation) said that ulcers that recur are especially evil.

Avicenna (*Canon*, Part 4) said that ulcers that refill with granulation tissues after they seem to have healed, are developing fistulas.

And Halyabbas (*Royal Dispositions,* Part 1, Sermon 8) said that any ulcer, simple or compound, that persisted for forty days was a fistula, but not a true fistula, as we will discuss later.

Avicenna, Part 4. said that indurated ulcers that show green or black matter are bad, because those are signs that the Natural Spirit has been destroyed (ie gangrene). He also said that the cool ulcers that are pale and have rolled edges should be treated with warm medicines. When they are pinkish and warm, use coolants. Simply touching them can be informative. Dryness and moisture are well-known signs, as are colors. When the malignant ulcers are accompanied by discoloration of the body as a whole, gray or yellow, that means that the liver and its blood are corrupt.

Note this: ulcers that follow other illnesses are not easy to cure. And ulcers that cause loss of hair in their vicinity are bad, but if the hair regenerates, that is a good sign. Hippocrates *(Signs of Impending Death)* wrote when a man has ulcers and abscesses in many places, he is moribund. And if a virulent ulcer reverts to laudable pus, that is a good sign, indicating that the patient's Nature is strong and that he will overcome the bad humors.

Prognoses

Avicenna's text, as follows, is a bit confusing. He said that ulcers at the ends of the muscles of the back, and ulcers of the thighs, the arms, and those that penetrate to the internal organs are dangerous. Furthermore, even those that we have cleaned perfectly and have applied incarnatives may produce bad proud flesh.[439] Round ulcers are slower to heal and may be lethal in babies. He recommended using a cautery to change their shapes.[440]

Also, ulcers in the extremities led to ulcers in the glands of the groins and axillae, especially in plethoric (ie obese) individuals. The bad humors traveling from their sources in the liver, et al, must pass through gland-bearing regions on their way to the ulcers, and some of it is trapped in those glands and form abscesses, especially when the patient is

---

[439]   Guy's repeated references to bad proud flesh describe the spongy granulation tissue that is riddled with pus, familiar to surgeons since the early times (LDR).

[440]   A side-side union of the surfaces of a wound or ulcer is resisted by the round contours. The surgeon incised the edges at the opposite ends of a diameter, with a knife or cautery,. That defined two opposed surfaces to bring together with sutures or bandages (LDR). Joubert added that the slow-to-heal rounded ulcer caused intolerable suffering that was lethal in babies (J).

plethoric.[441] It s a good sign when the medications seem to work well and do not harm the ulcers. But if you detect a bad response and the wetness (ie purulence) increases, the treatments should be changed to more potent desiccatives. But when those intensify the local heat and redness, add coolants. And if the additives do just the opposite, use warming medicines. When you observe softening, add astringents. When the medicine erodes and excavates, you must lessen its abstersives. Always be careful when using intense astringents because they may erode and increase the virulence of the pus in an evil ulcer. When that occurs, and you react with too potent detergents, the ulcer will deepen and become more hot, and become an abscess. The patient will feel the pain of the erosion into normal tissues.

In addition to all the bad problems with ulcers, consider the effects of the hot south winds that are so humid. Some say that is why ulcers are more difficult to treat in Avignon than at Paris.

The opposite holds for ulcers of the head where coolness and dryness are harmful to the brain.

Furthermore, be aware of the similarities of prognoses for ulcers are similar to those for wounds, and I refer you to the material in the Treatise on Wounds.

Treatments

The treatments have two aspects: for the ulcer qua ulcer, and for the ulcer with its causes, complications, where it is in the body, and other factors.

An ulcer as such needs to be dried (ie cleared of pus)[442]. Hippocrates said so as cited by Galen in *Therapeutics,* Part 4. Although the same is true for wounds, ulcers require much more desiccation, because they are much wetter. I have dealt with much of this in the Treatise on Wounds, especially in the section dealing with excavated wounds.

When we deal with the ulcer-complex, we must treat the causes and other factors that produced it or augmented it, as Galen stated (idem), as well as the ulcer itself. Refer to that subject in our Treatise on Aposthems, and in the Treatise on Wounds, in re complications.

The management of those factors has two aspects (idem.): getting rid of them and curing the damage they have caused. In other words, the twofold treatment is curative as well as preventive. That may be easy when the dispositions are minor, but when they are major, cicatrization of the ulcer may be delayed until the complications have been cured.

---

[441]   There was no concept of centripetal flow of lymph or venous blood to explain the association of ulcers of any kind or the abscesses in the regional lymph-nodes (LDR).

[442]   'Dryness': the surfaces of a wound will not cohere when they are 'wet' with pus. Therefore the surgeon emphasizes 'drying', eliminating the pus. However, surfaces that are too dry will not unite. Therefore, the surgeon may 'freshen' the surfaces by gently rubbing or scraping to moisten the pink surfaces with blood (LDR).

Therefore, the treatment of complex ulcers and their complications has three or four special elements. First is the life-style (ie habits). Second is to restore the normal balance of humors in the patient's blood, which are the antecedent causes. Third is to correct the complications and the associated abnormalities. Fourth is to eliminate the offending matter and allow you to treat the ulcer as an excavated wound.

The first and second elements are determined by the nature of the bad matter (humor) that is produced within the body. We use evacuations to attract the, matter from the ulcer: phlebotomy, purgation, diet, contraries, emetics, and other diversions. Also we deflect the inflow by applying bandages, epithemes, ointments of bol d'armenie, and other coolants and astringents. More of this can be found in the Treatrise on Aposthems.

Galen (idem.) wrote, "When bad humors end up as the matter in an ulcer, we treat as follows: Treat depletion (ie cacothymia) or plethora (repletion) as we do for aposthems, in two ways. When the inflow of humors is not excessive and the pus is not vicious, deflect and repercuss it, and apply coolants proximal to the ulcer. Also, you may apply bandages that contain repercussives,beginning below the lesion and wrapping over it toward the source of the inflow, using Hippocrate's method for bandaging fractures; such a bandage will retard the inflow. At the same time, apply the desiccatives which we use for simple ulcers and for wounds, but observe the differences. However, when the medicines fail to restrain the flux, try hard to find the cause (the source of the humor) and get rid of it."

If the fault is the weakness of the region in rejecting the rheum, treat it while treating the ulcer itself. If it is due to an excess (ie plethora) of blood and its defective humors, whether it is limited to the affected region or is throughout the body, treat it as such. If the incapacity is due to some local dyscrasia treat it as we describe below. But in all cases, local or general, disperse the matter as we described in the preceding Treatises.

The third goal, correcting the complications and the conjoint dispositions, is directed specifically at the abnormality in the ulcer. As Galen advised (idem.), treat one at a time with its specific contrary, rather than attack all the complications at the same time.

Treating The Dyscrastic Ulcer

If it is in the flesh and is dry and firm, and is shriveling, restore the tissue with repeated fomentations of warm water until the tissues redden and swell. Stop before your actions drive off the good humors that are attracted. Use water rather than wine because the fomenting agent must be wetter than the healthy flesh you are aiming for. But if the tissues are abnormally moist (ie edematous), treat with the contrary; fomenting with wine or vinegar with their drying qualities, or decoctions of drying herbs. Apply coolants when the dyscrasia is warm, and use warm things against the cool. Of course, all this repeats what we described in the treatments for wounds with abnormal temperatures.

For Painful Ulcers[443]

Galen said that nothing is worse than pain to interfere with your treatments for rheum[444] and other abnormalities. But, Avicenna advised us not to categorize an ulcer as painful unless the pain is severe and demands immediate efforts to relieve it before doing anything else for the ulcer. We should apply the emollients that we know are contrary to the ulcer, and which relieve pain. We have listed many analgesics in our Treatises on Aposthems and on Wounds. Use them here.

Ulcers With Abscesses

Avicenna warned us to do whatever we can to avoid this complication. Because it will delay all other treatments. When the prevention is not possible, treat the abscess first, as we described it in the Treatises on Aposthems and on Wounds.

Treatment For Contused Ulcers

Galen said to use suppuratives when crushed tissues are necrotic in order to convert them to pus, then to be treated as such. Avicenna advised us first to use softeners and moisteners, as we did for contused aposthems and wounds.

Ulcers With Superfluous Proud Flesh

Avicenna recommended that we should use potent medicines to corrode away the granulation tissues. First apply a coolant ointment and then corrosives such as we use to get rid of eschars. Then treat the exposed ulcer. Some good corrosives: troches of asphodel, the apostles' ointment, the Egyptiac, and others that we use to get rid of bad proud flesh.

Indurated Ulcers With Rolled Edges

Avicenna described these when the tissues around the ulcers are corrupted and are beginning to turn green and black. He began his treatments with scarification with cups, and with phlebotomy. Then he applied a dry sponge followed by desiccatives. When the evil spread, Galen (ibid., Part 4) discontinued the failed applications and proposed wide

---

[443] Guy's term is "doloreux". Our term, "Painful" obviously must be less than adequate to connote the suffering that Guy described. Dolorous meant relentless pain, sleeplessness, stench, uncomfortable dressings, itching, and the additional discomforts of the surgeon's topical medicines (LDR).

[444] Rheum included pus, serum, lymph, and any other discharge from the ulcer (LDR).

excision of the diseased tissues, or a long course of treatments with potent caustics. In such cases he insisted first to obtain the patient's consent, because he may prefer the long course rather than the operation, or he may choose the latter and be treated with a single, radical, surgical procedure, and avoid the corrosives.

Ulcers With Varices

Galen (idem.) said that we first should treat the varices with the methods we described in the Treatise on Aposthems, and then attend to the ulcer.

Ulcers With Corrupted Bone

Avicenna advised against hasty measures to remove the diseased portions of muscles and bone, and we will repeat this in the chapter on bones. It is better simply to incise the overlying tissues and expose the diseased bone as we will describe later for ulcers in the legs, and use both the knife and corrosives. After exposing it we can scrape away the diseased bone or cut it away. I quote, "To treat diseased bone, shave it away, or cut or saw it. Get rid of the corruption by repeatedly scraping until a healthy layer (ie periosteum) grows over the defect. That normal process can be assisted with medications." Avicenna preferred a plaster of aristolochium, iris, myrrh, aloes, the bark of opoponax, burnt cambil[445], pine-tree bark, et al. Mix all with honey.

Lanfranchi preferred applications of a red-hot cautery on the cleanly scraped bone. He said that the causes for corruption were moist, and the treatment should be dry; the cautery was better than corrosives. He applied the cautery above and below the diseased segment. Then he sprinkled it with warm oil of rosat.

My own practice after burning with the cautery uses the oil of rosat and egg whites for three days, and egg-yolks for three days, and then butter with rosat honey. Then I use detergents until I see granulation tissues on the bone. Then I apply incarnatives and consolidatives as powders or plasters.

If the disease penetrates into the marrow cavity, Avicenna insisted that we remove both the diseased bone and the marrow; He described a thirty-year-old man with a diseased femur. He used medicated drains until the soft tissues adhered to healthy bone. He marked the healthy margins where he could cut fearlessly. However, when the disease involves the head of the femur or the hip, you must abstain from such treatment, as is also the case with the vertebrae where the spinal cord is at risk.

---

[445] Cambil: a variously translated Arabic term. Some say it is a kind of Manna; others say it is Alkekeinji. Guy accepted Avicenna's claim that it is a red clay from the Middle-East or Athens. When it was rubbed in one's hands, it stained the skin. It was used as an ointment for the hads (J).

Ulcers Difficult To Cure Because We Cannot Determine The Causes

Avicenna said that these were neither purulent nor eroded nor undermined. They tend to spread on flat surfaces and to reopen after they have closed.[446] These call for potent desiccatives, such as flowers of bronze or iron, a collagen, colcothar, dragacanth, alum, galls, and repercussives that block the flow of bad humors into the region. Here Galen used waxes, ointments and powders (*Types of Compounds,* Part 4).

I use an ointment favored by Bruno: cadmia, alum, gelatin, flowers of bronze and burnt bronze flakes, cypress gum, wax, oils of rosat and myrtle.

The fourth element of treatment: Here we treat the ulcers that have been cleaned of all the bad matter as if it is an excavated wound. Galen remarked (*Therapeutics* Part 4) that all ulcers have exposed surfaces, and should be considered as erosions, and treated as such. He wrote (idem.), "There is nothing in the treatment of an ulcer that at the same time does not treat what causes it and makes it larger." He later added that after those dispositions were cured, we are left with a simple ulcer to treat as we have described.

To recapitulate, you will note how much the treatments for ulcer are like those for wounds and for aposthems.

---

[446]   I think that Guy describes basal-cell carcinomas (LDR).

CHAPTER 2

Purulent Ulcers—The Virulent And The Corrosive

These differ only in degree. At their onset when they do little more than discharge thin pus, they are called *virulent*. When that deteriorates and the pus is erosive, and the ulcer enlarges, or it forms an eschar it is called *corrosive*. When only its surface opening enlarges, it is called *ambulatory*. If it deteriorates further, it is called *invasive*, and then it may become a *lupus* or a *cancer*.

Cause

These ulcers derive from bad choleric humors which are bitter and erosive, the result of a defective desiccation. The same sort of humors may cause formis and herpes and other irritating pustules. The ulcer may be caused by the unwise use of irritating topical medications in the treatment of wounds.

Treatments

First attend to the patient's habits (ie life-style), and then to purgation, as we described for treating formis and herpes. Galen (*Therapy Contrary to Thessalus,* Part 4) wrote: "We often have seen what happens when someone tries to cure an undermining ulcer by bleeding. Perhaps an otherwise healthy person will scratch himself somewhere and a blister will appear and cause terrible itching. He will scratch and rupture the blister and an ulcer will ensue; it will drain bad matter and will erode—all of which will happen within three or four days. Then some Thessalan physician will advise an entirely bad treatment (ie only for the ulcer) whereas I recognize that the illness involves the entire complexion of the body, and I can tell what humors are at fault (ie excessive), not only from what happens in the ulcer but from the body's signs, and I will use the medicines designed to control them. As I have written elsewhere, and as did all the old-time Masters who used their common sense when they treated ulcers and other ailments,

I first attack the causes. It is foolish the treat the ulcer while the causes persist. Therefore, in all cases when the efficient causes persist, begin the treatment there." Galen added (*Therapeutics,* Part 3) that local applications of resolutives at the ulcer will act like cupping when the body is plethoric, and they will attract more of the bad matter than they can resolve. Therefore, before using them, evacuate the body or at least the affected region proximal to the lesion. Avicenna wrote (*Techni,* Part 3), "As general rule, first eliminate the efficient cause, then attack the dyscrasias that are the products of the malady."

As in the case of the Roman woman with herpes (ie a la Galen), we treat the lesion second. If it is warm, we cool it with dry and astringent desiccatives, and we irrigate the ulcer and its surround with alum-water; it acts as a repercussive and desiccative agent, according to Avicenna. Or we use decoctions of roses or iron-rust or a decoction of sedge, myrobalans, cypress, plantains, pomegranate-peel, et al. Then we apply a defensive of bol d'armenie around the ulcer. You may put into the ulcer a desiccative powder of litharge, tuthy, antimony, burnt bronze, coral, blood-stone, etc. Small pellets of sharpie coated with an ointment of litharge also are good; or coated with pompholyx as described in our Antidotary. Overlay the medicated ulcer with a pad moistened with oxycrat. Then wrap a pressure bandage such as we have described.

My own treatments for these ulcers use the washes alone, followed by placing a thin sheet of lead in the ulcer. I have coated it with some mercury and some plantain-water. Then I wrap my bandage. I have been pleased with the good results, but that will not convince the Idiots. However, I interpret Galen in his Book 9 of *Simple Medicament* to have meant lead where he wrote molybdenum. He used other terms that I fail to understand, but he explained the value of alum-water in Book 4. He describes the bandage admirably in the same book.

When a thick crust covers the ulcer in spite of the treatments, purge repeatedly and erode the stuff with corrosives, or use the actual cautery (which is the better way), or use a caustic powder of asphodels or quick-lime. Ink and sublimated arsenic (in small amounts) are useful, as we used them for esthiomene and abscesses. In those cases, apply coolants as defensives around the ulcer. If the lesion resists, you may have to extirpate it with a knife.

CHAPTER 3

Sordid and Putrid Ulcers

These ulcers are similar. A dirty ulcer with thick pus is called *sordid*. When also it shows necrosis, with shriveled tissues, and the stench is that of a cadaver, it is called *putrid* by default. And if it spreads under the neighboring skin it is called *esthiomene*, and that portends death.

Causes

The evil humors are thick and hot sanguine. The heat adds venomous qualities, as seen with carbuncles and anthrax, and in some aposthemes and wounds that have been badly treated.

Treatments

Establish healthy habits, and use evacuations as we prescribed them for carbuncles and crusted or putrid fistulas. Avicenna (Part 4) wrote that local treatments are improved after evacuations that clean the entire body, or at least the entire affected part. We use cups, leeches, epithems, and laxatives to prevent constipation. Improve the blood with good diets, etc. Then treat the ulcer itself.

1. Wash away the foul matter with honey-water. And use the apostles' or Egyptian ointments as detergents. Then, as Lanfranchi insisted, we mondify with the juice of aloin, rosat-honey, barley-meal, and myrrh. Always apply bol d'armenie as defensives around the ulcer, and cover it with stupes wet with oxycrat.
2. If the sordid mess has deteriorated and accumulates more putrid and corrupt matter, use irrigations of oxycrat or of soap or lye. Use plasters made with salt-fish, barley-meal, long (stressed by Theodoric) aristolochia, squills (in wine) and honey.

   In such cases, Avicenna used a topical ointment that he claimed was well tested and approved by Bruno: red dragacanth, quick-lime, alum, pomegranate-peel, frankincense, oak-galls, wax and oil. I add vitriol, colcothar and tragacanth, all boiled in vinegar to make a linament. And I use the bol d'armenie around the lesion, or stupes of oxycrat.

3. When things get too bad in spite of all the foregoing measures, you may have to extirpate the lesion with a hot cautery, or with caustics, or with the knife. Remove all the bad matter until you expose healthy pink tissues that bleed. The best caustic for this purpose is sublimated arsenic, as we use it for esthiomene and glands. Avicenna said that despite all our efforts, we may have to amputate a diseased limb in order to save the rest of the patient.

CHAPTER 4

Deep and Excavated (Cavernous) Ulcers

These ulcers have smaller surface opening than the defect in the depths; and there may be pockets in several directions. But the lining surfaces are not hardened or callous. In that way they differ from fistulas—a fact not admitted by the idiots who call all such ulcers fistulas. I shall explain later why that is not the case.

Cause

These ulcers begin as aposthems. Because trapped pus must be drained, and a deep wound cannot adequately be cleaned through a narrow opening at the surface, and a counter-drainage may be needed. Unless drained, the pus becomes corrosive and foul, and it prevents the surfaces from forming granulation tissues to consolidate the surfaces. The pus burrows and cavitates and the region cannot resist the influx of matter that is attracted from nearby tissues as well as from the rest of the body. So it is that an ulcer is created, one that is difficult to treat.

You can determine the size and the contours of the cavity with a silver or lead probe, or with a twig, or a wax taper, or by injecting some dye. The color of the discharge also will be informative. Is it pink (flesh-like) and thin; is it warm; is it white and thick; is it cool?

You can tell when an ulcer is closing by the appearance of even a small amount of laudable pus, the reduction in the size of a mass as the pain subsides, all of which is described in *Glaucon*, Part 2. Lacking any of the preceding signs, we know that the ulcer is not healing.

Treatment

Here also we emphasize healthy habits and proper purges, suited to oppose the offending humors, just as we described them for treating aposthems. Treat the ulcer itself with desiccative ointments and plasters, followed by incarnatives, applied on compresses, and lay on stupes wet with astringents, all wrapped with good bandages. Galen (*Glaucon*) favored the apostles', the black, and the diapalma ointments. When the topicals applied at the orifice cannot reach the bottom, you may remedy that default by elevating the limb and reversing the direction of the drainage.

But if that is not feasible, you may incise the opening, even to the full extent of the ulcer. Then you can clean and dry it, using packs and drains, as we used for treating deep and excavated wounds. I refer the Reader to those Sections. I suggest that you do not empty the pus before you make the incision; let what you see guide you. When the tract is narrow, insert a probe. Use only one well lubricated probe that can be introduced without pain, or use a probe with an eye-hole that can be threaded as a seton with a hempen or other thin cord, and push them through.[447] After easing the pain of the incision, use egg-white or other hemostatic agents to control the bleeding. Lubricate the seton and remove the needle; sew the ends or tie one end to another cord laden with detergent ointments that can be pulled into the tract. Use the medications for the ulcer and the surrounding surfaces that we described in prior sections.

When an incision for drainage is not feasible, Avicenna used a clyster-syringe (a la Albucasis) to irrigate with detergents until the returns were clear. Then he injected incarnatives.

Galen favored honey (*Glaucon)* at first, followed with wine and honey. To eliminate liquid matter, he preferred watered honey, and later on added wine to promote agglutination. He put the latter on a piece of new sponge. Avicenna irrigated with a lye-water or sea-water with alum when the pus was acrid; those waters were repercussive as well as detergent. Albucasis used the Egyptian ointment and added water and honey. Lanfranchi, Henry and others used watered honey plus a decoction of barley, lentils, and balaustium when the ulcer was hot and the virulent discharge was pink. When the ulcer was cool and the virulent pus was watery, they irrigated with wine and honey supplemented with a decoction of aloin, marrubium, pimpinell and myrtle.

As to the incarnatives, Avicenna insisted that they be thin liquids, although the detergents should be viscous and sticky. Some incarnatives were injectable and others were applied on drains and pads. I favor the apostles' ointment, as did Avicenna to which I add centaury; it works well. Like Galen, I add Illyric iris as a consolidative, followed with a flour of ers et al. As noted previously, we use plasters, and oiled stupes reinforced with diapalma or the black or red ointments with added galls, or we use honey heated with a powder of frankincense, myrrh, aloes, and some wine. We wrap bandages a la Galen and Henry. Although Galen's text is a bit confusing, I will try to summarize it, "After we have cleaned away the bad matter, we apply an incarnative plaster over the orifice and bandage it with a double-ended compressive bandage. We begin below the ulcer and lead to it, and we release the pressure after we pass it. We place a small pad with an incarnative ointment directly over the orifice, and continue wrappings until proud flesh appears, replacing the outer wrapping daily for three days. That exerts enough pressure to obliterate the bottom of the cavity." Many have questioned the use of a drain, as did Henry and my Master at Bologna, and we take care lest it be packed in too tightly. We may use tubes or a limp folded cloth to favor an easy escape of pus. To that end we lay a sponge atop the pads to soak up what comes through.

---

[447] Guy's instructions are confusing. He uses counterincisions in two ways. In one he makes a large incision to expose the interior of the ulcer for his topicals. In the second he guides a seton-bearing needle out a perforation and uses the thread to pull through a strip of cloth or a larger cord on which to isert his medications. He does not use the seton to saw open a larger opening between the ulcer and the needle hole (LDR).

CHAPTER 5

Fistulas

A fistula, as defined by Galen *(Unnatural Tumors)*, is a deeply excavated ulcer lined by a firm callus, which usually discharges large amounts of virulent pus. A fistula also is a long narrow serpentine sinus. The lining is indurated and it can be felt through the surrounding soft tissues, and it may discharge large amounts of pus, as if it is an abscess. Sometimes the sinus is plugged and the output is intermittent, according to how it is filled and emptied. Therefore, neither the virulent pus alone, nor the indurated lining differentiate it (ie from an excavate ulcer), nor does Master Arnold's claim that the discharge is persistent whether it be watery or viscous or foul. The fact that the discharge may stop does not mean that the fistula has healed. As long as the fistulous tract remains it cannot be cured simply by plugging the orifice. To be sure, you can dry it and plug it imperfectly, but only for a while. Neither Galen nor Albucasis treated it that way, and the latter said that a dried fistula may drain again after a few hours. Halyabbas, Bruno, Jamier, and the Four Masters agreed.

As to the induration, Henry agreed with Roger and Roland who insisted that successful treatment must consume the callus, and their Glossaters agreed.

Some fistulas arise in flesh, others stem from veins, nerves, or bones. Some are straight others are angulated or tortuous. Some have a single orifice, others have several. Some involve joints, eyes (ie lacrimal), or the throat, the chest, the abdomen, or the 'privates'. They differ, and call for different prognoses and treatments.

Causes

The causes are the same as those for the cavernous ulcers which engender all fistulas, but the discharge from fistulas are worse and do more harm. William of Saliceto said that their bad humors were from phlegm or melancholy that have undergone a combustion that led to acrid and venomous changes. Arnold said it is the natural humidity and the coolness of those humors that has been corrupted, and that is what causes them to engender fistulas.

369

Signs

The diagnosis is indicated by the leathery induration and the tubular configuration that gives them the name "whistle' (ie flute-like). The complicating features, such as the foul virulent drainage and the lack of pain (unless a nerve is the source) are diagnostic. When flesh is the source, the discharge is thick, sticky, opaque, and lumpy. As noted, a nervous source is painful, and the discharge is pale. A fistula arising from a vein emits bloody fluid with flecks (like wine lees). The discharge from bones is thin. The returns after irrigation and the color of the pus on the dressings help with the differential diagnosis. Cues may be taken from the locations and the timing. For example, a fistula that continues for more than a year comes from bone rather than from flesh or nerves. Later, we will discuss the lesions in bones that cause fistulas.

You can predict which fistulas will be hard to cure, judging from the depth of the source, the tortuosity, the involvement of bone, the chronicity, and the number of eroded side-pockets. Furthermore, beware of the difficulties when the fistula involves a noble structure, or enters the chest, the abdomen, the urinary bladder, the ribs, the vertebrae, or joints (especially the hands and the feet). Not knowing can make matters worse, as Albucasis noted.

Treatments

There are two categories: the General, and the Particular. The General, as in most maladies, has three elements: The healthy habits (ie Life-Style)), evacuating the evil matter, and treating the internal sources while drying the fistula to enable it to heal. First we seek a healthy regimen. Second we use evacuations suited to the offensive matter. All of this has been described in our chapter on aposthems, especially the cold ones. Third, we prescribe the most suitable beverages: agrimony, plantains, and olive-leaves, all minced and cooked in white wine. Offer a glass-full daily at dawn. Another good beverage is made from osmundia and centaury cooked in white wine. Like the first, it is useful for expelling bony sequestra.

There are three or four Particular Measures. The first is the enlargement of the external opening and the dilation of the tract of the fistula. Second, destroy the lining of the fistula by necrosis. Third, clean out the dead tissue. Fourth, encourage the growth of granulation tissue to fill and to heal the clean bed First, after tracing the full length of the fistula, insert a rootlet of gentian, aristolochia, dragontia, or bryony, or use a strip of sponge, twisted and moistened, cut to size. I do not use sambucus or ebulus twigs because they break when you try to remove them. Therefore, I suggest that you fasten a thread to whatever you insert, to control the depth of the insertion and to simplify the removal. Let it be long enough to pass through the entire length and of a diameter to fit that of the fistula. Leave it in place for twelve hours before removing it. When the tract is sufficiently enlarged, go on to step two. Second, use one or all three ways to destroy the lining of the fistula. Inject corrosive substances, or burn it with a hot cautery, or dissect it out with a knife.

William preferred the first method. He inserted a drain wet with asphodels. Roger used quick-lime, or a soap-lye, or arsenic (it never fails). However, when the fistula has side-pockets, you should dilute the caustics with wine or vinegar and inject the liquid through a tube that can get it into all the recesses. After the injection, plug the orifice to keep the corrosives within the fistula for enough time to do their work.

Arnold said that the tortuous and complex fistula cannot be completely necrotized unless first you soak it with acrid fluids, such as bile or niter. The best corrosive is the apothecary's aqua-forte (ie nitric acid). It mortifies and destroys all fistulas.

You may use a hot cautery to cut and burn the fistula. First insert a wood twig full length (ie as a dilator), as we did for cavernous ulcers. Then pour in egg-whites and enlarge the orifice by packing a drain. The following day cauterize the tract with the hot cautery or with a potential cautery of asphodels or arsenic. Avicenna used the sublimate of mercury. Maintain the action until none of the diseased tissue or callus remains untreated. When it is dry and has separated from the healthy flesh, remove it. Always apply coolants over the adjacent tissues. This method also is used to remove diseased bone.

Roger said that you will know when the corrosives have been successful when the non-ulcerated tissue is inflamed. Usually that will happen within three days. After a successful extirpation, mitigate the pain: at first with oil and egg-yolks, and then with butter or other fats, until the burning pain subsides and the eschar separates as the surrounding living tissue suppurates.

When the pus changes for the better, and the amounts are less, you will know that the fistula has been destroyed. As Arnold put it, "When the fistula emits laudable pus, no longer is it a fistula."

When you use a knife instead of a cautery, core out the entire tract, full length, leaving behind no remnants of callus or corrupted matter. You will know when you are successful; healthy granulation tissue indicates a complete cure (Avicenna).

When all else fails, you may call on St. Eloi, the popular favorite, even though Lanfranchi promised a cure with agrimony and salt. I cannot vouch for that claim. However, it may work in what Avicenna called a recent fistula. In such a case, the cure is effective if you set the scene for the herb by swallowing a spoonful of it while reciting the Pater Noster. So much, then, for the second measures.[448]

The third and fourth measures are detergents and incarnatives, as we used them for treating cavernous ulcers.

Palliation

When a fistula cannot be treated for cure, as when it is at an inaccessible site, or when it arises in a noble organ, or when it traverses or impinges on important nerves or veins,

---

[448] I assume that Guy wrote this paragraph with tongue-in-cheek. However, the Reader should be aware that Guy was a deeply religious man who would hesitate to offend Saint Eloi! (LDR).

or when the patient is too feeble to withstand the rigors of the treatments, or when he is terrified by the prospects and prefers to accept the problems of the malady, or when indeed the consequences of the treatment are worse than the fistula, as when the treatment of a fistula-in-ano can lead to incontinence of feces, then you must consider palliation alone. Emphasize healthy habits (and diet). Purge and divert the bad humors from the fistula to less vital parts. Clean the accumulated granulation tissue. Instill and fill the fistula with such familiar drying agents as diapalma and the black ointment, and apply them as plasters. Keep the orifice covered and forbid immersion in water. Avoid activities that cause pain. Avicenna and Arnold insisted that those long-term palliations of chronic fistulas must not block the escape of the discharges lest they be diverted or distributed into nearby regions.

CHAPTER 6

Cancerous Ulcers

A cancerous ulcer is rounded and it emits a dreadful stench. It has thick edges that are firm, nut-like, and rolled inward and over the excavated interior. It is dark purplish, and the surrounded area is mottled with veins filled with melancholic blood.

Avicenna said that the name cancer, has two sources: it digs in and fastens itself like a crab, and its contours and outstretched congested veins resembles a crab with its many legs. Henry added that it eats away like a crab devouring a dead fish.

There are three kinds of cancers determined by its appearance, by the matter in it, and by the part of the body that it invades.

First: Is it soft and movable and small, and cause little pain; or is it large, angry, and painful? Second, was the melancholy humor at fault a normal melancholy before combustion; or was it a corrupted combusted choler? Third: is the cancer situated in simple tissues, such as flesh, veins, nerves, or bones; or is it in a complex region, such as the face, where it is called a 'do not touch me' (ie noli me tangere), or at the hip where it is called a lupus, or in the center of the body where it is called a belt (by Roger). Bruno and Theodoric said that Roger's appellation was not used by the ancient masters.

Causes

A Cancerous ulcer may derive from a non-ulcer, or from an ulcer that has been irritated, or when an ignoramus incises a melancholic aposthem and converts it into a cancerous ulcer[449]. Then the wound deteriorates when it is treated with acrid medications. The combusted melancholic humors that caused the aposthem are combusted further and are agitated. Incision and drainage attracts more humors from everywhere in the body as well as from neighboring tissues. They suppurate and are overheated until they, too, are acrid and venomous. And that is how the evil is augmented and the lesion becomes a cancer.

---

[449] The 'ignoramus' erred in thinking the aposthem was an abscess when he incised it to drain pus, and he encountered solid tissue (LDR).

Other causes are the primitive humors that displace the antecedents, or join with them. That is a commonly held explanation.

Signs

The signs have been described: the hard rim, the large circular defect, the excavation and the invasion. The efficient causes and the complication are indicators: the terrible stench is impossible to describe in writing. Once you have smelled it, you can recognize the odor even from a far distance. When you wash it with very mild corrosives the matter turns ashy gray and viscous. It is irritated even by mild corrosives, and its evil character is worsened—as noted by Lanfranchi and Henry.

Diagnosis and Prognosis

The difference between the true cancerous ulcer and the non-ulcer is the issue. In addition to what was given above I emphasize its chronicity and resistance to treatment. Furthermore, Albucasis said that a large and long-standing cancer should be left alone, and he added that he did not attempt to cure such lesions if they had been treated by other physicians. Galen (*Commentary On The Aphorisms #6)* wrote that he would not treat such lesions unless he was forced to do so by entreaties and supplications. Furthermore, when a chronic cancer is embedded in a member and it invades veins, nerves, and bone, or when its penetration-in-depth is not known, or when it is in a place where the risks are great, or when the patient is feeble or fearful, he would treat to palliate rather than attempt to cure. The failed attempt leads to an earlier death of the patient, as was stated in the Aphorism. Galen restates what we wrote in our chapter on cancers. Avicenna said that the seeds of cancer (ie local metastases) persist as masses that do not ulcerate. Those represent the displacement of some of the cancer by failed treatments.

Finally, William of Saliceto said that cancer is a disappointing and regrettable malady, and that the more one touches it the worse it becomes. He advised even when examining it one should touch it lightly if at all; hence the term 'noli me tangere'.

Treatments

The treatments are General and Particular. The General measures are the usual three items, life habits, evacuations, and support for the vital functions.

The first two are managed by providing a good diet etc, and by prescribing laxatives used for treating melancholic aposthems. The third measures include beverages and herbs suspended from the neck. They support the patient's beliefs more than offer specific medications. Usually they are thread-like herbs such as ceterach, Herb Robert, and scrophularia, the so-called cancer-herb. All of them make good (palatable) beverages. The centinodia was favored by Arnold who also believed that crayfish were good[450]. Albert said that emeralds and sapphires were good against cancer.

Galen favored Theriac for treating Serpent Cancers (ie of Tyros) because it repercussed the venom of a cancer outward toward the skin

The Particular treatments are twofold. (*Glaucon*, Part 3). First is the complete extirpation of a cancer when that is possible. Second is the palliation when that is not possible, as we described above.

Extirpation is accomplished in two ways. First is by cutting with a knife or cautery. Second is by corrosion. Excision should not be performed inless one can completely remove the lesion and all of its roots. A failed attempt will make the patient worse than if he was left alone. After an extirpation, squeeze out all the melancholic blood from the surrounding vessels before applying a hemostatic cautery.

The second way to eliminate the entire cancer is with caustic medicaments. As Hippocrates (*Aphoirism #1)* wrote, as potent malady requires potent medicine. As we use it for esthiomene and glands, the best caustic is prepared arsenic. I will repeat the method. On Day One, Theodoric cauterized and extirpated cancers, lupus, esthiomene, the noli me tangere, fistulas, and other serious lesions. He was careful to limit his efforts to favored locations and to use only enough caustic to suit the need, and he coated the surrounding tissues with bol d'armenie. A successful removal exposed inflamed (red) flesh, usually after three days. He used mitigatives to lessen the pain, and later he attended to the removal of the eschar, as we do for fistulas. Pink granulation tissue appeared where once was dead eschar, and the pus was not virulent, and the stench was gone, just as after treating excavated ulcers.

Palliation

We emphasize relief of pain and caring for the patient's daily habits. We try to improve the elimination and diversion of bad humors. We use morel-water to cool and the white ointment with litharge, tuthy, burnt lead, and diapompholyx, and other metallic washes. Apply camphor-juice whipped in a leaden bowl. Bind on thin leaves of lead et al., as we used them for virulent ulcers and for non-ulcerated cancerous aposthems. Lead is important; do not neglect it. We use Herb Robert, scabious, chevrefuille, cerfueuil, bouillon, powdered human feces, and burnt aneth. Many use lupin to diminish the raging pain, apply the still-warm meat of a grouse. That last item has led people to call the cancer a wolf because it eats a hen every day, as does a cancer eat its host. Whatever may be the case, those things are tempered and they neither help nor harm.

Here Ends Doctrine One Of Treatise Four

---

[450] The 'river' crabs were not crayfish as some believed, including Rondelet in his *History of Fishes* Vol. 2 In certain rare places known only to a few persons—as described by Dioscorides—they were called the sea crabs but not crayfish (J). But I think Joubert was wrong, they really are crayfish (EN).

DOCTRINE II: ULCERS IN COMPOSITE STRUCTURES: EIGHT CHAPTERS

CHAPTER 1

Ulcers of the Head Called Moles or Turtles

H ere we will deal with methods of treating the ulcers, each differing from the General Measures described earlier, according to where they are on the body. We will describe the applications of the four categories of measures as we wrote in the sections on aposthems, as suited for the various parts of the body; we will begin with the head. If the Reader will review the aforesaid material about the four general measures he will find here supplementary material dealing with prognosis and treatments. As to the prognosis, we know that ulcers of the head that expose bone or the meninges, as happens with 'moles' and 'turtles', can be dangerous, especially when the ulcers are at the commissures. In such cases, Roger advised that we do not treat them for cure; we follow Lanfranchi's measures for palliation. However, whenever Roger had to treat for cure, he always scraped the roots of the ulcer away from the affected bone which he removed with the trephine and chisels, exposing the meninges for applications of detergents and incarnatives, using stupes and pads that have been moistened with rosat-honey et al., as we described them for head-wounds. My Master at Bologna did the same, and I follow Galen's (ie The Greek) treatment for fistulas coming from diseased bone behind the ears (ie mastoid osteomyelitis). Janvier, on the other hand, in his chapter on fistulous ulcers, applied this powder after removing the diseased bone: combusted aquatic apium, galls, sage, and myrrh. Then he overlaid it with plasters of the black or diapalma ointments.

CHAPTER 2

Ulcers of the Face

S ome of these ulcers erode, inflame and cause severe pain, and their virulent pus stinks. They are tenacious, difficult to eradicate, and are contagious. The more you disturb them the more they multiply, and that is the basis for their name, Noli me tangere (ie do not touch me). The feeble resistance of the facial tissues cannot reject the evil matter.

Treatments

As for managing the life-style and the frequent use of evacuants, we use what we described for aposthems (ie abscesses), choleric pustules, and virulent ulcers. We wash them with watered vinegar or alum-water, and try to dry them during several days with applications of ointments used for virulent ulcers. Jamier recommended applications of Theodoric's liniment, made from linaria, plantains, and rock-salt. To control the erosive tendency we lay a soft strip of watered vinegar or other cold decoctions around the ulcer, and renew it three times in twenty-four hours for three days. If some improvement is noted, we apply a detergent of honey, ache-juice, and barley meal. If we observe healthy granulation tissue, we apply the incarnatives which we use for virulent ulcers. However, if we make no progress, we go to the treatments for cancerous ulcer that were used by Roger and the Four Masters. Be very delicate in the use of caustics and the application of the hot cautery. The face is very sensitive and delicate and its bones are spongy and cartilaginous and are easily damaged. When the corruption penetrates them, they never will heal; I repeat what was written in the sixth *Aphorism*. And, finally, I recommend that the best corrosive for use here is a stupe wet with nitric acid.

Cancerous And Ruptured Blisters And Ulcers On The Eyes That Displace Uvea

When they are not caused by wounds, most ulcers of the eyes derive from aposthems, exitures, pustules, or blisters. I am not surprised that Jamier called all of them ulcers. Avicenna called them exitures, and Azaram called them pustules. Although those masters listed seven distinct types according to how deeply they affected the cornea, or their situation on the conjunctiva and their depth, they do not differ much in their treatments, which nowadays are not intensive, as stated by Lanfranchi. Nevertheless, I place them in three categories, as I did for ophthalmias: the small virulent type, the large cancerous, and the intermediate sordids. All of them are engendered by an influx of irritating and erosive humors.

Signs

All of them are painful and induce lacrimation and redness, as in the ophthalmias. Early on you can detect a small red spot somewhere on the surface of the conjunctiva when you examine the wide-open eye. If it is on the cornea, that tunic will be white and clouded. The reason for the different colors (ie red or white) was explained by Jesus. The cornea normally is black (ie transparent) and the whiteness here (as Gordon noted) is bedded in the cornea. If one applies consumptive medications, he will damage the eye and cause a permanent erosion at that site.[451]

Prognosis

The malignant ulcer penetrates the cornea, the uvea extrudes, and often the eye is lost. The ulcers leave permanent scars that cannot be cured because the cornea is a spermatic tissue that cannot heal by first intention (ie by replacing the lost part) and the scar itself is foreign (ie not corneal tissue). Corneal ulcers are not rare, and they tend to recur. We advise the therapists not to treat the ulcer per se when the inflammation, lacrimation, and headache are significant; the patient's suffering will increase while those complications are in force. Also, I suggest that the therapist review what we wrote in the chapter on ophthalmias; much there can be applied to these ulcers.

Treatments

Galen (*Composition of Medications According to the Sites of the Maladies*) wrote that we should simply modify the general measures for treating ulcers for use in the eye. Avoid all medications that erode when used as detergents, and do not fill the eye with them, and cause scarring. I prefer medications based on tuthy (ie zinc oxide) and use gentle irrigants that relieve pain while deterging, such as the juice of mandragore.

If the ulcer is sordid, we may mix tuthy with other detergents, such as ceruse and other metals. Also, attend to the four elements of treatment for ulcers in general: life-style, counteracting the influx of bad humors, evacuations, special topicals for the eyes, and treating complications.

---

[451] Joubert: The three causes for a white cornea are: 1. A permanent scar not affected by treatments or palliations. 2. A pustule on the surface that can be cured by eliminating the pus. 3. An ulcer caused by the use of overly potent medications. (J).

The first and second elements are those for ophthalmias with these additions: avoid reclining or sleeping with the bad eye down and thereby retaining the erosive pus in the orbit; avoid sneezing, shouting, and vomiting, all of which will bring more matter to the eyes, just when we are trying to diminish it and to reduce the pain.

The third element includes maturing pustules to 'point' with irrigations with decoctions of fenugreek or melilot. Jesus said that draining of the pus will advance the cure. The irrigations will deterge the ulcer and wash out the pus. Use a syrup of rosat as was especially favored by Rabbi Moses in Part 22 of his book. After it is clean, encourage the ingrowth of granulation tissue with the white collyrium or with opium mixed with mother's milk or egg-whites; it will relieve pain. When the pain subsides, add some cadmia (called luban) to the white collyrium, as advised by Avicenna. Galen also used a collyrium of frankincense (*Therapeutics*, Part 5). Jesus said that it will mature the thick pus. All these remedies are describe in our Antidotary.

At the last stage, a collyrium containing lead was used by Heben Mesûe, by Alcoin, and by Azaron. It consolidated ulcers in eyes. I use Rhazes' formulation: burnt lead, antimony, tuthy, gum Arabic, tragacanth, opium, and rain water.

The fourth element, treating complications, uses what we described for treating pain in ophthalmias.

Ruptured Cornea And Extruded Uvea

When the eroded cornea ruptures and some of the iris extrudes and pulls up the rest of the uvea, we follow Galen (idem.) by applying repercussives and astringents and wrap on compressive pads. The collyrium of powdered blood-stone, ceruse, cadmia, burnt bronze, wheat-flour, gum Arabic, tragacanth, opium, and a decoction of olive leaves.

When the extrusion is large, press it down with a lead disc. When the lesion is older than one and a half or two years, it cannot be cured. If you wish to improve the patient's appearance[452], tie off the extruded tissue with a silk thread and apply coolants and comfortives until the thread is cast off. Deal with the scar as we will describe below.

Lacrimal Fistulas: The Common Type Near The Nose

The fistula usually begins as a small abscess called a Garat, engendered by the same bad humors. They suppurate and remain long enough before rupturing to allow their corrosive qualities to increase. The lining of the ulcer and the fistula hardens and the bone within is contaminated and corrupted. The fistula works its way both inward and on the surface. Sometimes it reaches the nasal cavity penetrating both the soft tissues and the bone.

---

[452] This is cosmetic and not functional; the eye will not see (LDR).

Causes

The thick humors mature slowly and corrupt the tissues which attract more of the bad humors wbhich are erosive and nitrous; the ulcer then becomes a fistula.

The Signs

First there is a swelling; its lining tissues harden as the fistula erodes in depth. The serous pus is sticky and can be expressed The ophthalmitic eyes redden. Palpation (ie probing) gives the diagnosis: soft tissues and bone.

Prognosis

The thin fleshy layers do not readily heal, and the nearby sensitive eyes make treatments with topicals difficult. Sometimes the erosion at the draining orifice continues, and the lacrimation also is continuous, and the margins of the lids are permanently deformed.

Treatments (Two Parts)

The General measures are those for all fistulas. The Particular Measures have three elements: First, use repercussives, resolutives, and maturatives leading to our drainage of the pus. Second: Deterge the opened abscess. Third: necrotize the interior of the confirmed fistula.

The first measures include the medications used for ophthalmias: the repercussives, resolutives, and the suppuratives. The latter includes special plasters of barley-meal, sea-shells, saffron, aloes, myrrh, opoponax, and vinegar. The plaster also will act as an eruptive after the pus matures. When that fails, you must open it with care to avoid the duct (ie at the caruncle).

The second measures use three medications in the fistula after the interior has been emptied of pus with washes, pressure, and irrigations with rue-water. First is Avicenna's treatment: take floss from the core of the lower part of the stems of marsh-reeds and gather enough to fill the entire fistula. Cover the orifice with diapalma or a similar plaster, and change it daily. After everything within has been cleaned, the fistula will heal.

Second is Rhazes': introduce a collyrium of frankincense, sarcolla, aloes, sangdragon, balaustium, antimony, alum, bronze-flower, and rain-water. Avicenna added a decoction of galls. Inject two or three drops of the collyrium two or three times a day for one week, and after every injection the patient should recline on the side opposite the fistula. Rhazes said that this valuable collyrium will either cure or greatly ameliorate a fistula, almost as if it is cured.

Third is William of Saliceto's: dilate the orifice and introduce this green ointment as a detergent: bronze-flower, alum, and honey. Or he used a powder of asphodels followed by consolidatives.

If those treatments fail, we use third intention measures to necrotize the fistula. We enlarge the orifice and accurately determine the length of the fistula. Then, in most cases, we incise and cauterize or use corrosives.

We use a stiff-blade lancet for the incision, or a razor, and insert it full-depth (ie down to bone) through a cut that is as far away from the caruncle as possible. Then we stuff the wound with a drain soaked with egg-white. The following day we examine the exposed bone and cauterize as much of it as necessary with a round or claval cautery inserted through a cannula, that protects the eye, as did Alcoatin, and we cover the eye with a paste as did Jesus, or we slide the cautery along a grooved silver director[453] as did Theodoric. After the burn we mitigate the pain and treat to cast off the eschar and the bony sequestra.

If one uses caustics (ie in the fistula), cover the eye with a tent that is wet with coolants that protect it. However, I prefer the actual cautery, as did Lanfranchi. We can control the actions of a hot iron better than those of the caustics, and avoid injury to the caruncle.

After the eschar et al. are eliminated, allow the tract to close.

Heben Mesûe disapproved of poking an awl into the nose. He claimed that it was of little avail in healing a fistula because the perforation soon closed. I disagree. I approve of directing the drainage into to nose, as did Arnold, who also used head-purges (ie laxatives).

Nasal Ulcers and Polyps

Some ulcers deep within the nose are bare, whereas others have redundant tissue. In the first group there are virulent, sordid, and erosive ulcers. In the second group the redundant tissue may be soft and it may dangle on a narrow stalk, hanging almost free from the ulcer-base. Galen called them *Ozaenae* and Avicenna called them *Alharbat*. Other redundancies may be firm and solid and have broader bases. Galen called them *Polyps* and Avicenna called them *Cancers*.[454]

Their **Causes** are acrid corrupt humors that descend from within the head. They thicken as they are combusted and engender the polyps. In some cases the offending humors are cold and the results are the soft lesions. Galen (*Miamir,* Part 3) agreed that the ozaenae derive from the acrid and purulent humors, whereas the polyps derive from the thicker ones.

Some say that the polyps are shaped like fish ignoring the fact that the word polyp means many (poly) feet (pes). Avenzoar called them *Multipes*, because they are embedded in tissues that resemble those of the nose itself (Galen).[455]

---

[453] A'cullier' that resembled a grooved director as a guide, a half-shell made of bronze or silver (J)

[454] A modern Reader may be confused by Guy's deference to the ancient terminology, differentiating ulcers from polyps etc. The soft ozaena corresponds to what we today consider an 'allergic' polyp, and what Galen called a hard polyp we may interpret as a neoplasm without a stalk, closer to what Avicenna called cancer (LDR).

[455] Conceive of a clump of grass rooted in soil. Was the soil under the grass 'normal' or was it contaminated in some way? Hence Galen's use of the term 'resemble' (LDR).

Signs

The lesions can be seen with light directed into the nostrils with a so-called sun-mirror (a speculum), and, as did Halyabbas, you can touch them (ie with a probe). The polyp is different from the proud flesh lesion (ozaena), as noted by Avicenna and Lanfranchi, which is soft and it dangles. That ozaena has the color of lungs and is insensitive. Usually it appears after catarrhs (rheums). On the other hand, the polyp is a firm, dry, tender, dark-colored, malodorous, venomous lesion that is firmly embedded in the nostril. It begins as a pustule shaped as a chick pea and it enlarges and encrusts, and bit-by-bit it reaches the palate.

Do not under-rate nasal ulcers. They are the sources for polyps, and a polyp per se is pernicious, in the same class as hidden cancers which are incurable, and, Hippocrates said not to attempt to treat them. As Avicenna wrote, let palliation suffice, rather than use the knife or corrosives. But the non-polypoid nasal ulcer that has a healthy color can be treated without fear, as Bruno noted. But, as stated by Roger and many others, even some polyps can be treated. However, most polyps as we have defined them are of flesh that is not natural in the nose.

Treatments For Nasal Ulcers And Polyps

As always, we attend to life-style, to evacuations which act contrary to the acrid matter, which is melancholic, and we use, as did Galen, medicines that dry and fortify the entire head to eliminate its discharges from above into its lower region, just as we did for ophthalmias and excessive lacrimation.

Afterwards we apply medications to treat the ozaenas and the ulcers and try to dry the patient with medicines that have more than one property, repercussion as well as resolution. Galen (*Therapeutics,* Part 5) said that medications for the nose should be drier than those for the eyes and less dry than those for the ears.

Therefore, use the white ointment with burnt lead for virulent ulcers; Halyabbas said it was best for the nose. But if the ulcer is sordid and crusted, first wash it with wine and honey and with a decoction of camomille, melilot, nasturtium, hellebore, and myrrh, and, if necessary, with a lye. Then I deterge with the apostles' ointment and place in the nostril a drain wet with acoris that has soaked at length in an oil of juniper, and with added distempered scammony. What remains of the fluid can be introduced on a feather or a twist of cloth or a quill as used for writing. That collyrium is well-regarded.

The powdered dry residue of the collyrium can be insufflated and it will reach where other medicines fail. If it does not cure, use a troche of aldaran[456] or quick-lime (calidicon) diluted with a sweet wine or vinegar for use on polyps. Follow that with detergents and consolidatives.

---

[456] Aldaran: perhaps an ancient alder (alno). A hollow twig of alder wood stuffed with salt and charred before insertion (LDR).

When the ulcer is very painful, add small amounts of opium to the various ointments. When there is too much heat (ie inflammation), Halyabbas prescribed inhalations of rosat-oil or oil of nenuphar, and applied the following around the nose: sandalwood, memitte, purslane (ie pourpier), etc, with rose-water and vinegar.

Dry ulcers, rhagades and split lips are treated with wax, marrow from femurs, mucilage from quince-seeds, dragacanth, and almond-oil.

When the dangling ozaena is not foreign or cancerous, cut it away (Albucasis) while an assistant holds the head with both hands to expose the nostril to sunlight. Reach in and grasp (ie forceps) and pull forth the tissue and cut it away with a sharp, single-sided blade, and remove all that you can. Then gently scrape away that remains. When hemorrhage or swelling appears, treat it at once, as you have learned.

When the lesion is too high within the nose to allow you to risk cutting it, or if it blocks the passage through the bones—as you have determined by instilling vinegar or the like and it does not reach the mouth—or in school-age children, you insert a button-tipped lead needle threaded with a cord with a series of knots. Grasp the cord in the throat and pull the thread to-and-fro to saw away the polyps. Then attach a drain covered with the Egyptian ointment and pull it in place. Leave it until the remaining flesh has been consumed. You may cover the knotted thread with ointment during the procedure.

Some surgeons, including the Four Masters, slit the skin alongside the nose to expose the bone and to allow them and remove with cautery or caustics, the roots of the ulcer that they could not reach as above. Then they sutured the incision to close it.

I strongly advise that you do not close the incision until you control the bleeding and can see if the excision is complete. If you leave behind any bit of the lesion, it will recur and the operation will have been to no avail. Later, the wound should be sutured as we described for reconstructing a lip.

Others, such as Roger, used a cautery inserted through a cannula which will protect the surrounding tissues and minimize the pain. Drape the patient's head (ie cover his eyes) lest he see the approach of the hot iron and incite his fears and cause him to interfere. Then Roger always inserted a tent covered with a caustic. When the eschar came away, he treated the defect as a simple wound. In all cases he applied coolants, defensives, and sedatives around the nose and when called for he used lead tubes as cannulae to keep open the airway.

Nose-Bleeds

Galen (*Miamir,* Part 3) said that Heraclites of Tarentum removed the clots before inserting a tent covered with lycium ointment diluted with water. Then he pinched the nostrils with his fingers and held on until the bleeding stopped. Then he inserted a pack wet with frankincense and other medicines that he used for treating wounds. He mixed them with a decoction of shepherd's purse. Also he applied sponges wet with cold strong vinegar to the forehead, and he elevated the head, and he massaged the arms, upper thighs (ie groins) and scrotum, and he applied tourniquets at the knees and ankles. In that way he kept blood away from the nose. He provided cool beverages and he covered the nose and the mouth with cloths wet with cool rain-water.

Galen (*Therapeutics*, Part 5) avoided applications of astringents over the region lest they divert and harm the head. He preferred phlebotomy as the means of diversion, or cups placed below the ribs and behind the knees. He, too, used massages and applied tourniquets on the limbs.

Apthous Ulcers Of The Mouth

These ulcers are similar to those in the mouth. They are set far back or in the gums and may involve the jaw-bones. Galen (ibid. Part 1) called the superficial ulcers *Aphthous,* and Avicenna (Part 6) called them *Alcola.* Others have called them cancers when they are inflamed. The ulcers that invade the bones usually are fistulas, but are called hemorrhoids or figs when they involve only flesh.

Causes

They are the same as for ulcers in the nose except that here they occur mor often in children when the milk is bad and they have troubled digestion.

Signs

The lesions can be seen and touched, and their colors denote their origins. When they are red they derive from sanguine humors; when orange, from choler; when pale (ie white), from phlegm; when dark, from melancholy.

The ulcers usually begin as pustules, bothors, exitures, or abscesses in the mouth. Galen (ibid) said that a precise diagnosis is difficult because the region is a warm and wet place, and the lesions soon suppurate and erode, and the medicines do not remain where they are placed because they are washed away by saliva

Treatments

In addition to the measures we have described for treating nasal ulcers, we advocate phlebotomy from veins under the tongue, as we recommended for treating quinsy. Certain medications are case-specific. For virulent ulcers used such drying substances as diamoron and the juices of blackberries, and the green peels of walnuts. Galen (*Therapeutics,* Part 5) used the fruits of cypress trees, and Avicenna added lentils and sumach. Commonly used medicines are decoctions of plantains, roses, caprifolium. et al. When there is pus, use honey-wine, decoctions of chelidonium, reeds (ie cyperes), sedge, menthastrum, galls, saffron and myrrh. For corrosive ulcers, apply alum and vitriol. In *Miamir* Part 6, Galen described his use of lentils, and deer and veal marrow (ie from femurs). I add fruits that are astringent, such as quinces and medlars. Sometimes, I offer lettuce, endives, and purslane. I prescribe astringent mouth-washes of sumach and roses. Sometimes I make a liniment of diaphoretics, and I add vitriol and sour wine. When the ulcers are sordid, I add honey. When they are erosive, I add verdigris and enlarge the surface opening with oil and chalcitis. When the ulcer undermines, I add wax to the chalcitis.

When erosive and cancerous ulcers involve the gums, I rrigate them to wash out the debris, using vinegar with squills that have been boiled with olive-leaves. Another liniment contains alum, burnt salt, galls, pomegranate-peels, acorn-shells, cannelle, cloves, nutmeng, arsitoclochium, sage, roses, date-pits, and burnt cancerous leg-bones.[457] Grind all and add honey and vinegar. Also, the mixture may be applied as a powder (ie dry).

When all the foregoing applications have failed, apply troches of asphodels, calidicon and alandaron, and eau forte (ie nitric acid). Or you may have to use the actual cautery.

When the ulcer (ie fistula) between the teeth extends to or through bone, you must remove the teeth and enlarge the opening to allow you to insert a drop of eau forte or sublimated arsenic to eat its way down to bone. Use a silver or bronze cautery to burn away the diseased bone. Then treat the defect. If you cannot treat the fistula through the mouth, some surgeons make a counter incision below the jaw. That is a difficult task. The topicals are washed away by saliva, and you cannot identify where to site the lower incision by pressure from above (see *Prognostics,* Part 1).[458] And since severe pain accompanies the treatments, you first must relieve it with rosat oil applied both within the mouth and below. Galen (*Miamir*, Part 6) had his patient retain a mouthful of the oil of lentils: it repercussed without damaging or causing more suffering, and it resolved without necrosis.

After relieving pain and necrotizing the fistula; clean it well, irrigate it with wine and honey, and apply incarnatives. Follow with a decoctiom of frankincense and a liniment of aloes, myrrh, sarcocolla, mastic, frankincense, sangdragon, and rosat-honey.

When the hard cancerous lesions do not respond to curative treatments, go to palliation. The soft-tissue lesions can always be cured with excision and cauterization, as in the nose, and you can ligate redundant tissue (ie on pedicles) without causing bleeding or exciting fear.

Treat split lips with ointments such as we used in the nose, or use the oil of burnt walnut shells, an excellent medicine (Roger). Albucasis used a cautelary cautery to burn deeply when medications failed. Then he treated the wounds as such.

---

[457] Burnt Cancers: I assume that Guy meant powdered cancerous tibias that had been amputated, or incinerated tissues that had been excised. He offered no further definition. I have not seen it mentioned in any other medieval surgical treatise! (LDR).

[458] In Prognostis, 1, Hippocrates warned the physician against taking action blindly (LDR).

Ulcers Of The Ears

The varieties, the causes, and the diagnostic signs are categorized as for those of the nose and mouth. And in the same way we use very dry medications, as dictated by Galen (ibid. Part 5) where he described the treatments of a Thessalan sage. In *Miamir,* Part 3 he wrote, "The Glaucin[459] treats all recent and not very painful ulcers at the ears. Grind the herb with vinegar. They (Thessalans) also used diamyrrh and diacroca (ie from crocus). They treated painful ulcers with troches of andromache. For chronic otic ulcers they liberally applied iron filings which had been soaked in vinegar and set in bright sunlight or heated in a skillet. When called for, they irrigated the ear to clean it, using oxymel or wine with honey, or with rust-stained water. If, as needed, a fistula or the flesh of an ulcer had been burnt, they treated as above, and relieved pain as described for abscesses."

---

[459]   Glaucin was a topical based on the horn poppy, memitte, (LDR).

CHAPTER 3

Ulcers Of The Neck (Throat And Nape)

These are like all other ulcers, except for the risks involving the nearby veins, arteries, nerves, and the passages for breathing and swallowing. Ulcers at the nape are at risk for injury to the vertebrae and spinal cord.

These differ little from most ulcers as to diagnosis, prognosis, and treatments, as described in our treatments of wounds in these regions

CHAPTER 4

Ulcers Of The Shoulders And The Arms

T hese differ little from most ulcers as to diagnosis, prognosis and treatments, as described in our treatments of wounds in these regions.

CHAPTER 5

Ulcers Of The Chest.

U lcers on the surface are treated as in other regions. Penetrating ulcers are fistulas and cannot be expected to heal, and we resort to palliation which occasionally may achieve a cure.

Palliation includes a good life-style as we described for wounds of the chest, and treatments to localize pus and drain it before it interferes with breathing. That may require enlarging the ulcer to admit the nozzle of the apparatus for clysters, and the insertion of drains containing gentian; tie them to a cord at entry lest they fall in. Those maneuvers allow us to deterge and to irrigate with mellicrat (as used by Galen in *Therapeutics*, Part 5) and with honey-wine to evacuate pus when treating wounds in the chest. Then apply a topical of heated honey which acts as a detergent and as an attractive and as a treatment for non-penetrating excavated and undermined ulcer to attract pus from recesses. However, do not use corrosive substances such as verdarain in penetrating ulcers.

When the local treatments fail to empty the fistula, make a counter-incision between the fourth and fifth ribs as we have described. Then use the apostles' ointment or similars as detergents and keep the new wound open and allow the initial ulcer to heal. That may require that you consume the callus within the fistula with an actual cautery; do not use corrosives.

Many kinds of potions have been used for these cases. Henry recommended one that he had seen used successfully by a Master Surgeon (ie perhaps Hugh of Lucca) that was made from the roots of dipsaccus mashed with honey. Prescribe a dose about the size of a walnut, morning and evening. Averroes (*Colliget*, Part 5) said that dipsaccus is warm in the 2nd degree and dry in the 3rd. When taken with wine, it acts as a diuretic to evacuate pus, and it eliminates bad body odors from the armpits and elsewhere. It has another advantage, being tasty and pleasant to swallow.

CHAPTER 6

Ulcers Of The Abdomen

U lcers that do not enter the abdominal cavity are treated in common with others. The penetrating ulcer is a fistula and is treated as such. Use palliative applications and potions. Offer little hope for cures.

CHAPTER 7

Ulcers Of The Pelvis

These ulcers may appear only on the surface, or they may involve internal structures, or they may involve external appendages such as the penis and scrotum and the anal region.

Ulcers that affect the internal organs should be treated as we do the penetrating abdominal ulcers, although most of them are managed by physicians, not surgeons. Ulcers of the penis or the neck of the uterus (ie cervix, vagina, or vulva) usually are caused by friction or by local inflammations; they may be virulent, corrupt, erosive, or cancerous. Ulcers at the anus may simply be rhagades, granulatuons, warts, figs, and condylomas.

Causes

The ulcer may derive from corrupt humors, abscesses, improperly treated wounds, or uncontrolled scratching or rubbing.

Signs

We can see and palpate and use a speculum which Avicenna favored.[460]

Galen (*Miamir*, Part 9) and Avicenna (Part 3) agreed that treatments in this region are difficult because the parts are sensitive and are constantly wet with drainage of acrid choler. Therefore, the applied topicals do not remain in place long enough to work well, and are diluted by the discharges. The warm and humid region soon leads to corruption and prevents open exposure. Furthermore, the modest patient may delay exposing the ulcer to the surgeon until malignant changes have occurred.

The worst ulcers invade the muscles at the base of the penis (ie in the perineum) and around the anus. Other very bad ulcers invade deeply before they are discovered.[461]

---

[460] Guy's term is 'mirror'. However the speculum was not a mirror. Avicenna used a mirror with the speculum to reflect light into the vulva (J).

[461] Here and elsewhere Guy may have described Fournier's Disease (LDR).

When treating ulcers of the penis and the anus that are not badly inflamed, Galen (ibid. Part 5) recommended a softening cataplasm that also cicatrizes. For other ulcers he used drier medicines (ie to correspond), because the affected parts are drier than normal flesh, and the tissues of the glans penis are drier than those of the shaft. He cautioned us not to believe the false claims that their medicines can cure penile ulcers in three days. He said it is a sad and amazing claim made by those who have been educated in a heretical evil doctrine (ie Thessalon's).

Therefore, when there only small scabs and little inflammation, you need only to wash the lesions with waters of roses or plantains, followed with alum-water. Then we use the white ointment with camphor, or the bark of berberis, or balaustes, or powdered wild bedegar, all applied on soft thin cloths.

When new ulcers are virulent and erosive, aloes alone may be enough, or burnt lead, cadmium washed with wine, tuthy, litharge, and ceruse. The most potent are burnt bronze, pine-tree bark, and blood stone. Galen liked ashes of linen paper, burnt alum, and dry burnt melon. Avicenna used this potent desiccant and incarnative powder: tuthy, aloes, sarcocolla, frankincense, blood-stone, ashes of reeds, galls, balaustes, acacia, pomegranate peels, and bronze flower. He mixed all of them with oil of roses to make a paste. When the ulcer invaded the penile urethra, he injected it,

When the ulcers were chronic (ie old), corrupt, or cancerous, we wash them with Lanfranchi's collyrium of epithyme made from white wine, plantain-water, orpiment, and bronze flower, mixed with fluids. It absorbs, dries, and heals.

Troches of asaphodels and aldaron are stronger, and arsenic never fails. When the ulcer is malignant (ie necrotic), we cut it away and cauterize the bed, or use caustics (ie arsenic) placed between the necrotic and the live tissues, as we do for esthiomene. After the separation is complete, apply detergents before using incarnatives and consolidatives.

When the treatments cause bleeding, use our hemostatic powders or those of the Four masters, which are made from alkanna, burnt felt, and ashes of grouse-feathers. Remove the clots before you replace the arsenic; it will be impotent if it lies near an open vein (ie is diluted by blood). When the medications could not reach extensions of the ulcer,[462] the Four Masters advocated incising the overlying foreskin before applying their topicals. I disagree; the new opening will not heal and the prepuce will swell and form a mass under the glans that will be very uncomfortable. That outcome cannot happen in circumcised Jews. Galen (*Therapeutics,* Part 6) used this treatment in all cases whether or not he could expose the lesion, and he was especially attentive to the patient's discomforts to avoid burning pain: he instilled the populeum ointment with the juice of morels and a bit of barley meal and egg-white with the oil of violets (Roger's prescription). When he was

---

[462]   Here the ulcer is at the glans, and phimosis covers the balanitis (LDR).

concerned lest he cause bleeding (ie when separating the prepuce), he sat the patient in a tub containing a decoction of mauves, etc. Then he applied a defensive ointment to prevent suppuration, and coolants in an ointment of oxycrat and cool juices, placed over the penis and groins. Do not allow the swelling to obstruct the urethra; insert a drain made of wax or a twist of clith, and keep all in place in a sack-like suspensory tied to a belt.

Treat rhagades, split skin, warts, and redundant granulation tissue as we do at the anus, to be described.

When a swollen prepuce dangles below as a result of an incision (ie see above), simply ligate it and cut it off. If the wound bleeds, cauterize it. Incisions in the glans and the penile urethra may leak urine and heal with bad scars.

Hemorrhoids At The Anus

Galen said (*Aphorisms,* Part 6) that hemorrhoids are caused by ulcers or diarrhea.

They are swellings that may be painful and inflamed, engendered by humors accumulated in the hemorrhoidal veins. Lanfranchi insisted that we limit the term to those lesions. In our Anatomy we described five hemorrhoidal veins that reach the anus. The Greek term hemorrhoid means 'flow of blood', or flux as in Latin, that terminates at the anus, where the blood can purge itself, and bleeding can be either normal (natural) or abnormal. Natural hemorrhoidal bleeding has a different significance than menstrual bleeding, which represents a phenomenon that is important for our species. Here we have a mechanism that relieves individual persons of abnormal accumulations of melancholy and prevents many maladies. Nevertheless, Galen (*Maladies and Symptoms,* Part 6) said that all bleeding is abnormal, except for moderate menstrual flow. Later, in Part 2, he repeated that Nature normally does not need hemorrhoidal bleeding to get rid of bad blood.

Diagnosis And Prognosis

There are several kinds of hemorrhoids; we classify them as to their contents, their sites, etc.

As to the different humors in them, they vary as elsewhere, excepting choler. The lining of the veins is made from thick sanguine. Warts come from melancholy. Cystic (blisters) come from phlegm. Small varices come from weak humors, and they are named according to their appearance, as stated by Avicenna in *Canon,* Part 4.

Some hemorrhoids are intirely internal (ie occult) and others are external. Some are full of blood (ie clotted) and in others the blood is liquid.

Causes

Rabbi Moses said that a plethora of melancholic blood was the usual cause, but other humors that resembled melancholy were occasional causes.[463] A bad life-style led to thickening of the humors and their combustion. The weighted humors carried them down to the anus and congested the veins in the walls of the intestine, heating them painfully as they distend, and elongate, and cause them to leak blood. Acrid and irritating matter is engendered, to be treated with acrid medications, such as aloes, scammony, and others, as noted by Heben Mesûe.

Signs

You can see the protrusions and touch them, and, with the help of a speculum and mirror, you may examine the internal ones.

Usually the patient is uncomfortable and feels heavy in his pelvis and back. His face is discolored. Bleeding usually comes at intervals, monthly or once or twice a year. In general, moderate blood loss serves a good purpose for the patient, and it should be encouraged and not be restrained. It is prophylactic against depression, mania, strangury, and other melancholic illnesses. But when the bleeding is continuous, it must be controlled because hemorrhage may lead to dropsy and phthisis. We follow the advices of Hippocrates (*Aphorism* 6): "When one treats chronic hemorrhoids and he fails to leave one of them intact, the patient is at risk for hydrops and phthisis. Furthermore, do not neglect to treat the discomforts caused by hemorrhoids; an abscess may ensue and lead to a fistula."

Treatments

The General Regimen (Life-Style) of itself has three elements. 1. Avoid causes for a plethora of melancholy in whole blood. 2. Counteract the antecedent matter that engenders the bad melancholy. 3. Offer properly assigned antidotes to be taken by mouth.

The first is accomplished with the six non-natural things and their three annexes, all familiar to us for their excellence; they are praised by both Master Arnold and by Rabbi Moses. Patients with hemorrhoids should avoid fourteen foods (according to Moses):

---

[463] Rabbi Moses Maimonides: The Reader may be surprised at Guy's citations, written during the terrible years when hordes of Flagellants and other Millennarian revolutionaries were swarming through southern Germany and the Alsace, attacking and murdering thousands of Jews, Catholic Churchmen, Moslems, and other victims (LDR).

vinegar, fava beans, lentils, chick-peas, red cabbage, dates, plump (ie not dried) figs, beef and goat-meat, salted foods, meats from aquatic birds, heads of animals, old cheese, salty and spicy sauces, and all things that we already have described for treatments of melancholic aposthems. Also they should avoid constipating foods.

The second is accomplished with the dosing of diacatholicon, or diacassia, or with pills of bdellium as made by Rhazes: myrobalans, bellirica, chebules[464], serapinum, nasturtium, reglisse, bdellium, and the juice of leeks (or garlic). Dose two or three. Avicenna also used them for treating intermittent hemorrhoids.

The third is accomplished with this electuary: Indian myrobalans, bellirica, emblicus, decoctions of bugloss, roots of tassus barbatus, ginger, cannelle, galanga, nutmeg, frankincense, ammi, spikenard, squill, iron filings in vinegar, penides, sugar-loaf, and squinanthus.

Arnold's treatments had two elements: 1. Control the influx of humors. 2. Relieve discomforts.

Arnold's first treatment is accomplished in three ways. 1. Avoid spicy, bitter, and irritating foods that elicit inflammation. Control outbursts of anger, limit sexual activity, and strenuous exercise. Tasty and mildly astringent foods should be taken only after meals. If taken before meals, they may cause constipation; they include pears, quince, et al. Wheat, barley, pig-feet and ears, and strong sour wine, and rust-water are all good. During the summer Arnold prescribed morning and evening doses of a syrup of roses and myrrh, and a compote or jelly of quince. In wintertime he provided truffles and the roots of tassus barbatus cooked in rosat-syrup. 2. Apply astringent topicals. In summertime he made sachets of rose and myrtle petals, boiled in water, and squeezed damp before applying them. In the winter, he used crushed sage, well shaken in rosat-oil. He put the macerated matter in a sachet before applying it. In both seasons, the patient may sit on the sachets.

To restrict the flux from the liver into the hemorrhoids, Rhazes placed a plaster of spikenard over the liver, as described in his chapter on liver-diseases. Avicenna place cups on the shoulders and applied on the scarifications wet pads of hares' beard, spider-web, and a powder of frankincense, sangdragon, balaustes, etc., mixed with egg-whites on the scarifications, after removing blood from the openings.

We relieve pain with mitigatives as used by Master Arnold, selected according to the causes. When the pain comes from retained blood, it must be evacuated. At other times the pain is due to distension of an extruded hemorrhoid, or by inflammation, or by dryness, or by passage of hard feces.

When the cause is retention of blood, use one of two methods. One not only relieves pain, it cures the hemorrhoid by removing the blood in the way Nature does it, by bleeding. Use one of three techniques. 1. Use a phlebotomy knife or a lancet. 2. Apply leeches set with a reed. 3. Use caustic medications. The best corrosive for this purpose is the milk squeezed from a fig-tree leaf. Rub it on the hemorrhoid until it opens. Or you may use a slice of onion, or lay on a bit of cotton cloth or fluff covered with aloes dissolved in

---

[464] Guy listed them as a varieties of indian myrobalans (LDR).

beef-bile, or use Avicenna's medicine of colocynth pulp and bitter almonds, applied for one hour, repeated five times. But when that was too slow, Arnold advised phlebotomy of a vein on the dorsum of the foot or from behind the knee, removing three ounces. Failing that, he bled from a basilic vein.

To treat only for relief of pain—not for cure—use softening fomentations that gently calm the inflammation and resolve the superfluous blood. Foment in two ways. 1. Heat the medicines in a pot placed under a perforated seat. The anus will receive the fumes that are boiled out below. 2. Place the medicated fluid in a flat dish on which the patient can sit and immerse the hemorrhoids. Or you can soak a sponge with the fluid and press it against the anus. Or you can fill a sack with the medications and apply it. The medications are to be boiled in the water: leaves of cynoglossus, mauves or guimauves, violets or parietory, and fenugreek.

Rhazes preferred boiled and diced white onion with melted butter made from cow's milk, applied warm. Avicenna preferred melilot and peeled lentils mixed with egg-yolk and rosat-oil and duck-fat, with small amounts of saffron and opium.

Halyabbas used this plaster: camomille, melilot, leeks, and roots of guimauve minced and boiled at length in water until macerated. Grind the pulp with egg-yolks in a mortar before adding fenugreek, linseed, bdellium, and the fat of grouses.

Rabbi Moses' remedy consisted of heated skimmed butter ground in a lead mortar while exposed in sunlight until it turned dark. He claimed that it is a wonderful mitigative that he called lead-juice. Avicenna mixed some oil of chrysomeles.[465] with added bdellium. In some cases he put in some hen- or duck-fat.

William of Saliceto described this hemorrhoidal ointment: rosat-oil, ceruse, litharge, opium, and mandragore-bark.

However, when the painful hemorrhoids defy all these treatments, use Alexander's Ointment, which I have used also in treating tenesmus and all manner of painful anal disorders.

Lanfranchi said the same for his liniment: frankincense, myrrh, lycium, saffron, opium, egg-white and mucilage of psyllium and rosat-oil. He put it on drains which he inserted into the rectum, and he placed moistened packs over the anus.

When the discomfort is due to external excrescences caused by the extruded hemorrhoids, treat as follows: For Warts: use use egg-yolks with oil of violets during the summer. In winter-time mix with oil of almonds or with butter or some mucilage. For flat warts use desiccatives that are not erosive, such as a powder made from leaves of tassus barbatus or of plantains, and roots of burnt reeds, and ceruse and litharge. You may make an ointment from the powder.

When the medicines fail, remove the warts. If they are old you should erode them with aloes mixed with figs, as above. During those treatments the patient must remain sober. Do not attack all the warts at one session; proceed stepwise.

---

[465]   Chrysomeles: apricots, so-called Armenian apples. Avicenna that the oil made from the large seeds was useful for relieving hemorrhoidal pain (J).

Master Arnold used mild corrosives that are salty, such as rock salt, crystals of niter[466], or a lye made from burnt grape-vines mixed with honey. Other surgeons, including Rhazes, Avicenna, and Halyabbas used more intense medications such as troches of diabardich (ie verdigris) or calidicon. Roger used an erosive ointment.

Avicenna, Albucasis, and Bruno, and their followers used a cold knife or hot cautery to remove surface warts, and they everted the anus by placing a cup over the opening to expose internal warts, or they used their fingers with or without a cloth. Master Arnold ligated some of them and let them drop off, or he sawed them with a thread.

When inflammation causes pain, Arnold said that irrigations may suffice, using warm water that had been boiled with the cold seeds—cucumber, zucchini, and purslaine, or he added egg-white. Or he used cold water with a mucilage of psylliuim, or he anointed the hemorrhoids with populeum ointment or Galen's wax ointment.

When the pain is caused by the passage of hard feces, we use as a softener an ounce of cassia taken before meals. After defecation the patient should sit in a tub containing a decoction of mauves; then anoint the anus with warm rosat-oil. All of this has been well tested.

Figs At The Anus, Atrices, and Condylomata[467]

These lesions at the anus, penis or vulva are cured (a la Theodoric) with minced millefuille, parietory, and salt applied daily, or can be ligated, cut away, or cauterized with the iron or with caustics, as we treat hemorrhoids. Attend to the pain.

Anal Fistulas

Some anal fistulas may enter the rectum, some of them as far up as three fingerbreadths; others enter right at the anus. Fistulas that do not enter the bowel may enter the buttocks,

---

[466] Sal or sel nitre: sagimen nitris is sagepenum, an umbellifer resin, also called salt of glass, axunge nitri, or 'glass-fat' (J)

[467] A 'fig' is a soft spongy, gray or pale, irregular clump that dangles from a slim pedicle—the resmblance is the source of the name. Usually it appears in the rear (ie between the buttocks). Sometimes it emits a whitish discharge. Some are difficult to remove, those which derive from melancholy. Others come from contaminated phlegm.

An 'atrice' resembles a mulberry and is an excrescence derived from thick choler and can be very tender. Its color varies from red to violet, covered with little papules like the berry. Arnold called it attrite, and defined it as proud flesh that can come forth anywhere in the anus.

A 'condyloma' derives from thick melancholy, and is a firm excrescence that grows at the anus or vulva. Its name comes from the Greek word for a joint (condyle) (J).

the tissues of the pelvis, the bones of the coccyx, or reach the bladder or the base of the penis. Each type calls for its own type of treatment.

Causes Of Fistulas

One may infer that the usual causes are hemorrhoids, or abscesses, or poorly treated wounds. When one delays treatments of such maladies, he allows pus to remain in the warm and humid region, and it soon corrupts, erodes, and complicates by allowing the fistula to worm its way.

Signs Of Fistulas

These include the obvious manifestations of the causes as well as the induration, lumpiness, and swelling near the anus. It may drain or not, emitting watery or serous virulent pus.

The depth of penetration can be determined with a malleable lead probe or with a twig of parsley, or with the central stem of a leaf of mauve or periwinkle. A communication with the rectum will emit fecal matter and flatus, and the probe will enter the rectum and be felt with a greased manicured finger. When the fistula involves the sphincter muscles, the patient may be incontinent of feces, and the sphincter will not respond by squeezing the examiner's inserted finger.

A connection to the bladder will cause dysuria. Involvement with bone will be detected with the probe.

Avicenna and Lanfranchi agreed that a fistula that communicates low (ie at the anus) is not threatening, and can be neglected except for attention to cleanliness, We use medicated cloths or fluffs of cotton. Also, we use Rhazes' collyrium as for lachrymal fistulas, and the black plaster, which is more potent. When the patient is not bothered by it, you may leave it to serve as an emunctory, as do hemorrhoids. Remember that drainage from such natural emunctories should not be obstructed.

Albucasis said that a fistula to the bladder or coccyx should not be treated because the treatments may be harmful both to the patient and to the surgeon's reputation. The twists and turns of those fistulas will prevent medicines from reaching their goals. Therefore, simply palliate.

Furthermore, a fistula that destroys the sphincter muscle cannot be cured, because the treatment will make the incontinence worse. Here, too, we palliate.

However, Rhazes said that a fistula that does not damage the sphincters, or one that does not enter the rectum too high, can be treated without such concerns.

The treatment has two elements. The General Regimen is what we prescribed for most fistulas, in the chapter devoted to the topic. The Particular Treatments follow here.

When a non-penetrating fistula comes from an abscess in soft tissues, enlarge the external orifice with a drain wet with gentian, and incise into the abscess and cauterize the cavity with actual or potential cauteries. Be aware that the region requires more coolants and defensive than elsewhere. Both Bruno and Theodoric preferred the actual cautery because it does not increase the burden of bad matter.

Rhazes insisted that a communicating fistula must be treated only with setons. First he used a sickle-shaped—lancet to drain an abscess if one existed, followed by incarnatives, after the setons had cut through. On the other hand, Bruno and Theodoric said that one can prevent an abscess by keeping open the tract to allow free passage of feces.

Albucasis described his technique for passing a seton. He inserted a malleable needle threaded with three or four strings. His well-lubricated fingers in the rectum bent the needle and directed it through the anus to remove it while leaving the threads. He retied them tighter every day until they cut through the mass of tissues. Meanwhile he applied sedatives. If the patient could not tolerate the painful procedures, Roger tied a strip of clothe wet with a corrosive to one end of the strings and pulled it through and tied it loosely. Then he applied mitigatives (ie to the strip).

The sickle-lancet was used by Albucasis as follows: He pulled the loop of threads to create a curved tract through which he passed the knife to pull it through (ie the blade was on the concave edge) My Master used as his guide a half-split reed, to function as a grooved director to lead in a hot cultelary cautery to burn through the tissues that contained the fistula. Later he used detergents to remove whatever eschar was present, and he applied the incarnatives used by Roger.

Although Bruno and Theodoric cauterized their incision and destroyed the callosity of the fistula, I see no purpose for that; the callosity is a minor factor. It is better to leave it, even to favor it and allow the entire tract to be lined by scar, including the edges of the rectal wall, thereby to prevent the painful accretion of feces against bare tissue.[468]

Rhagades

This is a condition of cracked skin at the anus, penis, and vulva. It calls for a soft diet and fomentations with decoctions of mauves and their roots, and linseeds. Also, we apply Rhazes' ointment as improved by Lanfranchi with rosat-oil, wax, ceruse, wheat-flour, tragacanth, opium, camphor, and egg-whites.

The dressings that hold the medications in place are suspended in panties, or, trousers, and sacks (ie for the scrotum and penis).

---

[468] Note that the threads were used as a seton only in the one method in which the threads cut through and the tract healed behind it. That is the purpose of the seton-technique. In the other two methods, using the knife and the cautery, the threads were used as a drawstring to establish a curved path to guide the tip of the sickle-shaped lancet, or to pass the grooved director for the cautery (LDR).

CHAPTER 8

Ulcers Of The Thighs, Legs, And Feet

The treatment of these ulcers is as elsewhere, excepting the bandages; here begin at the knee to wrap the thigh, and begin at the ankle to wrap the lower leg. Rest (ie with the leg elevated) is important here to counteract the descent of humors into the lower limbs. Cancerous ulcers in the thigh were called lupus by Roger and by most lay folk, and were called cancers when in the lower leg. Lanfranchi called them esthiomene, differentiating them from mal-mort (ie venous stasis ulcers) which he called vilaine rogne (dirty mess), which we will discuss later. But, as Galen said, let us not waste time with terminology.

Cancerous ulcers should be treated with alum-water and plantains. I follow Lanfranchi and advocate complete excision with a cultelary cautery, and I incise the round ulcer at both ends of a diameter to allow it to be closed as a linear defect. Mitigate the pain of the cauterization with egg-whites and rosat-oil. Let it suppurate to assist in the separation of the eschar with a plaster based on celery. Apply a defensive ointment of bol d'armenie around the lesion itself as part of our usual practice.

When the underlying bone is involved, you will see the damage. Roger applied a protective paste around the lesion, or used a waxen cloth or diachylon, or a cool adherent plaster, before burning away the overlying flesh with a caustic left in place during an entire day or over night. Then apply some egg-white and rosat oil over the dead tissue and proceed to separate it with butter and minced cabbage. Then scrape the exposed bone to remove all the disease and see the healthy bone. You may have to use the hot cautery to get rid of all the corrupted matter. What remains will be an ordinary excavated ulcer. However, when the bone is diseased through-and-through, discontinue your operation, because the lesion is incurable.

As always, use corrosives and caustics gingerly; they are dangerous substances. Use them as we would have them used on ourselves.

HERE ENDS TREATISE IV

HERE BEGINS TREATISE V

Figure 5.
A Consultation About A Fracture
From A 14th C Manuscript Of Guy's Surgery, fol. 80, #396.
Bibliotheque Nationale, Paris

RUBRICS FOR TREATISE V

DOCTRINE I: ON FRACTURES:
EIGHT CHAPTERS

CHAPTER 1

The Treatment of Fractures: General Considerations[469]

As Galen described it in Greek (*Therapeutics,* Part 6), a fracture is a disruption of continuity in a bone; in other words, it is a wound. But in French we use different terms—a fracture simply is a broken bone, no matter how it is caused, with or without an open wound. That emphasizes the differences in immediate causes. Galen (idem, Part 4) specified other differences as to treatments: the complexion of the patient the affected part, the complications, and the late effects (ie after restoration).

In light of the above, he classified fractures as simple and as compound (idem. Part 6. A simple fracture may be across the long axis (ie transverse) or in the long axis (ie vertical) when the shaft of the bone is split. Furthermore, as Lanfranchi wrote, the fracture may be completely through the bone or it may be a partial crack. And further, a simple complete transverse fracture may have only two pieces or it may have several fragments, including spicules. Sometimes both of paired bones may be fractured.

As to compound fractures, some occur with open wounds, and others cause excessive pain, or cause much swelling, or an aposthem (ie either an abscess or a mass of edema). In some cases the the fragments override and cannot unite on-end, as in the nose, jaws, clavicles, or arms. Each calls for its own kind of treatment.

The **Causes** are what break bones: falls or blows.

The **Signs**, as Halyabbas wrote in *Royal Dispositions,* Part 1, are obvious. Palpation reveals a separation of the fragments and one can see the deformed contours. Both Rhazes and Avicenna described palpable crepitus, tenderness, and the unusual mobility of the distal part of the limb. Knowing the cause will help with the **Diagnosis** (Avicenna). A split in the long axis produces an abnormal swelling (Lanfranchi) when other signs may be absent (Halyabbas). However, always there is pain, and a loss of length in the limb. Other signs will be described, case by case.

---

[469]   Guy has taken most of this Doctrine directly, often verbatim, from Albucasis *On Surgery and Instruments.* (see Bbibliography) And there is much room for confusion! (LDR).

Prognosis

Avicenna's comments were in General; Special cases will appear in other chapters. A complete transverse fracture may heal poorly because it may be difficult to restore an end end-to-end apposition for the separated fragments, especially when they over-ride. That occurs more often when the bone is not one of a pair, as in the case of the thigh bone, or when both paired bones are fractured. Fractures near joints may be difficult to reduce because a proper reduction cannot be maintained by bandages, and the result will be stiffness and reduced range of motion. A fracture associated with a twisted limb can produce a troublesome swelling that must subside before you can reduce the fracture.

A compound fracture with a wound is not easy to treat because you must leave a gap for drainage and require bandaging and splinting that expose the wound, and they may fail to stabilize the reduced fracture. The fracture then may heal slowly and imperfectly. Furthermore, the union will be unstable because the space between the fragments will be filled with non-bony tissue.[470] In such a case the traction needed for reduction may be excessive, and that may cause a spasm (ie convulsion).

The times required for healing (ie bony union) differ according to the sites. A cranial fracture needs about thirty-five days; a fractured nose needs eighteen; ribs need twenty; other times will be described later. The time will vary with the age of the patient (Jamier). Both Avicenna and Halyabbas (Sermon 9 in Part 2) mentioned other causes for delayed healing: the use of too hot water for the embrocations, too much motion allowed too early in the course, too thick blood (ie poor nutrition), too tight bandages that restrict the inflow of nutrients, and the presence of detached bits of bone. Furthermore, as Avicenna noted, the union may be less secure in persons with choleric complexions, in persons convalescing from other ailments, and in older people. And Albucasis said that decrepit people may suffer permanent non-union.

A good prognosis for recovery may be offered when the reduction restores the proper length of the limb, when the reduction relieves the pain, when the wound is puffy, and when signs of inflammation subside after the initial repair.

Treatments in General

The four General Measures are those we have described for all wounds, as proposed by Galen (*Therapeutics,* Part 4) and by Avicenna (Parts 1 and 4). First: reduce the fracture. Second: secure the reduction with proper bindings. Third: favor the appearance of callus. Fourth: treat the complications.

Before discussing the four Measures (ie Intentions), I will list six requirements.

---

[470]   Interposed muscle, etc, (LDR).

The First Requirement has Ten Elements:

1. Do your work in a suitable place.
2. Have capable assistants.
3. Have at hand a good supply of egg-whites and rosat-oil, cloths of various sizes, all wet with the medications,[471]
4. Have a supply of binding cords, suitable sizes (width and length) of single and double-armed bandages. All are to be moistened and softened with oxycrat—to be molded to the contours of the part.
5. As Halyabbas advised, have enough thin cloth stupes and oakum, moistened as above.
6. Have a supply of smooth light weight splints of juniper-wood or wood that is used to make scabbards, or of horn, of iron or of leather. They should be long enough to extend at least three or four fingerbreadths above and beyond the fracture, or longer as needed, but they should not impinge on and harm a joint. The splints should be thicker at the center than at the ends. You should have enough of them to surround the limb; each should be about one fingerbreadth wide, and each should be wrapped with a moistened cloth.
7. Have short tubes to engage the cords that bind the splints. Each tube will be a small windlass to adjust the tightness of the application. A slim rod will pass through all the aligned tubes and secure them.[472]
8. After the reduction use a cradle and slings suspended from above to support and immobilize the limb.
9. The mattress and the cot should have openings through which the bed-ridden patient can defecate.
10. A sturdy cord suspended from above should be a handle for use by the patient as needed when he wishes to shift position, as when he is at stool.

The Second Requirement deals with the reduction:

You will have one of your two assistants grasp the end of the limb, and the other will hold proximal to the fracture and distract the site and prevent jagged fragments from piercing through the overlying soft tissues. If simple manual traction cannot distract the fragments, you may use thongs and instruments designed by Hippocrates, and later by Galen. I believe they

---

[471] I call attention to the frequent use of egg-whites, especially in bandages. The moist bandage will conform to the region while it is wrapped, and when it dries it will be rather stiff. The application served as does a light-weight plaster cast in today's armamentarium (LDR).

[472] See Plate. V in The Glossary of Instruments (LDR).

consisted of a windlass and columns, as described by Albucasis[473],and with pins to hold the position of the windlass (Lunel). While the traction is in effect, the surgeon will manipulate the fragments and bring the limb to its proper length, as compared with the normal side, as insisted by Galen.

The Third Requirement deals with the wrapping to maintain the reduction, thereby to relieve the pain and discomfort. Apply a bandage to suit the case, and cause no added discomfort. Nothing is more harmful than a too-tight and painful wrap. I have seen many sloughing infections (ie esthiomene) caused by such. Rhazes offered the same warning.

On the other hand, too loose a bandage will not hold the bones in place. Although it is the opposite condition the tight bandage, it can lead to the loss of a limb. The middle course is well tolerated, as Rhazes described it.

Hippocrates used three bandages to immobilize the fracture. The first is centered over the fracture and proceeds proximally; it will block the inflow of the humors attracted to the site. The second is unwrapped distally; it will press out the abnormal accumulation. The two work together to protect the fracture from inflammation. They are applied with multiple layers, (ie turns) and both end over normal tissues. Then use a third wrap to hold the pads over the fracture. Galen used waxed pads rather than rosat oil. Also, as in his treatment of wounds, he used sour dark red wine. Rhazes began to tighten the bandages after the seventh day, and was sure to have the center tighter than at the two ends. He claimed that will prevent an abscess proximally and yet permit the flow of nutritious blood to the distal part of the limb.

In this *Third Requirement* you have two goals: relieve suffering, and restore a normal limb. A sign of success is a well-shaped comfortable limb, and the patient will be pleased with what he sees, and he will be comforted further with the cradle and slings, and all the devices that hold the limb securely while it heals.

The Fourth Requirement uses slim splints to prevent motion especially during the initial seven days, until after the period of greatest risk for formation of abscesses. Use enough splints from the very beginning, to hold until callus forms and secures the union. Avicenna urged us not to be impatient. In that regard, Galen (idem, Part 6) said that a slender limb after seven days indicates freedom from infection, and we may replace the multiple splints with a few wider slats. But if edema persist, it will be wise to continue the more secure supports, although with less pressure.

The Fifth Requirement deals with the resumption of motion after we are sure, after ten, fifteen, or twenty days, that the fracture has been properly reduced and that there is no suspicion of complications. Rhazes said that the longer the wait, the better the result.

---

[473] A sturdy post was fixed in the ground. The patient was set athwart the post in his groin when the fracture was in the femur, or in the axilla when the erect patient's fracture was in the arm. The thong was attached to a windlass (LDR).

However, when there is any question that a good reduction was not sustained, as you will determine by your inspection before six to ten days, before callus has formed. Then you can decide if the damage can be repaired. When the patient suffers bad pain, or there are signs of an abscess, or the signs of esthiomen, you should inspect by the third day. Galen (I admit that the citation is unclear and the original Greek is uncertain as to when to unwrap the bandage) said that Hippocrates ordered the removal of the bandages on the third day, and if things were not right, or he saw pus, or other complications, he said not to replace them.

On the other hand, if there are no bad signs or symptoms before the seventh day, you may remove the initial dressings, although there is no harm in leaving them until a dirty limb needs cleaning. That was the practice of Bruno and Theodoric, and others.

The Sixth Requirement is the inducement of good callus (also called porus) which usually appears on the tenth day. That is when you should improve the diet. Galen offered nutritious foods, including plump meats that refresh the body and engender benign viscous humors which in turn engender callus. He mentioned broths of rice and wheat, and the cooked feet, stomachs and heads of animals, and thick somewhat vinegary wine. He avoided diluting the broths, and searing the meat and burning the blood. He avoided weak thin wine, garlic, onions, mustard, spica, bile, and sexual activities. Rhazes kept the diet thin during the first few days and avoided wine. He kept open the bowels with laxatives, and he used phlebotomy to prevent swelling (ie edema). When the initial period at risk for such was past, he allowed the patient to return to his usual diet.

Now having described the six Requirements. We will return to the

Four Goals (Intentions) of Treatments [474]

First we reduced the Fractures by traction and gentle manipulations to elevate or depress the fragments to properly align the ends of the bones.

Second we apply proper bandages, just tightly enough. The techniques vary. Some authors recommend anointing and wrapping the fracture and delaying the application of splints until the fifth or seventh day. Others apply plasters and pads and stupes, as did Theodoric, or moist cloths and splints as did Master Peter Arelata. Those methods are risky. In the case of delayed splinting, the limb may twist. In the other, a heap of pads may interfere with accurate bandaging.

I tend to a middle course, following the leads of Galen, Albucasis, Avicenna, and Halyabbas in three stages. 1. I relieve pain. 2. I promote the formation of callus. 3. I return the comforted limb to its normal functions.

First Stage: After the fracture has been reduced, and while the assistants continue the traction, I either wrap a long wide moistened bandage directly over the site, as did Roger, or I first apply a cloth or a stupe as did Lanfranchi, not too thick lest it interfere

---

[474] Most of this section deals with transverse fractures (LDR).

with the good wrap. Both (bandage and stupe) have been moistened with a mixture of egg-white and rosat-oil. I begin the wrap over the fracture and go distally and proximally to cover a good length of healthy limb, careful not to cause pain, although I wrap more tightly over the fracture. Then I apply a sheet of felt or folded cloth, both moistened, to cover and protect the entire part of the limb that will be enclosed by the splints (ie see footnote 469). I bind the wood splints with cords or leather strips and adjust the tightness. Then I put the patient in a firm bed with the supports described above. On the following day, when needed, I bleed the patient who can tolerate it, and I provide a thin diet. I avoid laxatives for a few days, especially when the fracture is in the lower limb. The initial dressing remains in place for ten or fifteen days, unless the reduction had not been perfect, or something elseintervenes, leading me to remove the dressings as early as the third or fourth day. At that time I make whatever repairs are called for, such as resetting the fracture, etc.

The Second Stage: After twelve or more days, when the callus is forming, as indicated by freedom from pain, the absence of swelling, and a limb with a healthy color, I remove the initial bandage and wash the limb with warm water, and correct any malposition. Then I apply a plaster of wheat-paste, red powder, and egg-white smeared on a cloth, and I bandage over all. Then I reapply the splints et al., but less tightly (ie restrictive). I improve the diet and prescribe laxatives if needed. That routine is repeated every seven to nine days until the callus secures the reduction, as your hands will reveal, and the edema is gone, and all good signs appear at the expected times.

Now we begin the Third Stage: Wash the limb with salted wine every three days. The wine is reinforced with a broth of roses, aloine, and oak-tree moss. After every wash, wrap the leg with a moistened bandage that encloses two or three splints. Use common sense in gradually returning the limb to normal activity. Ease the way with applications of dialthea and oxycrat.

The Fourth and Fifth Stages deal with such complications as persistent pain and swelling (ie edema or infection). For pain remove the bandages and apply sedative topicals containing lanolin, vinegar, et al., and re-bandage over the sedatives, and do not include splints. When itching is a problem, remove the bandages and bath the limb with salt-water and anoint it with the white or the populeum ointment.

When dealing with a fracture compounded by an open wound, gently remove the spicules of bone, as we described in the chapter on wounds of bone. When pus appears, allow it to escape (ie through a gap in the wound), and use detergents as needed.

When the formation of callus is deficient, attract more with massages, irrigations, and plasters of resin covered with bandages that are not tightly wrapped. On the other hand, when the callus is excessive, compress it with a snugly bandaged lead plate. When the union has been faulty and the initial callus is less than six months old, you may be able to soften it with daily baths and plasters containing guimauve, for fifteen days (Jamier). Then, with attached extensions (ie for leverage) lay the limb over your knee and break it. Then reset it. At times, when the callus is soft, you may be able to re-align the bones without re-fracturing them. I have seen that accomplished with traction after applying resin. When the callus is old and firm, all the experts advise us to leave it. Wisdom outweighs valor. Halyabbas wrote that life with a limp is better than death from complications. However, when you cannot resist the pleas, Avicenna said to incise and expose the callus, cut it away, separate the fragments, and re-set them. When the limb is atrophic, the risks for infection are greater, and esthiomene may ensue. If the limb survives, treat it as we do gout and other diseases of joints. All the foregoing material deals with transverse fractures.

Now, in respect of the axial fractures, the treatments in general are similar, excepting the need for more compression over the fracture-site, to restore the separated fragments.

CHAPTER 2

Fractures of the Cranium and the Nose, Jaws, and Face

C ranial and nasal fractures were discussed in the chapter in Treatise III "On Wounds". As to fractures of the mandible, Halyabbas, Albucasis, and Avicenna all instructed us to reduce them by inserting fingers in the mouth. A proper reduction will align the teeth and will restore the bite. Then fasten the teeth of the broken anterior segment to adjacent teeth in the healthy part, using a well-waxed thread or a silver or gold wire. Then apply topicals on stupes and pads of folded cloth atop which you lay a splint of heavy leather. Wrap the entire region with a long bandage that begins under the throat, passes over the jaw, returns under the ears and behind the back of the head and returns to the front where it is tied. Repeat the scheme several times until it holds the jaws firmly closed. The patient is fed broths until he is allowed to chew; that may take about twenty days, according to Avicenna and Albucasis.

CHAPTER 3

Fractures of the Neck and the Vertebrae

P aul of Aegina wrote, as cited by Avicenna and Halyabbas that vertebral fractures were unusual except when the bones were atrophic; Albucasis disagreed. Theirs was not a real dispute because Paul and Halyabbas were discussing fractures of the bodies of the vertebrae whereas Albucasis referred to fractures of the processes (ie spines and wings). Whatever is the case, when the injury is in the neck the most serious damage is suffered by the nerves that leave the vertebrae; they transmit the paralysis: in the arms when the upper vertebrae are involved, and in the lower limbs when the lower vertebrae are damaged. In any event, death may ensue (Avicenna) and you must prognosticate the outcome. When the patient is incontinent of urine or cannot initiate urination, the injury is lethal (Albucasis), and you need not attempt to cure the patient. But if that does not happen, your treatment with applications of rosat oil and cooked egg-yolks should contain analgesics and other substances that will reduce the swelling (ie edema or infection). After relief is obtained, apply comfortive and desiccative plasters. Limit motion with bandages. Place the patient at rest, to lie on the uninjured side.

When the tail-bone is fractured, insert the thumb of your left hand in the rectum and reduce the fracture as well as you can with your right hand. Use splints and plasters and bandages.

CHAPTER 4

Fractures of the Clavicle[475] and The Scapula

The angulated fractured clavicle may be depressed inward or it may jut outward. An inward depression may not be easy to reduce, whereas an external projection may be reduced by gently distracting the arm while you press the bone into its proper place. In either case, after the reduction, apply a stupe or a multi-folded cloth and a wood or sole-leather splint, about two fingerbreadths wide and eight widths long, and place a ball of wool fleece in the armpit and bind the dressings in place with a long bandage that passes under the axillae, making enough turns to hold well. Suspend the affected arm in a sling hung from the neck. Check the dressing daily to that it is not too loose.

Reduce a depressed fracture by pulling back both shoulder while you thrust your knee in back between the shoulder blades, and you work at the fracture with your hands.[476] Or, as did Avicenna, place the patient supine on the floor atop a pillow or a rounded block between his shoulders, and press down the shoulders while you manipulate the fracture. If that fails, glue a leather strip over the site, as we do for depressed fractures of the ribs. When it adheres, pull it away with a jerk; that may reduce the fracture. If the patient's breathing is impaired, and all else fails, very carefully insert a hook under the bone, avoiding the pleura. Use it to elevate the bone. Then dress with plasters and bandages as above. The fractures will not heal before twenty-four days (Albucasis).

---

475 The *Furcule* consisted of both clavicles and the attached manubrium. As described here, the clavicle is bent and the two fragments are not separated (LDR).

476 The text is unclear. An assistant is not mentioned. However, the surgeon cannot retract the shoulders while he tries to grasp and pull up the inwardly bent clavicle (LDR).

CHAPTER 5

Fractures of the Humerus, The Forearm Bones, and The Hand[477]

The Arm

Usually the arm is bent outward when the **humerus** is fractured. The special treatments vary. Of the two methods recommended by Albucasis for reduction and restoration, I find one of them hard to accept; simply he bound the hand of the fractured limb to the opposite shoulder. The other technique is better. One assistant holds the shoulder, another gently applies traction at the bent elbow while you use your hands to reduce and align the fragments. Then apply four to six splints and bind them in place and suspend the arm in a sling and let it hang free against the belly. You may apply compresses and stupes when you change the dressings every four to seven days, until the union is secure after forty.

Forearm

Sometimes only one or both bones are broken. A fracture of the larger bone (ie ulna) requiresd more force and is more deforming. Everyone agrees that you manipulate the bones Ie one or both) while two assistants apply traction at the elbow and at the hand. Dress and wrap as we described in the General Chapter.

When only one bone is broken you need less splinting, but you will need five or six splints when both bones are broken. Then suspend the arm from the neck in a sling, allowing it to rest at the belly. Expect union after thirty days.

The Wrist and Hand and Fingers

The carpal bones in the palm seldom are broken or displaced. When that happened, Albucasis placed the open hand palm-down on a flat surface and pressed the bones to

---

[477] The palm of the hand and the wrist are taken with the term 'hand'. Perhaps fractures of the distal ends of the radius and ulna were included with 'wrist'.Albucasis, who was Guy's model in his descriptions of fractures, made no mention of the typical deformities described so many centuries later, by Colles(LDR).

realign them. Then he used plasters and splints as indicated. He filled the palm with stupes and dry cloths.

Splint a broken **Finger** by bandaging it against the adjacent intact ones. Change the dressings after four days if suspect. Union will occur in twenty.

CHAPTER 6

Fractured Ribs And Other Bones of the Chest

Although they share in the General Matters, the fractures of the clavicles, ribs, and other bones of the thorax differ from others as to causes, signs, diagnosis, and treatments. The ribs, like the clavicles, may be bent inward or outward. The fragments of rib fractures seldom come apart.

Signs

Unlike the clavicles, the signs of rib fractures may include those of pleurisy, and painful respiration. Coughing may raise bloody sputum. Those signs bode ill.

Treatments

Halyabbas, Avicenna, Albucasis, and many others treat fractures of ribs in the same way. The first three apply oily wool fleece and pads, and wrap bandages. Roger reduced bent fractures with his fingers that were covered with glue, after he warmed the patient in a bath or sat them near a live hearth, and then applied the apostolicon. Jamier did the same, and added a plaster of honey, cumin, laurel berries, pouliot and costus during the first four days. Theodoric and his Master Hugh of Lucca used the same topicals.

Bruno was different. When the broken rib was bent outward he pressed it in and applied a plaster and splints. When it was slightly bent inward he fomented it, as did Avivenna, with lanolin and oil. But when the concavity was deeper he incised and elevated it (ie with a hook, or with his fingers). William of Saliceto used manual reduction and plasters of egg-white and wheat-flour and other sticky substances. Lanfranchi used Roger's methods, and he ordered the patient to cough while he was working with his fingers to bring out the depressed bone.

As for me, I first attend to the General Measures, using phlebotomy, laxatives and diets and beverages that dissolve coagulated matter (ie blood clots, edema fluid) such as broths of chick peas, as did Bruno. When the bend is outward, I press it in and apply plasters of egg-white, wheat, agglutinatives, and stupes, and I bound on a thick strip of shoe-leather with a long bandage. Later, after it is stable, I soften the region with dialthea and oxycrat. When the rib is bent inward, as did Roger, Jamier, and Lanfranchi, I set the patient in a warm bath or near a warm hearth, and cover my fingers with some terebinth or other glue and set them to adhere over the depression, and I lift them carefully when the patient coughs or holds his breath (ie valsalva maneuver), all the while palpating what I lift. If necessary, I make an opening (ie for a hook) as did Avicenna. I apply rosat oil for three or four days to relieve pain and discouraging infection; then I add stupes of egg-whites. The bandages should only be snug enough to hold the in place the medications. Later, I use a plaster of bean-paste and honey. Finally, I soften the region with dialthea and oxycrat. I replace the dressings every five days, and expect healing in twenty.

CHAPTER 7

Fractures of the Pelvis and the Femur[478]

T he pelvic bones are seldom fractured, except for the pelvic brim, although the bones may separate[479] or the brim may be broken. Sometimes the thighs may be driven upward toward the abdomen and cause much pain, and cause numbness and shortening of the limbs [480] Reductions are difficult and require much traction on the thighs while the pelvis itself is fixed. Reduce as best you can and apply plasters and bandages.

When the femur is broken powerful traction is needed (Avicenna). Give due attention to the cause and to the General Measures. Know in advance that some lameness is almost inevitable. The technique is similar to that used for a fractured humerus, but here even stronger traction is used. Bind straps above and below the fracture, to be pulled by the assistants. After the reduction use strong bandages and six or seven splints. William of Saliceto made the exterior splints longer and thicker. Albucasis used only three. He flexed the knee and fastened the foot to the buttock thereby to lessen the need for splints. I do not like that maneuver.

Several ways to manage the patient after the reduction have been proposed. Roger, Albucasis and William all used a flat bed and restricted the limb with bolsters and blankets; I think that is inadequate. Others, including Master Peter of Arles[481] surrounded the limb with a sort of basket of reeds wrapped and sewn in a shroud which was tied to the bed. Others, Avicenna and Bruno, used two long (to the feet) splints secured with bandages. Lanfranchi and other more recent surgeons set the leg in a cradle. All of the methods

---

[478]   Here Guy includes the femur at the hip-joint as part of the hip. The Reader should interpret the text as to what is femur and what are pelvic bones (LDR).

[479]   I assume that Guy (as did Albucasis) referred to separation at the symphysis pubis rather than dislocation of the femoral head or to inward displacement of the acetabulum and its contents(LDR).

[480]   Here Guy describes fractures at the femoral neck or at the trochanters (LDR).

[481]   Peter Of Arles (Arelata) was a respected contemporary of Guy, who cited him as Master Peter, to be distinguished from Peter Argentiere, (another contemporary) and Peter Argelata (late in the 14th C). See Nicaise's Introduction (LDR).

serve to maintain the reduced fractured thigh at rest and prevent any motion or bending at the knee. Roger insisted that the fractured leg be bound to the healthy one, placing the patient in a narrow bed with a hole through which he could defecate while supine. To insure the immobilization, he bound the leg to the bed in three or four places and he tied the foot to a post at the end, to prevent the patient from pulling toward the head of the bed. Theodoric did the same.

As for my own practice: I use full-length splints and either a basket of reeds or a cradle. I attach a cord to the ankle (ie foot) and lead it over a small pulley and attach a lead weight. That maintains the length of the limb as reduced, or if there is any shortening it may correct it gradually. I change the dressings every nine days or longer during the fifty days needed for union.

CHAPTER 8

Fractures of The Knee, The Leg, and The Feet

The **Patella** rarely is fractured although frequently it is dislocated. However, if fractured, use your fingers to assemble the fragments (Halyabbas) and restore its shape. Then use plasters and a disc of leather as a splint. Bandage all.

The Tibia may be fractured alone or with the **Fibula**. When alone, it is displaced to the rear. When the fibula alone is fractured it can be displaced forward or inward. A tibial fracture is more serious because it is the mainstay of the leg. Even more serious is a fracture of both bones. Both Albucasis and Halyabbas treated the fractures as they did fractures of the forearm bones, except that here they placed two splints between the bones, the full length of the leg, and they used a cradle. I do the same, and when necessary, I use a traction-device as for a fractured femur.

The **Calcaneus** is a very hard bone surrounded with fibrous ligaments. It seldom is fractured.

The **Forefoot** bones (tarsus), like the heel bone, are seldom fractured. When such occurs, any treatment is difficult and is subject to serious complications. Albucasis placed the foot on the ground and stepped on it to reduce the fractures. Then he applied plasters and wide leather splints. He filled the arch with padding as used in the palm of the hand.

The **Toes** are treated as we do fractured fingers (Halyabbas).

DOCTRINE II: DISLOCATED JOINTS
EIGHT CHAPTERS

CHAPTER 1

Dislocations In General

A vicenna and Albucasis called them *deflorations or untyings*. A dislocation is a displacement of a bone from its natural position in a joint. Avicenna (Part 1) added the description of four kinds of joints. One is a saw-tooth connection, as in the commissures of the cranium. A second is an embedding, as the teeth fixed in the jaws. A third is simple contact, as is the meeting of the parts of the sternum. The fourth is when one bone is bound within a cavity: shallow as in the vertebrae or a deep cup as in the pelvis for the head of the femur. His term 'untying' refers only to these four structures. It does not include unusual mobility or enlargement of the opening. Lanfranchi's definition included enlargements.

A dislocation may be complete; a true dislocation is when one bone completely departs from the joint. A partial (imperfect) dislocation is less than total, what Avicenna called an extortion, as when one bone twists and stretches the bindings but does not empty the joint. Galen called it 'stretched ligaments'. Avicenna used that term twice in Part 4 of his *Canon.*

Dislocations occur in four directions: anterior, posterior, superior, or inferior. Some are simple and others are compounded by fractures, wounds, severe pain, and abscesses. In some cases indurations persist.

The **Causes** may be external, such as falls, blows, or unanticipated over-stretching. Internal causes may be slippery humors in the joint.

The **Signs** are clearly noted at the site: a deformity, a bulge, or a hollow. Other signs stem from the injury: pain, limited mobility and function, especially when compared with the normal limb (Avicenna).

Diagnosis and Prognosis

Hippocrates and Galen (Aphorisms, Part 6) said that once dislocated a recurrence is commoner because slippery matter has accumulated to favor it. The limb may cramp and be stiff unless the region has been cauterized. Hippocrates' words are unclear and Galen tried to explain them. Albucasis taught how we should cauterize with a circular iron.

Avicenna declared that a dislocation is more dangerous and more difficult to treat if compounded by a wound or an abscess. In such cases it will be better to leave the dislocation untreated, and Galen agreed. An old and indurated dislocation may be impossible to cure.[482] Do not hasten to take on such cases.

Furthermore, the dislocations vary as to the involved joints. One may dislocate easily and be easy to reduce, as at the knee (ie the patella) by softening the ligaments. Another may resist reduction, as at the elbow, the mid-foot, and the toes. Some are middling, as at the shoulder and the hip. And any dislocation where the margins of the joint have been broken may be very difficult to reduce.

Finally, you may know when your reduction is successful when you feel a pop when the bone enters the joint, when the normal contours have been restored, as compared with the healthy side (Jamier).

Treatments

I suggest that you review the 'Requirements' discussed in the Doctrine on Fractures, and then review the Four General Elements: 1. Reduce the luxation. 2. Immobilize after the reduction. 3. Prevent swelling (ie edema or abscess). 4. Treat complications.

First: The reduction is obtained by traction and manipulation, pressing the dislocated part into its proper place. Do it gently and as painlessly as possible.

Second: Anoint the region with rosat oil before bandaging. Apply a moist cloth and cover it with stupes or pads of folded cloths moistened with egg-white, and, when needed, with a leather splint. Wrap all with a long and suitably (to fit the joint) wide bandage that is moist with oxycrat. Set the limb at rest in a comfortable position for four to seven days. At the second treatment, wash the region with warm water. I remind you not to do that at the first treatment because warm applications will increase the swelling (Avicenna). After the wash (ie second treatment) apply a plaster of wheat-flour, red powder, and egg white. The second bandage should be tighter than the first (Halyabbas).

The third measures include phlebotomy, laxatives, and a healthy diet, very light at the beginning, and will increase as the pain subsides and the swelling abates. All of this is the same as the measures we use in treating fractures. In time, use comforting washes with decoctions of roses, aloin, and white-oak tree galls. Apply cloths wet with oxycroceum as you increase the range of motion towards normal.

The fourth measures befit the complications as they appear. Bad pain should be relieved before you begin the reduction, because the traction may induce a convulsion and other bad effects. Lay on wool-fleece wet with warm water and oil before you perform the reduction.[483]

---

[482] Such as a congenital dislocation of the hip (LDR).

[483] Guy seems to overlook his warnings to avoid warm applications or embrocations when treating recent dislocations (LDR).

When it is complicated by an open wound, first reduce the dislocation, and use sutures if needed to close the wound. If the wound suppurates, leave a gap for drainage. If the complication is a fracture, whenever possible first reduce the luxation, and then treat the fracture. If that sequence is not feasible, reduce the fracture and await the formation of callus before you attend to the dislocation.

When the dislocation is old and indurated, use frequent washes with a decoction of mauve and guimauve followed by inunctions of dialthea and diachylon, and cover the region with unwashed (the lanolin is retained) wool fleece that is wet with a mucilage containing the bark of the roots of guimauve that you have minced and boiled to make your ointment. After the region has softened, you may be able to reduce the luxation and restore the joint, and, even then the mobility may be partly or completely restricted. Then use the treatments which we will describe in the chapters on disorders of joints in Treatise VI, and in the Antidotary.

CHAPTER 2

Dislocated Jaws

The mandibular joint may be loose, when the ligaments are relaxed, or it may be in spasm, or it may dislocate. If the luxation is to the front, the jaw remains open when the ligaments are loose. When it dislocates to the rear the opposite occurs, and it is not due to looseness (Avicenna), and the jaw is clamped shut, not like the clench during a convulsion.[484]

Signs

Both Avicenna and Halyabbas said that induration soon appears when the luxation is not promptly reduced, and that fever, pain, an accumulation of choleric humors, and other serious complications will follow, and the patient will not survive beyond the tenth day.

Treatment

Reduce a posterior luxation as follows: While an assistant holds the head, the surgeon inserts his thumb in the mouth and his fingers under the jaw, or uses a wood rod if he cannot get his thumb in. Then he pulls the jaw forward while raising the chin and sets it in place just below the ear. An anterior dislocation was reduced by William of Saliceto and Lanfranchi by placing a strap under the chin and placing a rod between the upper and lower teeth at the rear of the mouth, as far back as [possible. An assistant standing behind the patient holds his knees against the patient's shoulders and pulls up and back on the strap. The surgeon guides the mandible into the joint (a la Jamier). That will reduce the dislocation with a little help from The Lord.

Then apply plasters, etc., and bandages as used for fractured jaws. Change them every four days for about twelve. Put the patient to rest with his head on a firm pillow. Feed him liquids and foods that do not need to be chewed.

As noted, when the reduction is delayed at length you must soften the induration with warm water and oils, et al. before attempting the restoration.

When serious complications appear, such as pain, shave the head and anoint the back (occiput) and behind the ears and on the neck and in the armpits with rosat oil.

---

[484] When the convulsion is epileptic the clench is temporary. Guy does not consider tetanus (LDR).

CHAPTER 3

Dislocations of the Neck And Back

Vertebral dislocations may be total or partial, deviating to the front, back, or the sides. When it affects the upper (ie cervical) vertebrae it causes quinsies and scrofules, as were described in Galen's *Affected Parts,* Part 4. Sometimes the luxation is in the lower vertebrae, and when it involves the middle sector it produces a gibbus.

The **Signs,** are obvious to the eye and the hand.

The **Prognosis** always includes the risks and the limited expectations for a cure, because all the nerves at the neck come from within and they cannot be repaired. The prognosis also should state that dislocations of the cervical vertebrae interfere with swallowing and breathing, whereas dislocations of lower vertebrae impair micturition and defecation.

Treatments

Albucasis, Halyabbas, and Avicenna all dealt with them at length. In brief, we treat cervical dislocations by traction on the head with your hands or with a head sling, as did Jamier. He put a twig of quince-wood between the upper and lower teeth and pulled the chin, ears, or hair while he set his heels to push away the shoulders. And he pushed in the protruding bones to relocate them.

He reduced lumbar dislocations with traction on the ribs with his hands or with a strap attached to a windlass or over a pulley. He pushed in posterior protrusion with his hands, or by stepping on them, or by using a plank as a levered wedge.[485] He applied comfortives and softeners. He persisted for five days until the reduction was assured after twenty.

We shall describe hump-backs caused by humors in the chapter on disorders of the back.

---

[485] Bruno of Longoburgo described this curous method in detail, as coming from Hippocrates. See Bruno in the Bibliography, p. 47 (LDR).

CHAPTER 4

Dislocations of the Shoulder

Although Avivenna wrote that there were only two kinds of luxations at the shoulder; the common one, into the upper axilla (ie the tickle-region), and the less common anterior dislocation, Albucasis described another rare dislocation on the chest wall, and he said that a superior luxation was prevented by extensions of bone.[486]

The **Signs** common to all types are the obvious bulges and hollows, and limitation of movement, partial or total. Other signs are the hollow above the depressed shoulder, the palpable egg-size mass in the axilla, and the hand cannot touch the head. When the mass can be felt in front (under the pectoral muscles) the upper axilla is empty and the victim cannot reach behind to touch his back. When the luxation is to the rear, the arm is held at the side.

An accurate diagnosis may be obscured by inflammation, edema, and by the twisted arm. That was Rhazes' prognosis when he treated the daughter of the king of France.

Treatment

There are five ways to reduce an inferior dislocation. The first is easy. Gently raise the arm while you push the head of the humerus with your palm or with your shoulder. Then lower the arm while holding the bone in the socket.

The second technique requires more force. Place a firm ball in the armpit and pull a strap (ie folded towel) under it or push it with your heel while you lower the arm.

The third is even more forceful. Place the ball in the axilla and suspend the arm over a rod which two assistants lift while you pull the arm downward.

The fourth arranges the patient so: he stands on a foot-stool and places his arm, with a ball in the axilla over a rung in a ladder. You hold his arm down while an assistant removes the foot-stool. The patient's weight will reduce the luxation.

The fifth uses traction on the hand, and I will not describe it, although Halyabbas, Albucasis, Avicenna, Bruno, and Theodoric included it with the others. I follow Roger here

---

[486] The coracoid and acromion processes of the scapula (J).

in most of his methods: he used a fist, or a ball, or a board instead of a ladder, but used the foot-stool.[487] Jamier used only the ball and the foot method. William and Lanfranchi used the ball and strap (ie towel) method.

Other types of dislocations were reduced by traction with a strap or towel and manipulation.[488] When the induration of a chronic dislocation does not yield to the attempted reduction, he applied softening topicals for a long time before he repeated his efforts.

After a successful reduction, we return to our General Measures. As did Roger, I release the bandages if pain persists for three days. Then I apply stupes soaked with egg-whites, and lay on a plaster of wheat-flour and red powder. When I replace the bandages I put a wad of cloth or charpie in the armpit and secure it with a two-end bandage that is five finger-breadths wide and at least as long as full arm-spread. I place the center of the bandage over the wad and unroll one end over the shoulder and the other under the opposite axilla They meet at the wad. I repeat the turns and make it snug before I sew the laps in several places. I support the arm in a sling for nine days or until the local inflammation (ie swelling) abates. The treatment lasts twenty days. I change the dressing every nine days, when I apply dialthea on a cloth or as a plaster.

Although the clavicle can dislocate at the sternum, that cannot happen at the shoulder where it is bound between the two bony projections from the scapula. The treatment for medial separations already has been described.

---

[487] See Bibliography: Roger, p 112 and Roland p. 75 (LDR). The 'fifth' method was used for treating dislocations of the shoulder in children (LDR).

[488] Joubert described Lanfranchi's method for reducing an anterior dislocation. He placed a towel around the chest, under the axilla and behind the back at the shoulders, both ends of the towel were to be pulled by a strong assistant. He fastened a strap over the bent forearm near the elbow to be pulled by another assistant. While the two applied traction he used his palm to press the head of the humerus toward the joint. When he sensed that the bone was where he wanted it the assistant at the elbow released his pull and allowed the bone to settle into the joint. (J).

CHAPTER 5

Dislocations at the Elbow

A vicenna described them as minor and major, and Albucasis and many other authorities said that the former were anterior, and the latter were posterior.[489] However, Roger was concerned only with the anterior because they were more frequent.

The **Signs**, other than the obvious deformity, are the lateral angulation of the forearm at the elbow and the restricted mobility such that the hand cannot touch the ipsilateral shoulder.

The **Prognosis** recognizes that the elbow does not dislocate easily and that the reduction is difficult because of the complicated bindings that join the three bones. The radius rarely dislocates because its range of action is limited (ie rotation). The ulna dislocated more frequently because it cannot rotate.

Treatments

Avicenna reduced the anterior displacements with two maneuvers. He pushed the patient's palm towards the opposite shoulder while he manipulated the bone into place. For posterior luxations he pulled at the hand and with greasy fingers he pushed the protrusion to reduce it. Roger used a stirrup (ie see below) for traction on the fore-arm; he used it with his heel. Lanfranchi used it also for anterior dislocations. He treated posterior dislocations by attaching a weight to the extended forearm (hand) for traction, to be carried about by the patient.

As for me, I reduce the anterior dislocations as I do the knee, and I follow Avicenna in treating the posterior luxations, as did the Romans and the Bohemians.

The stirrup and the heel are used to pull the forearm away from the humerus and allow you to slip the humerus into the fossa of the ulna. The stirrup is a long looped strap placed over the middle of the forearm. The other end of the loop is near the floor. The surgeon sets his foot (near the heel) in it; it can press and control the downward traction while an assistant holds the humerus to prevent its dislocation at the shoulder, until the forearm can be flexed.

When you use the knee the lower loop is controlled by it.

---

[489] When the person stands and the palm sits against the body, the thumb is anterior. Therefore the radius is the anterior bone of the forearm, and, the ulna is posterior. When the elbow is flexed, the radius is higher than the ulna (LDR).

Whatever method is used, traction on the forearm is important. Jamier used a portable weight for the purpose.

As to the immobilization after the reduction: Wrap as you do for a dislocated humerus at the shoulder, but keep the arm in a sling while you gradually, over the course of fifteen days, increase the range of motion. Change the dressings every four days.

CHAPTER 6

Dislocations at the Wrist (Hand)

The wrist bones are easily displaced[490], and easily reduced if treated early. They can be shifted in all directions but more anterior or posterior than lateral.

The **Diagnosis** is easy to see.

The **Treatment** consists of traction and manipulation with pressure on the abnormal projections (ie eminences). When that failed, Albucasis placed the hand (palm) on a flat surface and pressed it with his own. Stabilize the repair with splints and bandages, changed every four days until secure after twelve. Use softeners as usual to lessen residual induration.

---

[490] This seems to contradict what Guy wrote about fractures at the wrist. Perhaps he considered a bad sprain a dislocation (LDR).

CHAPTER 7

Dislocation Of The Femur at The Hip

L et us first review what the Authorities had to say. Albucasis described three kinds: inward (ie anterior), outward, and posterior. Bruno, Theodoric, and Lanfranchi all accepted Avicenna's description of four kinds, of which the outward was the most frequent. William of Saliceto disagreed, claiming that a posterior luxation was more common and that external dislocations never occurred because the femur was bound to the hip within the joint, and that the treatment involved only the anterior and the posterior types. He ignored the others.[491]

The **Signs,** other than the General, were best described by Avicenna. When the dislocation is anterior, the bad leg is elongated and rotated such that the patient cannot set his foot flat or the leg parallel with the other. The head of the femur is visible and palpable in the groin

Treatment

Although the Masters described many techniques, all of them share two elements. Albucasis declared that all types of dislocated hips can be reduced by traction of the upper body against traction on straps attached to the thighs and pulled by a windlass, supplemented or replaced by traction on the hip and the lower thigh. While the traction is in force, it will lift the patient from the ground and allow the surgeon to move the thigh in all directions so to reduce the luxation. When the traction has freed the femoral head, the surgeon can press it out of the groin and manipulate the shaft of the femur from the knee so the head will enter the socket.

When the dislocation is to the rear (into the buttock) and the traction is effective, you can feel the head of the femur and angle it into the joint by abducting the thigh.

A successful reduction equalizes the length of the legs, as noted by Roger, and Jamier, who copied Roger in most matters. Sometimes I use this device for traction. The patient lies supine on a table longer than him, with posts secured in the ground at the ends and sides. I pass a heavy towel or table-cloth through his crotch to extend over his belly and

---

[491] See Bibliography, William of Saliceto's *Surgery,* p.174 ff. (LDR).

under his back toward the shoulders and I fasten the ends around a post at the head. I bind another towel around the thigh and the entire leg down to the ankle and fasten the ends around a post at the foot. By twisting the looped towels with a rod I shorten them and distract the dislocation. When the traction is effective, I restore the joint.

To maintain the reduction, I bandage as follows: I begin over the bulge and cross to the other side and pass it around the back (at lumbar level). When the luxation is anterior, I place a moist cloth or stupe over the groin and hold it with the bandage that passes over the normal side, holding the thighs together. When it is dry,I lay a strip of leather from the foot to the shoulder and a long splint along the side to the heel. When it is dry, I fasten it to a cord bound to the length of the lower leg and lead it over a pulley and tie a three- or four-pound weight. I change the dressings every five days during the thirty-day time for healing.

CHAPTER 8

Dislocations of the Knee, Patella, and Foot and its Parts

T he knee is easily dislocated[492] A dislocation may result from a misstep, or a skid, or a jump. The luxation may occur in all directions other than forward, where the patella and its attachments are a barrier.

Treatments

Treat the Dislocated **Knee** by seating the patient on a bench with his legs dangling. A strong man swings the leg gently to and fro and then suddenly and forcefully pulls the leg and extends it and returns the knee joint to a normal position. Then bandage it.

The **Patella** is reduced by placing the patient supine and filling the popliteal hollow with a wadded cloth and splinting the leg and thigh to prevent flexion until the repair is secure. Avicenna insisted that the return to normal should be gradual. However, Lanfranchi and Jamier kept the patella fixed in place (ie between the femoral condyles) by flexing the knee and binding the leg to the back of the thigh for one hour.[493]

Dislocations of the **Foot and the Toes**. The instep is easily dislocated and easily reduced. However, when the several small bones are displaced in one or more directions, the normal arrangement is not easily restored. The **Diagnosis** is clear: a misshapen foot, pain, and limited movement.

The only suitable treatment is gentle manipulation and traction to accomplish as what is possible. Then bandage and splint. Re-dress every five days for thirty or forty, hoping that the no serious defect persists, and that The Lord will forgive us our defects.

HERE ENDS DOCTRINE II OF TREATISE V
AND TREATISE V.

---

[492] A true dislocation of the knee usually is caused by a devastating force, and is an unusual occurrence. It is clear that here the Ancients and Guy described a locked knee due to injuries or fragmentation of a cartilage, or the so-called internal derangements, such as torn ligaments (LDR).

[493] I assume that is a misprint, and that the treatment lasted longer than one hour; perhaps it was one day (LDR).

HERE BEGINS TREATISE VI
MISCELANEOUS SURGICAL MALADIES
TWO DOCTRINES

RUBRICS FOR TREATISE VI

TREATISE VI:
MISCELANEOUS MALADIES

---

494 Items in Italics are not dealt with in this Treatise VI (LDR).

DOCTRINE I: HAS EIGHT CHAPTERS

CHAPTER 1

Gout And Other Painful And Indurative Disorders Of The Joints

G out, or Arthritis, refers to painful joints caused by humors accumulatinmg in them. Galen (*Commentary on Aphorism #6*), while discussing Podagras in Eunuchs, wrote,

"The malady is caused by a humor flowing into a joint. In that case the arthritis is called podagra." Albert of Bologna copied Halyabbas (*Royal Dispositions,* Sermon 9) stating that arthritis is an aposthem, meaning a swelling, involving joints. On the other hand, many authors, including Rhazes (*Painful Joints*), wrote that arthritis could develop from a bad complexion (ie disposition) and not be due to a humor. But Avicenna said that such rarely occurred, and that when it happened, it was not a real gout. And Galen (*Miamir,* Part 10) wrote, "The excess of matter that causes arthritis, sciatica, and podagra in its 'rheumatizing', it accumulates and fills joints and the ligaments around them, and affects all the nerves and causes pain". He repeated the claim in his *Commentary on the Aphorisms* where he described podagras in general.

Gordon said that Gout and Arthritis are the same; Gout being the name of the flux, and Arthritis, referring to the word for joint, Arthron. When it involves the hip it is called Sciatica, and elsewhere it simply is Arthritis, as Galen had written.

Chiagra really is not arthritic, because it is caused by phlegm, as we described it in the Treatise on Aposthems. The same holds for swollen knees as caused by indigestion in children (Galen, idem). Children do not get podagra. We should have concern for the terms, because they lead us to different treatments, determined by the causative humors, except in the case of sciatica. Some painful arthritides are cool gouts, others are warm. Sometimes the pain makes the patient groan, and he is disabled, as described by Paul of Aegina, and sometimes it is mild and tolerable and is easily relieved.

Causes

The General Causes are those for Aposthems (rheum). The Special Causes are the Primitive (ie underlying), Antecedent (ie immediate), and Conjoint, as we described them in that Treatise. The principal sources for phlegm are the brain and the stomach; for choler and the others are the liver and the veins. Avicenna said that the causative humors usually are engendered in the second and the third digestions, and the joints are the primary recipients.

439

Avicenna described three kinds of causes: the material or efficient (called the instrumental). The second enlarge the passages for easy entry of the humors. The third are the patient's own contribution: the feeble resistance against the inflow, either his natural tendency for gouts, or the effects of an injury (a fall or a contusion), or the effects of bad habits (life style).

As to the first, the instrumental, we will refer to Hippocrates' *Sixth Aphorism*, where he wrote that women get podagras as a result of defective menstruation; the bad matter is retained and causes it. Galen agreed (*Commentary on the above Aphorism*).

The second causes are the enlargers, as are demonstrated by the eunuchs and children who get podagra, as a result of sexual activities[495] that dilate the openings which admit the bad humors.

The third are described in Galen's *Commentary,* where he wrote that podagras will not occur unless the feet and other weak joints lack resistance to what comes from a source such as the brain of a person with epilepsy.

Galen *(Miamir)* explained that the passages referred to above sometimes admit sanguine humor[496], but more often it is phlegmatic or choleric or a mixture of the two, or sanguine plus them. Avicenna said that melancholy seldom was involved, because splenetic and melancholic persons seldom are rheumatic.[497] Avicenna cited Hippocrates to that effect. Still less frequent causes are corrupt humors; and least often, according to Rhazes, were humors in their normal proportions. A normal joint suddenly becomes symptomatic and that suggests that the humors are gaseous, as it is claimed in the Papal Regimen[498]. Usually the offending humors are mixed at random, and are not single. A raw (a single bare) humor usually cannot penetrate the joint unless it is mixed with choler, as Rhazes claimed (*Divisions*). Yet we treat the single humors, as described for the aposthems, and in that way we get at the composites.

Signs

When a gout agonizes the patient, no other sign is necessary. The distinguishing signs of cool versus the warm gouts are those for the aposthems. Galen added eight that identify their specific matters: the colors, what we can feel, changes in nearby structures, the prior life-styles, the complexion, the age, the weather, and the region of the body.

---

[495] Sodomy and masturbation (LDR).

[496] The major humor in normal blood, which contains a proper allotment of the four humors (LDR).

[497] We are reminded that catarrh was phlegmatic, and that a prime source was the brain, whereas melancholy came from the spleen (LDR).

[498] The Papal Regimen was a set of rules of health for the Pope. It was erroneously attributed to Guy. We know that it existed before Guy came on the scene as a papal physician-surgeon. See the final paragraph of the section on Biography in Nicaise's Introduction. Guy cited it frequently (LDR).

Avicenna added even more: the different kinds of pain, their duration, their tendency to recur, the urine (ie urinoscopic), and other excretions.

Therefore, Gordon's practice was first to examine the affected site: was it red, tender, warm? What were the patient's personal habits, his complexion, his age, and other particulars as suggested by the degree of the patients's warmth—the gout probably will be warm, and cool when the patient's complexion is cool. Not infrequently, all these signs are not enough for a precise identification of the offending humor or composites, as is desirable when selecting the correct medications, and where to apply them. Both Rhazes and Avicenna agreed in that.

They also agreed that gout usually begins as a podagra in the large toe or near it. Sciatica usually appears at the hip and spreads to the ankle. Hippocrates declared (*Aphorism #6)* that all podagrous maladies subside within forty days. And Galen said that fourteen days are the duration of fleshy phlegmons[499], and forty days for phlegmons of nerves, although the fleshy substance is thinner than that of ligaments (ie nervous tissues). The bad humors in flesh may form local masses or they may be diffused; not so for matter in ligaments. We consider it a good sign when a varicose swelling involves a gout, and a bad sign when it is absent.[500] Therefore, the humors eventually leave the nervous tissues and diffuse into the surrounding flesh

As in other aposthems, disorders of the joints go through four stages and end with a good resolution or a stiff joint. They do not suppurate as do other aposthems. (Rhazes in *Divisions).*[501]

Hippocrates (idem) said that arthritides are more common in the spring—the result of the accumulation of humors engendered during the winter, and in the autumn—the result of the degradation of the humors during the summer when the passages (ie into the joints) are dilated. Nevertheless, a gout may appear during the winter when cold causes the humors to combine, and in the summer when the heat separates them (Hippocrates, *AphorismI, #3).*

Avicenna said that painful joints are inherited because the semen is part of a person's inborn complexion.[502] Furthermore, the painful joints serve to accumulate the matter that they fail to repercuss, and it could spread to and harm the principal organs. So it is that a gout may spare the body of some harm (Galen, *Therapeutics*, Part 7, and *Health,* Part 4).

Sciatica is the worst arthritis (Avicenna).

---

[499] Edema in soft tissues, usually after a contusion or a dislocated joint (LDR).

[500] Joubert explained: Here 'varices' does not mean the common varicose veins of the thigh and leg. It refers to the congested venules around a joint, perhaps not seen individually (J). The observation meant that the vitality of the region was good enough so to enrich the region of the sick joint (LDR).

[501] Joubert explained: ;They are not true abscesses as can happen with an aposthem in a muscle, as Galen defined them in *Unequal Temperatures*. The humors that cause a gout are not spermatic as are those in aposthems in muscle, which lend themselves to drainage. The tendons, ligaments, and membranes around joints are dry and do not suppurate. They usually heal by resolution (J).

[502] Joubert explained. Semen is a benign product of the third digestion, as occurs in all organs. In this case the spermatic organs are the testicles. It is their special function in the same sense that the special function of the kidneys is to attract the watery or serous matter from the nutritious blood that goes to all parts of the body (J).

Painful joints often cause fever, and in reverse, fever and colic can cause painful joints (Avicenna). And when a painful joint lasts for a long time, the limb becomes thin, weak, and stiff in extension. When there is a gibbus, or when the joint is wounded, both easy to see, it means that the joint's natural resistance to influx is even weaker.

In summary: When a sick joint persists in a limb, even when the cause has been eliminated, the result may be permanent. When its normal functions have been impaired for a long time, the limb will remain weak, and recurrences of the gout may be frequent.

Rhazes reminded us (*Divisions*) that painful joints sometimes lead to asthma, paralysis, apoplexy, emotional depression, and sudden death.

Treatments in General

The Physicians (ie the Academics) agree to two elements in the treatment of arthritides, and we add a third, the Papal Regimen. First is prevention; Second is dealing with the humors already in the joint; Third is the return to a healthy life-style after the ailment is controlled.

Preventive Treatments: Three Types

First: deter the formation of the humors at fault. Second: eliminate the bad humors before they get into the joint. Third: repair the damage. Treat according to the type of combustion:
- A. avoid what makes the bad humors by applying the six non-natural things and their three annexes, as we commonly use them. They are: air, food and drink, inanition and repletion, sleep and wakefulness, the various states-of-mind (ie emotions, etc.). We use external things such as baths and the provision of a healthy general ambience.
- B. Use phlebotomy and laxations.
- C. Apply comfortives and desiccatives.

Specific Curative Treatments: Four Types

First: The Diet should be thin and spare. Second: Evacuate the antecedent matter with repercussives and resolutives. Third: Eliminate the conjoint matter with repercussives and resolutives (ie evaporatives). Fourth: Treat the complications.

Rehabilitative (ie Resumptive) Treatments: Three Types

First: Good habits and a gradual return to a healthy life-style and diet. Second: Get rid of residual traces of bad matter with oral diuretics and theriacs. Third: Baths, Inunctions with fox-lard, et al., Comfortives and Analgesics. All of them usually are prescribed by the Physicians, and the Surgeons only apply them. I am very careful about my tasks as a surgeon; whatever else is described here is for the sake of completeness.

Details of The Preventive Treatments

A. Master Arnold offered ten aphorisms about the use of Non-Natural Things to prevent the formation of Cold Matter, in respect of the first category noted above:

1. Provide cool or warm air as required for the comfort of feet affected by podagra.
2. About meats and beverages: Avoid piglets, riverine fish, and old grouses as harmful to joints in the lower limbs.
3. Avoid fish such as eel and mullet.
4. Do not mix animal milk with wine or meat.
5. Drinking when not thirsty makes joints worse.
6. Too big an appetite for food, and too much eating are bad.
7, Too much ambulatory activity is bad, whereas resting feet elevated is good.
8. Sleeping supine for too long spells is bad.
9. Angry outbursts excites gouts in hands and feet.
10. The overuse of the arms and legs in unnecessary activities is harmful.

B. In this context, Galen said (when commenting on Podagras in Eunuchs) that gourmandizing, high-jinks, indigestion, laziness, omission of customary laxatives, and prolonged coitus all can induce gouts.

Master Arnold's details in re the second category: He agreed with Hippocrates (*Aphorisms*) who advocated phlebotomy and laxatives in the spring and autumn at the very onset of the arthritis, and that the methods were those we use for aposthems. But Arnold prescribed his own laxative: an electuary of powdered tragacanth, a comfit of quince-pulp and sugar, white ginger, hermodactyl, crushed carthame, diagridum, turbith, manne-seeds, foam from rosat-honey, and sugar-loaf.

C. Arnold in re the third category: He used comfortives for joints taken from what Hippocrates listed in his *Aphorisms*. Alum-water for the feet; barely cooked sage taken after a meal to relieve nerves and joints; flowers of almond and myrtle, camomille, melilot, roses, acoris, ground-ivy, nutmeg, and various clays; raisins and and moss make a good foot-bath for podagra; vervaine and chelidonium are good diuretics (Rhazes and Halyabbas both highly recommended diuretics for arthritides that purge the superfluidities after the second and third digestions. The improvement is soon effected). I used chick-peas as a diuretic for the Pope. Other comfortives include electuaries, troches, plasters, et al.

Details Of Four Curative Treatments

1. The Life Style (Habits)

Avicenna eliminated wine and he limited meats for all cases of warm gouts, He prescribed mead with diuretics; I agree to use the usual ones at the onset, especially for gouts of the upper limbs. When the patient cannot abide total abstinence, I allow small amounts of well watered thick wine. Later on he can take thin wine. Early on he may drink barley gruel etc. as we prescribe it for aposthems

2. Diversives and Repercussives Against The Flux

I prescribe emetics and irritating enemas containing Herb Benedict, and laxatives containing diacarthame. I bleed from the opposite side when the patient is robust and plethoric. That works better when performed at the onset, as did Avicenna. Every diversive medicine works wonderfully if given early in the course., especially when the gout is due to choleric and phlegmatic matter. Arnold described phlebotomy. I bind the limb, but not on the lesion. A proximal tie blocks the inflow.

3. Dealing With Conjoint Matter

As we do for aposthems, we should use the repercussives from the onset (except in cases of sciatica) to prevent the return of the offensive matter to the involved joints, where they can harden and resist the resolutives. During the phase of augmentation, add resolutives, a small amount at a time, until they come to equal the amounts of the repercussives during the phase of stasis. Erlminate the repercussives during the phase of decline. All of that scheme was Galen's. Although the familiar remedies work well, as described in the Treatise on Aposthems and in the Antidotary. Hwever, we will repeat them here to round out the chapter. Often we alternate thew medicines in use, because it is natural that the effectiveness of a remedy will vary from time to time.

TOPICAL REMEDIES

The Repercussives for Cool Matter. Two Kinds

Avicenna's: A plaster of juniper, cypress seeds, icinerated bones, alum, tragacanth, fish-glue, Rhazes': A liniment of oil nard, styrax, myrrh, aloes, acacia, and a decoction of galls.

Three Kinds of Resolutives For Cool Matter

Avicenna's: A plaster of warm beef turds (to which Halyabbas added goat's), and combusted cabbage leaves, honey; or add a decoction of camomille, melilot, aneth, marjoram, and and centaury to make a stupe.

Rhazes' (*Almansor*): A plaster of ammoniac, bdellium, styrax, old wine, fenugreek, linseeds, and oil of costus.

Dino's version: aloes, myrrhe, salt, saffron, lupin-flour, bran, honey, lye-water. A fourth may be made by adding alcohol and Holy Oil, and distilling it (see our Chapter on Paralysis).

Repercussives For Warm Matter

Avicenna's: A stupe of waters of endives, roses, plantain and morel; a decoction or oil of sandalwood, rosat oil, washed wax; a mucilage of psyllium, egg-white.

Rhazes': A plaster of sandalwood, bol d'armenie, memitte or glaucium, pomegranate-peel, iron-rust, sour wine, rose-water, coriander juice.

Another from Rhazes: A comfit of combusted bones, hermodactyl, wheat-flour, ceruse,
Dino's: A comfit of roses, barley-meal, lentil-flour, oxycrat, rosat oil.

Three Resolutives For Warm Matter

Avicenna's: Aloes, myrrh, saffron, cabbage-water (or endive), barley-meal.

Galen's (*Miamir*, Part 10): Aloes, centaury, pumice, feather-alum, frankincense, myrrh, opium, mandragore, sweet wine and milk. Apply it with a feather.

Rhazes': Psyllium-mucilage, linseeds, fenugreek-flour, and oil of camomille. Rhazes fomented with warm water before anointing, and kept it in place for ten hours.[503]

Resolutives For Mixed Humoral Causes

These require mixtures of medicines, sometimes composed by trial and error (ie conjecture). However, here I will describe resolutives prescribed by the academical Doctors.

1. Rhazes' (*Divisions):* Rye-bread crumbs, egg-yolks, and saffron
2. Avenzoar's: Barley-meal, ashes of combusted cypress nuts, water, and oil.
3. Albucasis' (*Antidotary,* Part 23): A plaster of fenugreek-flour, linseeds, camomille, bdellium, liquid amber, ammoniac, galbanum, oil of lilies, and vinegar.
4. Avicenna's: A plaster of fenugreek-flour and oxymel.
5. From the *Papal Regimen*: Urine boiled with salt and sambucus, and thickened with rosat-oil.

I will mention another one used by the lay surgeons (empiricists, or 'experimenters')[504]: An ointment of snail-shells, meats of snakes, frogs, tortoises, foxes, and bats, and others. All are boiled in brine and mashed with the lards and salted anew, and set in a perforated

---

[503] Guy's readers knew that all applications—plasters, stupes, inunctions, etc.- were held in place with light bandages, unless more bulky compressive or splinting dressings were called for. Here the instruction to leave an inunction in place for ten hours inferred that a bandage was used (LDR).

[504] Although the use of animal-parts was not frequent among most of the medieval surgeons other than Theodoric, Guy takes many examples from the more complicated remedies accepted by the physicians. Indeed, all of them were used by the Arabs, and before them by the ancient Egyptians, as will be noted later in the Antidotary in Guy's Treatise VI (LDR).

earthenware jar which fits inside another, and heated in an oven. The liquid which seeps into the outer jar is collected for use.

Composites

As used by Galen (*Catageni)*: An ointment of blood of frogs and tortoises with the oils of the roots of wild cucumber, marjoram, and alkanna, and wax, mixed with terebinth, galbanum, bone marrow of deer, and balsam.

An ointment of Heben Mesûe: A gutted fox placed in an earthenware jug with brine, wine, oil, sage, rosemary, juniper, aneth, oregano, and marjoram and cooked until the flesh comes away from the bones. Mash the flesh and use the expressed fluid.

Rhaze's ointment: Place seven bats in a pot, cover with water and boil down by half, and add rosat oil and willow-tips.

Halyabbas bath: A decoction of turnips, leeks, onions, eruca, cabbage, fennel, and celery. My own substitute is a wash with turnip-broth (also used by Isaac).

Thadeus of Bologna's Ointment: A plucked and gutted fat goose stuffed with the meat of a fat tom-cat, ordinary salt, saltpeter, sel ammoniac, rock-salt, and alum. Add euphorbia, asafetida, and castoreum. Instead, you may stuff the goose with ive, dove's-foot, crow's-foot and hermodactyl. Theodoric added parietory, rue, marrubium, roots of wild cucumbers, ivy-leaves and gum thereof. The distillate (ie seepage) was very effective against cool arthritic matter.

Rhazes' Plaster used against warm matter: An ant-hill, including ant-eggs, barley-meal and bean-meal, roses, egg-whites and yolks.

Special Remedies *For Sciatica*

All of the aforementioned substances may be fortified with mustard, yeast, et al, to attract the bad matter from deep in the joint. For the same purpose you may apply cups and counter-irritants (ie reddeners), vesicants containing garlic, cantharides, lupin, marrubium and psyllium. In addition to those caustics, you may apply actual cauteries: olivaries and circulars. Place them around the joint but not centered over it,; Albucasis' description will come later. Let the blisters drain for forty days, which is the usual duration of the gout, keeping them open with cloth drains, and leaves of cabbage or ivy until the joint heals.

Here we finish our discussion of the Third Goal of Treatments.

4. Dealing With Complications

Pain

We use two methods: Treat for cure with evaporatives (ie resolutives). Palliate with narcotics. Treating for cure may exacerbate the pain, therefore, it may be better to combine the two. For that purpose we use four remedies:

First is Rhazes' and Avicenna's, which they took from Galen's *Miamir:* A liniment of white bread-crumbs mixed with milk to make a paste to which they added opium and saffron. They reapplied it frequently.

Another: A waxy plaster of rosat, opium and saffron.

Lanfranchi added wheat-flour, camphor and rose-water.

A liniment from *The Papal Regimen*: The tips (ie pods ?) of white poppies with their seeds and coats, barley-meal and hyoscyamus seeds. Add fenugreek, flea-bane, linseeds, vinegar, rosat oil, and egg-white. Apply it cold on a cloth and replace it when it is warm.

Also, we follow Hippocrates and Galen (*Aphorism, #6*) and frequently douse the joint with cold water. By numbing it we dull the pain.

Induration

Although we may not be able to cure it when it has become very tough, we accept Ovid's advice; when we cannot resolve the induration of a podagra we always can improve it with softeners and common resolutives, as we describe them in the chapter on sclerosis and in the Antidotary. Rhazes (*Treatise on Joints*) described an evaporative made by heating marchasite to red-hot and staunching it in vinegar and making a plaster by adding ammoniac, opoponax, galbanum, bdellium, sulfur, niter, mustard, pyrethrum, and litharge.

When the induration causes stiffness (ie spasm) we use this ointment: Bdellium, vinegar, warm wine, honey, frankincense, opoponax, ammoniac, myrrh, oil if camomille, fats from hens, geese, eels, and calves.

Another: Diachylon (large and small) of Heben Mesûe.

Another from Galen (*Simple Medicines)*, Aged cheese cooked in salt-pork broth. Vigorously rub the skin with gypsum to soften the induration. Rhazes added some nasturtium.

Induration As After Fractures[505]

Here, as it is after other conditions that affect the joints and nerve-bearing regions. it is not easy to cure, especially when the joint is dry, stiff, and little used, and when it will not redden after massages, and when the region is numb and you note that it is very weak and will react sluggishly. Sometimes you will help by incising the indurated tissue (Avicenna), followed by applications of softeners, and resolutives (ie evaporatives) such as hot blood-stone staunched in vinegar. Also use our familiar resolutives in plasters and ointments described in the chapter on sclerosis. That is a good resource for all such maladies. But if the matter is scanty and dry, proceed as follows: First passively move the limb, flexing and extending it, while applying watery mucilages and decoctions of

---

[505]   Induration after dislocations and around joints that were inflamed usually at first was edema and fibrosis later on. The authors did not differentiate, except in certain cases of edema, as after a dislocated shoulder. They treated it with 'touch cauterization' (LDR).

the bark of the roots of guimauve and elm trees, camomille, melilot, linseeds et al. Also use decoctions of the heads and the feet of sheep with moss, or use the warm blood of an animal.

Then set the patient near a warm hearth and apply this ointment: Dialthea, oils of laurel, mastic, and lilies; musk, ben, and nutmeg; fats of ostrich, eagles, eels, mountain rats (ie marmots), ducks and chickens and donkeys; marrow from a deer's femur, or a calf; bdellium, lanolin, liquid amber, castoreum, and wax.

Then apply this plaster: Wax, resin, diachylon, donkey-fat, labdanum, lanolin, galbanum, opoponax, ammoniac, bdellium, styrax, calamite, mastic, sarcocolla, and wine. Add fats of bears, ostriches, eagles and eels; scum of oils of lilies and Terebinth; flours of fenugreek, linseeds, and saffron.

All medicines similar to diachylon are useful, as is seven-time distilled human blood, as was favored by the alchemists (ie apothecaries) and Henry. Many other suitable medicines are in the Antidotary.

Also, refer to suitable bandaging, instruments, and mechanical devices.

CHAPTER 2

Leprosy, A Disease Of The Body As A Whole[506]

Leprosy, as defined by Galen in Parts 2 and 6 of *Maladies and Symptoms,* is a major impairment of a person's metabolism (ie assimilation) such that the entire body is diseased. I add that the error of metabolism is immediate, whereas the error of digestion in the liver is mediate, where blood is formed. Avicenna said that the liver was the very first efficient cause. When the liver's heat is defective in its formation of blood it becomes the primary source for defective melancholy (ie the humor). When the defective blood[507] undergoes a third digestion in other organs, they are debilitated by the by the bad, cool, and dry complexion of what they receive, and it cannot change its color to the normal color of the organ, which is the uniform red of flesh[508]. Instead its color is that of dark[509] and putrid granulation tissue. So it is; when the result of the error of digestion is disseminated, its effects are dire (ie hectic). Galen concluded (*Natural Faculties*) that when it occurs on one place, the result is hydrops; when it affects the general metabolism, the result is leprosy.

Leprosy is both a primary disease or a complication, and Gordon agreed with Avicenna in deciding when to treat it. It is a disease of itself, consemblable, and recognizable as such. It is Consemblable because it has a cool dry complexion wherever it appears in the body. It is Official because it corrupts the appearance in typical fashion. It is Common because as a general default of continuity it is an aposthem. It is a Complication because that is its fate.

A leper is called such because his nose comes to resemble that of a Hare (ie lepus), one of its principal and distinct signs. Or it may be called Wolf (lupus) because it devours whatever part of the body that it affects, as a cancerous lupus (Haly Abbas, *Royal Dispositions*, Part I, Sermon 18). And Avicenna said that it is a cancer of the entire body.

---

[506] Guy's term was Ladrerie, which connoted miserable poverty (LDR).

[507] Be reminded that blood is a composite of the four humors: phlegm, choler, melancholy and sanguine. The last named predominates in normal blood, and lends its name to the final product, *sang* (blood) (LDR).

[508] The normal product of the liver is nutritive blood, which is converted to tissues when it is distributed throughout the body (LDR).

[509] Hence the term *melan-* (LDR).

The varieties and differences of leprosy are named for the various bad humors and their principal signs. Although Halyabbas (idem) as taken from Galen, said that there are two kinds of leprosy, based on two kinds of badly combusted choler, our present assessment defines four kinds, because any of the four humors can be combusted into melancholy.[510] Elephantiasis from melancholy, Leonine from choler, Tyrian from phlegm[511], and Alopecia from sanguine. However, they seldom occur singly; more often they are composites, as are most aposthemes. Therefore, we give them the names of animals.

Causes

The three cause of leprosy are primitive, antecedent, and conjoint.

The *Primitive Causes*, according to Avicenna, are corrupt humors, direct contact with a leper, rotted food, and a bad inheritance. All of them are exacerbated by superfluous melancholy, as from hemorrhoids[512], defective menstruation, quartan fevers, a weak spleen, and an overheated liver.

The *Antecedent Causes* are the badly combusted humors converted to melancholy.

The *Conjoint Cause* is melancholy exacerbated by everything above. As we described in the chapter on aposthems, melancholy has two forms, Natural and Non-natural. However, leprosy is not caused by primary natural or non-natural melancholy; rather it comes from products of combustion. Avicenna said that the bad melancholy is amplified wherever it goes in the body, or when it lodges in only one part. Wherever it goes it is further combusted and causes fever. When it is not combusted, it causes morphia in the skin and leprosy in the flesh. Galen said (*Maladies and Symptom,* Part 6) that when it is localized, it causes cancer, warts, et al.

Signs And Diagnosis

To understand the signs for diagnosis (according to Master Jordan of Montpellier), one should know the two elements of Leprosy: the *underlying disposition*, and the *actions*

---

[510] Joubert explained: Really there is only one leprosy, with one final humor, with different names. Elephantine because the victim's legs are swollen and uneven; Leonine because the disease is invincible; Satyric because he is hideous. All of them were described by Aretaeus of Cappadocia. Alopecia is a specific name for a disorder of the hair, and it is not leprosy. Leprosy was commonly called ladrerie, or St. Lazar's Disease. (J).

[511] Joubert esplained: Tyrian is a copyist's error for Satyrian. Alopecia also occurs in the fox (αλοπψξ) and in fish.

The Arabic word Tyros meant serpent, and they use the word to mean baldness, which is caused by contact with any snake (οπηυξ in Greek, when the hair-loss is limited to the area of contact (J).

[512] As noted earlier, hemorrhoids were important routes for evacuating faulty blood from the liver. Mild hemorrhoidal bleeding was healthy; too little as stated here, was implicated in leprosy (LDR).

of the humors. The disposition favors the occurrence of leprosy, its susceptibility to the primitive causes and their accompaniments.

The actions are the dissemination of the bad melancholy throughout the body. The action has four phases: onset, augmentation, stable-state, and declination leading to death.

At *Onset*, when the damage begins to affect the internal membranes (ie the muscles beneath the skin) the diagnostic signs are subtle until the malady approaches the surface, and then returns to affect the internal structures more intensely and to destroy them.

The ever-increasing external signs are indicators of the *Phase of Augmentation*

During the *Stable State* the lesions ulcerate and the diagnosis is obvious to the physicians.

During *Declination*, the final phase when the body-parts are destroyed, the diagnosis is obvious to everybody.

The signs of the underlying Disposition are the same for all types of leprosy. Other signs derive from the Actions. Persons who are disposed to leprosy exhibit a sick color, morphia, skin-folds, foul excretions, and others mentioned in the section on causes.

Signs of *Actions* are either *Unique* or *Equivocal* (Shared). The *Six Unique Signs* that occur only with leprosy vary in intensity. They are: rounded eyes (ie sunken) and ear-lobes, loss of eyebrow hair and the brows appear knobby, dilated nostrils that are narrowed above, ugly lips (ie distorted), hoarse nasal voice, foul halitosis and body-odors, and an ugly, beastly, Satyr-like stare.[513] Galen added (*Maladies and Symptoms,* Part 2) that the bridge of the nose collapses, the lips fatten, and that the ear-lobes are floppy thin like an elephant's, and pointy like those of the satyr.

The *Seventeen Equivocal Signs* shared with other serious illnesses are: Firm and nodular subcutaneous tissues, especially near joints; the dark skin-color of morphia; loss of hair, replaced by fuzz; atrophy of muscles especially noted in the palm of the hand; numbness and cramps in the limbs; deep wrinkles, scabies, rosacea, and ulcers over the body; sandy irregularities under the tongue; under the eyelids, and, behind the ears; pins and needles and burning discomforts anywhere in the body; goose-bumps in the skin when exposed to cold air; when sprinkled with water, the skin feels oily; lepers react to treatments with fevers; they are quick to anger; they suffer sadness and have bad dreams; their pulses are weak; their blood is dark, leaden, cloudy, ashen gray, sandy or lumpy[514]; the urine is dark, or white, thin, or flecked, as with cinders.

After your examination, be very circumspect about your diagnosis. A wrong judgment will either relegate an innocent patient or will put the populace at risk by exposing people to a leper, because the disease is contagious. Therefore, the physician must review his findings and reexamine the patient, and ponder at the signs, and separate the unique from the equivocal, and not offer a diagnosis based on a single sign especially the uniques, when several should rule.

---

[513] The Saton or Satyr is not the Greek faun. Here it is a corrupted biblical name for a mythical beast of the desert (LDR).

[514] References to blood refer to the inspection of the products of phlebotomy as caught in basins : the clot, the foam, and the serum here called the urine (LDR).

First pray for Divine assistance to ease your doubts that your diagnosis is not based on hunch, and that there be no doubt that the facts are real. When you find a patient to be a leper, you consign his soul to purgatory, and the world will abhor him, but not the Lord Jesus who loved Lazarus, who was a leper, more than the others. If the physician declares that the patient is not a leper, he will live in peace.

After his prayers, the physician must determine that the patient's answers to his questions were true. First he should direct his inquiry to those whom he knew were disposed to leprosy. Had there been lepers in their families? Had the patient had sexual relations with a leper? Did he have problems with hemorrhoids or she with menstruation—retained or too little? Was the life-style attuned to a melancholy? What other illnesses had he suffered? Ask about his friends: their habits, their wits, their ambitions, dreams? Does he itch or have prickly or burning discomforts? Feel his pulse. Then bleed the patient and examine the blood: its color, black or gray? Pour off the serum. Is it (ie the clot) granular or lumpy—both are reliable signs? Test some of it in another bowl: does it salt dissolve it; does the serum (ie 'urine') freely mix with vinegar? Does it sink or float like flour when dropped in a basin of holy water?

After the tests, carefully reconsider the patient's appearance when he is up and about (ie after the bleed), and again the next morning when he brings his urine.

The physician thinks about everything he has seen, and the following day He should reexamine the urine and study it for any hint of leprosy, or a proclivity for the disease. Is it colorless, thin, or gray—all are signs of leprosy? He will study the patient's face, his eyebrows, the hair, any signs of inflammation? Any small pustules? The eyes themselves are special: is the cornea cloudy or white (ie scarred)? Is the nose twisted, are the nares large, are there ulcers inside? Are the ear-lobes rounded and shrunken[515]? Is the voice hoarse and nasal? Are the lips and tongue ulcerated or bloody? Are they granular? Is he short of breath, and is the breath foul? Is his body horribly misshapen? Of all the above, the facial signs are the most reliable.

Leave the face and examine the body. Its color: dark like morphia? The subcutaneous tissues: are they firm and depleted, lumpy, especially near joints, thrown in folds? Does the skin itch, or feel stringy; is it ulcerated? Does it have goose-bumps? Are the muscles atrophic? Are the limbs easily fatigued? Can the patient feel a pin-prick on the heel or leg and can he tell you where it is? Sprinkle water on his body: does the skin feel oily? When you sprinkle salt, does it adhere to the skin?

Diagnosis And Prognosis (Judgment)

After the physician has reexamined the face and its expression, he turns away and weighs all the signs and the results of his interrogations, and the consistency of the signs.

---

[515] This is not consistent with Guy's earlier description of the ear-lobes as being thin and floppy, and pointed (LDR).

If he thinks that some of the equivocal signs disagree with the signs of a leprous disposition, he will gently tell the patient to follow a healthy regimen, and he will consult with some Physicians (ie the Academics). The alternative would be to declare him a leper then and there.

If the equivocal signs outnumber the uniques, the common diagnosis is Cassot or Capot, a less threatening condition if the patient follows a healthy life-style and has good medical care and keeps to a small farm or an isolated cottage, and does not become involved with society and engender cases of leprosy.

But when the unique signs prevail, you must gently and sympathetically isolate the leper and consign him to a leprosarium.

When the patient is declared to be healthy (ie non-leprous) he should be absolved and carry letters from his physicians directed to the church officials (ie Rectors).

The identifying signs for the humors are described above in the Treatise on Aposthems, with special attention to the maladies associated with elephantiasis which are worse than the others.

In Summary: The diagnosis of leprosy, an evil disease, labels it as hereditary and contagious, and almost impossible to eradicate after it has reached a stage when a diagnosis can be made. As Avicenna wrote, the treatment of leprosy is like treating a universal cancer, whereas only localized cancers can be cured. All we can do is to prevent it or to palliate it. Furthermore, the prognosis is worse for the leonine and elephantine varieties, which derive from the worst humors. The other varieties of leprosy are more gentle and treatable.

Treatments

There are three elements: prevention; elimination of what already is in effect, before it is fully established; palliation of the full-blown disease.

Three Preventive Treatments

First: A. Try to prevent the bad combustion of the faulty humors. Second: B. Repercuss what has been formed. Third: C. Amend the complexion of the liver and the rest of the body.

A. This employs the six non-natural things and the three annexes.as general measures: attend to the ambient air, foods and beverages, tending to moderation.
B. Use purgatives according to the season, but during the Spring and Autumn use all of them containing diacatholicon or pills of fumitory. Also, use phlebotomy and provoke bleeding from hemorrhoids. Apply cauteries in the antecubital and popliteal hollows.
C. Provide electuaries containing diarrhodon from monasteries, and epithems good for the liver.

Curative Treatments; Four Elements

The four measures are applicable when leprosy can be diagnosed, but before it is full-blown. First: A good regimen that will moderate the intensity of the bad humors. Second: Evacuate the combusted humors. Third: Improve the condition where damage has occurred. Fourth: Treat the complications.

A. *The Good Regimen*: uses the six non-natural things and the three annexes, that control the temperature and the humidity
B. *The Evacuations:*: Use phlebotomy, laxatives, caput purges, baths, cups, massages, and other resolutives.
C. *Repair The Damages*: Use snake-meat, potions, and confections of gold, sedatives for the heart, natural remedies.
D. Treat *complications* as they appear.

Palliative Treatments: Three Elements

When the diagnosis of leprosy is not in doubt, use 1. Humectants internally to protect the uninvolved parts of the body. 2. Comfort the heart and other principal organs to keep them functioning. 3. Protect what has not been deformed.

The First uses milk, chicken-broth, and other humectants.

The Second uses Galen's Electuary, called the Laetitian., and monastic diarhodon.

The Third applies cauteries where indicated, and uses gums and others medications to improve the appearances

Remember that most of these substances are to be used by the physicians rather than by the surgeons, except when they are called upon for manual actions. I will limit my descriptions to the essentials in eight sections to follow: diet, bleeding, laxatives, caput-purges, baths, inunctions, epithemes, embrocations, et al., the use of serpents, cauteries, complications.

1. The regimen for melancholic aposthems also suits leprosy. I refer the Reader to that section. Avoid coitus and other agitating activities as Avicenna dictated. He also recommended milk in the diet because it improves the halitosis, and, when drunk after evacuations, it improves the hoarse voice. The milk should be fresh from the cow, and in amounts to tolerance. It is good source of nutrition. As the disease goes on, one may forego the milk.[516] As time passes the regimen can be reduced to what the hectics can accept.

---

[516] Joubert explains: This passage is a scribe's error, and makes Guy contradict himself, or perhaps he had been misled by a bad translation of an edition of Avicenna that cites Paul of Aegina (J).

2. On Bleeding. Finding veins in lepers can be a difficult task unless they are fully replete or they hold their breaths.[517] Small veins can be bled by applying cups over scarifications on the buttocks or between the shoulder blades. You may incise veins on the face near the nose. That may succeed because the bad humors have already diffused from the veins into the nearby tissues. Halyabbas wrote (*Royal Dispositions*,Part 1, Sermon 4) that after the diagnosis is confirmed, you can accelerate the outflow from jugular veins (ie external)[518], and from two veins behind the ears, and from the mediastinal veins. One should stop the bleed when weakness appears.[519] However, Rhazes began his bleeds from the large dark vein in the antecubital region in the right arm. He repeated from the left after a brief delay. He claimed that he had cured a young leper on whose face some nodules had appeared and whose hair had fallen. The appearance was obvious, and his treatments were phlebotomy, laxatives, a decoction of epithyme, and pills that purge black bile. He bathed him frequently and fed him humectants. He put him to bed for a few days and then resumed a series of laxatives. During five months he purged him forty times. His hair began to grow, the eyes and the color of the face improved, and a state of health returned. He left him for a six months, limiting the laxation to milk and a healthy diet. The cure was complete.

3. Laxatives: First use this syrup of fumitory to digest faulty humors: Fumitory, bugloss, tendrils of hops, scabious, lapathum, venus-hair fern, adianthum, polytric, fresh endive, scolopendre, chicory, reglisse, melon seeds, ozelle, anise, cuscuta, flowers of roses, violets, borage, bugloss and epithyme, polypode, pomegranate wine, vinegar, agresta, and sugar-loaf. Then evacuate the digested matter with repeated doses of mild laxatives and diuretics, with additions from the list above, and senna, plums, tamarind, and cassia fistula. Add no vinegar. Dose it twice weekly and perhaps make it more acceptable with an electuary of roses or a syrup of myrobalans as described by Heben Mesûe.

A strong laxative is Avicenna's pill: Yellow myroblanas, chebules, aloes, scammony. Stir continuously while you add water or the juice of fumitory.

---

[517] Valsalva maneuver (LDR).

[518] Joubert explained that the fear of uncontrolled bleeding from the external jugular veins (ie the guides) or from cutting too deep, led to several methods for control other than by ligation. The method of Halyabbas was favored. The patient turned his head as far as possible and the surgeon incised the exposed vein. After the bleed, when the patient returned his head forward the incision in the vein was beneath a flap of skin, pressure on which could easily control the outflow, or a small pad could be bandaged over it. The method was described in detail by N. Carpe (J).

[519] Joubert explained: The Weakness or faintness was either heart failure or the loss of blood to the head, manifest by the pallid face (J).

A very potent laxative is Avicenna's Heirarufinum, Hieralogodon, the Theodoricon (see Antidotary). Add colocynth for tart flavor or give an electuary of rose-water with each dose.

Decrease the intensity of any purge as you observe the decline of phlegm and choler, according to the season, and the status of the patient.

4. Caput Purges: After the evacuations (above) prescribe caput purges containing a decoction of myrobalans, chelidonium, nasturtium, staphisagre, pyrethrum, nutmeg, peppers, euphorbia, scammony, and an electuary of rose-water. Administer one drop in each nostril and close (ie pinch) the nostrils.

5. Vapors, Baths, Massages, Inunctions, et al.: After #3 and #4 above, expose the patient to steam containing vapors of the herbs listed.

Shave the head and rub it with and wash it and the rest of the body with this decoction: fumitory, parelle, scabious, camomille, melilot, staphisagre, mustard, long peppers, nutmeg, sulfur, niter, aloes, and orpiment: all cooked in vinegar and water.

After the rub-down apply an ointment based on the blood of hares. Then dose a theriac with wine. When the blood has dried, return the patient to the vapor-bath where you wash him with a decoction of the oil of the roots of lilies and arum.

Then apply this ointment: the yellow, the white and the populeum ointments, snake fat, and the oils of roses and myrtle,

6. Serpents as Remedies: Avicenna wrote that the flesh of the Tyrian Viper was one of the better medicines for leprosy. Galen (*Simple Medicines,* Book 11) described five examples.

According to Gordon, the best vipers come from arid places and had dark backs. Tie off the head and tail and bind the snake to a slim rod. Cut of the head and the tail and let the body bleed-out. Then skin it and wash the flesh with warm brine and fresh wine. The serpent's flesh has more uses than we can list here, other than for treating leprosy

Boil the snake until the flesh separates from the bone, in water containing fennel, anise, soft bread, hard biscuits, and some salt. The decoction is a beverage, and the meat can be eaten.

When the serpent flesh has been prepared, mince it with a chicken wing, sugar, and ginger until it is a white pudding. Or you may prepare the meat with powdered ginger, coriander, and saffron to make a paste. Or make a paste with ginger, nutmeg, and sugar and use it as an electuary. Another: at grape-harvest time, cook the live snake with epithyme, senna, polypode, anise, fennel and aneth. When broth is clear (ie the sediment has separated) pour the broth into another jar. Serve it two or three times a day. Henry placed the decapitated body in a distilling vessel (an alembic) with water, and used the drip in a wash.

Be aware that when used, the remedies will cause the patient to swell (ie subcutaneous edema), and the crusts to be shed, and some of the skin to slough, as part of the treatment. Use the snake remedies until the patient has scintillating scotomas and blurred vision; then discontinue the applications.

To sum up: Salted viper flesh is useful as a beverage and food. In Arabic they were called Bederasuli and Alfelude.

7. Cauteries: Cauteries should not be used before the other remedies have been tried, even to treat edema and corruption. Although Albnucasis had seventy kinds of cauteries, I use only the pointed and round ones, applied in the antecubital and popliteal hollows, on the dome of the head, in the groins and axillae, and at the neck where I make a perforation for a seton. I use potential cauterize (the eruptors) under the chin and elsewhere on the neck. You will do no harm to apply cups before using the cauteries.

8. Treating Complications: Leprosy causes many complications that require treatment, such as morphia, deep folds, erosions, and scabies. All will be discussed in the following chapter. Others are nodes, glands, tuberosities, ulcers, and corrosions, all to be discussed. Alopecia, pustules, and collapsed nostrils will be in Doctrine II below. Shortness of breath and halitosis have been noted above, and others will be dealt with as they appear in the text.

CHAPTER 3

Morphia, Scabies, Erosions, Lice, Nits, And Lesions Resembling Leprous Skin

A ll these maladies are alike because they are cutaneous: the morphia, albaras, algades, algases, (the latter three have been called rosacea), freckles, sang-mort (bruises), gouts (also are rosacea), scabies, heat-rash (Persian fire), scurf (or dartre). All of them occur in various sizes, locations,and colors, and contain various humors. Yet we physicians have had many different opinions, each his own. A good example was Halyabbas who said that leprosy was a white albaras (a morphia). However, most of us agree that when they are only on the surface, and are not ulcerated, and are smooth, and are dark, they are to be called morphia. When they are white, they are albaras; when reddish they are rosacea (couperose); when large they are pannus; when small, they are freckles. But when they are not smooth and have irregular surfaces, or are ulcerated, we call them either scabies or Persian fire. Both Henry and Lanfranchi made much of the differential diagnosis.[520]

Therefore, we shall consider all of the non-ulcerating lesions as a single Category, as Morphias. When they ulcerate, the category will be Scabies (rognes) or Scurf (dartres). In this chapter we will deal with general matters. Later we will deal specially with the lesions as they affect the face.

---

[520] I think that Guy was unclear as to Henry and derogated him. Here is some of what Henry wrote about this topic. "In respect of these lesions and many others, the authorities contradict each other as to definitions, treatments, etc . . . . I have yet to find two of them in complete agreement . . . . It is almost as if they agree to disagree . . . ." (Treatise III, Doct. I, Ch. 15). "I have searched all the authoritative texts and the practica, and I have perused them thoroughly, only to discover that no two of them agree in all points as to causes, diagnosis, symptoms, and treatments . . . . Furthermore, no one has offered a treatment that some other author or practitioner has not disagreed with or called it disgraceful.". (ibid., Doct. II, Ch. 6.) See Bibliography. (LDR).

Morphias

Morphia is a defedation.[521] It is a flat macule that has been called leprosy, although there are two more familiar terms, the black and the white. White morphia is caused by debased phlegmatic humors, and the black by the melancholic, as claimed by Galen in *Maladies and Symptoms,* Part 6.

Signs

The signs are self-evident, but the prognosis is difficult.

Prognosis

The highly respected Gordon described a chronic morphia as very large, and which does not redden when rubbed, and neither bleeds nor weeps edema when pin-pricked. It is either incurable or highly resistant to treatment. When it is less advanced than what we defined above,there is hope for cure.

Treatments For Black Morphia

Adopt the regimen (diets et al.) prescribed for early leprosy, and, as did Avicenna, bleed copiously and purge with laxatives specific against melancholy in order to evacuate those humors, as we described for leprosy. The best laxatives are based on a small amount of milk with epithyme, followed by a full cup of milk. Rhazes also washed the lesions with milk and applied a plaster of radish-seeds, eruca, a lump of soap, and vinegar. Then he put the patient in a tub bath before washing the lesion directly.[522]

---

[521] Defedation meant soilage. Morphia derives from an Arabic term taken from the Greek μορφοσ, which was a synonym for αλφοσ. When the lesion appeared in the skin, it became 'white morphia'. According to Bazin, white morphia and the alphos represented the first phase of leprosy. Today, perhaps we would call white morphia 'vitiligo'. Many other lesions, in Guy's time, whether alike or different were called 'black morphia (EN).

[522] Guy's term is 'epitheme" which means a local irrigation, not necessarily with 'epithyme', which is another name for the herb cuscuta, a dodder, A confusion of terms may have led to some errors by scribes as well as by generations of translators, including the present one (LDR).

Halyabbas made a cataplasm with minced onions that were exposed in sunlight. Gordon rubbed the spot with a dry cloth and then applied a paste of epithyme, red orpiment, and the juice of fumitory. On the following day he washed it with a decoction of bran. Jamier rubbed the morphia with memitte.

Roger prescribed this ointment: tartar, soot, niter, sulfur, orpiment, rock-alum, hellebore, juice of fumitory, abrotonin, parelle, pig-bread, soap, and oil. Later he washed the region (ie epithemed it), using blood produced by scarification at the site.

William of Saliceto applied cantharides, yeast, vinegar, and cashew-honey, as did the Glossators of Roger. When blisters appeared, he laid on cabbage leaves until the eschars sloughed. When the lesion was deeper, he applied arsenic with the others. When the subcutaneous tissues were clean, he consolidated the defect with a yellow ointment fortified with litharge.

Treatments For White Morphias

We use the same measures described by Avicenna for phlegmatic aposthems, avoiding phlebotomy, and evacuating the phlegm with irritating laxatives containing colocynth or Rhazes' cochia pills.

Rhazes then massaged the lesions with a liniment of thapsus, garance, hellebore, mustard, and radish-seeds, while the patient sat in warm sunlight. But Avicenna used salsola and cooked cabbage until the site ulcerated, Then he applied pitch, wax, terebinth, the ashes of nutshells, the blood of young pigeons, and the oil of henna. He used it until the lesion healed and it had the color of normal skin.

Theodoric's recipe used what a Lady therapist of Pisa devised for treating all kinds of morphias: First he washed the site repeatedly with cold water. Then he applied this plaster: ashes of an incinerated snake in a covered pot, burnt litharge, galls, roots of flammula, bits of old shoe-leather (soles), incinerated dark feathers of a grouse, arsenic, quick-lime, mercury, and vinegar. He used it as an ointment, and applied it two or three times or oftener. Each time he washed it off in a tub bath. Before the patient stepped into the tub, he applied a depilatory of lime, arsenic, vinegar, and water. However, I believe it would be better simply to use the medicated bath and to apply the ointment afterward.

If any or all the above measures fail, the Physicians treated with cantharides and arsenic (when the lesions were deep and dark). They do not use cauteries or scarifications, and they do not leave behind the scars of the treatments. Avicenna declared that they were worse than the morphias.

Avicenna used a tincture of litharge, lime, galls, henna, and tragacanth when all else failed. He ground all of them with honey and dark vinegar and used it as a liniment.

Impetigoes, Serpigo, And Assaphati[523]

All of these are skin lesions, varying in depth, that end as shallow ulcers. Avicenna said they all were very much alike, and he called them pustules. They begin small and smooth, often in clusters. Then they become small crusted ulcers that shed flakes (ie the crusts). Sometimes they are in clear view, but at other times they are occult (ie as in the mouth). Those that are fixed are called assafati and impetigo; those that are movable (polypoid) are called serpigoes, which are called dartres or Persian fire in street-language. Some are moist, others are dry.

Causes

A bad corrosive substance mixed with salt-phlegm is a special cause for moist lesions. The dry lesions derive mostly from melancholy. The faulty matter is driven to the surface where it corrupts the skin. The moist matter is very fiery and erosive, as note by Theodoric and the Glossators. Often the lesions appear on the face and on the heads of babies. They are more common during the winter, when the cold contracts the pores and retains the matter in the skin (Theodoric). When they occur during the summer, it is due to excessively hot weather (Jamier).

Treatments

The evacuations are the same as for aposthems and choleric and melancholic pustules, just as we treated aposthems, leprosy, and morphias. However, Avicenna specially eliminated sweets, such as dates, and bitter, and spicy, and salty foods. He used baths etc. to normalize the body's humidity.

As to local applications for recent lesions, we use fomentations with warm water, and moisten the lesions with juices of purslane and cucumber, and flea-bane. And we use the saliva of a man taken on his awakening (in *Simples,* Part 11) mixed with the sour juice of lemons, the gum (ie Arabic) with vinegar, and mustard. Heben Mesûe used oils of wheat, eggs, serpents, and ginger. The common folk prefer an oil of tartar.

Roger preferred soap with the juice of chelidonium mixed with the white ointment. The Glossators said that the cosmetic result was the prettiest of all. The Faculty at Montepellier preferred a wash of rose-water, lemon juice, and powdered sulfur kept in a

---

[523]   Impetigo: The word derives from 'impetus' or 'eruption'. The connection is vague. The Latin authors used 'impetigines' to include a group of crusted, dry, and chronic lesions that corresponded to the Greek λευην of Galen and the 'mentagre' of Pliny. The Arabs and other medieval authors followed (EN).

basin exposed to sunlight for twenty hours. The Bolognese prefer fresh milk mixed with vinegar and litharge, then distilled it and mixed it with brine. At Paris they use the white ointment containing litharge, oil of tartar, and lemon juice and pulp.

The Treatment of Assaffati

The Ancients (Avicenna) preferred an ointment made from cimolean clay, incinerated melons, colocynth-pulp, and vinegar.

Theodoric's ointment: roots of parelle, aged salt-pork lark, vinegar, and mercury tempered with saliva. Boil off half the liquid before adding the mercury.

Roger's ointment: tartar, burnt lead, soot, incinerated melon, pyrethrum, and juice of cyclamen.

Henry observed the treatment of dartre by a physician at Paris, who used this ointment: crushed ginger-seeds, water, fresh pork-lard, and terebinth. All are melted, then cooled. Pour off the water and beat the mass in a mortar until it is an ointment; then add the sulfur.

But Rhazes wrote that the ancients needed leeches and rubbing to evacuate enough blood to allow the affected tissues to resolve and regain their normal appearance.

And Avicenna said that the best of the ancients used a corrosive to erode the disease and expose healthy tissue. Then he treated the ulcers with ointments—especially the white ointment with litharge—until they healed.

Scabies And Itching[524]

These all are ulcerating lesions of the skin that itch and form scales and crusts. Some are virulent and purulent (Gordon).

Cause

The offending matter, according to Avicenna, is sanguine humor mixed with choler that was converted to melancholy, or with phlegmatic sal niter. The first engenders dry scabies, the second causes moist scabies. Therefore, there are two kinds, with which I include pruritis. When Nature expels the bad matter from within the body toward the skin, some of it persists beneath the surface. When it is very thin (ie not visible), it causes itching; when it is thick it causes scabies (Halyabbas, Part II, Sermon 18). The bad matter

---

[524]   The Latin word 'Scabies', according to Pliny, was a synonym for the Greak πσορα. In southern France it was called 'rogne' or 'gratelle', or 'gale'. Halyabbas and Avicenna thought that it began in the interdigital webs of the fingers as tiny pustules. Later, Avenzoar (12ᵗʰ C) thought he saw small lice, but he could not establish them as a cause for scabies. Bazet thought that Guy related them to 'gale' but confused them with other lesions. Here, Guy called lice 'Cyrons', as will be noted later (EN).

appears more often in persons who eat too much, who eat spoiled meats that are too salty, bitter, or spicy (Avicenna), Those persons garb themselves with their soiled clothing even after a bath. They work long hours, and they drink undiluted wine.

Signs

The small pustules that itch at the onset and then ulcerate are scabies (Halyabbas). Their color, heat, and severe itching indicate which humors are at fault, and a simple examination can tell which type may do some good. Nature's way is to repercuss bad matter toward the skin, as Galen wrote in *Therapeutics,* Part 4.[525]

Prognosis

Bad scabies is difficult to treat and may not be curable. The prognosis also should note that scratching the scabies can cause ulceration, scurf, and malodorous lesions. Furthermore, scabies is a contagious disorder.

Treatment

All of these maladies are treated in the same way, as to diets (ie life-style) and evacuations, except that Avicenna recommended special laxatives for scabies, based on chelidonium. Also, he and Rhazes emptied the scabies with prescriptions of aloes, endive, and fennel, dosed three times daily. They used laxatives to empty the intestine, and they bled robust patients. Avicenna also advocated the use of cups at the buttocks when treating bad scabies.

After the evacuations, our physicians use fumigations with herbs steeped in the juices of fumitory, as we described them in our chapter on leprosy. While they are in the steam-baths, they massage with Avicenna's prescription for shampooing and bleaching the hair: mauve, white blett, parelle, oxalis, ache, bran, flours of lentils, rice and fenugreek, cooked melon, oil, vinegar, and pomegranate-wine. After the fumigation they dose a theriac or a red troche. The patient will sweat when he is put to bed for sleep, and his skin will improve when he is anointed with the oils of violet, roses, and almonds, and with vinegar and the juice of pomegranates.

As for treatments of humid scabies, Rhazes and Avicenna preferred a plaster of mercury tempered with saliva, cadmia of silver, wild olives, hellebore, salsola, litharge, and vinegar, to be left on through an entire night. In the morning, they bathed the patient in a tub and rubbed the lesions with vinegar and green moss from an oak tree. Then they washed him with warm water, then with cold water; then they applied rosat oil as he exited the tub.

---

[525]   The pores of the skin, sweats, and scurf are natural avenues for excretion of superfluidities (LDR).

Avicenna treated dry and pruritic scabies with doses of sour cows' milk and baths in warm water, and inunctions of materials made with cold oils containing the juice of celery, rose-water, endives, vinegar, aloes, sel armoniac and alum. Medicines special for itching are minced poppies in vinegar, and a wax containing opium.

Rhazes used this ointment for scabies: borax, costus, salt, hellebore, styrax, vinegar, and oil. He made a liniment of them and washed the patient with it while he lingered in a tub-bath. Afterward, he washed him with water that contained vinegar that had been boiled with roses, myrrh, red sandalwood, and alum. It relieved the itching.

White ointments with litharge are in common use. Other familiar ointments are used for both kinds of scabies. Galen taught some of his fishermen-friends (*Simple Medicines*, Bk. 9) to mix sulfur with oil and honey or terebinth as a great treatment for scabies; they succeed because they are both repercussive and resolutive. Terebinth mixed with fresh pork-lard and sulfur and some mercury is another good remedy.

Theodoric made an ointment of roots of parelle, meadow-enula, asphodels, and cicuta, all broiled or boiled in water, minced and mixed with old pork-lard.

Henry's plaster: laurel oil, pork-lard, green wax, frankincense, tempered mercury, common granular salt, fumitory, and plantain. Some of it can be sipped. When he added some soot in vinegar, he used it for all these disorders.

Master Dino's ointment consisted of parelle, scabious, chelidonium, enula, fumitory, oil and salt. Boil off half the juices and add wax, and melt to be mixed well. For extra dryness, he added vitriol. For salt-phlegm he added ceruse, litharge, burnt lead, ashes of sarment, and more vinegar.

Master Pierre de Bonant's ointment for salt phlegm scabies: juices of chelidonium and hedera, and pork-lard. Boil off the juices and add mercury, After the inunction, apply a leaf of lapathum face down, or of a lily.

The Saracenic ointment for treating scabies, mal-mort, and salt-phlegm scabies, evacuates by inducing salivation and sweating when applied only on the lower legs and forearms while the patient sits in direct sunlight or near a fire. He must avoid cold air. The ointment contains euphorbia, flea-bane, mercury, and old pork-lard. Apply only once a week.

Be aware that the over-use of mercury can be harmful to the vital organs, and can harm the teeth, and the gums. Avicenna insisted that the applications containing mercury must be made as far away as possible from the stomach and other vital organs.

Henry said to wash the teeth and gums with wild mint, chevaline, aneth and camomille, or with a water of morelle.

On Lice, Nits, And The Like

We all know about lice. They are engendered from the same matters as the other lesions in this chapter, but they are less harmful. Deterioration is slower and virulence and putrefaction are rare, because the lice are living things and are in The Lord's Dominion. We are not concerned here with their creation and we will leave that topic to the Physicians.

But it is enough to know that after they arise within the body they move to the surface, where they dangle. People spread them by copulation, unclean habits, and by wearing contaminated clothing.

Their **Sign** is their color (Gordon)

Prognosis

Gordon also said that when they multiply too profusely at their origin within the body, they tend to become morphia, and leprosy. However, if the patient has tough skin it will not succumb, and it will digest the matter. When that digestion is faulty, leprosy follows.

Treatment

First cleanse the interior of the body with evacuations, bleeding, and laxation. For laxatives use hierapicra and others that eliminate the putrid humors. Improve the patient's bad habits and prescribe medicines that kill lice, such as cooked garlic and mountain-calamint (Avicenna) and other local remedies.

Steam-baths and tub-baths that contain water in which alum, salt, blett, tansy, cypress, pine, calamint, lupin, and flea-bane. Apply ointments containing bastard saffron, radishes, sumach, and roots of oxalis.

Here is a special ointment spread on a greasy sash of raw fleece: oil, wax, and mercury. It will kill lice and prevent others from reproducing

Nits are tiny animals who tunnel between the skin and the underlying flesh, especially in the hands of basket-weavers. Treat them by washing in brine or a fish-broth, or with a juice of hedera and vinegar plus aloes, etc.

The Variolas and the Roseolas, the Night-sweats, and Prickly-heat are tiny pustules in sweaty regions. Urticaria are small nodules that itch when a person is overheated and he scratches. All of these maladies are problems for the physicians. When called on, we surgeons treat them as aposthems. The residual marks (scars) will be treated as described below as for pox-scars on the face.

CHAPTER 4

Enlarging And Fattening The Body

The business of treating obesity and inanition (asthenia) of the body as a whole belongs to the Physicians. However, we surgeons are called on to deal with enlarging and diminishing the parts (limbs), and we shall limit the following to those matters.

Galen (*Therapeutics,*Part 14) defined enlarging and diminishing. A person can become so obese by overeating fatty foods, that he can barely walk, and he cannot reach to clean his anus. His belly is so protuberant that he cannot remove his own shoes. His breathing is impaired, and he is called Fatty. But when the person is at low ebb and depleted, as in atrophy or phthisis, the condition is called Dry. Occasionally, dryness can involve only a part of the body.[526]

Causes

Therefore we seek the principal causes for the two affections, as Galen called them in *Maladies and Symptoms*: Overloads or Deficiencies of matter, and we easily can recognize those features in the obese and the starved, in the body as a whole or in a part. In *Conservation of Health* he commented on the strength and weakness of the resistance, and of the nutrition of both types. Avicenna (Part 4) added such external causes that lead to dryness as consumption of poorly nutritious foods, and too little rest. Further on he said that those who abjure exercise tend to deteriorate and be feeble (ie hectic). In other words, they over work, rest too little, are quick to anger, are over anxious, keep late hours, ignore hunger, and sleep on hard beds.[527] As to an affected limb: if you bind a limb with a tight cord, or allow cold or heat to constrict the pores, or avoid moisture, or do not attend to pain or inflammation at joints proximal to an affected part—just as we described for podagra and deformed joints (ie gibbosities), or a badly reduced dislocation—the distal limb will be affected as we described in previous chapters. The passage of nutrition to the distal limb will be obstructed, and its powers of attraction will be impaired.

---

[526] Guy's text is unclear as to what are 'enlargements' and 'diminutions'. When he describes the affections involving the body as a whole and refers the matters to the Physicians, he describes obesity and emaciation. When he describes affected limbs, he refers to edema and atrophy (LDR).

[527] Guy described what today we call a hard-driven, anorexic, ascetic, tense, duty-bound character (LDR).

Prognosis

Hippocrates (*Aphorism #6*) wrote that athletes who consider themselves to be in top form, and believe that they observe the most healthy habits, cannot be too sure of themselves, because their peak status cannot last forever. They deteriorate, they limp, and their veins begin to bulge. Galen (*Techni,* Part 2) said that once dry (ie emaciated) the depletion of the important solid parts of the body is very difficult to remedy, and when far advanced, it is incurable. And in *Therapeutics,* Part 7 he noted that the condition of excess humidity is easier to treat than dryness, although when it is excessive, it is more a threat to life than is aridity. Hippocrates said the same (*Aphorism # 2*); the very obese die sooner than the very emaciated. Galen, in his *Commentary* on that Aphorism explained that the obese are cool and their blood vessels are constricted[528]. The patients lack blood and are dispirited, and whatever of their Natural heat remains often is corrupted. As a result, as Avicenna (Part 4), wrote they are more apt to suffer apoplexy, paralysis, irregular heart-beats, diarrhea, halitosis, syncope, and bad fevers. They do not tolerate hunger or thirst. Therefore, it is better to limit solid food and to let them sustain themselves on their own supply of humidity (ie body-fat); limit them to watery and oily stuff.

Treatment Of The Superfluidities Of The Body As A Whole

We have two goals: First: diminish the excess of blood (ie sanguinous humor). Second: resolve the accumulated conjoint matter; and weaken the patient's avid attraction for it.

Galen described the first goal (*Therapeutics*, Part 4). Limit the diet and prescribe diuretic and sudorific medicines such as rue and its seeds, round aristolochia, gentian, pouliot, centaury, combusted snake-meat, and salt. Rhazes added vinegar. Galen also used laxatives to produce diarrhea of phlegm, as did Halyabbas in Part 2, Sermon 1. Galen also advocated vigorous exercises and periodic fasts.

Galen's second goal was achieved with plain water or medicated baths containing flower of salt, and sea-water. He anointed with irritating oils of wild cucumber, gentian, aristolochia, etc. He allowed no food during the baths, and he insisted on fasting before and exercising after the baths.

Treating A Swollen Limb

Refer to the section on chiragra and elephantiasis in The Treatise on Aposthems, and note the additional measures. The goal is to drive the matter (ie edema) into other regions by applying weights[529] and by bandaging.

---

[528] Guy's reference to constricted blood vessels in the obese explained the difficulties the phlebotomist had in finding a vein hidden in fatty or edematous tissues (LDR)

[529] 'Applying weights': A common practice was to lay a disc of lead over a recently dislocated shoulder to reduce the edema (LDR).

Treatments For Shrinkage And Drying of The Body

I will briefly describe three items: First: Engender a good supply of nutritive blood. Second: Attract that blood into the depleted flesh. Third: Increase the nutritive value of the blood, and let it be retained.

The first is accomplished with the juicy (ie fatty) foods we use to treat the hectics, as described in Part 14 of Galen's *Therapeutic*: A regimen of thick wines, juicy meats, gentle exercises, and gentle massages, all of which are contrary to the causes.

The second measures are used when the patient refuses vigorous slapping[530] to provoke swelling. As described in *Conservation of Health,* Part 6, we massage the body gently until the skin reddens. Then we massage more firmly, but still with moderate intensity. We exercize the patient, followed by a moderately long bath. Then we anoint with oil before offering food. Galen (*Therapeutics,* Part 14) preferred feeding the patient before the bath. If any obstruction occurs,[531] treat it at once with capers in oxymel at the beginning of the meal, until the heavy feelings are dissipated.

The third (ibid. Part 6): Feed foods that warm the flesh, yet do not evaporate the new blood which will be the basis for growing more flesh Use ointments of oils that favor retention, such as an oil with melted pitch. If the patient's age permits, add cool baths to the treatments. All of these measures will help.

Treatments For Shrunken Limbs

When dealing with limbs that are not easily nourished, when the cooling has been excessive (as used above to ablate the cause for the emaciation) and causes pain and contraction of the limb, Galen (idem, Part 14) anointed the part with thapsus and honey. Another very good remedy uses a wax to attract blood to the region where it is applied. Make it of black pitch alone or with resin; melt it and smear it directly on the skin or apply it on a cloth. Use it only once during the summer, and twice during the winter. Leave it on for three or four days, or longer if tolerated. If you use a plaster of tar, first massage the region, foment it, and gently tap it with strips of wood until the tissues begin to swell. Then stop at once before you cause the tissues to resolve the attracted matter. An hour later, tear off the tar plaster quickly and with force. Then anoint the irritated site with a sticky oil or wet it with cold water. After an hour of repose, repeat the process. Continue until the limb has been cured. While that treatment is in progress, exercize the limb by

---

[530] Light flagellation with thin slats covered with pitch was used to produce local edema (LDR).

[531] Canappe thought that this meant biliary colic. Mingelesaulx thought that it meant intestinal blockage (EN)

However, I think that Guy described an anorexic patient gagging at the food (LDR).

having the patient tug against resistance or lift weights. Tie a band around the opposite limb to divert the nourishing blood to the feeble one.

Other measures were used by Rhazes, Halyabbas, and Avicenna: foods, confections, electuaries, beverages, enemas, baths designed either to enlarge or to shrink. We shall leave them to the Physicians who use them, rather than to the surgeons.

CHAPTER 5

Falls, Blows, Extensions, Drowning[532]

A lthough we have discussed wounds and contusions in soft tissues, nerves, heads, and eyes in general, we have not dealt in detail with falls and blows and similar injuries.

Avicenna said that Falls and Blows are injuries that disable the victim by contusion and attrition, and often with associated fractures and dislocations. The Glossators stated that they all differ as to the site of the injuries. A fall to the ground or being struck by a falling rock involves the entire body, whereas a blow is an injury caused by another object coming at a part of the body. Other authors defined a fall as a contusion of the body. Others defined a fall as an external and a blow as an internal injury.

An Extension is an injury cause by pulling at the body as by a rope or a chain. A Submersion is suffocation by fumes or water.

Many disabling injuries may be caused by Falls and Blows, including actual incisions in the muscles over the heart and abdomen, leading to sudden death or to impaired defecation or urination, causing emesis, hemorrhage, and shortness of breath, or loss of voice or speech. All those serious complications are life-threatening results of divided nerves, muscles and blood-vessels, and they cause much suffering and disturbance of vital functions. And it follows that the larger the body (ie the target) the more serious is the injury. Furthermore, Avicenna wrote that glands are numerous after blows and falls[533], and they should be treated as we described for such.

The **Prognosis** is similar to that for large wounds.

Treatments For Falls And Blows And Extensions

In general, these treatments are like those for dislocations, fractures, and contusions, as we described them in the assigned chapters. As usual we put them in four categories:

---

[532] Much of this chapter deals with the surgeons' treatments of the victims of torture and hanging. The Inquisition had been in full display for more than a century, and the relentless pursuit of so-called heresies was relentlessly cruel, especially in southern France. The Reader will note that Guy avoided any statements about those activities (LDR).

[533] I assume that 'glands' means regional lymphadenopathy (LDR).

healthy habits; evacuations and diversions; prevention of further accumulation of the bad matter (humors) that cause swelling; resolution of what already has accumulated.

First we provide a liquid, but wine-free, diet, without meat. On day one the patient fasts. On days two and three we offer small feedings, delaying longer if we observe an increase of swelling. Then we expand the diet to include chick-peas and goose berries (ribes) and other items that are easily digested.

Second: We evacuate with phlebotomies, mild laxation with cassia fistula, enemas, etc. Rhazes used rhubarb with rosat-syrup.

Third: We provide this beverage, beginning at the very onset of the treatments: bol d'armenie, nutmeg, terra sigillata, in a decoction of plantains, dosed every morning for five to seven days. Avicenna used this plaster: beans, rice, bol d'armenie, sumach, aloes, alum, diluted quick-lime, and egg-whites. Rhazes' ointment has been used frequently: oils of rosat and myrtle, and powdered myrtle seeds.

Fourth, we use a beverage after the first day: powdered rhubarb, costus, roots of garance, centaury, and aristolochium. Mix with sweet vinegar, and a decoction of anagallis. Dose it every morning for nine days. Anagallis was William of Saliceto's favorite consolidant, also called pimpernel.

Master Aimeri d'Alais (Americus d'Alesto) use another beverage: wine, honey, cooked roots of osmondia or the feucher fern, and arum.

Some surgeons use fumigations and baths as often as three or four times a day. The water contains consolida, osmundia, sanamunde, benoiste, morgeline, sambucus, artemisia, aloin, roses, camomille, melilote and the dust found beneath hay-stacks. You may make an ointment of it by adding honey.

William of Saliceto and Henry used this ointment: Wax, resin, terebinth, oil, and frankincense.

However, when the site is badly smashed, apply a paste of bean-flour cooked with oxymel and saffron. Avicenna used this: Oxycroceum, or the Apostolicon, and leaves of elder-treet, artemisia, cypress, and tamarisk, all mixed with sour mik.

Some surgeons bury the victim in a mound of fresh warm horse-manure where he sweats. Instead, Halyabbas and Avicenna wrap him in warm fleece of a recently flayed two-year-old ewe, dusted with table salt. If possible, they leave it on until the next day, when the victim may have recovered. Galen witnessed such and reported it in *Medicaments, Part* 11.[534]

When the injury affected the abdomen, Avicenna applied a plaster of wild pomegranates and raisins mixed with labdanum, roses, spikenard, mastic, eupatorium, and oil of lilies.

---

[534] The Reader will observe many examples of what were Guy's anachronisms in his citations. Here we know that Galen came many centuries before Halyabbas and Avicenna (LDR).

Treatment For Extensions[535]

Most surgeons follow Halyabbas methods. First he used his hands and feet to remold the victim's deformed body[536] after applying a cloth wet with cold water. After the restoration, he was treated by sweating. If that treatment failed, the operator did as follows:

When the victim was stretched by falling (ie hanging) and has lost his voice and cannot speak intelligibly, be suspicious that his brain or anoher vital organ has been damaged. Rhazes said that a clear mind is a sign of a healthy brain. Furthermore, you must determine if the patient is alive. Feel his pulse, shout at him, tug his hair, pinch his nose, look at his pupils to see if they are dilated, place a wisp of cotton fluff at his mouth and nostrils, or a cup of water on his chest and note if they move. Dust pepper or euphorbium in his nostrils to make him sneeze. If he is alive treat first by rubbing his limbs with vinegar, salt, and rue. Make him sneeze by tickling his nostrils with a straw to expel whatever blood is inside. When enough vitality has been restored, go on to the usual treatments with evacuations, etc. If a brain-injury was the cause for the unconscious state, treat him as we described for contusions of the head. If he is dead, let him remain at peace, and do not attempt to treat.

If the victim simply has fainted, tilt him (ie in bed) head down-feet up until his edema subsides. Have him gargle with vinegar that was boiled with peppers, and later steam-bathe him daily with a decoction of chick peas for several days. After his body is somewhat restored and he reacts well, let him gargle with oil of violets, and be steamed with warm water in which wheat flour has been boiled, and let him eat warm and tasty foods.

---

[535] Racking or hanging (LDR)

[536] Feet: He trod on the displaced vertebrae of a victim of torture (LDR).

CHAPTER 6

Burns From Hot Water Or Other Hot Things

Painful blisters are the result of those encounters. The pain upsets the patient, and he is ill-disposed. The suddenly produced blisters attract humidity (water) to collect beneath the skin, which is a barrier to its escape. Arrested beneath the skin, it elevates it and makes blisters and edema. Sometimes it fills the body and causes general edema, aposthems, and malignant ulcers. If the burn is mild, the body will show no damage.

Treatment

Treatment of the burn-site has three elements: prevent blister; cure what is there; close and heal the burn

First use available refrigerants. Rhazes applied a cloth wet with rosat-water or chilled it with snow. Often that is all that is necessary. If you are delayed or the burn is too large, bleed from the opposite side and feed a meal of cold liquids. Avicenna applied rosat-oil with well-beaten egg-yolks, mauve-leaves, blet, lentils, roses boiled in sugar-water and rosat-oil, or bol d'armenie with vinegar, terra sigillata, or clay with litharge and ceruse. Her also advocated the use of a decoction of endives and morel to prevent blisters. Halyabbas used a water of olives. Theodoric bathed the whole limb in vinegar-water. Galen used a wax of with rosat oil. Roger preferred a washed oil with lard and the leaves of sambucus. The Four Glossators used the populeum ointment with egg-yolks. For relief of pain, they added opium. Alexander approved of that.

Second: eliminate the blisters by shearing them off with a scissors or a knife.

Third: we use our familiar desiccatives. Rhazes' was the white ointment with egg-yolks and camphor, or an ointment of well-washed quick-lime: seven rinses or as often as is necessary to reduce its potency, as we will describe in the Antidotary.

Avicenna had a long recipe: dry cow-turds, pine-tree bark, litharge, ceruse, bol d'armenie, washed quick-lime, tuthy, burnt lead, iron filings, camphor, all mixed with rosat oil, and marrow or lard of deer. The compound was for relief of pain

In line with the foregone, use dove-droppings wrapped in a cloth and incinerated. Add rosat oil to the ashes. Cooked garlic is good for the ulcers (ie the open blisters). If they fail to heal, treat them as we do the malignant ulcers.

CHAPTER 7

Leeks, Warts, Horns

According to Halyabbas (Part 1, Sermon 8), warts are firm small pustules in the skin. Some of them are called Leeks (ie poireaux) because of the resemblance to the plant with its many layers seen when transected, some are called Clavals because they have a single claw-like root, and others are called Corns because they have a hard surface suggestive of a finger-nail and are usually seen over the joints of the toes (Avicenna). Galen called them achrocordons, warts, alphas, and leuces, all of them are multiple and are unnaturally hard (*Therapeutics*, Part 14).[537]

Their **Causes** (idem, Part 2) derive from matter that is contrary to nature, that arrives in the skin (from within the body)—the efficient cause—and persists, as is their character, as Avicenna said of all pustules and pimples (papillomas). Their matter is thick and melancholic, or of salt-phlegm converted to melancholy. On the other hand, Halyabbas said there always were two types of humors, melancholy mixed with undamaged phlegm.

The warts appear anywhere on the body, more frequently on the hands and feet, and they grow. They do not form from the blood that spills and stains the skin during a phlebotomy, as is a belief of the common folk. They also believe that a large wart can convert the nutritious blood that it attracts to form more warty tissue and allows it to multiply.

Treatments—Three Goals

First: purge the phlegmatic-melancholic matter from the body. Second: keep the matter from increasing. Third: extirpate what is there.

Halyabbas accomplished the first goal with a decoction of epithyme and agaric. Avicenna used phlebotomy, and established a healthy life-style to accomplish the second goal. And the third goal was achieved in two ways: with medications and with surgery.

The **Medications** were chosen to resolve and to dry. Rhazes rubbed the lesions with the leaves of the caper plant and moist carob leaves. Avicenna used the oils of pistachios

---

[537] I think that Guy's Verrucae and leeks were our common warts, and the clavals are our papillomas, and horns are our common corns (LDR).

or a decoction of leeks and sumach. William of Saliceto used the bulbs of squills after he softened the warts with warm water. Halyabbas favored plasters of goat droppings with vinegar, or of nigella and vinegar, or of vinegar and salt, or of minced cinqfoil and moss. Henry used leaves of rue and millefuille, and minced Robert's herb bandaged on the warts. He said that the lesions were ablated within three or four days. He also used as a wash applied two or three times a day for six days, water that had been in a lead bowl in which he had soaked red snails for four days. All the lesions were painlessly uprooted.

The **Surgical Measures** of Jamier consisted of applications of corrosives: yeast tempered with flour, tempered capital ointment, or milk of figs. Avicenna used a milk of tithimal with an oil of cashew-nuts. Halyabbas used quick-lime and soap as a corrosive.

One can scratch open the warts with his finger-nails or a lancet, or use cantharides, and apply a small amount of arsenic in the scratched openings. It never fails. You may use Francis' method and apply a bit of the alchemists' eau fort (ie nitric acid) with a cotton-tip reed or pen-pont. Rhazes ligated the base of a papilloma with a silk thread or a horse-hair, and cut it away., and touched the stump with a live cautery or with a corrosive medication. Be sure to protect the surrounding skin while you destroy the roots of the wart. Or do as Galen and use a hollow goose-quill to core out the root with some surrounding adherent flesh by twisting and turning the cannula.

As for a corn on the foot, Henry advised that we do what a Parisian shoemaker once did to my toe without permit use a knife to scoop out as much of the corn as possible. Then lay on a perforated metallic or leathern disc with the hole the size of the corn. Place a bead of live sulfur in the opening that is set directly over the lesion and cover it with a film of wax, and leave it in place until the corn is destroyed. Avicenna had described the same method (Part 4).

You can ignite the tip of a twig of wood and bring close to a wart which will shrivel and die. That works quite well in a cluster when you attack one wart at a time. Then dress the area with melted butter until the shreds come away.

CHAPTER 8

Amputations of Superfluous or Gangrenous Limbs

G alen (*Maladies and Symptoms,* Part 1.) said that such limbs either are natural, such as a redundant sixth finger or toe, or they are unnatural, such as goiters[538] or gangrenous limbs.

Causes

A Natural superfluous part is a redundancy of normal tissues endowed at the time of conception (idem, Part 2).

There are three causes for a necrotic limb, as we described them in the chapter on esthiomene: a dying person: drying of only a limb (ie lack of blood to the part); blockage of the source of nourishment (so-called suffocation).

Signs

A redundant part is self-evident, whereas the signs of infarction are those a dead body, as we described the signs of death in Chapter 4.

Galen described death by poisoning in Part 6 of *Internals.* The victims seems to have been healthy and well nourished before he suddenly died, as if he had taken an overdose of a poisonous drug. It may have made him livid or dark, or varicolored or pale. The body stinks of putrefaction. Lacking such signs, we attribute death to the progress of a disease.

Prognosis

If you do not get rid of a corrupted limb, it will spread. When gangrene of a distal part of the limb spreads to the thigh or to the arm, it is beyond cure (Albucasis) and the patient will die. Do not attempt to treat what should be left to God and The Saints.

---

[538] Goiter is Guy's term, and I do not understand why it appears in the context of this chapter. Perhaps it was a scribe's error when writing 'gout', meaning an infected joint. Goiter is not mentioned elsewhere, whereas gangrene passing beyond the knee or elbow was lethal (LDR).

Treatment Of Redundant Fingers And Toes

As Galen insisted (*Techni*, Part 3), the only good treatment for redundant digits (natural tissues) is amputation. Getting rid of the digit by any other way is as difficult as wanting to form a digit, and that requires a miracle. Galen wrote (Therapeutics, Part 14) that it is easy to lop off an entirely fleshy part, but less so if you must divide bone (Halyabbas, Part 2, Sermon 9.)

The technique: cut the soft tissues down to the base of the digit and disarticulate and remove it. Control the spurt of blood with applications of the red powder with egg-white. Then treat the wound as such.

Some surgeons, as did Avicenna, cauterize the wound with boiling oil. He said that it was a sure way to prevent infection and to avert hemorrhage, and it breeds good granulation tissue and a tough scar.

Amputating A Dead Limb

When the treatment for esthiomene fails, including repercussives and scarifications, Albucasis and Avicenna agreed that you must amputate through proximal healthy tissues in order to save the patient's life.

Choose the level of the amputation as follows: When the necrosis reaches almost to a joint, incise and disarticulate at that level using knives, etc.; a saw will not be needed. However, when you have enough healthy tissue below the joint, amputate there, proximal to the gangrene. Distract the soft tissues with a cloth saturated with analgesics, and divide the bone with a fine-tooth saw.

Here are the details: Wrap separately the healthy proximal limb and the distal dead part with a gap between. Two strong assistants hold steady while you incise in the gap down to the bone which you denude of attached soft tissues. Cover the exposed flesh to protect it while you saw. After you cut through the bone, apply a red-hot cautery to the stump to stop the bleeding, or use boiling oil. Then bandage the stump tightly. Later, treat the open wound as such.

You may use the red powder with egg white and other familiar hemostatics later on to control residual bleeding,

Some surgeons, as did Theodoric, used soporific medicines to dull the pain, such as opium, the juice of morel, hyoscyamus, mandragore, hedera, cicuta, and lettuce. Soak a fresh sponge with the juices and allow it to dry in sunlight. When needed, wet the sponge with some warm water and place it over the patient's nose. He will nod and then sleep before you start your operation. Later, apply another sponge wet with vinegar over the nose to awaken him. You may apply the juices of rue and fennel over the nostrils, mouth and ears to awaken him.

Other surgeons have the patient drink opium; that is a dangerous action, especially for the young. I have heard that sometimes the patient becomes combative and savage, or manic, and sometimes dies.

As for me: when dealing with dead limbs, after failing with incisions proximal to the advancing necrosis, and scarifications and arsenic, and with applications of bol d'armenie, etc., I wrap the necrotic part of the limb and expose the joint and the healthy proximal tissues so to be observed by the attending Physician who will see that we are cutting through healthy tissues. Almost inevitably whenever one amputates, there will be accusations and second thoughts by the patient.

Embalming Cadavers—Two Methods[539]

Rhazes used this method to temporarily preserve a dead body. First inject the rectum with several doses of strongly detergent substances, as colocynth and red borax, with the body tilted head down. Then press the abdomen to expel the fecal matter. Then inject, as clysters, aloes, myrrh, acacia, ramich, pomegranate-peel, cypress nuts, nutmeg, sandalwood, aloe-wood, cumin diluted with vinegar, and rosat water. Then plug the anus with cotton batting and cloth strips, all wet with the foregone substances, and wrap the lower torso with many layers of wet bandage. Then inject mercury into the nostrils, ears and mouth to prevent liquefaction of the brain.[540] Then wet down the body and anoint with a resin of cedar (ie black tar). As far as possible, cover all the pores and body-openings with bandages.

Another way: Use many layers of cloth impregnated with the black tar, colophonic resin, frankincense, mastic, styrax, gum Arabic, tragacanth, and borax powder. Wrap every part, body and limbs, separately: the legs up to the buttocks, the arms up to the shoulders, the body up to the head. Sew the plies with sutures soaked in the tar. Place the arms across the chest and tie the legs together at several levels. Dust the powder over all, fill the hollows with wet stupes, using the substances of the clysters, Then rewrap over all and sew the seams with a stitch that runs contrary to the inner ones. Then wrap with a waxed cloth, and sew it with tightly knotted threads. Then wrap with a mesh before enclosing the body in a sheet or in a box of lead which you seal by melting the seams with a hot iron. Place all in a casket of cypress wood or walnut which you bind shut with six metal straps, for carrying. Some surgeons cover the entire apparatus with a cow or horse hide.

The second method: Soon after death, open the abdomen and eviscerate it. Fill the cavity with the foregone substances and a large amount of salt and cumin. Close the incision with sutures and wrap as above. If you must save the viscera, empty the gut and clean it. Place it in a lead jars which you put in the casket.

---

[539] The surgeons had accepted this onerous task since ancient times. The methods described by Henri de Mondeville varied according to the status of the dead person, and the length of time the embalmed body had to be preserved for exhibition. He had embalmed Kings. See Bibliography. (LDR).

[540] Those openings were emunctories, that is, they were directly connected with the brain (LDR).

The first method is preferred for emaciated bodies during cold weather. The second is preferred for plump bodies during the summer.[541]

Rhazes had a neat way to avoid gassy distension of the cadaver. He place the body face-down and bored openings into the cavities with an awl or a metal cannula and allowed the accumulated gas and liquid to escape. That method was used by Jacques the Apothecary at Paris who had embalmed several Popes.

To permit the face to be exposed for up to eight days, and yet not need to prevent putrefaction of the body, Rhazes simply sprayed the face with salted rose-water or a balsam. There are many choices, but I have found that few of them are reliable, and so too, Henry.

Many the Lord preserve our souls with his own Balm of Pity. Amen.

Here Ends Doctrine I of Treatise VI.

---

[541] I presume that the embalmers took into consideration the weight of the cadaver when engaging in the various ponderous maneuvers (LDR).

HERE WE BEGIN DOCTRINE II:

SURGICAL MALADIES OF PARTICULAR PARTS OF THE BODY THAT ARE NOT ULCERS, OR DISEAES OF BONE: EIGHT CHAPTERS

CHAPTER 1

Maladies Of The Head

O ther than the disorders-in-common, there are several that especially affect the head: Tinea, Calicia (Falling Hair), Canicia Gray Hair), Alopecia, and Other Disorders of Hair. We will take them in order.

Tinea[542]

Halyabbas (Part I, Sermon 8) said that tineas were small ulcers in the scalp, crusted vesicles of several kinds.

First are the Favus-like which drain a heavy thick matter.

Second are the Fig-like which contain seeds, and are spherical with a red tip, like the cognominal fruit.

Third is Amedosa. It is moist, like subcutaneous flesh, and has smaller openings for drainages than the favus-like.

Fourth are the Uberous which resemble the nipples of a woman's breasts. They are pink and they exude pink, or blood-colored fluid.

Fifth are called Lupinosa because they resemble the lupin plant in color and shape. They have covering membranes (ie skins) and they cast off flakes. That group includes the Furfurous, with bran-like scurf shed from the head. They do not ulcerate.

---

[542] From ancient times, Tinea included all kinds of crusted and flaking lesions in hairy skin. The Latin name was *Porrigo*. Celsus did not differentiate impetigo.

The Arabs included lesions caused by parasites that infest hairy skin. Avicenna, Avenzoar, Rhazes, Halyabbas, et al. called them *Sahafati, ot Safati, or Albathine,* which were contagious disorders that caused loss of hair. They described two kinds: the moist was similar to impetiginous excema; the dry was the same as favus (ie tinea).

The word *Tinea,* entered the European literature in a work by Stephen of Antioch, who translated Halyabbas' works in 1127.

Guy described five varieties of tinea: t. favus, t. ficosa, t. amedosa, t. uberosa, and t. lupinosa. We can only identify t. lupinosa as what we call tinea (favus). Perhaps we can accept t. amedosa as a fungus of hair. Guy's t. favosa is a pseudo-tinea which today we call impetiginous excema. (EN).

Galen (*Miamir*, Part 1)[543] described only three varieties of tinea: achorous, favous, and furfurous. The achorous was called such because it consists of small nodules and tiny ulcerations which emit a slightly viscous fluid. The favous resembled them but had larger openings that drained thicker, honey-like, matter. The furfurous (dandruff) discharges dry, bran-like scurf when the hair on the head is ruffled

Avicenna seems to have renamed the achorous as assefati, ameda, and tyrean (ie snake-like) because it shed its scales. He called the dry flakes furfur. However, he repeated his wise advice that we not devote our attentions to names, except to proudly show off how much we know.

And this is what really we should know: Tinea, in every-day usage, is a scabies-like malady of the head, with scales and crusts. It moistens and it depilates. It is ashen-gray, it is malodorous, and it disfigures the scalp. It derives from corrupted humors; it may begin in utero, or appear later in life in persons with bad habits.

Signs

Some are natural, others are non-natural, and others are contrary to nature. Galen wrote (idem) that they begin as erosions that itch and are scratched off, and they form small nodules (ie scars) in the emptied erosions.

Prognosis

It is not easy to treat and cure, especially when it is chronic, callous, and scaly, and it eats away the hair. Roger said to ignore it when it is extensive, rather than to treat to no avail. Furthermore, the treatment depilates and leaves a nasty scabbed scalp as an ugly source of embarrassment. Jamier used the term tinea as a pun for tenir (ie to hold), and he made another pun with the name of a worm (ie teigne) that invades the bark of trees. Tinea may appear after a large invasion by lice and it may lead to leprosy.

Treatments—General And Particular

The General Measures include the diets and evacuations that we use for aposthems, leprosy, and scabies. Although those may be sufficient, Halyabbas applied cups on the upper back (ie neck) and used phlebotomy from a cephalic vein when he could find it, or from veins below the ears (ie external jugulars), where the Saracens (according to Gordon) used scarifications to treat maladies of the ears. Galen preferred his own laxatives rather than our routines; his were pills of aloes colocynth, scammony, and cabbage.

The Particular Treatments have two goals: to correct the disfigurement of the scalp caused by the tinea, and to promote the regrowth of the hair.

---

[543] Nicaise re-identifies the *Miamir* as Galen's pharmacopoeia. The term itself is a Latin version of the Arabic translation of Galen's book (EN).

When the tinea is new, the First Goal was accomplished by William of Saliceto by repeatedly washing the scalp with water and vinegar containing fumitory, blet, and camomille. Then he rubbed the scalp with alum and wine-lees (tartar), followed by an inunction with a liniment which we described for dartres and scabies. Avicenna used a decoction of willow-leaves. Galen (*Secrets for Monte*)[544] used an ointment that he claimed had been successful in many cases of tinea, ulcerating scabies, assafaty, serpigo, fallen hair, and lice. The recipe is extensive: Galls, rue-seeds, red arsenic, both kinds of aristolochia, sel armoniac, oven-soot, sulfur, bitter almonds, colocynth, roots of capers, leaves of fig and olive trees, roots of marsh-reeds, verdigris, alum, memitte, myrrh, aloes, frankincense, cow-bile, black tar, and enough wine vinegar to make a paste with the consistency of honey when mixed in sunlight.

Galen also used this recipe for treating achorous and favous tinea (*Miamir)*: litharge, rue-leaves, staphisagre, vitriol, vinegar, oil of myrrh, cimolia, and, ashes of paper.[545] Galen said that when he could not find papyrus growing nearby, with which to make his medication, he sometimes found unused paper in the patient's home. He asked for a lamp and burned the paper, mixed it with vinegar and anointed the affected part. He requested a visit by the patient at his own establishment on the following day. He then could see if the flesh[546] was firm; if so, he left the medication in place. When he anticipated an early cure, as was his usual practice, he saw no reason to change.

Gordon favored this ointment, which I, too, have found to be effective: dark and pale hellebores, live sulfur, ink, orpiment, litharge, quick-lime, vitriol, alum, galls, soot, calcined tartar, tempered mercury, and verdigris. Powder all and mix with juices of scabious, fumitory, borax, parelle, vinegar, and old oil. Boil off the fluid and add liquid pitch and wax. It will cure tinea, scabies, stasis ulcers (ie mal-mort) and most other skin lesions.

When the tinea was furfurous, Galen (*Easy Treatments)*[547] washed the head with brine or with a decoction of lupin, or a juice of purslane, or blet, or wild cucumber, followed by an inunction of staphisagre and soap. Avicenna favored a gum of rue. Halyabbas used rosat and vinegar. Rhazes got rid of the scurf by keeping the head cleanly shaved, and by applying

---

[544] Not identified (LDR).

[545] Joubert explained: Dioscorides described a kind of paper made from papyrus; that is known by only a few today (1569), and Pliny described how it was made. Both Galen and Dioscorides used the ashes of burnt paper as a desiccative when treating undermining ulcers. Excepting that use as a primary, it was used as a vehicle to retain other medications for topical applications, as today we use cotton fluff or shredded cloth, or pads and drains. It was used to stuff and dilate the openings into fistulas, as today (ie 1569) we use medicated strips of sponge. Dioscorides described the methods. As such it was used as a dry medicine called *Diacarthas,* of several kinds, described in his *Compound Medicines,* for use in putrid ulcers. Thirteen medications are listed, and Number Seven is from a physician named Appelle and is the only one we now call Diacartha (J).

[546] Firm flesh: This describes clean granulation tisse, in contrast with a spongy matter that weeped pus when it was pressed (LDR).

[547] Unidentified by Joubert and Nicaise (LDR).

ointments during the night, and eliminating them in the morning with warm water. If that failed, he mashed chick-pea-flour, seeds of guimauve, and vinegar, and applied it for three days.

A potent medication is made with the flour of chick-peas, with fenugreek, bran, borax, sel de niter, mustard, guimauve, vinegar, and water. Wash the head with it once a week.

Chronic, indurated tinea needs potent medicines to be applied over scarifications (Halyabbas), which are bleeding. First lay on a cataplasm of lupin-flour boiled in vinegar, or with epithyme or cantharides, as follows: cantharides, sulfur, nut-shells, mustard, myrrh, honey and vinegar. Keep the plaster in place for one full day. Then cover the head with beet-leaves and warm cabbage for four days, until the edema subsides and the scalp no longer is malodorous, and the skin and the subcutaneous tissues are clear.

If the tinea persists, use corrosives to clean out the corrupt matter, and let the lesions heal.

Roger and his Glossators and Jamier treated chronic tinea by depilating with medicines or by smearing tar to form an adherent cap[548], or they used forceps. Then they irrigated with vinegar or sea-water or baby's urine. Roger then applied this ointment on the bare hairless skin: bear-fat and cow-fat, rat-droppings, juniper-berries, burnt spider webs, liquid resin, and lamp-oil.

Jamier used this ointment: old pork-lard heated with oxalis-water, juices of abrotonin, mint, fumitory, parelle, rocket, mercuriale, alum, vitriol, gum of incinerated hedera, soot, roasted salt, tartar, and aloes.

Be aware that an angry inflammation may be provoked by any of the very potent medications. Mitigate it with the oils of roses or myrtle, or with the Appollonium ointment. This formula was describe by Galen (*Miamir*): shake and stir a large amount of good oil in a lead bowl using a lead rod, until it thickens and darkens. Then mix in litharge and ceruse, enough to make an ointmeent. It is an effective mitigant for ulcers and cancers that have been treated with corrosives, and for other malignant lesions, not only of the head. It also is effective for relief in cases of chaps and anal fissures, and for painful ulcers anywhere on the body. We try to promote the regrowth of hair with medicines that attract good humors. They are described in the next section.

Alopecia, Calvicia, And Loss Of Hair[549]

Galen (*Miamir*, Part 1.) used the term alopecia to include total loss of hair, change of color, etc.—both are seen in leprosy and alopecia. In order to provide a complete discourse, Galen reviewed the sources (origins) of hair.

Hair begins as an interior vapor, and the body resolves it by emitting it through the pores of the skin, where it is condensed on contact with the outer air. Galen (idem, and

---

[548] When the 'cap' was forcefully pulled off, the hair came with it (LDR).

[549] Joubert's term was *Pelade*. The Greek term was οφιασισ, denoting a kind of alopecia where the hair fell out in clumps, like a snake molting. The word pelade today refers to a group of disorders with various origins. Bazin stated that pelade was ophiasis in Greek and *arca* in Celsus' Latin; Chambord said that Celsus' term meant alopecia in general. Here Guy includes all types of alopecia, including ophiasis (EN).

Commentary on Temperaments) said that the process resembles the sprouting of plants from below the ground.

There are four sources, as taken by the School of Montpellier from Part 2 of Galen's *Temperaments*. The efficient source is the Natural Heat of the body which is only moderately elevated so to vaporize the affected humors. The Natural Source is a dry vapor. The Formal Source is the porous structure of the skin. The Final Source is the need for hair, both as a useful as well as a decorative material. Therefore, the hair must grow so to please—elegant women do not want hair on their chins or to be bald. And that explaims why the hair easily comes away from the scalps of old people and others who are feeble. In some the hair is curly, and it comes in various colors, and in some places it is scanty. And there are many more problems with the hair that can occupy the surgeon other than just the loss of it.

Causes Of Many Disorders

Galen said (idem) that the subject resembles the farmer's dealings with his crops. When they are too dry, they shrivel; at other times too much moisture is the problem. So, too, the growth of hair can be corrupted by a lack of nutrition. It is a cause for loss of hair (ie calvicia) followed by alopecia. Avicenna said that other causes were loose pores, or when the pores were too constricted, as caused by cold weather, or when the skin was replaced by scar, as when an ulcer heals.

Galen said the term alopecia is derived from the Greek name for a fox (αλοπεξ), and the term *ophiasis or tyrian* describes the spread of baldness, as if a snake was burrowing under the skin.

Signs

Alopecia is visible. The effective humors at fault show their colors, and can be inferred from the general appearance of the person and by knowing what their habits had been.

Prognosis

A natural depilation occurs in eunuchs and in women; they do not sprout beards. The same occurs when normal skin is replaced by scar, and in old persons and in hectic maladies, when the capacity for hair-growth is lost, or when someone's natural complexion lacks vigor. Hippocrates (*Aphorism, #6*) said that hairless persons cannot develop large varices, and when a bald person forms large varices, his hair will return. In his *Commentary* on that aphorism, Galen said that such baldness was not a true alopecia which is caused by corrupted humors; in this case it was caused by a drift of normal humors away from the scalp toward the lower body, and not by a general depletion. How else can we explain why that kind of baldness is incurable? Furthermore, eunuchs are not born bald, any more than congenital stammerers become so by an overabundance of humidity in their brains.[550] Galen (*Miamir*) noted that we cannot cure the baldness of a scalp that does not redden when it is rubbed, and that the degree of redness that appears in a rubbed bald scalp is a gauge of its curability.

Treatments of Alopecia—General and Particular

The *General Measures* include the diets and evacuations that are suited for the offending humors, as we use them for treating tinea. But after purging the body, Gordon, as taken from Galen's *Miamir*, insisted that we especially purge the head of phlegm with caput-purges and gargles and with such anti-phlegm medicines as Diacastorium pills tempered with marjolaine.

The two goals of the *Particular Treatments* are to correct the defects in the scalp, and to attract healthy blood to be converted to hair.

The first has two elements that apply to different stages of the malady. Early in the course (like Galen, *Therapeutics*, Part 14), we try to reject the bad matter and to change the scalp's attractive powers. That is, we weaken its previous bad tendencies, and use our familiar repercussives to control simple calvicia (shedding). Heben Mesûe washed the scalp with a decoction of roses, myrtle, and venus-hair ferns. Then he shaved the scalp and epithemized with the juices of the tips of myrrh plants and wild olive trees, dry rose petals, aloin, and vinegar. He boiled off half the liquids, cooled it, and added labdanum. Two days later, he added oil of myrtle and sour wine to make it as thick as honey. He sweetened the odor with musk and galls. He also used the liniment as a plaster or as an ointment, applied daily for three days. After every application he washed and dried with a rough cloth. He repeated the process until the hair was restored.

We follow Galen (*Miamir*, Part 1) and use the following for early cases. First we shave the scalp and rub it before applying an ointment with moderately attractive powers. He said that there was no better medicine for treating falling hair (calvicia) than a mixture of labdanum and oil of pistachio; in a rare case it may be better to use oil of myrtle. You may add some nard during cold weather and in the winter. He added that sometimes purges alone will work, and no local treatments will be needed. He described a successful treatment (ie only with purges) of a young athlete.

Rhazes used this liniment: venus-hair ferns, leaves of myrtle, bark of pine trees, burnt and minced aloin, labdanum, myrrh, frankincense and oil of radishes. He applied it at night and washed in the morning. Galen (ibid) cited Archigenes for this ointment: juniper-seeds, labdanum, aloin, venus-hair ferns, oil of myrtle, and old wine.

When treating long-standing baldness, first resolve (evaporate) the bad matter that has enfeebled the scalp. Galen (*Therapeutics,* Part 14, and *Miamir,* Part 1) banned the use of hot or very warm substances lest he dry or burn the skin. A thin medicine can

---

[550]   The causes are not acquired (LDR).

penetrate the skin down to the level of the hair-roots (ie now hairless). Therefore, first shave the the scalp and wash it with a decoction of camomille, aneth, stoechas, and fresh thapsus. Avicenna replaced the last-named with a gum of wild rue. But he agreed with Galen that the hair-roots should be penetrated only with moderately warm oils, such as from juniper trees, or old ricinus, or with onfacium.

After the thapsus-mix Galen applied mustard, nasturtium, sulphur, meerschaum, niter foam and seared niter, the two hellebores, rue, laurel-oil, roots and bark of marsh-reeds, liquid cedar-resin, rat-bile, and bear-lard.

Rhazes treated loss of beards as well as scalp hair by rubbing with a rough cloth until the sites were red. Then he replaced the cloth and rubbed with an onion until the place was red and hot. He awaited a full day before repeating the maneuver, and again on the third day. When blisters appeared he applied goose-or hen-fat, and avoided rubbing for several days. When hair appeared, he re-shaved it and renewed the rubbing-routine, and anointed with a decoction of venus-hair ferns, camomille, and oil of ben. He boiled off the water and used the oily residue. He also used this ointment: meerschaum, borax, burnt sulphur, a gum of wild rue, euphorbia, staphisagre, and cantharides; he applied it on the rubbed surfaces. Then he treated the fresh blisters in the usual way. When the site was inflamed (ie hot) Gordon used a compound of the powdered shells of almonds, hazel-nuts, and almonds, and the withered 'skins' of roasted goat droppings, and made a paste of it with honey and vinegar, or he used it as a liniment after shaving and rubbing.

When those measures failed to redden the site, Avicenna applied leaches and cups and pricked many places in the scalp with needles, and repeated the applications until he saw a revitalized scalp. Then he proceeded to the next elements.

The second goal is achieved with the medicine of Philagrius (4th C) which later was copied by Heben Mesûe: the flesh of snails, leeches, honey-bees, and wasps, with burnt salt. He put all in a jar with perforations, and set that in a glass bowl. After a day, the moisture leaks from the inner jar. He saved what collected in the glass bowl to use for an inunction to treat calvicia.

Rhazes instead used ashes of avrotonin, old cantharides, labdanum, burnt hazel-nuts, and gall-musk applied on the rubbed scalp.

Heraclides of Tarentum, as reported by Galen in *Miamir* Part 1, used this fomentation to treat longstanding alopecia, after shaving and rubbing with fig-leaves: sea-urchins and their shells, green galls, bitter almonds, bear-fur (? a fern), adianthum, roots of marsh reeds, and fig-leaves. He incinerated all and added rat-droppings, vinegar, resin, and bear-lard Another is listed in a chartulary[551]: juice of vitriol, powdered incinerated leeches, green lizards, moles, and soles of old shoes, with verdigris and roasted pork (ie drippings) and honey.

---

[551] Chartularies were collections of many kinds of documents. Here Guy probably referred to an old document, perhaps in a monastic library (LDR).

On Dyeing White (Gray) Hair (Canicia)

First let us state that we deal here only with non-natural (ie premature) gray hair, not the natural graying that comes with aging.

Causes

Avicenna said that the immediate cause was the corruption of an excess of the phlegm (the humor) which engenders the matter that weakens one's natural heat, as occurs with chronic illnesses, with stomach disorders, with weighty worries, in those who shampoo too frequently, or wear head-covers too much of the time.

The **Signs** are manifest.

Prognosis

We can consider premature white hair as a bad sign, perhaps as a harbinger of death (Gordon). Galen (*Miamir,* Part 1) said that the common practice of dyeing gray hair to black—usually with cold and astringent medications)—threatens a woman's health, and puts her at risk. Not only do the medications over-cool her head, they may cause apoplexy, or epilepsy, or may damage her lungs and lead to phthisis.

Treatments—General and Particular

The General Measures include an adjusted diet (ie life-style) and the evacuations which we use for treating phlegmatic aposthems. We use the Saracenic tryphere and and myrobalans, as did Halyabbas., both of which slow the progress of the graying in the hair of oldsters. Avicenna took an opposite route and simply had the patient fast until noon every day.

The Two Particular Measures: preparation of the hair for dyeing, and the actual process. As his preparative routine.Gordon repeatedly washed the hair during several days with a lye made from the ashes of incinerated cabbage-cores and some alum to act Avicenna used henna and some roots of yellow and black indigo which had been mashed by trodding. The second measure uses substances chosen for their colors. Galen chose cedar-resin for his black dye, used with or without oil; it is cool and dry. The peasants and shepherds of remote parts of Asia add an ointment of liquid pitch to stain even the roots. Those darkeners are benign; their astringency lends an adhesive quality that binds to the roots.

Archigenes described a darkener which Avicenna, said was the best of all: He boiled the roots of the caper bush in the milk of a woman or of a female ass, and reduced the volume by two thirds. He appled it at night. Sometimes he used a dog's urine kept for five or six days. Or he used the bark of a plum tree cooked with honey, or the black centers of red poppies mixed with the oil of myrtles. All were used as inunctions or as cataplasms.

490

The author of *Easy To Use Medicines*[552] favored an ointment of oil that was whipped with a lead rod in a lead bowl. Rhazes used a liniment of galls boiled in oil with henna, burnt bronze, tragacanth, and rock salt. First he washed the greasy gray hair or beard with warm water and dried it by toweling. After it was dry he covered the head with beet or elder leaves for six hours. Again he washed, noting the good effects of the dye on the gray hair. Avicenna darkened and kept some color in the hair not yet grayed by age: dark myrobalans, emblicus, galls, labdanum, and the seeds and the leaves of myrtle. He soaked all in oil for three days, and heated it to thicken it. In a later chapter he said that an added bit of clove intensified the darkening and protected the brain.

To Redden Hair

Galen attribute this formula to Archigenes : an ointment of wild rue, and niter-water will turn hair yellow. To make it blond and curly, he washed it with the foam of niter, myrrh, roots of asphodel, and heated pure wine. Rhazes and Avicenna both used a lye made from the ashes of grape-vine wood mixed with minced lupin., and allowed to set over night before adding myrrh, and roasted henna. They poured off the liquid and added wine and used it to wash the hair. They repeated the process several times in order to dye the hair red.

The women of Montpellier add flowers of stoechas and broom. Those of Bologna add shavings of box-wood, and lemon peels. The Parisian women add the roots of gentian, barberry-bush, and the flowers of carthame.

Cleansing Shampoos

The cleansers are lyes, especially those made with the ashes of grape-vine wood, wine lees, egg-whites, and soap. Another uses a double soap: Saracenic (soft) and French (hard). The Saracenic soap is made of capitel and olive oil. The French is made of capitel and ram-lard. Capitel lye is made from the ashes of beans and quick-lime; The ingredients are mixed with water as we do for any lye, and the filtered decanted fluid is the Capitel

Depilation

Galen as did Criton, used this depilatory: yellow arsenic, quick-lime, and amydon (the flower of silver).[553] Mince all and mix with water and boil. The decoction is read for use when it defoliates a feather dipped into it.

---

[552] Attributed to Appollonius or Dioscorides (EN).

[553]   Joubert claimed that Guy misinterpreted amydon, which really is a kind of wheat, and is not litharge (J).

Rhazes': yellow arsenic mixed with water in which some lime (chalk) had set for six days, during which the lime was replaced three times daily. During that time the mixture was set in sunlight. It is ripe when it defoliates a dipped feather.

Avicenna's: quick-lime, arsenic, aloes, and warm water for dilution. Ameliorate the bad odor with the musk of galls.

The common folk use five kinds depilation (Henry): First, they cut the hair with scissors. Second, they shave the region. Third, they pluck some hairs by the roots with forceps. Fourth, they smear on naval tar (with fingers or cloth). Fifth, they use psilothrum or another from the above list. They foment the scalp with the steam from boiling water and apply a thick liniment of psilothrum before they apply the tar. They leave on the tar as a cap long enough to say a Miserere. By that time the hair comes off quite easily when you rub lightly with warm water. Then apply cool rosat oil. If the skin has been singed, use the white ointment.

To Prevent Regrowth Of The Hair

Galen's recipe: sea moss (ie a mollusc), marsh apium, blood of a sea-turtle and of bats, ant-eggs, and a gum of white and dark bryony, and nettles.

Avicenna and Rhazes used hyoscyamus, opium, psyllium, vinegar, cimolia, lead-ceruse, and alum. Mince all and add a water of white hyoscyamus.

Others have used iron-filings boiled at length in vinegar.

CHAPTER 2

Various Matters Concerning The Face And Its Parts
In Five Sections

Section I: Cosmetic Matters In General

Herein we deal with Natural and Contrary-to-Natural matters. We try to preserve the beautiful natural features and to improve what is unsightly. For example: preserve and augment whiteness of the skin, or redden it when such is desired. Galen (*Miamir*, Part 1) explained the differences between cover-ups (ie face-paint, etc.) and true embellishments, although both are parts of surgeon's practice. No generalizations fit everyone, and the outcome cannot always be guaranteed, as affirmed by Master Raimond de Molieres of Montpellier. Whereas true surgical embellishment is an acceptable practice, simply applying cosmetic preparations does not pass that test. I quote Galen, "What simply makes a face beautifully attractive is not for me to prescribe for the asking. Whereas, I can advise honorable ways to deal with the signs of deterioration that come with aging, and to eliminate what offends her husband."

To Improve The Color Of The Skin

In the Arabic version[554] of Galen's *Commentary* on Hippocrates' *Aphorism #1,* he showed that the color of a person's skin indicates the predominant humor that comes to the surface; a healthy pink indicates good sanguine humors; a poor color comes from defective humors; dark skin from melancholy; pallor from phlegm; yellow from choler. In addition, Avicenna led us to consider external influences: heat, insufficient bathing, use of too much vinegar, and drinking contaminated water; all darken the skin. Cold, sexual activities, emotional disturbance (ie sadness), and chronic illnesses lead to pallor. Ingestion of too much yellow stuff, such as cumin, ammi, and salted meat leads to yellow skin. Jean of St. Amand followed Avicenna in

---

[554] Guy reminds us that the earliest Latin translations of Galen's works came from Arabic translations before the 12th C. They were rather crude and inaccurate. The same material was Latinized directly from the Greek late in the century. It became the elegant European resource for Galenism through the later centuries, and largely replaced the earlier version (LDR).

claiming that good internal sources provide a good skin-color, and a bad color indicates external sources.The good internal sources are those that engender a balanced, thin blood[555].

When treating skin color, a good diet includes soft-boiled eggs, sweet-smelling wine, good meat (ie not fatty), foods that increase the volume of nutritional blood and abet its distribution through to the surfaces. They include the limited use of figs, cloves, and saffron. Other substances purify blood, such as the lesser tryphere, myrobalans, et al.

In general, the externally applied substances that favor a healthy color are attractive[556] and detergent: liniments and embrocations of flours of wheat, bean-pods, chick-peas, ers, barley, tragacanth, radish-seeds, amidon, and rice. Others were used by Rhazes, Halyabbas, and Azaram to make face-creams: flours of chick-peas, beans, barley almond-shells, tragacanth, radish-seeds, all mixed with milk, applied at night and washed away in the morning with warm water and a decoction of dried violets and bran. Roots of lilies, narcissus, iris and arum as troches may improve the results.

Avicenna's face-cream: soap, ammoniac, water, frankincense, mastic, niter, and honey. Cook in a glazed bowl.

Theodoric's whitener: ceruse diluted with water and exposed in sunlight for a month during which the water was decanted and replaced daily. He used the ceruse to make a liniment with pearls, crystal (powdered glass), niter, borax, camphor, sarcocolla, myrrh, and mercury sublimated with saliva. All were ground on a marble plaque, and the powder was set aside. When called for, he made troches with it by adding rose-water. He warned the therapist to avoid bad reactions by diluting the liniment with an oil of tartar before applying it.

Other recipes for face-creams: Yellow ointment. Goat-fat (from the omentum) and dove-fat. Both of them stink and are seldom used.

The Use of Cosmetics

First steam the face. Then wash with a weak soap and warm water in which you have boiled the flowers of beans, or lilies, or nenuphar, or elder, or bryony; or use filtered milk instead of water. Dry the face and apply the cream, to remain all night. In the morning wash with a water of bran or of violets, and cover the face with a cloth for brief period. Then, to color the cheeks, wet the cloth with alum-water diluted with scrapings of bresillet wood.

To remove unwanted hair and prevent their regrowth, use the methods we described in the previous chapter.

To Remove Macules, Freckles, And Pannus

First anoint the face with an oil of tartar containing wheat-flour, or use Henry's French oil: calcined tartar, mastic, camphor, and egg-whtes. Add rose-water and filter.

---

[555] "Healthy Blood" is a properly balanced mixture of the four humors (LDR).
[556] "Attractive:" bring the desirable humors from within the body into the skin (LDR).

Avicenna's Diachylon: litharge heated in old oil and cooled by adding a mucilage of fenugreek, mustard, bdellium, and myrrh. Reheat it while beating it vigorously.

Another good recipe is called Virgin's Milk: litharge tempered with white vinegar and filtered through felt before adding brine.

Rhazes and Avicenna both declared this recipe to be the best for clearing macules: mercury with enough minced almonds to hide the mercury. Add minced melon-seeds. Apply at night for seven days. Wash with warm water every morning.

When all the foregone fail, use the prescriptions in our chapter on morphia.

Treatment of Discolored Bruise-Marks On The Face And Elsewhere

Galen (*Miamir,* Part 5) fomented with a sponge soaked in dilute warm vinegar. Then he applied a decoction of ammi, hyssop, milk, and wine.

Henry applied the juice from the roots of grape-vines mixed with hepatic aloes, three or four times a day. He declared success within two days when removing the black-and-blue discoloration.

Rhazes waited until the local pain and heat subsided before applying leaves of cabbage, radishes, or mint (the best). When that failed, he used a liniment of yellow arsenic, lapis lazuli, frankincense, ammoniac, and the juices of coriander and celery.

Avicenna favored aloin and honey, or an ointment also used by Dino: marjolaine, yellow arsenic, camomille, oil, and wax. He also used the diachylon.

Popular remedies use fomentations of rose-water and fenugreek; or as a plaster smeared on a wine-wet cloth; or mixed with bean-flour and oxymel. The most rapidly effective is ceruse mixed with rose-water or with the white ointment.

Treatment of Pox (Variola)—Its Marks and Scars[557]

The General Treatments usually are conducted by Physicians, and include special diets and evacuations.

Four parts of *The Particular Treatments* are Surgical: 1. The Attraction to the surface of whatever bad matter lingers within (ie after the Physicians have finished their work). 2. The Protection of internal and external structures from further involvement with pox. 3. Rupture the pustules when they are 'ripe'. And fourth: Heal the pox so to leave less disfiguring scars.

---

[557] In this section, Guy limits his discussion to the cosmetic effects of variola and does not hint at the dire effects of the disease itself. I assume that he avoided any conflict of interests with his physician-colleagues. The term 'variola, was used since the 6th C, and the disease was clearly and minutely described by Rhazes around 900, one of Guy's favorite authorities. Another of his authorities, Gilbert, described it in 1225. (see Bibliography, Garrison pp 114, 120, 121, 145 (LDR).

First: Prescribe special beverages of lentils, figs, and saffron, and wrap the body in a red sheet to protecy it from cold and wind.[558]

Second: Defend the eyes with a collyrium of rose-water and saffron; the nostrils with sour wine; the throat with diamoron; the lungs with diatragacanth; the intestines with troches of spode. Etc.

Third: Open the pustules by snipping off enough of the tops to prevent their closure.

Fourth: To avoid scratching, cover the pox with a mustard plaster containing the flours of lentils, beans, lupin, ers, litharge, ceruse, and aloes. After the pox have dried, apply an ointment of litharge, ceruse, cadmia, oil of lilies, and the fats of donkeys or grouse.

Rhazes used this liniment on the face: thick litharge, roots of marsh-reeds, old bones, flours of chick-peas and rice, clean melon-seeds, ben, costus, and a mucilage of fenugreek and linseeds.

Treat the scars of pox as we described in the chapter on flesh-wounds.

Couperose (Roseacea) and Acne (Facial Pustules)

Avicenna called the abnormal redness *Albedsanen*; it resembles the red skin that is an early sign of lerprosy. Usually it first appears on the nose or on the knobs of the cheeks. At times, the entire face is involed, and was called *Butizaga* by William of Saliceto. At times there is no inflammation (ie heat), and at times there are pustules, some of which are crusted, and are called *Assafati*.

The causes are salt-phlegm and other combusted humors. At times their emanatiuons are foul and toxic, and the inflammations may ulcerate the tissues around a pustule.

Signs

The colors the shapes, and the virulence (inflammation) indicate what humors are at fault, and whether they are natural or contra-natural.

Prognosis

It is a contagious malady in its early stages, as is leprosy. Rhazes said that it multiplies more rapidly during the winter and when the weather is cold. Avicenna said that cold represses the vapors. It is not easy to treat because the face readily accumulates the thin

---

[558] Joubert described the reasons why the common folk used the red wrap—a sheet or a sash—and covered the sick-bed with red items. They believed that the patient should not look away from red things. Indeed, if he saw only red, the recovery would be hastened. On the other hand, those who bleed excessively during phlebotomies should avoid looking at red objects (J).

Both Gilbert and Gordon ascribed the emphasis on Red to the practices of Arabic quacks (Garrison, ibid. p 156) (LDR).

causative matter, and it is too feeble to reject it, as is true for ulcers in general, especially in older people.

Treatments

The General Measures, as usual, involve diets, evacuations, and the diversion of the bad humors, according to the various types. We use various rubbings, cupping, cantharides, applications of corrosives between the shoulders, at the back of the neck, and under the chin. Bleed from veins on the forehead and the nose, and apply leeches at those sites. Laxatives include electuaries of rose-water and others that are suitable. The diet is kept thin or minced. Abstain from strong undiluted wine. Avoid spices, tangy foods such as garlic, onions, peppers, mustard, rue, aromatic elder, salty fried, roasted, and broiled meats, and melancholic and gassy foods. Offer cooked foods that are not too cold or too hot. Avoid constipation. Sleep with the head elevated, and only for short spells, all of which is the same as the general regimen for infections, choleric pustules, burns, and the early stages of leprosy.

The Particular Treatments: Early in the illness, cool the inflamed sites and use repercussives as drying agents. For longstanding rosacea, bolster the resistance of the affected region with resolutives. At first use alum-water with verjus, juices of poplar and plantains, egg-whites. Filter it as we do we in making rose-water. Lay on a soaked cloth and renew it frequently. Treat the chronic case with fumigations of steam containing camomille, melilot, roses, violets, and water-lilies. Afterwards gently clean the region and anoint it with the white or the yellow ointments; they contain mercury, sulfur, alum, and the oil of tartar.

William used this: sour lemon-juice thickened with ceruse,and mercury tempered with saliva. Boil it down to the consistency of an ointment.

Theodoric used this troche: borax, flours of chick-peas and beans, camphor, honey, onion-juice. Dilute as desired.

Another recipe uses fresh cow milk rose-water oil of tartar, wheat-flour, and other substances that we use for dartres, and are in common facial cosmetics; mix them with cantharides and lard. Overlay the purulent field with beet-leaves. Then use desiccants and consolidatives.

Section II Disorders Of The Eyes, As Yet Not Dealt With

Galen (*Maladies and Symptoms,* Part 3) said that all disorders of the eye interfere with vision, and the oculists [559] have said that some maladies involve the entire eye, such as

---

[559] From ancient times certain famous physicians devoted themselves to treating disorders of the eyes, especially among the Arabists; among them were Jesus, Acanamose, Alcoatin, and Azaram (LDR).

aposthems, inflammations, and impaired movements, whereas other maladies affect only parts, such as the eyelids, cornea, the optic humors, and the visual sense. I quote Galen (idem, Part 40), "In all senses, vision as an example, there are three sets of symptoms. One involves the affected structures, such as the crystalline humor (ie the lens). A second set involves the sense itself, which derives from the brain, in this case, via the optic nerve. The third set includes all the supporting structures." He added that,some disorders involve only simple tissues; and others the entire complex structure (ie organ) is involved. In most maladies both the simple tissues as well as the structures are involved.

Although many maladies can occur in any part of the body, those which affect the eye are special, and they call for many different kinds of treatments. Jesus said that some disorders of the eyes are commonly occur elsewhere in the body, such as bad complexions, aposthems, and wounds Many are limited to the eyes as are lacrimation, cataracts, and affections of the membranes of the eyes. Jesus counted ninety-two, Avicenna forty-eight, Alcoatin fifty, Azaran sixty, Acanamose of Baldac sixty (in his pharmacopoeia, where he credited Galen with one hundred-five), Almansor sixty five[560], Bien-venu and Peter of Spain[561] did not provide counts. We shall choose from those vast collections and eliminate what was discussed in other parts of this treatise, such as aposthems, wounds, and ulcers. We will consider other surgical maladies here and reserve others for later, those which are hidden and escape examination

Causes

As we have for most ailments, we classify causes as primitive, antecedent, and conjoint. In addition, Galen (*Interiors* and *Affected Places)* listed Sympathies and Idiopathics—that is, Compassions and Tendencies. Compassions refer to maladies that come to the eyes from other parts of the body (ie accompaniments), and Tendencies refer to matters proper to the eyes (ie idiopathic).

The causes derive from humors that are engendered by unhealthy habits, and by the intrinsic weakness of the eyes that allows the entry of the bad matter. We can say that diseases of the eyes are caused by what comes to them and by what happens when it gets there, as we stated in our discussion of ophthalmias and inflammations (aposthems).

Signs

The signs are apparent for lesions that are exposed. For others we must hunt for them as noted in Galen's *Afflicted Places*. We will get to them later.

---

[560] Almansor is the title of a treatise by Rhazes. It is not the name of a person (LDR).

[561] Peter of Spain (really of Lisbon) (1215-1277) became Pope John XXI. Wrote the *Book of Eye-Diseases* (LDR).

Prognosis

Treatments are difficult; the eyes are complex structures, as we described them in our Anatomy. Furthermore, they are very sensitive (Galen, *Therapeutics,* Part 13), and the medications require utmost care *(Miamir,*Part4). Acanamose said that ocular ailments almost always are complicated, as when scabies occurs with sebel, and the multiplicity adds difficulties (Galen idem Part 7*)*.

The Five Goals Of The General Treatments

The First Goal includes eight concerns. The Second deals with evacuating the antecedent matter. The Third deals with provisions for a healthy life-style (ie habits) that will impede the production of the corrupt humors. The Fourth eliminates the bad matter (ie conjoint). The Fifth deals with complications.

The Eight Items of the First Goal:

Item 1. When the head is sick along with the eye, treat the head first, as was the teaching of Avicenna and Jesus, as set forth in their doctrines for ulcers and ophthalmias.

Item 2. Galen insisted (idem, Part 13) that we do not treat the eye until both the body and the eyes have been purged.

Item 3, All manual operations on the eye must be performed delicately, and as painlessly as possible. Jesus insisted that an eyelid should be folded up or unfolded gradually, not suddenly. Albucasis added that when we undertake a painful or disfiguring procedure, we should stop until the suffering is appeased before we complete what we set out to do.

Item 4. Operations should be performed under clear skies and when there is no wind. The patient's head must be secured motionless. Albucasis held it in a trough or he held it between his knees or fastened on a metal table (Halyabbas)[562]. After the procedure, put the patient to bed in a dark room and cover his eyes with a dark soft cloth.

Item 5. Re-apply small amounts of medicines frequently rather than a large amount at one time.

Item 6. All medicines that are put into eyes must be pure and be finely minced or powdered, lest they do more harm than good.

Item 7. Before introducing any medications, wash the eye with warm water and blot it dry, Gently wipe it clean of accumulated matter with a cotton-tipped quill or probe.

Item 8. Only a surgeon who is quick-witted and ingenious should undertake operations on the eyes. He must have good vision and steady hands. He must have observed operations by other good surgeons. He must have at hand the necessary equipment: hooks, needles,

---

[562]   Guy's instruction for stabilizing the head during operations for cataract are different. See below (LDR).

probes, scissors, spatulas, lancettes, and small razors: all of them clean and keen. He should have both single and double-edged knives. He should have a good supply of cotton, egg-whites, rose-water, and clean cloths for bandages and wraps. He must always be prepared, although the operations are not frequent occurrences (William of Saliceto).

All the eight items are part of the First of Five Goals of General Treatments. The Second, Third, and Fifth Goals are as we described them in our chapters on aposthems, ulcers, and ophthalmias. The Fourth Goal will be discussed piecemeal, with each malady, as below.

Four Disorders Of The Entire Eye

I. Lacrimation and Rheum

Tears flow from lacrimal ducts, chiefly through a tiny opening at the end of the hairline, as described by Bien-venu. Drainage (ie not tears) also occurs when the eye is inflamed. Avicenna described a disorder when there is a continuous drainage of watery matter coming from the brain, to be discharged via the eyes.[563] It reaches the eyes through veins, internal and external (Jesus).

Causes

In these cases the head and the eyes are replete, and the matter can be diagnosed by touch. When it is cool, its cause is a cool humor; when warm and prickly and it irritates the nearby skin, it means that the humor is warm. Whenever an eye is inflamed, there is an ophthalmia. A shrunken or incised caruncle can be seen and that will provide your diagnosis and prognosis.

Prognosis

When the Natural parts are involved: When the caruncle has been incised or a bit has been cut away, the lacrimation cannot be cured. When the interior of the eye is involved, a cure is difficult to obtain. However, you may be able to palliate with our desiccative powders.

Treatments

The General Measures are those we apply to ulcers, ophthalmias, and rheum. They are attention to good diets (ie life style) and the use of purges. In addition, Bien-venu and Acanamose applied dialibanum in the evening. Usually I use diversives including touch-cauterization on the dome of the head. Galen (*Therapeutics,* Part 13) sometimes bled from arteries in the temples when treating internal sources (ie rheum) or from veins

---

[563] This is the counterpart of catarrh discharged via the nose and pharynx. Rheum is the common term, distinct from pus (LDR).

when the causes were external (ie irritants, etc.). Albucasis and Halyabbas described at length their difficult-to-perform procedures, and I will delay reporting them except to comment that much of it is in our chapter on ophthalmia.

We also use touch cauterization in the axillae and apply setons at the back of the neck; I have found it useful. We sedate and dry the brain (ie the source of the rheum) when the cause is cool, and use liquid-amber. When the cause is warm, we use roses and camphor. The Glorious Avenzoar, as did his own father, laid sachets or mustard-plasters on the occipital prominence (ie inion), containing cinnamon, cloves, mace, peppers, lemon-peel, waters of roses or mint (according to the weather and the patient's complexion), and Rabbi Moses (*Aphorisms*) approved. Do not omit applications of cups and divert the matter with astringents, as we described them in our chapter on ophthalmia.

The Particular Measures for Treating Warm Causative Humors include Jesus' collyria for treating lacrimation: washed blood-stone and marcasite tuthie, pearls, memitte, and aloes. Some experts add a paste of the hard centers of combusted myrobalans (ie chebules) and coral, and perhaps a bit of pepper.

Avicenna and Heben Mesûe preferred this collyrium for treating lacrimation: concentrated (ie by boiling) juice of pomegranates, aloes, memitte, lycium, saffron, and musk. Place all in a glass jar and expose it to sunlight for forty days.

Another, favored by Arnold, is in the Antidotary.

Others are the white collyrium with climia, etc.

For treating Cool Matter: Azaram Galaf (*Antidotary,* Part 21) used this collyrium to treat lacrimation, wet eyes, and sagging eyelids: heads of vipers combusted with salt, antimony, tuthie, verdigris and camphor.

Another: The Basilicon

Another: the common collyrium

Another: borax to clean away the sticky matter that is produced when verjus was used to treat lacrimation. Then add sumach, myrobalans, sel armoniac, and table salt.

Another: Have the patient sniff onions and eat mustard to cause tears to flow and wash the eyes; it is a well known remedy, and allows the brain to clear itself of rheum.

II. Enlarged Eyes, and III. Small Eyes[564]

Avicenna said that there are three kinds of enlarged eyes. One is the inflation of congesting atter in the orbit. Another is the result of pressure on the surface, called suffocation by Jesus. Another is the result of severe headaches, or when the patient strains when vomiting, or is constipated, or during hard labor at child-birth, all in persons whose retentive eye muscles are weak.

Sunken eys have opposite causes: excessive evacuations, the use of consumptives when treating hectic fevers, sleeplessness, sadness, convulsions.

---

[564]   Large: Exophthalmus and Proptosis. Small: Sunken Eyes and Narrowed palpebral fissures LDR).

Treatments

The General Regimen for enlarged eyes; diets and purgation are as we described for ophthalmias, including cups at the inion. We bandage a cloth over the eyes, wet wth astringent juices: of plums and olive leaves, or Albucasis' liniment of acacia, frankincense, and sarcocolla. Or we lay on a small disc of lead or other substances that we use to treat extruded uvea and optic uilcers. All of them work well Also, we wash the face with cool brine, as praised by Jean. Avicenna's favorite was a plaster of bean-flour, roses, frankincense, egg-whites, incinerated date-pits, and spikenard.

For Treating Small (ie sunken) Eyes we use our regimen for feebleness, and we foment the eyes with warm milk and sugar-water, and we anoint the head with oil of violets. Also we massage the head and repeatedly apply a plaster of wheat-flour, saffron, violets, milk and beef-bone marrow.

IV. Squint (Strabismus)

Strabismus is crossed-eyes.

The crystalline humor (ie the lens) shifts with the eye, to the side or upward or downward. Galen (*Maladies and Symptoms,* Part 3) wrote, "When the shift is to one side, there is little or no harm; but when it is up or down, sometimes there is double vision, because of a defective fusion of the images at the crossing of the optic nerves." Avicenna said that it was due to a defective function of the visual sense (*Naturals,* Treatise I).

Causes

The cause for squint may be external, as when a baby stare too long at his fingers (ie held close to his face), or is attracted by light from a lamp or a picture on the wall on one side of the cradle; all are matters for concern by the wet-nurse. The cause may be internal, after a convulsion or a paralysis, or an epileptic fit, when the eyes are contorted.

Prognosis

Childhood strabismus is treatable early, but when it lasts until adulthood it is incurable.

Treatment

In babies, squint can be treated by exercizing the eyes to look at the opposite side, at a candle or something that glints, or at a brightly colored object. Jesus made a collyrium for an eye wash, a decoction of lung-wort. Avicenna used blood from a turtle.

When the adult's strabismus is of recent origin, as after a convulsion or paralysis, treat it when you treat the underlying disorder.

Now we come to the Treatments of parts of the eyes.

Twenty-four Disorders Of The Eyelids

I, II, III, IV. Scabies et al.

Although there are many disorders of the eyelids, they usually have much in common: heat, redness, heaviness, swelling, verdiginet[565], puffiness (silac), xeres (small red nodules), styes (formi), ulcers, et al. Most of those features accompany scabies, which also causes stinging, and produces sandy excrescences in the inner lining of the lids, which induce lacrimation and inflammation (azaran). Scabies[566] appears in four phases, graded by their severity, as noted in our chapter on ophthalmia.

Scabies is **Caused** by salty and nitrous matter that first appears as erosions, followed by ophthalmia-like ulcers (Alcoatin). The treatments are described in the chapter on scabies.

The Signs have been described, and the diagnosis is clear when you turn up the lids. Depress the outer surface with a flat probe or a round penny, and lift by tugging the hair.

The Prognosis is guarded because scabies has so many variant forms and complications, as noted by Rhazes. We will comment further in the section on sebel.

Treatments

The General Measures include diets, purges, diversives, comfortives for the brain, and interrupting the flow of rheum from the brain.

The Particular Measures use irrigations and fomentations with rose-water containing some iron rust, or rose-water boiled with lentils. Also we apply egg-whites with rosat-oil and the juices of poplar berries or endive. Those substances are chosen to eliminate erosions (Alcoatin). Bruno preferred rose-water with white wine and a small amount of couperose or verdigris. Alcoatin added alum and saffron.

Bien-venu use this collyrium: Alexandrine tuthie, sugar, powdered dried rose-leaves, all heated in a large volume of good wine until it is reduced by half. Save the liquid for use ad hoc. The recipe is improved by adding small amounts of antimony and burnt bronze.

Avicenna washed the eyes with a sponge wet with warm water, and then anointed the lids with aloes, lycium, memitte, saffron, and a water of morel.

William of Saliceto covered the lids with this plaster: powdered rose-petals and camomille, boiled in wine, and thickened with egg-yolks.

Avicenna used a plaster of mauves after cleaning the purulent lids with sugar. He added tuthie to Bien-venu's plaster. And he used Jesus' recipe to comfort the eyes: incinerated blood-stone, bits of samphire, and seeds of gallitricum.

---

[565] Verdiginet: Perhaps a erroneous copy or an Arabic term. Not identified (LDR).

[566] Scabies of the eyelids probably was trachoma (LDR).

Rhazes used this blood-red collyrium: blood-stone, incinerated verdigris, coral, pearls, indian salt[567], gums Arabic and tragacanth, clean myrrh, sandragon, and sulfur. Use old wine to make a paste or use as a collyrium.

For severe ocular scabies Avicenna turned back the eyelids and rubbed them with powdered meerschaum and fig-leaves, or with sugar (a la Alcoatin) or with Jesus' red collyrium, or with Rhazes' green ointments (white or dark) for sebel and ongle. The recipe includes copper-flower, incinerated colcothar, red arsenic, borax, powdered meerschaum, and ammoniac, all dissolved in the juice of rue.

If that failed, he shaved off the subconjunctival 'sand' from the undersurface of the lids, using a knife or a spatula, as did Rhazes. He filtered the water and added a small amount of wine, and cumin-water, and applied the filtrate on the bared surface of the lids. Then he applied the yellow powder that we use for ophthalmia, and then in the winter, a plaster of bitter almond. In the summer he used the marrow of beef-bones and oil of violets. When that stirred up too much inflammation, Jesus mitigated it with blood-stone, avoiding wheat-flour and the white collyrium. After all that, I have nothing more to add.

V. Drooping Eyelids (Ptosis)

This section deals with what Jesus called a lengthening of the upper lid that cannot voluntarily be lifted, and sometimes is inverted and allows the eyelashes to press against and injure the cornea.

He said that the **Cause** was a defect of the palpebral muscles.

Treatments

The General Measures, as for apostems, involves diets and purges. The *Particular Measures* used by Jesus for recent-onset cases were applications of medicines that dry and stiffen the lids: memitte, aloes, saffron, acacia, myrrh, and a broth of bran.

In chronic (longstanding) cases, and when the foregone treatments for new cases have failed, he operated; Alcoatin described the four ways:

The First Way: He incised of the skin of the upper lid with a lanceolate (ie myrtle-leaf) incision, and removed enough skin to shorten the lid. His technique: He stretched the lid with his fingers and caught the upper edge of the incision with three threads, each about a palm-width in length, introduced on needles, or he used a three-pronged hook, and dissected under the skin to allow him to use a scissors to excise enough skin to shorten the lid. Then he closed the defect with sutures and applied the red powder with

---

[567]   Indian Salt: Galen and Dioscorides called it a sugar, a small amount of the sap of marsh-reeds. It was imported from India, hence the name. Today (1563) we use it in sugar-candy, and other things that contain what the ancients called Indian Salt (J).

egg-white on the suture-line, and covered it with a bandage. Follow-up treatments of the wound used diapalma.

A Second Way: He inserted two thin metal or wood rods to undermine and elevate the skin. When he tied the ends of the rods together tightly enough to isolate the segment of skin between them, the segment would die and slough away

The Third Way: He used a curved red-hot cautery to cut away a segment of skin. The defect would heal by scar and close it.

The Fourth Way: He used caustics as potential cauteries, applied on a shaped ellipse of paper-ashes (ie paper maché) wet with the caustic. He held the mold in place until the patient no longer felt the destructive heat. He treated the eschar with butter until it came away and was replaced by scar.

VI. Short and Everted Eyelids

Opposite to the above, the lids do not close a gap that exposes the cornea; it resembles a hare's eye (Halyabbas). Avicenna said that the condition may be due to a weakness of the muscles that move the lids.

Treatments

In addition to the usual *General Treatments*, with a diet regimen and purges, the *Particular Treatments* aimed to relieve the dryness of the (ie exposed) eye with humectants. Jesus moistened the lids with oils, douches, and fomentations, a mucilage of fenugreek, and milk. Avicenna favored chicken-fat. Failing that, Alcoatin incised the lid and spread open the incision and inserted a drain to widen the gap while it healed with granulation tissue covered with skin. When the granulations were excessive, Jesus applied corrosive powders, the green collyrium. Others use a hot cautery, which I approve, because it avoids the risk of contact with the corrosives.

Failing that, put traction on the lid with two or three hooks, or as above, insert threads with needles. With the lid on stretch, incise with a knife or scissors and control the bleeding with cotton fluff or cloths. Prevent early union by scar with applications of a water of chewed cumin (Haly-abbas). Apply egg-white in the eye. Later use the red collyrium,and as in treating scabies, and the yellow powder as in ophthalmia. Be careful not to cut the cartilage with the knife or cautery; it does not unite solidly.

VII. Sticky Lids

The lids can adhere to the surface of the eyeball as well as to each other. The adherent matter may appear after finger-nail scratch, or a sebel, or from superfluous granulation tissue, after rubbing the eye to relieve itching, by irritation from infolded eyelashes produced by compressive dressings that restrict eye-motion, as used after operations, or as a reaction to irritation by medications such as salt, cumin, etc.

Treatments

Jesus separated the lids to admit a probe under the lid. While elevating it he inserted a quill to complete the separation, taking care not to injure the cornea, and allow the escape of uvea. After the separation he instilled cumin-water with salt. He placed a strip of cloth beween the lids, as did Alcoatin, and that allowed him to introduce a cotton-tip applicator with egg-white and rosat-oil. Two or three days later water he used the green collyrium, then the red, and then the yellow powder.

VIII. Redundant Eyelashes That Prick The Eyee Engendered By Corrupted Humors

The General Treatments are the usual twofold measures.

Six Particular Treatments: First: Pull out the hairs with fingers or forceps.

Second: Dry the exposed hair-line with Jesus' collyrium of vitriol. That medicine also is good for treating silac, scabies, ungula, pallor and burns: gum Arabic, dragacanth gum, socotrin aloes, verdigris, red and green arsenics, colcothar, flakes of burnt bronze, three peppers, blood-stone, wheat-flour, roots of garance, sangdragon, acacia, tuthie, lycium, incinerated galls, cadmia of silver, ceruse, myrrh, sarcocolla, ammoniac tempered in rue-water, and lemon juice.

Third: After removing the hair and turning back the lid, cauterize the hair-pores while an assistant holds the head motionless. Then lay on a cotton pad wet with egg-white and rosat-oil.

Fourth: Removal of the eyelashes with their roots: Evert the lid and penetrate it with a needle at the base of the root, threaded with fine silk or a woman's hair, leaving a loop on the side opposite when you remove the needle. Catch a couple of lashes in the loop and extract the lash and its root when you withdraw the loop. Repeat the action as often as needed to get rid of the offending lashes, but do not use the same needle puncture lest you make it so large that you will fail to engage the lash. Cover the hair-pores with some glue to prevent the regrowth of the lashes.

Fifth: Stiffen the non-natural inwardly bent lashes with a glue of mastic, frankincense, aloes, sarcocolla, tragacnth, and egg-whites. Apply it until it adheres to the straightened lash.

Sixth: Shorten the lid as we did for the sagging lids; that may evert the lashes.

IX, and X. Loss Of Or Blanching Of Dark Eyelashe And XI. Lice

The Cause for loss of eyelashes may that of alopecia, or it may be inflammation or ulceration or the kind of scabies engendered from salt-phlegm. Jesus and Alcoatin **Treated** it with a paste made from the ashes of incinerated pits of dates and spikenard, applied with a cotton-tipped quill or probe; it is a potent and reliable medicine.

Avicenna used several others, one of which is a paste of sloe and antimony. He applied it with his finger or a twig.

To restore the original dark color of pale eyelashes, treat them as we did white (gray) hair, with an ointment of oil or of goose-fat that is ground in a lead bowl.

Lice in the lashes are treated as we have described. Here we emphasize washing with sea-water, brine, and sulfur-water, and an ointment of alum, staphisagre, aloes, oil, and a vinegar of squills.

XIII, And XIV: Indurations (Hard Nodules), Lupus, And Hordeolum XV, XVI, And XVII: Chalazion (Gresle), Silac (Puffiness), Xeres [568]

The first group are the kinds of nodules and phlegmatic swellings that we dealt with in our chapters on gland and lupus. Here we may be able to soften and resolve them with applications (ie compresses) of warm water and diachylon, or with a recipe of opoponax, serapinum, ammoniac, and vinegar (Jesus). If the lesions persist we may be able to cure them by rubbing them with a narrow spatula until they bleed, followed by desiccants and treatments for wounds.

If a lesion is large, find a spot where it is most prominent and incise over it, from inside or outside, parallel with the skin lines, and as far away from the eyelash-roots as possible (where healing will be unkind (ie scar). The incision should be centered over the eye to avoid injury to the lacrymal apparatus at the medial or lateral corners. Peel out the nodule intact if possible, and suture the incision when necessary to close it. When you cannot enucleate it, consume what is left and deterge the defect with the apostles' ointment or a powder of asphodels. Then use egg-whites and the red powder, followed by diapalma as a cicatrizer,

XVIII And XIX: "Mulberries" And Warts On The Lids

These are excrescences on the skin, derived from tears. When flat and firm they are warts (verrucae), and when they are soft and dangle, they are "mulberries" (mûres) They resemble the'figs' and the warts at the anus.

Treat them by excision or hot cauterization, and avoid the risks of applying caustics near the eyes.

Sections XXI To XXIV Aposthems, Fistulas, And Ulcers Of The Lacrymal Apparatus, And Redundant Granulation Tissue (ie At The Caruncle) And Eliminating It Without Causing Uncontrolled Tearing[569]

All of this has been dealt with in previous chapters.

---

[568] These terms are vague, and have been variously defined. Indurations or dartre probably meant sclerotic nodes. Gresle probably was our chalazion, a meibomian gland. Silac probably meant puffiness, although some authors claimed that it was a mild form of scabies in children, and that Xeres was scabies of the eyelids—possibly an adult form of silac (EN).

[569] Section XX is omitted, unexplained (LDR).

Thirteen Disorders Of The Conjunctiva

Section I. Ungula (Pterygium)

Setting aside many of the maladies of the conjunctiva (including ophthalmia, wounds, subconjunctival bleeding) that we have dealt with in the chapters on aposthems, and in the preceding chapter on the eyelids, we now arrive at Ungula (ie pterygium) and Sebel.

Avicenna described ungula as a flat (ie a pannus) excrescence on the surface of the conjunctiva that grows over the eye from the lacrimal caruncle to reach the cornea at the iris. It usually arises at the lacrimal corner, and only occasionally from the lateral corner. Rarely it may come from below or above. Acanamosale of Baldac said that there are four kinds of ungula, but Alcoatin said that there were only two, fleshy and nervous. Albucasis said that 'nervous' meant membranous, thin like the peritoneum, and that it included a fatty type (his number three), all of which really are blemishes. It (ie the fatty type) can be pale like a clear humor, or cloudy. One cannot slip a hook beneath it and pick it up to cut it away, as we do for the others. Avicenna described several colors: orange, red, and pale tan, verging on clear. Furthermore, he said that when they first appear and are thin, one can easily free them from the conjunctiva. When they are chronic and tough, the dissection is more difficult.

Bien-venu said that their **Causes** were thick and fatty phlegmatic humors that were the result of unhealthy life-styles (regimens).

The Signs are clearly visible. The fact that they are not conjunctiva itself is clear when you separate the ungula by lifting it with a hook. And it is not part of the caruncle. Furthermore, the ungula remains pale when it is released after being put on a stretch, whereas the caruncle regains its redness.

Prognosis

Jesus said that when one lifts the ungula before cutting it, it may tear apart, especially at the corneal end. And when you cut it away at the lacrymal corner you may damage the caruncle and cause uncontrollable lacrymation[570], or you may cut a vein and cause bleeding that is difficult to control by ordinary means (Acanamosale). As to that, Jesus advised dissecting and transecting with great care, even to leave behind some of the ungula to be consumed with medicines, a bit at a time. However, you must get rid of all of it, lest it recur.

Jesus described two elements of **Treatments.** The usual *General Measures* Are those which we prescribe for cool aposthems and for cataracts (in a later section).

---

[570] Although some of the special oculists, such as Jesus and Alcoatin mentioned the lacrymal pores at the medial ends of the lids, just beyond the eyelashes, the clear anatomic distinction of lacrimal ducts and the fleshy caruncle came in a later epoch. For our surgeons the fleshy-fatty caruncle was part of the lacrimal apparatus, and when it was damaged, tearing was inevitable (LDR).

The Particular Measures for treating recent-onset thin ungulas are softeners and resolutives, and detergents. We use fumigations and irrigations with warm water. We use Avicenna's abstersive collyrium: burnt bronze, colcothar, bull-bile, and a small amount of honey. Jesus used a rosat-like collyrium (ie sief) for ungula as well as sebel, scabies, vestiges of ungulas, lacrymation, and dim vision: washed blood-stone, burnt bronze, cadmia of silver, indian salt, borax, verdigris, black and white long peppers, meerschaum, socotrin aloes, spikenard, cloves, ginger, myrobalans. All of these thirteen simples were crushed and mashed and put in wine and a water of fennel.

Alcoatin's collyrium for the same purpose is described in our section on scabies, and his vitriol collyrium is in the section on eyelashes.

Avicenna said that the best treatments for a longstanding indurated ungula are either to scrape it away, or elevate it and separate it, especially when it comes away easily by scraping or by elevation. If not, your operation may cause damage. There are two methods, or three if you include the fatty blemish describe by Albucasis.

We will describe Jesus' method first: When the lids are held wide apart (not everted), catch the ungula with a hook at its center and lift it. You may need two or three hooks. Then slip the flat blade of a narrow lancette, or a fine quill, under the ungula. You may first make a small nick alongside the ungula through which to admit the knife or quill. Then deftly dissect the ungula away from the conjunctiva over the sclera, and cut the film[571] along the side the ungula with a scissors, carefully avoiding the cornea and the caruncle. Then instill salt and crushed cumin, repeatedly until the eye has healed. Also, use egg-whites and oil to prevent swelling, and encourage the patient to move his eyes to prevent adhesion to the lids. Perform the exercises two or three times a day for three days. Afterward get rid of the remnants of the ungula with collyria and the familiar blemish-removers.

A second technique is that of Albucasis and Avicenna, who adopted the use of a hair from Halyabbas, Alcoatin, and Bruno. As above, with the eye fully exposed and the ungula caught and lifted with a hook, pass a threaded needle through the lesion and knot the thread at the ungula. Then pass a second needle through another puncture, under, not through the first, if possible pass it beneath the entire ungula. The needle is armed with a thread of fine silk or a woman's hair. Then gently pull at the ends to saw the ungula away from the eye, first toward the pupil and then toward the caruncle, and cut across the ungula with a scirros. The followup treatments are as above.

A third maneuver uses a razor, as did Albucasis in treating fatty ungulas, and when removing spots. We will describe it later.,

We already have described, in the section on eyelids, how to eliminate proud flesh from the conjunctiva and the caruncle.

---

[571] The reflection of thin conjunctiva that passes over the ungula (LDR).

II. Sebel[572]

Avicenna defined Sebel as a flat excrescence on the eye, appearing as veins in the conjunctiva and cornea, appearing as a smoky cloud. It is accompanied by itching, tears, ulcers, thickened lids, and scabies, all of which obtund vision. He said that there are two types, arising from internal or external veins.

The **Causes** are repletion of the head and lack of resistance (ie feebleness) of the eyes.

The Signs are described above. Its matter and its appearance were noted in the chapter on ophthalmia. Jesus said that when large, red, and tender veins appear on the forehead above the eyebrows, or when you see a stye on the lid, the cause of the sebel is external. And when the veins and the tissues around the eye (ie in the orbit) are red and there is repeated sneezing, especially when one looks at the sun or at a lamp, or when that causes throbbing pain in the eye, that indicates an internal cause.

Rhazes gave the same **Prognosis** as for scabies. He said that scabies and sebel are difficult to cure. Avicenna said that sebel was inherited and carried from one generation to the next. He also prognosticate that sebel caused the lids (ie the gap) to be narrowed and the vision to be impaired.

Treatments

The General Measures are as for ophthalmia, ulcers of the eye, scabies, and lacrimation. I refer the Reader to those chapters.

The Particular Measures are the medications for ungula, as used and tested by Avicenna, Alcoatin, and Azaran. Also they used an collyrium made of sea-shells and fresh hens' eggs. It was set in vinegar for ten days and evaporated in the shade. Then it was made into a powder.

Other recipes used by various authors for sebel, ungula, scabies, and lacrimation contained the red, green and rust-colored collyria; or the basilicon powder. Jesus' powder contained peppers, ginger, seedless yellow and Indian myrobalans, socotrin aloes, meerschaum, minium, canelle, cloves, and sel armoniac.

Ten Disorders of the Cornea

I. Blemishes (ie Macules)

In earlier sections we discussed some of the many disorders of the cornea: hypopyon, pustules, ulcers, and rupture. Here we will consider blemishes (ie taches) and cataracts.

---

[572] In Guy's time, Sebel was neither ungula nor pterygium. Guy followed Avicenna in distinguishing Sebel from Ungula. Joubert said that Sebel simply was a collection of natural veins, abnormally distended, lying in the conjunctiva and coloring it red, and causing lacrimation and itching. William of Saliceto said that sebel was a collection of arteries and veins enclosed by a thin membrane (EN).

First let us agree that a blemish—Jesus called them pustules—is a remnant, a small scar (ie point). It is a pale clouded speck, as if a bit of cloth was deposited on the surface—what Albucasis called a fatty ungula, and Bien-venu called a cloudy fleck—or a bit of scurf or a freckle, or a chip of barley-bran. The treatments vary only with the size of the blemish.

All of them are pale and lie on the surface; they neither penetrate nor do they cause a lump; all of them derive from the cornea. Some lie in front of the pupil and impair vision; others lie at the periphery and do little to impair vision. Some are tiny and involve only the most superficial layer, and others go deeper in the cornea. Some are entirely flat. While others have some irregularities of the surface, as scars of pustules that retain some of the matter.

The Primitive Causes are aposthems, wounds, blows, frost, or burns. **The Antecedent Cause** is the thick humor that invades the eye. **The Conjoint Cause** is the persistence and induration of the invaded matter, or the scar of a corneal ulcer. That information lets us infer that there are two kinds of corneal macules: one is a deposit on the surface and the other is a scar.

The Signs are what you can see on the surface, different from hypopyon and cataracts, and a white lens, all of which are behind the cornea. They are not like pale corneal ulcers or reddish conjunctival ulcers. Most blemishes are flat, unlike pale pustules that project above the surface and contain pus. If the source of the causative humors is internal(ie the brain), it enters via the veins in the conjunctiva around the cornea. The scar—blemish is oval, that of the ulcer it has replaced, or of other lesions that heal by scar (ie the external causes).

The Prognosis is bleak, because a corneal scar cannot be eliminated. The more one tries, the worse is the scarring. Furthermore, old scars in older persons, according to Alcoatin, that have affected the entire cornea are incurable because the cornea is a spermatic tissue that cannot be regenerated; it is replaceable only by scar, which is a foreign tissue. However, you may be able to improve its appearance and its color.

Treatments

The General Measures, which are diets purges, etc. are those that we prescribe aposthems and cataracts. However, Jesus said that laxatives are unnecessary unless the eye is reddened and the veins in the conjunctiva are dilated, as indicators of the cerebral source of the blemish.

The Particular Measures for fresh and thin macules includes licking them, especially by a woman, or by fomentations (Avicenna and Alcoatin) with warm water or by immersing the face in a basin, or by using collyria containing centaury with honey. Dioscorides and Gordon favored the juice of polygonum, commonly called bird-tongue.[573] Bien-venu made

---

[572] Joubert's Comment: In our (1569) apothecaries' shops the 'bird tongue' is the seed of the ash-tree, the centinodium or corrigeola, or renoueé in French. The same holds for 'swallow-tongue. It is not chelidonium (J).

Note the association between the herbs named by birds' tongues and the use of the woman's tongue (LDR)

his nabatine powder from a sugar-candy called cassonade; he claimed that it softens and cleans the macule; he called it 'the clarifier'. Fumigations with the smoke from aloe-wood will comfort the eye. Another good detergent (Rhazes) is meerschaum, sarcocolla, and mouse-droppings. Jesus added eggshells.

The English Rose[574] used ceruse, bronze flower mixed with white wine, juice of rue and chelidonium, mixed in a copper pot. After it sat for a full day and night he filtered it through a cloth and used the liquid as a collyrium.

Acanamose used a powder of Alexandrian tuthie, camphor, ginger, and pepper.

When the blemish was large and old, and its veins indicated an internal source, the treatment after the General Measures, may include cutting some of the veins in the conjunctiva, or some of the measures used for ungula. However, blemishes from external causes need more intensive measures. First mitigate with a softener such as a decoction of barley-straw, violets, camomille, melilot, mauves, and fenugreek, all concentrated by evaporation. Follow that with a detergent powder of sepia, white ginger and pepper. Grind it fine and apply a bit directly on the blemish with a narrow flat probe. Shut the eye at once and rub the lid over the spot, with the ball of a finger. In addition you may try some of the medicatuions we use for ungula, sebel, and scabies.

The purpose of the following treatments is to darken the white blemish. After the eye-bath use the green collyrium, as we did for scabies, and then some of the musk-confections. Jesus used this powder: lizard-feces, niter, meerschaum, ostrich-egg shells, whole pearls, tuthie coral verdigris, and musk. Grind them fine (ie as a mildly corrosive abrasive). For the same purpose, Rhazes, Avicenna, and Azaram used a confection of massacumie[575] which Lanfranchi said was the best of all. The recipe was taken from Heben Mesûe, and was called 'badly cooled 'glass by Dino: lizard feces, meerschaum, borax, powdered sugar. Grind fine and put it in water in which you had boiled fennel, chelidonium, and acoris. Boil it down to three ounces and grind it again, and let it evaporate before using it as a powder.

If the blemish is inflamed or contains some matter, use our medications for treating pustules, hypopyon, and cataracts. First try a balsam, and, as advised by Alcoatin, add some powdered gold. Also try droppings of swallows in honey (a la Azaram, and Jesus).

When the blemish is fatty and cloudy (ie Albucasis' third type of ungula), he scraped it with a razor several times. He held open the eyelids and scraped lightly as he did for ungula. He follow each episode with sedatives and collyria and repeated the treatments until he have removed all of it, unless pain and swelling intervene. After a delay, he returned to the treatments as soon as feasible.

---

[574] John of Gaddesden, fl. 1330 (LDR).

[575] Joubert explained massacumie: Other authors called it varnish, such as used on glazed pottery. It included the shells of marine animals and materials used in making glass. The common term is masse cuite (cooked mass), reduced to 'massacumie' by the author of the *Pandects*. Gordon called it Marcia cocta, and he used it to make counterfeit pearls (J).

If all of the described treatments have failed, the Masters advise only to dye the macula with this collyrium: galls, acacia, and calcanthum. Or: pomegranate flowers, vitriol, acacia, gum Arabic, galls and antimony, grind all, and add the juice of poppies to make a collyrium.

II and III. Cataracts And Gutta Serena

A cataract is a flat blemish that obstructs the pupil and blocks vision. It derives from a moist foreign matter that accumulates over time and is congealed by the natural coolness of the eye. The moist matter may derive from the humors of the eye itself (ie aqueous or vitreous or crystalline), especially the aqueous (ie albugineous) as described by Galen in *Maladies and Symptoms*, Part 4. That is the **Primitive Cause.** When it comes from the stomach or the brain it is a vapor, and it liquefies when it reaches the eye. That is **The Antecedent Cause.**

The congealed matter accumulates between cornea and the uvea (Jesus) or between the aqueous and the crystalline, as noted by Galen *(Usum . . . .* Part 10). I will not argue which it is.

However, the cataract develops in three phases, each with its own name. At the onset it is called 'Imagination' or 'Fantasy', because the patient thinks that he sees things that are not there. In mid-course it is called 'Suffusion', because it is like a water-fall, sometimes only as drops, or the patient sees a watery cloud. Finally it is called 'Cataract' when it completely blocks the vision, as if one stands behind a mill-race or in a heavy rainfall and cannot see through it.

Galen (idem) and Avicenna said that cataracts vary as to the amount, the kind, and the quality of its substance.

As to quantity: It may be large and cover the entire pupil and completely obstruct vision. At times it covers only a part, differently shaped as in phases of the cusped moon, around an open center as a window.

As to the kind of substance: At times it is fragmented and the bits seem to move and the patient can detect shadows and light, as if a cloth hangs before the eye. At times he may see through specks or strings or hairs of light that rise or fall or pass from sided to side when the matter is disturbed. Sometimes the pieces are so large that he sees nothing.

As to the quality (ie appearance) of the matter: Sometimes it is like ashes—sometimes blue, sometimes pale, sometimes orange, or black, or flaxen. Avicenna, listed six colors by dividing the white into pearly and chalky. Bien-venu counted seven by adding green. Alcoatin counted ten, adding reddish, silvery, and glassy, Jesus counted twelve, adding quick-silverish and livid. Acanamose counted only four.

The Primitive Causes of cataracts are falls, bruises, fevers, headaches, severe cold weather, and weak (ie non-resistant) eyes. **The Antecedent Causes** can be bad fumes, vapors, and liquid humors coming up from the belly, the result of over-eating and bad digestion. **The Conjoint Causes** are the substances of the eyes themselves (ie humors and membranes).

The Signs are what we have described as their appearances. A Cataract differs from Gutta Serena in that the former blocks the pupil whereas in the latter case the pupil is empty; the name describes a cloudless sky. The blindness is due to a blockage in the optic nerve (*Internal Evils*, Part 4). If one claimed that the nerve was normal, he would have to blame the blindness on an invisible black cataract, as was Bien-venu's incorrect opinion.

The signs of a cataract before it can be detected at the pupil are dimmed vision and distorted images and mirages. Galen described the three indicators of the sources of the primitive causes (ibid). First: When the source is from the sick stomach, it affects both eyes; when the source is within the eye, only one eye is affected. Second: When the vague distortions of vision persist for several months and nothing is detected at the pupil, the source is the abdomen. Third: When the fantasies end or are intermittent at the times when the digestion has been corrected with laxatives, but the disturbed vision returns when the digestion again is disturbed, then the cause is from the abdomen. But, when the visual symptoms persist without interruptions in a person whose diet-regimen is normal, including the use of laxatives, then the source is not communicated to the eye from below.[576] Galen claimed to have learned this from letters sent to him from patients he had treated in other regions. The signs that indicate that the source is a disorder of the brain are frenzied fevers, scotomas, severe headaches, and inappropriate self-control and displays of emotion. (*Maladies and Symptoms*, Part 3).

Prognosis

Gutta Serena is incurable because it is an obstructive malfunction of an optic nerve. We must not attempt to treat for cataract when the pupil does not react when the other eye is shut, or when the affected eye is rubbed, or pressed, or receives exhalations blown from the examiner[577]. Other reasons to avoid operations are when the eye is totally blind, and the blindness has persisted for years, and the patient is too old. Do not try to needle the eye, and fail.

The prognosis is poor when the cataract recurs soon after having been couched by rubbing; one may infer that there is nothing to hold it down. If he has had some return of vision (can see shapes), and the recurrence of the cataract has not been long enough (Acanamose said up to five years), and the eye is tender, and you cannot see the cataract, all are contraindications to needling or to another attempt to couch by massage. In such cases the needle will penetrate an uncongealed cataract, as if it is water. But indications for needling are when one can see the cataract as a whitish thing that returned after the iris had been dilated[578], and the patient can still see some light or forms, and he is willing to undergo the prodedure. Furthermore you have ascertained that the signs of gutta serena are absent, when a needling would not overcome an obstructed optic nerve and fail to enable patient to see. Bien-venu described two kinds of gutta serena, the black and the

---

[576]   'Communication'. Galen conceived the vagus nerve to be the connection between the head and the abdomen through which humors and vapors could travel in both directions (LDR).

[577]   Perhaps a 'medicated exhalation'. See below, the section on Treatments, (LDR).

[578]   The compression-massage operation has failed to clear the pupil and is described as a failure to dilate it (LDR).

orange, and that the pupil always was dilated (ie clear). Alcoatin advised us not to operate for cataract when the patient has bad eyes (ie inflamed), when he has headache and painful eyes, who coughs and sneezes or has a dripping nose (ie rheum), or is vomiting or has some other annoying infirmity that will cause him to move during the operation and cause the couched cataract to come back up at once. Jesus and Alcoatin agreed that a cataract that is the result of a contusion is not a favorable candidate because the normal humors of the eyes (ie aqueous, etc.) are either expanded or diminished, and the patient's blindness is not just due to the cataract, and a successful needling will not correct it.

We cannot be certain about the substance of the cataract, and we cannot be predict successful responses to our medications, and we cannot guarantee a successful outcome after needling when it is not clearly indicated. Galen (*Miamir,* Part 4) said that although the expectations of the medications may be great, the results may be nil or meager. The same certainly holds for the results of the needlings of the eyes of fashionable people who succumb to the false promises of the nomadic operators who use instruments.

On the other hand, Avicenna claimed that when treating the early stage of cataract in an intelligent and compliant patient a proper regimen with abstinence and the prescription of laxatives, and the use of delicate resolutive collyria, may yield great improvement. But when the cataract is firmly established, surgical treatment is the only successful way.

The operations should be performed under clear pleasant skies, and when there is no wind. Jesus preferred a northern rather than a southern light, in mid morning.[579], during May or September. Acanamose added: when there are no clouds, no thunder, no torrid heat or chilly cold to upset the patient.

The needle for couching is called Elmadec in Arabic and Acus in Latin. It must be slim and extend about a thumb-nail length beyond the handle, which should not be heavy and should be shaped for easy manipulation. Bien-venu preferred silver, whereas Acanamose preferred gold. I use both kinds rather than iron, because they are malleable and not easily broken.

Treatments

The General Measures[580] are a diet regimen and certain purgations. The diet changes from a pretreatment to a postoperative variety. Before using any medications, emphasize a diet that favors the six natural elements and the three annexes which reduce the body's heat and dryness, and attend to the air (ie general milieu), foods and beverages in proper amounts, between repletion and inanitition, and attention to his emotional status, to baths and other daily habits as regulated by day or night. Those are the guides for a lifetime

---

[579]   The 'Tierce hour'. The canonical hour of mid-morning, about 9:00 AM (LDR).

[580]   Most of this section deals with 'Internal' cataracts. Guy uses 'water' and descending water' when referring to the rheum coming to the eyes as the basis for the cataract. He uses 'humors' when referring to the normal fluids of the eye, aqueous, crystalline, and vitreous (LDR).

of health. Galen wrote a book about the regimen (*The Thin Diet)* emphasizing the diet. Arnold also wrote a book to guide the King of Bohemia, and I call your attention to it, as well as to what we wrote in the chapter on cool aposthems. Here I shall interrupt our treatise, and repeat much of it.

I have often said that we should avoid three kinds of foods: First: The moist and the raw because they engender phlegmatic humors. Second: The heavy and the gassy which offend both the head and the stomach. Third: The astringents, because they challenge the stomach and loosen the bowels, especially when taken after other foods. And when the air is cool and humid, poorly baked and unleavened bread, legumes, cabbage, cheese, fruits, fatty and heavy meats, and fats. Fish and unclean water cause strangury and indigestion; they are avoided by most physicians.

Rhazes was especially averse to onions, garlic, mustard, eruca, and leeks; they harm the head and dull the vision, and they cause hot gas; Avenzoar agreed. What diet is good? Abstinence, especially at the evening meal; sobriety as to beverages, and include fennel in the diet. Democritus and Avicenna said that the venomous animals that live above-ground as well as the blind ones that live below the ground during the winter, will eat it when they emerge in the spring and rub their eyes with it to see.

The herb euphrasis (adhil, or eye-bright)) is a wonderful stimulant (Heben Mesûe*)* for vision. Avenzoar favored a broth of turnips in which headless doves had been boiled, and eating beef-liver, and using the gravy as a collyrium. It moistens the eyes that suffer from night-blindness (Galen, *Miamir,* and *Medicaments*, Part 11). Avicenna added long peppers and sel niter. Rhazes advised the patient to exercize his eyes by reading small letters and by examining pictures to strengthen his vision Avenzoar advised immersing the eyes in water that had been boiled boiled with saffron.

Avicenna advised bathing in a clear stream and repeatedly immerse the open eyes, and stay for one hour, especially in the morning during the summer. I prefer to pour warm water, mimicking a water-fall-rather than to immerse in a cold stream. I use an orange or green glass jug. As for the exercize, I instruct the patient to stare at the eyes of a wild donkey to disperse his own cataract.[581]

Most authorities used a metal plate as a mirror to shine light into the eyes. Arnold said that the greenery of a living thing, the transparency of water, the shiny colors of gem-stones, and the clear sky above all are comforting sights and resolve the water (ie the cataract).

Massage the feet and scratch the head with pine-resin to divert the rheum from the head away from the eyes. Have a child, chew some fennel seeds or other tangy herb, and blow into the patient's eyes. That will digest and consume the water (ie the fluid of the cataract), as experience has shown.

Other decoctions of seeds help to get rid of the superfluous water (ie the cataract). Taddeo prescribed a troche of seeds of fennel, anise, ammi, mountain-ash, ginger,

---

[581]   I assume that staring into the shining dilated pupils of an angry donkey was an adjunctive treatment by suggestion (LDR).

cubebs, cloves, peppers, nutmeg, roots of chelidonium, euphrasia, rue, betony, and royal haste, or he made a powder or an electuary, to be taken dry, morning and evening for the same purpose.

To evacuate digested matter use hierapicra or cocchia pills, or another suitable laxative. Then purge the head with diacastorea tempered with the juice of marjoram. Avicenna said that we should use a special purge of the stomach at the onset for patients whose cataract was preceded by visual aberrations (ie fantasies). Then purge the head with gargles, head-purges and chews, all of them potent and dosed several times. He added that it is a good practice in general to clear out the belly first when treating all maladies of the eyes.

The diet regimen changes for patients with established (ie mature) cataracts, when the diet described above has failed to change the course after a cataract was suspected, and has reached the stage when an operation will be indicated. Now let him eat fresh fish and moist meats to prepare himself for the operation.

The General Regimen after a successful operation includes seclusion, quiet, and rest in bed with the head elevated. The diet should be spare and not require chewing, to include broths, soft eggs, and water (Jesus) or sour wine (Acanamose).

The Early Particular Measures before the cataract has congealed includes liquefiers, incisives, and consumptives. Avicenna used fennel, honey, and oil as mitigatives, expecting that the oil would have balsamic effects. Galen (*Miamir*, and *Therapeutics* Part 14) used a collyrium of myrrh, frankincense, galbanum, saffron, et al.

Rhazes used a collyrium of the bile of cranes, falcons, eagles[582], cows and one or two others. Let them evaporate in an exposed copper bowl and add colocynth, serapinum, euphorbia, and the waters of fennel and rue.

Jesus favored a collyrium of cow-bile, asafetida, balsam, concentrated by evaporation in a glass bowl. He also used a collyrium of borax, juices of fennel and rue, and the basilicon, Master Peter of Spain's (ie later a Pope) water, and others that improve vision.

The Particular Measures for Mature Cataracts: When the patient is deemed suitable for needling, after he has been purged and bled (if the surgeon wishes) and he has been treated with restrictive plasters pressed against his temples and forehead to divert the inflow of humors away from the eye. He is young, and is a willing patient. He is healthy and not over-anxious. The weather is clear, the hour is 9:00 AM (ie tierce), the moon is increasing and is not in Aries (ie zodiacal). The opposite eye is covered with a bandage and he is seated on a narrow bench, his legs straddling it, as if he was in a saddle. An assistant sits behind him and holds-tight his head. The surgeon has prepared himself by chewing seeds of fennel, garlic, or other aromatic herbs. He sits astride the same bench, slightly higher (ie on a pad). The patient thrusts his hands under his own thighs and the surgeon's knees snugly embrace those of the patient. He will use his left hand to needle the right eye, and his right hand for the left eye. He hold open the lids on the operated side and gently exhales three

---

[582] Birds have keen eyes! (LDR).

or four times into the open eye, to warm the cataract and to loosen it. He aks the patient to turn his eye inward to stare at his own nose, and to hold it so. Then, with a prayer to God, the surgeon introduces the needle far enough under the conjunctive so he can see it clearly, with care to avoid veins. He directs the point of the needle through the cornea so he can see the tip, which he pushes into the center of the pupil, or just above-center. He catches the cataract and pushes it down and holds it there long enough to recite the Pater Noster or the Miserere three times. If after release, the cataract rises, he repeats the maneuver, taking care not to tear the iris or touch the lens.[583] When it remains below and there is no need to re-couch it, he withdraws the needle by reversing the directions of the insertion.

You may make the patient happy by briefly exposing the operated eye and covering the other side, and asking, "What do you see?" Offer God's blessing, and introduce some egg-white and bandage both eyes to keep them at rest. Were he to move the healthy eye, the other side will move as well. Put him to rest in a nearby bed, and follow the regimen described above. Deny him food for one or more days. Then he may eat as before. Feed two meals a day with his eyes covered for nine days. Then gently wash the eyes with cool water and allow the patient to move them, briefly at first. Meanwhile, if the cataract had come up or there is pain, relieve the pain, and re-couch the cataract through the original puncture if possible.

Jesus and Avicenna favored displacing the catarct to lie between the cornea and the uvea (ie in the anterior chamber). I agree with Alcoatin and Bien-venu; that was unwise.

Albucasis and Avicenna both described an ancient operation in which the puncture was made with a hollow needle, from below, allowing the 'water' to be sucked out. I think that, too, was a bad maneuver; the aqueous humor will escape along with the water of the cataract, and simply will compound the original malady

Maladies Of The Interior Of The Eyes That Interfere With Vision

Impaired vision may be due to defects of parts within the eye.

The First Source may be the pupil: too large or too small. Avicenna called the dilated pupil Alintizor. Galen (*Maladies and Symptoms,* Part 4) said it was nearly always harmful (ie for vision), whereas the too small pupil, called Constricted, was praised in babies, but later on it was blamed.

Second may be the aequeous humor (ie albugineux). It may be too dense or too thin, or be discolored and cause distorted images and cloudiness.

Third is the lens (ie crystalline humor). It may suffer the same abnormalities as the aqueous humor, or it may be displaced.

Fourth is the vitreous humor, Its defects are the same as the aqueous.

---

[583] Guy's cataract was a membrane in front of but not part of the lens. The word itself defined the concept of a liquid, a 'water' falling from above over the front of the lens, indeed a cataract was a waterfall. Guy often used the word 'waters' (eaux) instead od cataract. This translation uses cataract rather than 'eau' when in context (LDR).

Fifth: The Visual Sense itself may be defective. One may be able to see large and dense objects in the distance but not up close. Or he can see large and small things up close but not when afar unless they are large and dense. Or when they are dense but small, he may not be able to see them at all. Another defect is Nyctalopia[584] when one sees poorly after sun-down (Rhazes). The opposite is called Alihahan, when one sees better at night.

Sixth is obstruction of vision by or in the optic nerve.

Causes

An accumulation or a deprivation of the substances of the various internal parts may cause defective vision. Or the defect may be a faulty communication between the eyes and the brain, or the stomach, or with the body as a whole, as seen when the visual defect follows strangury, indigestion after eating heavy and gassy foods—as noted as causes for cataract—or when the patient sleeps too much, especially after meals, or sleeps with his head flat, or when he sleeps while wearing his shoes[585], or keeps to himself in dark rooms, or exposes his eyes to cold, to dust, or to fumes, etc.

Sometimes, the cause may be unanticipated complications of treatments: too much laxation, or cuppings at the nape. Other potential causes are too frequent coitus, or too much toil, or too much tearful sobbing, or prolonged illness, or prolonged open-eyed exposure to flames or to bright sunlight or moonlight, or over-use of salt, peppers, strong spices, et al.

The Signs of weak vision are obvious to the patient, but the sources must be researched.

Treatments

Most treatments are prescribed by the physicians, and there is little need for manual procedures. However, in *General*, when the loss of vision is an unexpected complication, we can help the patient in the prescriptions for rest, relaxation, diets of moist food, baths, and fomentations of the eyes and upper face with warm water or milk. We can urge abstinence when there has been abuse, and we can administer purges as we do for cataracts. We can use collyria to clear the eyes, composed of the bile of rapine birds and balsam. Here, I will conclude the doctrine with three recipes that I use for many maladies in addition to poor vision. All were taken from Heben Mesûe.[586]

---

[584] Nicaise noted that so-called Nyctalopia was called Hemeralopia in his time, and that it described blindness in day-light (EN).

[585] This brings to mind a caveat when I was a schoolboy, not to wear overshoes when in school lest that lead to impaired vision later in life (LDR).

[586] Heben Mesûe (ηεβε =young) an 11th C herbalist-pharmacologist, from Baghdad, d.1015. This younger Mesûe is to be distinguished from M. The Elder (9th C), who was John Damascenus, and M. the third (13th C), a surgeon. All wrote in Arabic, although H.M was a Christian. Nicaise noted that Guy cited Heben Mesûe sixty times (LDR).

The First recipe is a pleasantly flavored electuary (alfarti) that relieves the body of superfluidities, especially the head. It comforts persons who have poor vision or other sensory defects. It slows the appearance of gray hair, and retains the aspect of youthfulness. It contains: powdered peels of myrobalans, chebules, and emblicus, all ground with the oil of almonds that have been washed with sugar-water until their bitterness has been ablated and dried. Then add clean white turbith, mastic, reglisse, ginger, galanga, canelle, wood of aloe, cloves, peppers, cubebs, mace, the fuzz of spikenard, seeds of fennel, powdered sugar, pine-seeds, and foamed honey. Dose one-half of the electuary after midnight once or twice a week.

The second electuary improves vision: euphrasia, seeds of rue, sesali, calamint, pouliot, ginger, cubebs, nutmeg, crystal (ie ground glass), pearls, esula, mastic, incinerated snake-meat, saffron, balsam, and sugar-loaf. Dose half of an electuary every morning.

The third is an irritating collyrium to sharpen and preserve vision: juices of fennel and rue, pomegranate wine, honey, long peppers, socotrine aloes, sal armoniac, and tuthie. Grind all and put it in a glass jar to be set in sunlight for three months. Then discard the lees and keep the rest.

Also, you may use a decoction of fennel, rue, chelidonium, euphrasia and vervain. That was Peter of Spain's (ie later Pope John XXI) recipe.

As a last resort, you may use spectacles made of glass or beryl.[587]

Section Three Maladies Of The Ears

Deafness[588]

Just as maladies of the eyes cause disturbed vision, so do disorders of the ears cause impaired hearing: either a loss of it or the distortion of the sounds. Galen (*Maladies and Symptoms,* Parts 3 and 4) said that the ear is a nervous conduit for the sense of hearing, and its functioning parts were the canal, the ear-lobe, and the agent of hearing itself (ie the nerve) that goes to the brain.

---

[587] Nicaise explained: Beryl was the gem-stone, similar to emerald, described by Marbode, Bishop of Rennes (d. Sept. 11, 1123) in his work on gems He wrote that beryl helped weak vision, a text that is not easy to interpret. Littre wrote that beryl was used as a magnifying glass or in spectacles. Trotula (12th C) also raised the question of glass spectacles in her *de Mulierum passionibus.* And Gordon (*Lilium medicinae)* mentions glass and beryl spectacles (EN).

[588] Brugsch said that the Egyptians during the reign of Rameses II- the putative father of Moses, during the 14th C BC- was the first to mention disorders of the ear. Later the topic was dealt with by Hippocrates, Appollonius, Galen, Celsus, Alexander of Tralles, Paul of Aegina, Rhazes, Albucasis, Serapion, and William of Saliceto. He said that Pierre of Cerlata was the first to mention an ear-speculum, and that maladies of the outer ear were described during the early medieval years. Guy used the speculum and described disorders of the ears as connected with maladies of the brain (EN)

The various maladies include dyscrasias-especially related to cold (Avicenna)—wounds, ulcers, blockage by an abscess or by liquids or gas or pus, or blood or by detritus (ie ordure), worms, warts, seeds, ashes, or by a congenital web (see Fn 592).

The obstruction may enter from outside the ear, as spider, lice, pebbles, nuts, seeds, powders, or water.

The Primitive Causes may be falls or bruises, or bad habits. **The Antecedent Causes** are liquid or gassy humors. **The Conjoint Causes** derive from the ears or from anywhere in the body, especially the head and the stomach. As for maladies of the eyes, those for the ears may be compassionate or communicative.

The Signs

Some signs are visible to us, and others must be inferred from the various kinds of deafness, Some have lost low tones, some high. Some persons are slightly deaf or are slowly losing their hearing (*Miamir,* Part 3). Some hear strange sounds such as dripping water or rainfall, or trumpeting noises, or continuous tinkling. et al. Some hear distortions of normal sounds.

The causes are clues to the symptoms. Examples: When there is an abscess, there will be fever and pain, and bursting discomforts, sluggishness, and throbbing, and chills. When the matter is warm, the patient will feel prickling and the deafness may be preceded by bleeding, or by a choleric fever, while the pus accumulates.(*Aphorism #4*). When the cause is gassy, the ears will ring. When there is an ulcer, there will be pain and itching. When there is a wart or some foreign material in the ear. The patient may know what it is, and you may be able to expose it by tugging the lobe to enlarge the opening to the canal for you to insert a speculum and use bright sunlight.

When the ear seems intact, and deafness is the only symptom, you may infer that the nerve is at fault. But when other parts of the face are awry, we know that the brain is affected, and the ear canal is healthy. Avicenna said that nerve-deafness was the diagnosis when it was not of sudden (ie recent) onset and it has persisted after a good evacuation and a proper regimen.

Prognosis

When the above status (natural causes) is confirmed, we must seek causes of blockage of the outer ear, especially when the impairment has lasted longer than two years. If we encounter a hard mass (ie not an abscess) or a hard scar, the condition is incurable.

When the deafness is a recent complication and the cause is identified, it may be cured. Hippocrates (*Aphorism #4)* wrote that when the deafness was a complication of diarrhea due to choleric humors, that it will end when the diarrhea does. Galen explained that when a diarrhea was accompanied by high fevers, the choleric humors were wafted upward to the head.

Treatments

The General Measures are the diet-regimen and purgatives and the relief of earache, as we described for treatments of abscesses of the ear.

The Particular Measures are twofold. First are the *Eight Requirements*.

First: Do not attend to the ear, especially with irritating and painful substances before completing a purgation of the body.

Second: Be gentle, especially when working within the ear, and do not injure the membranes (See Fn # 592).

Third: All applications within the ear should be tepid: neither too hot nor too cold.

Fourth: All medications must be liquid so they may easily escape.

Fifth: Leave nothing that you introduce into the ear longer than three hours.

Sixth: After introducing your medicine, have the patient recline with the healthy side down. Cover the medicated ear with cotton fluff or wool fleece.

Seventh: Introduce no medication to follow another until the first has been completely removed. Use a cloth or cotton-tipped probe to swab it out. The patient may help during the cleanout by coughing and sneezing.

Eighth: The surgeon must possess the special instrument for removing foreign material: curved hooks and tubules for suction and fumigation. He will need supplies of wool and cotton fluff, sponges, bits of cloth, glue, etc.

The Specific Treatments

Refer to the chapters on Abscesses and Ulcers.

When cold and wind cause the hearing-loss, fumigate the ears through a funnel and cannula, producing the steam in a kettle with a narrow spout. Galen and Appollonius used bull-urine with myrrh as the fumigant; they covered the kettle with a cloth hood to direct the fumes only to cannula, so not to injure the external ear or its region.

Most physicians use another fumigant of white wine boiled with rue, calamint, hyssop[589], marjoram, centaury, betony, leaves and seeds of laurel, stoechas, fennel, spike-nard, et al.

After the fumigation, Appollonius introduced into the canal a mixture of goose-fat, beef-bile, and oil of laurel. More commonly used is a warm oil containing camomille, aneth, bitter almonds, costus, nard, radish, ash-tree leaves, and filtered water. Others prefer a decoction of joubarb fortified with falcon's bile, balsam, and the pure juice of radishes.

Rhazes and Heben Mesûe used this collyrium: pulped colocynth, aloin, aristolochia, costus, borax, and castoreum. They added cow-bile as a liquid and used the oil of bitter almonds as a diluent.

---

[589] Nicaise spells it Hysop, which is lanolin, the grease of lamb's wool, which is not an appropriate addition to these and similar recipes. I believe that the intended substance is 'Hyssop', which is the herb hasca, an aromatic mint. Threfore, I shall use hyssop where I deem it proper (LDR).

[590] I assume that the crunching sounds were meant to stimulate hearing (LDR)

Avicenna used a troche of castoreum, niter, hellebore, and radish-juice.

Alexander had the patient walk on a sandy ground[590], and Avicenna urged the patient to shout, for the same purpose. On the other hand, Galen (*Miamir,* Part 3) said that ears that have suffered pain need rest (ie not a walking exercise), and others have said that they need active exercise and to be restless in bed.

When the impairment is caused by an accumulation of ordure (ie wax, et al.), soften it with warm honey, and remove it with a curved stylet. Then insert a strip of cloth wet with honey and the oils of camomille and spikenard, or with nasturtium and borax.

If the cause is water trapped in the canal suck it out through a funnel, or use a smaller cannula in children. Then introduce some oil of sweet almonds, or apply it on the surface of a cannula or on a wisp of charred cotton (while warm). The heat will favor the discharge of the water. Or insert a strip of sponge attached to a sturdy cord, and remove it after it has absorbed the water.

When an embedded foreign body is lodged within—a pebble, a nut, an insect, etc.—Albucasis first introduced some oil of violets, and provoked sneezing or coughing, or had the patient spit and jump on the foot of the affected side, and he slapped the ear while tugging at the ear-lobe to enlarge the orifice.

Failing those actions, he put some glue on the tip of a probe and placed it against the surface of the object, to be withdrawn when the glue adhered. Failing that he tried to grasp it with a hook or forceps.

Needing yet more, he sucked at it through a cannula surrounded with oil and wax (ie to seal the cannula).

As a last resort, to avoid an inevitable abscess or a convulsion, we make a semilunar incision at the base of the ear, deep enough to reach and dislodge the foreign body (a pebble). Then we suture the incision and treat the wound.

When the object was a seed or any object that can swell, and Albucasis had failed with the simpler measures, he morcellated it with a narrow spatula (ie a chisel) and removed it piecemeal with a forceps.

When the object is an insect or worm, and he could not dislodge it alive, Albucasis (*Divisions*) instilled a water of aloes or the juice of calamint or aloin or a strong vinegar. Then he removed the dead object with an instrument, or filled the ear with water or hair, as did Bruno, or with brine, as did Avicenna.[591]

When the obstruction is due to an exposed wart or polyp or granulation tissue, I chisel off the polyp with a spatula. A dangling wart or redundant tissue may be ligated or sliced

---

[591]   Guy's citation of Bruno was inaccurate. When the object was a louse, Bruno introduced some lamb's fleece. After an hour, when the louse had attached itself to the fluff, he removed it. Or he filled the canal with water (to drown the insect) and had the patient shake his head vigorously to expel it (J).

[592]   I doubt that Guy meant the tympanic membrane. He made no mention of it or the ossicles in his Anatomy, nor did any of the writers before him, from Galen on (LDR).

off after catching it with a hook. Or you may use a hot cautery or corrosive. After any of the above, insert a drain covered with the green ointment or some other mild corrosive.

When the lesion is within the canal (beyond easy view) the treatment is difficult. Albucasis used a radial cautery with care to avoid the nerves. [592]

Section IV. Disorders of the Nostrils

Many diseases of the nostrils are destructive or disfiguring, and they interfere with the sense of smell, and with breathing, which are their prime functions. Other dysfunctions are complications rather than primary, such as dyscrasias, ulcers, foul odors, shriveling, abscesses, obstruction by swelling (accumulated humors), or granulation tissue, or polyps, runny-nose (coryza), and epistaxis (bloody nose). Some of the maladies arise in the nose; others begin elsewhere. Most of them belong in Medical Treatises, and many will be described in the later chapters. Here we shall limit the discussion to cathesial obstruction[593] and foul odors.

Cathesial Obstruction

Avicenna said that it is engendered by humors, granulation tissue, and crusts that collect in the space between the nose and the throat (ie the nasopharynx). That makes the patient hawk sputum and obstruct breathing when the mouth is closed. The patient snorts and he may be nauseated.

The Particular Treatments are fumigations and aromitizations, and snuffing up a water in which were boiled camomille, melilot, hyssop, marjoram, et al. Also we use a funnel to introduce via the nostrils small pills of diacastoreum with added juice of marjoram. Or we use Rhazes' confection that he designed specially for this purpose; it was praised by Heben Mesûe; nigella set in vinegar for three days and evaporated. The dry residue is powdered for use in a collyrium. Make a confit with old oil. Clean the nostrils with water, and instill drops into the nostrils when the head is tilted back and the mouth is filled with water. The patient sniffs strongly and moistens the interior of his nostrils. Repeat the treatment three times daily for three days. After each treatment, rinse the mouth with warm water. If the maneuver causes pain or inflammation, introduce some oil of melon-seeds, and splash the face with warm water. You may fortify the collyrium with colocynth, hellebore, cyclamen, and sel ammoniac. You may prescribe confits of crane-bile and camel-urine for the same purpose.

Avicenna said that he often had to remove the accumulations and crusts by sawing them off with a knotted string, as we described the method in our chapter on ulcers of the nostrils. Or he scraped them with detergents placed on the knots. Sometimes you will bring out so much bad matter as to dismay the patient.

---

[593]   Cathesial: coming down from above (ie the brain) (EN).

When nothing else works, dilate the nasopharynx by packing it through the nostrils with thin strips of cloth wet with gentian or the juice of acoris-roots, as did Master Pierre de Bonant.

Foul Odors

This refers to malodorous exhalations from the mouth (halitosis) and the nose. In may arise on site when there are corrupt tissues or foul ulcers in the nose, in the gums, or from decayed teeth or adjacent structures. Sometimes the stench comes up from the stomach, or out of the brain or from within the chest, as Galen describes in *Maladies*.

Signs

When the chest is the source of the foul breath, the patient also coughs up and spits pus. When from the stomach, the foulness may be intermittent, before or after a meal. When from the brain, one can smell it when the patient's mouth is full of water.[594] Furthermore, the foul odors were continual and disturbed the patient.

When the engendering humor is warm, the stench is hot and bitter; when the humors are cool, the odor lacks those qualities.

Prognosis

When the stench is offensive even to the physician, and when the disorder is caused by constriction of the filtering channels, it is incurable, especially when the nose is collapsed.[595] The same holds when the odor is that of rotten fish. And when the patient is feverish, the outlook is bad.

Treatments—Two Elements

The General Measures are diets and purges, suited to the corrupted humors at fault. In general it is better to abstain from all things that favor suppuration, such as moldy bread and decayed meats which are mushy, pale and glutinous; avoid poor wines, especially when they are sweet and dark; avoid fish and dairy foods, soft fruits and vegetables such as melons, cabbage, spinach, fleshy beans; broths, soups, garlic and onions, all of which cause bad breath.

---

[594]   The brain communicated with the nasopharynx via the olfactory foramina and passages through the cribriform plate, et al. The brain drained itself of rheum (catarrh) and other emanations, including the foul ones, as is here. Some of the latter came up to the brain from the stomach, via the vagus nerves (LDR).

[595]   Guy implies that the cause is leprosy, where the collapsed nose was a prominent symptom (LDR).

All sour things are good, such as vinegars and pomegranate-juice. Dry meats are good, such as from partridges and small birds. Favor things that dispel vapors (ie belching) after meals (as deserts), such as quince, pears, and prepared coriander (Arnold). Rhazes said that parsley was best of all when eaten frequently. Also good are twigs of myrtle ground with dry raisins (Avicenna). Sage and marjoram are among the principal herbs. Good aromatic sweets were made by Heben Mesûe; they taste like our rosat, and were dragees made from seeds of cloves, vinegar, canelle of Aleppo, souchet, iris, leaves of nard, aromatic rose petals, lemons, myrtle leaves, aloe-wood, sandalwood, and tablets of rosat-sugar.

Good habits as part of the regimen are sobriety, limited sleep, ambulatory exercise, and frequent baths. Use myrobalans laxatives, and phlebotomy when needed. As an additional measure to divert the humors and favor resolution we use touch-cautery over the coronal suture of the cranium.

The Particular Measures vary according to the timing and causes of the foul odors. When a gangrenous limb, an infection, or an ulcer are causes, treat them. The same holds for excrescences, pustules, and obstructions. Treat as described in the chapters devoted to them.

In all cases, clean and deodorize the nose and the mouth and the throat. Heben Mesûe first washed and rinsed the nostrils with scented wine, then he used this recipe: a decoction of myrrh, souchet, Roseau, aloe-wood, roses, myrrh, and rock-salt. Rhazes used gallie (a musk), spikenard, cloves,and the urine of a donkey. Then Mesûe inserted strips of cloth wet with the following paste: hierapicra, aromatic reeds, myrrh, cloves, leaves of spikenard, squill, canelle, and wine. Avicenna used palliative pills when no other medication is effective: cloves, galanga, pyrethrum, aloes, mustard, musk, camphor, and wine. Every morning insert a pill into each nostril and hold two in the mouth (ie not to swallow before it melts).

Gordon made this paste to put on the strips for insertion into the nostrils by adding to the foregone: marjoram, basil, marjoram, canelle, aloe-wood, styrax, calamite, amber, and rose-water.

One may chew souchet to disguise the breath-odor of wine, or chew rue to disguise the odors of onions and garlic (Rhazes).

We dealt with polyps in the nose, and with epistaxis in the chapters on ulcers.

V. Maladies Of The Mouth

Galen wrote (*Maladies and Symptoms,* Part 4) that the parts of the mouth, as it is for other regions, each has its own symptoms. The mouth has two principal parts and actions: the tongue for tasting and the teeth for chewing. The other parts are subsidiary. The mouth's functions are controlled by the brain, and its symptoms vary in three ways: degrees of feebleness, absence, or corruption. As is the case for all simple, organic or mixed structures, the symptoms vary in intensity, especially as to debility, ablation, and lack of changes. Galen, Avicenna, and Avenzoar all agreed in that assessment.

The Causes sometimes are specific, sometimes in common, as elsewhere in the body, all to be described.

The Tongue

The functions of the tongue are impaired by dyscrasias, ulcers, apothems, edema, ranula, spasms, and wry-tongue, softening, and stammering.

Swelling and Enlargement

When caused by warm humors, prescribe a healthy regimen and proper laxatives (cocchia pills). Galen (*Therapeutics*, Part 14) prescribed gargles of lettuce-juice, and that may be all that is needed. But when more is called for, divert the humors at fault by cupping at the nape, and by phlebotomy of veins under the tongue.

When caused by frothy cool humors, Rhazes and Avicenna tried to eliminate them by inducing salivation and by rubbing the tongue with sal ammoniac and onions. Then they rinsed the mouth with vinegar, and added ginger and pepper as resolutives. Also, they dehydrated the head with medications that desiccate, as we use them for rheum.

Ranulas and Excess of Sublingual Tissues[596]

Avicenna described an oval mass under the tongue that limits its motion, that resembles the tadpole of a frog. He treated it with applications of astringent resolutives such as pomegranate-peels, oregano, and salt. In children he used combusted vitriol, hermodactyl, and egg-whites, to be retained under the tongue. He reported successes.

Failing that, he went to more intensive measures: rubbing the mass with sal ammoniac, bronze-flower, and with troches of aldaron and calidicon, with care that the medicines were kept in the front of the mouth.

Failing those measures, we operate. Albucasis' technique: Expose the open mouth in bright sunlight. When the mass is very hard and dark (black or brown) and is firm and insensitive, it is cancerous. Leave it! If it is pale and soft, catch it with a hook and dissect away the covering membrane (ie mucosa) with a spatula, and shell out the mass, sponging away the blood. If the bleeding is vigorous, apply zegi, but do not leave the patient. Control the bleeding and finish the job, and repeatedly irrigate the field with vinegar for a day or two. William of Saliceto used wine as the irrigant, after heating it with myrrh.

Spasms and Tongue-tie

A spasm curls the tongue back towards the base, and disables it.
The Cause is a congestion, an inanition, or a retraction.

---

[596]   Guy is not clear. The sublingual mass may be a benign or malignant neoplasm of a salivary gland or a cyst. Guy quotes Albucasis almost verbatim (see bibliog., Albucasis, p 298) (LDR).

The Treatment for congestion is with purgations, both general and special. The General uses cocchia pills. The Special uses head-purges, chews, mouth washes with calamint, vinegar, mustard, fenugreek, peppers, pyrethrum, and others such as we prescribe for incontinence (uncontrolled fits) and paralysis. We may apply an evaporant to the neck: camomille, melilot, stoechas, and aneth.

The treatment for dryness or inanition uses humectants and a diet of rich foods; gentle mouth-washes, and inunctions of the back of the head with oils of nenuphar and lilies, and fomentations with warm water and milk.

Treat tongue-tie by dividing the restraint, and applying pads moistened with vitriol for several days to prevent recurrence. If nearby veins make you reticent, do as Avicenna and pass a threaded needle under the restraint, knot it and let it slough through on its own. Lanfranchi used a red hot cautery as the cutter.

Paralysis and Stammering

Although paralysis of the tongue may follow a convulsion, ulcers, or other disorders of the tongue, more often it is the result of a local paralysis due to a blockage and accumulation of humors in an obstructed nerve, or in the muscles of the tongue.

The Cause and **The Signs** are those of paralysis. Also, there is uncontrolled salivation and inability to initiate speech or to pronounce words. Galen said (*Aphorisms*) that stammering is a kind of lingual diarrhea.

Prognosis

When stammering is a complication of paralysis, the prognosis is as described for that. Inasmuch as fever can cure a paralysis, we treat stammering with humectants. But when the paralysis and stammering are longstanding, we may not be able to cure it completely. When it occurs in children, often it improves during adolescence.

Treatments

In addition to *The General Measures* used for treating paralysis, there are *Three Special Measures* and special goals, all described by Heben Mesûe.

First: Divert the matter with irritating enemas, massages, and cupping over the nape of the neck.

Second: Relieve the congested head with desiccative plasters applied over the entire cranium. Use mustard, dove-droppings, millet seed, roasted salt, laurel-berries, anise, fennel, peppers, cloves, et al. They will comfort and dry the rheum. Apply touch-cauteries on the ribs and over the vertebral spines in the neck. Halyabbas applied this plaster on the nape to comfort the nerves: camomille, melilot, marjoram, ginger, mustard, pyrethrum, laurel-leaves, opoponax, castoreum, wax, and the oil of elder-berries. He also used an inunctions of costus, nard, rue, castoreum, and Holy oil (very important). Also he applied an oil of terebinth on the back as in his treatment of paralysis; here it had a special purpose.

Third: Consume the conjoint matter with gargles and mouth-washes, and rubbing the tongue with what follows, graduating the intensity. Begin with oxymel and squills, as an excellent gargle. Add stoechas or domestic hyssop as comforters, and add the shells of capers, pyrethrum, ginger, and the three peppers.

Evacuate the phlegmatic humors before prescribing the gargles. Use tongue-rubs with sel ammoniac, ginger, and onions to improve the effectiveness of the mouth-washes.

Heben Mesûe prescribed this gargle to resolve the phlegm at the back of the tongue[597]: oregano, marjolain, hyssop, pyrethrum, ginger, three peppers, canelle, costus, and nigella. He added oxymel and squills to make a plaster. He made a tongue-rub of it by heating it with wine, and he use it as a daily gargle.

Lanfranchi claimed that he had restored a woman's normal speech with a paste of stewed figs honey and euphorbia. He placed small amounts (the size of a small pea) under her tongue.

Rhazes used this paste for treating paralysis and clumsiness of the tongue: sal ammoniac, pyrethrum, staphisagre, mustard, pepper, and acoris. He rubbed it on the tongue and placed it beneath, several times a day.

Halyabbas used this for the rub: hierapicra, mustard, and pyrethrum. He also used Dioscorides' recipe and added acoris to soften a rigid or paralyzed tongue: sage, rue, calamint, rosemary, seeds of basil and wild centaury (the latter being the most important.) Also, he prescribed pills of castoreum, asafetida, and terebinth to be placed under the tongue. Also he made candy pills of anacardus, diacastoreum, and theriac.

We urge the patient to make long speeches and to rub his own tongue with rock-salt. Avicenna encouraged children to pronounce words.

The Teeth In General

Halyabbas, in his *Royal Disorders*, Part 1, Sermon 9, listed five or six disorders of the teeth: toothache, decay, congelation, numbness, bad odor, loss of teeth, and looseness. The duration of each is less than that of aposthems, although, as Avicenna wrote, there are similarities, as to decay, and suppuration of the gums, as taught in modern schools. Are the teeth themselves sensitive structures, and can they feel pain? Galen (*Miamir*, Part *5)* and Avicenna (*Canon*, Part 3) both said yes. But Halyabbas insisted that was not so for the tooth itself; sensitivity is a property of the nerves that come to the teeth via the third nerve. Galen also said (*Usum.,*Part 7) that the teeth themselves cannot form aposthems, and that their sensitivity comes from the gums and the nerves thereof.

---

[597] Be reminded that Guy called it rheum and catarrh elsewhere in his text, and said that it came from the brain into the pharynx (LDR)

The Causes of the ailments and the damage are their intrinsic bad complexions, their lack of attachments (ie solutions of continuity) and abscesses. Therefore, some causes (ie conjoint) are in the teeth and their attachments.

Sometimes the causes are tied to the brain, the stomach, and to other organs. Therefore, the causes may direct our treatments. The Primitive Causes may be falls, blows and bad diets. The Antecedent Causes are superfluous humors. The Conjoint Causes involve the inherent quality of the teeth and their attachments.

The Signs are obvious. One may see eroded, pierced or discolored teeth. They may be overly sensitive to heat or cold. The pain may identify the source. (Galen, *Miamir*, Part 5). Some other information will tell us what to avoid and will be helpful: the weather, the prior life-habits, the patient's situation—all of which was described in our chapter on gout.

Prognosis

Of all Man's ailments, toothache is the most grievous. Obvious inflammation of the jaw (ie external redness and swelling) may be a good sign in a case of toothache; it indicates that the pus has left the nerve and ligamenst and has spread to the soft tissues (ie where it can be drained), as we explained in our chapter on gout.

Treatments

The General Measures include an improved life-style, and proper evacuations. There are six elements involved in the life-style. First: Avoid things that provoke suppuration, such as fish and dairy foods. Second: Avoid extremes of heat and cold in the medications, especially when one follows another. Third: Do not chew hard things, such as bone. Fourth: Avoid foods which you know can harm the teeth. Fifth: Care for your teeth gently rather than roughly. Sixth: Brush your teeth with honey and burnt salt. If you add vinegar it will do everything that is called for (Halyabbas, *Royal Dispositions,* Part 2).

The laxatives should be suitable for the case. Phlebotomy should be from a cephalic vein or from veins under the tongue. Use massages, cups and head-purges as diversions. Dry the rheum and comfort the head, as we have instructed more than once. Resolve the phlegmatic humors with pyrethrum, mastic, et al.

The Particular Measures 1. Although the physicians have deferred the removal of teeth to barbers and special tooth-extractors, the most reliable performers are the physician-surgeons. 2. The physicians in charge of a case should be aware that there are several techniques for carrying out their requests, including mouth-washing, gargles, chews, filling teeth, evaporations, inunctions, massages, head-purges, instillation of drops in the ears (ie for treatment of dental maladies), as well as the manual operations. 3. Albucasis insisted that the dentist have all the necessary equipment: razors, rasps, spatulas (straight and curved), simple elevators or double-armed, gum dissectors, toothed forceps, probes, cannulas, augers, and files, and others when needed.

Toothache In General

When toothache is a complication of another malady, treat that first. If an abscess in the gums is a cause, get rid of the pus and divert it (ie repercussives). If the matter is warm, use cool astringents, such as vinegar-water swished in the mouth, or waters of roses or plantains fortified with some camphor. Another wash: oil of roses or myrrh or onfacium. After the early stages, add resolutives to the foregone, as mastic and dry raisins. If those medication fail to relieve the pain, add some opium or other narcotics.

When the bad matter is cool, at first use rosat oil with mastic. Later, add styptic wine, and then add a decoction of hyssop and calamint.

When the matter suppurates, use decoctions of raisins, figs, and seeds of flax and fenugreek.

When the cause of the pain is in the root of the tooth and involves the nerve and the ligaments, and there is pus, purge and resolve it with the foregone recipes. When it is warm, especially use rosat oil with camomille and aneth. When it is cool, use the oils of ben and nard. If there is gas, dissipate it with a decoction of cumin, laurel-leaves and seeds of rue, galbanum, and serapinum. After the pus has been properly eliminated by the opposites, and the pain persists, even after the application of salt, alum, and the ashes of galls, and after a dry affectioin has been humidified with ram's lard, then you should apply boiling oil with a cotton-tipped probe, directly on the tooth, repeatedly. Or use an actual cautery, or remove the tooth with your instruments.

Lest the Reader be dismayed by the multitude of medications that have been advocated for treating toothaches, I shall set apart here those that I have used with success as resolutives and alterants, and a group used as stupefacients.

Galen (*Miamir*, Part 5) said among the medicines for toothache, the resolutives and the repercussives should be very potent, and their major component usually has been very strong vinegar. They would be of little use if the vinegar injures the teeth; but the vinegar loses its vehemence when it is properly mixed with other substances. Avicenna (*Treatment of Headaches,* Part 3) said that the cooling effects of vinegar can easily be tempered without losing its penetrating and incisive qualities, when used for treating cool matter. But there is no better treatment for warm matter. Galen agreed (*Simple Medicaments* and *Miamir)*. Archigenes, cited by Galen idem. Part 5) placed vinegar at the top of his list for treating toothache, using it warm and mixed with galls.

Recipes For Treating Cool Matter in Toothaches

1. Parietory and mercuriale (moderately combusted), salt, burnt alum, oregano, iris, peppers, pyrethrum, costus, mustard, sesali, hyssop, dry mint, deer or bull horn, amome or cinnamon. Grind all to a powder and rub it on the gum over the root of the tooth. The gum should be moist.

2. Rhazes uses a medicine that warms. Seeds of purslane, coriander, and sumach; seed covers of lentils, yellow sandalwood, roses, pyrethrum and camphor, add the juice of morel to make troches. Apply one after moistening it with rose water. When the matter is cool, Heben Mesûe put the troche over the root. Also he used a theriac made from five things: pepper, asafetida, opium, myrhh, and castorem. He sweetened it with honey.

For warm matter

Halyabbas used vinegar with rose-water, or sumach with camphor. For cool matter, he heated vinegar with snake-skin and added ginger, pyrethrum, ginger, and salt.

Alexander used this confection based on garlic: garlic, frankincense, and myrrh. Add wine and boil down to the consistency of honey. It should be swished in the mouth.

Galen and Heben Mesûe attested that if one held a minced garlic root in the palm of his hand and it caused pain, then the garlic would be a good remedy for toothache. Avicenna used vinegar boiled with colocynth, aristolochia, pyrethrum, asafetida, mustard, shells of capers, menthastrum, nigella, and soap. He applied it on the tooth with a warm hard-cooked egg-yolk and warm bread, and then he sipped hot water.

Avicenna also used a resolutive (ie evaporative) plaster two hours before or four hours after a meal. It contained salt, millet, mauve, aneth, camomille, and seeds of fenugreek and flax. Also he used fumigations of seeds of colocynth, mustard, onions, and rue. He approved of Rhazes' treatment of toothache by instilling this mixture into the ear on the affected side: oils of almond, elder, castoreum, and others.

Now, as to narcotics (stupefacients): We use them when badly needed. Avicenna placed the following on the tooth: seeds of white hyoscyamus, opium, styrax, galbanum, peppers, asafetida, and boiled wine to make candies.

Another: Opium, castoreum, and rosat-oil, instilled into an ear.

Another: Roots of marjoram and hyoscyamus decocted in wine, swished in the mouth.

Another: A potion of narcotics used by Philonium (1st C) that is to be retained in the mouth long enough to induce sleep when the pain subsides.

Another: Avicenna said that swishing cold water may relieve the pain.

Weak and Loose Teeth

Sometimes a fall or a blow may loosen a tooth (a Primitive Cause). At times humidity (ie accumulated humors) may soften the nerve and ligaments (an Antecedent Cause), and sometimes dryness of tissues due to inadequate nutrition, erosion, or retraction of the gums may be at fault. When dryness and bad nutrition are causes in old or phthisical people, treatments will fail, whereas in others resumptives may help when the patient is careful about chewing, hard foods, and if he limits talking, and he avoids touching or wiggling the loose tooth.

When a contusion loosens the tooth, after the customary phlebotomy, laxation, and reduction of salivation with mastic and pyrethrum, Galen followed the advice of Archigenes, and applied on the gum that overlies the root of the loose tooth a recipe of alum, frankincense, cannelle, and seeds of cypres.

Rhazes used this: balaustes, roses, gallie-musk, souchet, sumach, and alum, to be rubbed on the gum. Elsewhere, he rubbed the gum with troches made from acacia, hypocystus, mirobalans, and a 'sprinkle' of vinegar.

When all the foregone fail, you may save the tooth by fastening it to a healthy neighbor with a gold wire (Albucasis). If the loose tooth falls out, you may replace it or make a substitute from cow's bone. When you fasten it securely in place, it may serve well for a long time.

Corruption (ie Cavities), Worms, Erosion And Separation Between Teeth

The General Measures of diet-regimens, purgations, and comfortives for the brain are as we described them in the section on toothache. *The Particular Measures* include mouth washes with hot water or hot wine containing mint, saffron, calamint, peppers and pyrethrum. Then fill the cavity with gall-musk, souchet, mastic, myrrh, sulfur, camphor, wax, sal ammoniac, asafetida, et al.

Failing that, make a small opening in the side of the cavity with a scissors or file through which to eliminate food that can be trapped in the cavity, and to admit a cautery. When necessary, remove the tooth with care not to splinter it; if the forceps breaks the tooth, the roots may remain. Then pack the gap with cloth or cotton fluff.

If a worm had been in the cavity[598], after the mouth washes as above, stuff the tooth with pills made of small seeds of leek or onion, hyoscyamus, mashed with cow-fat to make small pills. Use them one at a time.

Treating Dirty and Discolored Teeth

After the General Measures, use mouth-washes of wine heated with menthastrum and peppers. Rub this powdered dentifrice on the teeth: squid-bone, white sea-shells, porcelain, pumice, ashes of horns, niter, alum, rock-salt, burnt sulfur, iris-roots, aristolochia, and ashes of reeds.

Master Pirre de Bonant used a wash of sal ammoniac, rock-salt and sweet alum. He made a powder and set it in a glass jar and added water. He rubbed it on the teeth with a piece red cloth,

When the foregone fails to remove hard stains, scrape them off with files and spatulas.

---

[598] The 'worm' decribes the impacted detritus in a large dental cavity. When it was removed or expelled intact, it may have appeared as a small curled mass. The Ancients made of it, dead or alive, a worm (LDR).

Numbness and Congelation[599]

Request the patient to retain in his mouth and to swish warm wine or hot water. Then rub the teeth with burnt salt. Then apply warm roasted walnuts and hazel-nuts and other warm things (ie pastes), and ask the patient to chew things that are like leeks ond purslane.

Extraction of Teeth

When medications fail to save a tooth, you must remove it. First identify the one that hurts lest you remove a healthy tooth. Hold the patient's head between your knees and dissect the root free of the gums and other tissues, and wobble the tooth to loosen its roots, so that nothing will be left behind to cause trouble in the eyes or jaw-bone. Then grasp the tooth with a forceps and remove it, root and all. Use a forceps shaped like what a cooper uses when wiring a barrel. Or use a simple elevator or fork to dislodge and expel it. If some of the root remains, remove it, piece by piece with instruments. Then rinse the mouth with salted wine containing alum and vitriol to arrest the bleeding. Finally, induce granulation tissue with wine, myrrh, and frankincense.

When a tooth is abnormally large, simply file it down rather than remove it.[600]

The Ancients used many medicines designed to remove teeth, and eschew the use of instruments. They included tithymal, pyrethrum, roots of mulberry and caper plants, and yellow arsenic, ram-fat, nitric acid (ie eau forte), and the fats from tree—and forest-frogs. The medications were applied to the gum over the root of the tooth. The results seldom were mentioned.

Disease of The Lips, Gums, and Uvula

The disorders of the lips and gums include nodes, aposthems, pustules, cracks (ie chapped lips), and ulcers. All of them have been described elsewhere in the text.

Here we will deal only with the disorders of the uvula that can interfere with breathing and swallowing.

Inflammation and Prolapse of An Elongated Uvula

When the mouth is wide open and the tongue is depressed and the sun shines brightly you can see the fleshy mass. The late Grecian writers called it a Column (Cionis); we call it the Uvula, a small grape. If the 'grape' enlarges and its 'stalk' narrows it may dangle and interfere with breathing and swallowing. Simple inflammation is more common.

---

[599]   I think congelation meant insensitivity to cold and heat (LDR).

[600]   Althogh Guy copied Albucasis in most of this chapter, he abbreviated the Master's detailed instructions and descriptions. See Bibliog., Albucasis, p 280 ff (LDR).

The Cause is a warm or a cool humor descending from the brain, as does rheum.

The Signs are as we have described them, visible through the open mouth. The uvula is red when the faulty humor is warm, and it is pale when the humor is cool and the uvula is not inflamed.

Prognosis

Hippocrates (*Prognostics*, Part 3) wrote that an inflamed uvula (when the swelling is not limited to the grape) is more dangerous when you incise to drain pus, and your incision bleeds; it can choke and kill the patient. Not so, when it is pale or cyanotic and the swelling does not involve the slim stalk. Then you may cut it off with little risk. Albucasis warned us to keep your hands off when it is dark, hard, and insensate, lest we encounter or provoke a cancer at the site. An increasingly enlarged uvula cannot be cured with medications alone and should be amputated before it suffocates the patient without warning. Before that happens we must balance the risks, and choose the least perilous. That is what Aristotle said, and Galen, too, when he discussed the same dilemma in treating an ulcerated penis (*Therapeutics*, Part 4). When we can find a way to cure a threatened patient, although the route may be devious, we choose it no matter the condition of the patient.

Halyabbas told us not to remove the entire uvula and thereby add another malady affecting the patient's lungs, because the uvula serves five functions, as Galen defined them in *Voices* and *Usum* . . . Part 11.[601]

Treatments

The General Measures—diet, laxatives, and repercussives—are what we described for treating quinsy. We add desiccatives for rheum and to shrink a dangling uvula. Roger and The Four Masters treated all children with a penny-size bit of red cloth moistened with pitch, frankincense, and mastic placed on the fontanelle at the dome of the cranium. In adults Heben Mesûe suggested pulling at the hair as if to separate the scalp from the bone, as he did when abating rheum, and applied a hot cautery on the dome. Women try to elevate the uvula by pressing the throat upward.

The Particular Measures include medications and operations. As to the medications: Rhazes treated warm matter with gargles of rose-water and vinegar; or he lifted the uvula with a spoon containing a powder of roses, sandalwood, balaustes, and camphor. When the faulty humor was cool, he prescribed gargles of almuni[602], and with a vinegar-syrup containing mustard, sel ammoniac, and aloes. Roger added cannelle, peppers, pyrethrum, galls, and balaustes.

---

[601] Does the author here confuse the uvula with the epiglottis ? (LDR).

[602] Joubert cited Albucasis for Almuni. I cannot find it (LDR).

Galen (*Miamir,* Part 6) recommended a very good recipe taken from Asclepiades for treating a dangling uvula: handfuls of dry rose-petals, celtic spikenard, some soil around the plants, swallows' nests, myrrh, and unripe (green) galls, Make a powder and blow it on the uvula through a tube, or apply it on a finger or on a spoon. The medicine also acts as a diuretic and a repercussive,

Many remedies for treating quinsy are useful here.

The operations with instruments remove the uvula in three ways. In the first Albucasis seated the patient so that sunlight illuminated the open mouth when the tongue was depressed by a flat blade. He hooked the uvula and used a blunt-tip scissors or a sickle-shaped lancet to cut it off. Afterward the patient gargled with water and vinegar. He controlled the bleeding with applications of galls, aloes, and vitriol. If more was needed, he cupped at the nape of the neck, as did Avicenna, and he prescribed troches of amber in a water of plantains. He placed the patient face-down to let him spit out the blood. A little blood should be retained (ie to clot).

A second operation divides the uvula with a hot iron cautery. Heben Mesûe trapped the uvula in a lateral opening near the end of a fenestrated cannula,. He introduced the cautery through the tube and cut off the isolated uvula. Then he cauterized the stump.

In the third procedure Albucasis used a potent caustic to remove the trapped uvula through the fenestrated tube. He applied the corrosive with a cloth or cotton-tipped probe, using nitric acid (ie eau forte) or a caustic made from quick-lime and soap (ie lye), or with tempered arsenic. He held it against the uvula for a half hour, until the entire uvula was destroyed. Then the patient gargled with rose-oil or rose-water. The slough came off within three days. He was very careful not to allow the patient to swallow any of the caustic, or to let it touch any of the nearby tissues.

In any case, after the removal, treat the wound with warm frankincense and myrrh.

Albucasis described a fourth method, fumigating with vinegar containing calamint, hyssop, rue, amome, and camomille, directing the fumes at the trapped uvula through the fenestrated tube that entered the kettle through a hole in the lid. That is a rarely used method; it depends on resolution (ie absorption of the damaged tissue). I will leave it at that.

Inflamed Large Tonsils and Sore Throat

Here we use the same treatments that we described for treating the uvula and quinsies. When they fail we use the knife to prevent the advance of a disease that could impair respiration and swallowing. As Albucasis had warned us in the section on the uvula, do not invade a tonsil that is dark (brown or black), hard and insensitive. When it is pale and soft it may be removed.

Seat the patient to expose his open mouth to the sun. Hold his head between your knees, open his mouth and depress his tongue. Hook the tonsil and draw it forward and incise its membrane and dissect the tonsil from the surrounding tissues, working from side to side, using a blunt-nose scissors or a sickle-shaped lance, and remove it. Immediately, have the patient gargle with rose-water and vinegar, as we did after removing the uvula.

When Swallowing Is Obstructed

When an incompletely swallowed object is caught in the throat and you can see it, depress the tongue and grasp the object and bring it out.

When you cannot approach it directly, try to push it through the esophagus with a curved lead rod (Albucasis). When that was not feasible, he fed the patient sticky soft fruit, hoping to lubricate the object and carry it down. And he helped by having the patient gargle with hot wine and decoctions of figs, and by anointing the neck with warm oils of violets and almonds, and butter. Failing that he had the patient swallow dry soft bread, or water-cabbage (ie naveau). Failing that he provoked vomiting by administering water containing minced nasturtium, hoping to dislodge and expel the object.

Failing that, he tied a chink of raw meat or moist sponge with a long strong cord, and had the patient swallow it half way. When he briskly pulled it back, he hoped to bring up the object with it and the other materials. When the object is hard and large, he squeezed the shoulders and slapped the back.[603]

If, by chance, the patent had swallowed a leech that fastened itself in the throat, give him garlic and a strong vinegar to gargle or swallow. If the leech can be seen, remove it with a forceps or fumigate it with asafetida, or trap it in the fenestrated cannula and burn it with the cautery.

---

[603]   This was Albucasis' and Guy's version of the Heimlich Maneuver! (LDR).

CHAPTER 3

Disorders Of The Neck And The Vertebra Of The Back

Quinsy and Goiters were discussed in the chapter on aposthems, and Uvula was discussed in the previous chapter. Here we limit ourselves to the vertebrae of the neck and the protrusions of the vertebrae in the back, the so-called gibbosities. We mentioned them in the section on the chest.

A gibbosity is a protrusion of vertebrae that makes a person hump-backed, and limits his mobility. **The Primitive Cause** may be a fall or a contusion. However, the problem really is a dislocation of one or more vertebrae. Its **Internal Cause** may be an abnormal amount of slippery and viscous lubricant humor, or of a gas that inflates the joint from within, or an aposthem that displaces the vertebrae, or by pernicious coughing, or by a dry surface that attracts the bones.

Signs

The patient can tell you if trauma was the cause. When dryness was the cause the entire body has a shrunken appearance. When the coughing (consumption) is the cause, a fever came first. When the hump is due to lubrication, the patient's diet has favored moist and soft things. An aposthem as the cause is tender and inflamed (ie hot). It is a mass of gas, tightly inflated but can be indented and is tender but not feverish.

Prognoses

Hippocrates (*Aphorism # 6*) said that a patient whose hump was the result of coughing and asthma will die before puberty [604]. Galen's *Commentary on the Aphorisms* explained that weak and small chests can not withstand the stresses of the coughing and they give way as a hump not only in young persons, but in all asthmatic persons with weak sunken chests.

---

[604] Guy's comment here explains that he has used the more reliable translation of Hippocrates' *Aphorisms* taken directly from the Greek, rather than the Arabic version that claimed death did not occur before adolescence (LDR).

The prognosis is poor unless the protrusion is treated early; the deformity spreads downward and inward involving the pelvis and hips, pressing on nerves, even to cause paralysis.[605] Avicenna noted that the thighs of such victims became atrophic because the deformed bones compressed the veins that carried nutrition to the limbs. Rhazes said that when such obstruction was complete, a cure was impossible.

More about this topic will be found in the chapter on dislocations.

Treatments

When treating the 'dry' variety, use moist foods and beverages; baths and humectants; enemas containing the oils of violets and almonds; decoctions of the roots of guimauve, mauve, linseeds, milk, and broths of tripe, and other things that we have prescribed for treating hectic fevers and dry convulsions.

When the hump is firm, treat it as we do sclerosis.

When coughing is the cause, use amelioratives.

When the hump contains raw humors or gas, *The General Measures* first attend to the dietary regimen and laxation that we use for treating paralysis and convulsions caused by humidity (ie accumulated phlegmatic humors) as found in some aposthems and in gout. *The Particular Measures* include the resolutives and the comfortives, which also are warm astringents, such as cypress nuts and leaves, laurel and juniper leaves, aromatic roseau (reeds), enula and acacia. All of them are used as embrocations, ointments, and plasters.

Avicenna prescribed this plaster: juniper, sehan (Serapion called it aloin, Rhazes called it stoechas), enula, pyrethrum, cassia-bark, cypress-roots, marjoram, cardamon and acoris. Boil all in water and oil until the water is gone. Add more of the same herbs, water, and oil, and reboil. After the second 'boil-off' add castoreum, euphorbia, and ammoniac. Fortify it with rue, sysimbra, spikenard, acoris, styrax, and bdellium. For even more potency, use wax and terebinth for the plaster.

After a bath and embrocations Albucasis manipulated the bones by covering them with oil and pressing them inward. Then he applied a plaster covered with a disc of lead and a splint, and wrapped all with a tight bandage. And, when all else had failed, he cauterized the circumference of the hump.

---

[605]   The descriptions fit the osteoporosis of elderly women, and the progressive scoliosis that begins in the child (LDR).

CHAPTER 4

Maladies Of The Shoulders And Arms

The shoulders share the maladies of the arms and hands. Occasionally we encounter redundant fingers, and we described the operations to remove them in an earlier chapter. Sometimes the fingers are webbed. We treat them with the knife and use dressings to keep them apart until the incisions heal.

Fingers and Finger-Nails

The nails may be torn or they may be injured by contusion, and may trap foul clots beneath them. Other lesions are splits, discolorations, and blemishes.

Avicenna treated the torn nail with leaves of myrtle and pomegranate, and appled the basilicon ointment, and used lenitive waxes and greases. Clots and pus under the nail are treated with goat-lard and sulfur (Halyabbas). When that failed, he carefully slit the nail and released the material (also Avicenna).

They treated deformed and eroded nails with cataplasms of diachylon, oil of almonds, mastic, oil of ben, and dry seedless raisins (Halyabbas). Avicenna used onion or squills fried in onfacium.

Rhazes treated stained and spotted nails with eruca and vinegar, Avicenna used fish-glue with linseeds, nasturtium, and red arsenic.

When all failed, and the patient's distress was inconsolable, he abraded the nail (the surface) with oil containing opoponax and serapion.

Halyabbas used a paste of oak-tree sap, cantharides and thapsus, and added arsenic for more potency.

I use a spatula as a chisel and scrape the nail.

When a nail was shed, Avicenna used a small copper or silver cap (not an air-tight)to cover the finger tip until a new nail appeared after about one month.

CHAPTER 5

Maladies Of The Chest And The Breasts

D isorders of the breasts are the only ones special to the surface of the chest. They include hard masses, such as aposthems and caked milk, which we described in our *Treatise On Aposthems*. Persistent and inadequate lactation are disorders treated by physicians. Here I will repeat Galen's *(Easy Remedies)* recipes.

To attract milk: a beverage of sweet wine containing radishes, fennel-roots, and bran.

To reduce an annoying rapid growth of the breasts in an anxious young woman, he prescribed massages and tight binders and washes with cold water and vinegar, often with added clay and mill-dusts. He increased the potency with alum, galls, and pomegranate-peels. Sometimes (also used by Rhazes) he made a broth of powdered cumin in water and vinegar and wet a cloth, applied on the breast, and wrapped it with a tight bandage, alternating three-day applications with a paste of roots of lilies, honey and vinegar. The cycle was repeated three times every month. It worked well.

When a man's breasts enlarged, he incised above and below the nipple, or made a cruicate incison, and scooped out the underlying fatty tissue, and sutured the incisions, and treated the wounds as such.

When a nipple is inverted and cannot serve to nurse a baby, he applied a cup or a hollowed nut shell or sucked at the nipple at the end of a tube until it was everted.

541

CHAPTER 6

Maladies Of The Abdominal Wall

The abdominal wall can be considered as a structure in itself, and it has no unique disorder (ie of its tissues) other than an everted umbilicus caused by ascites, which we described in the chapter on hydrope. The everted navel really is a ventral hernia, and will be discussed below in the section on omental, intestinal, watery and gassy hernias. There may be hematomas, so-called aneurysms, following ruptures of arteries or veins.

The Causes are those already described for hermias and aneurysms. And the same holds for **Signs.**

Prognoses

Do not treat an aneurysm with a cautery and frighten a patient, and that caveat applies to treating all herniations through the abdominal wall at the navel. I advise you to treat them with plasters and bandages rather than use the knife. In that region the muscles do not cover the opening, and an incision through the peritoneum will allow an extrusion of intestines that will be difficult to reduce through a small opening. I cite Galen's *Therapeutics*, Part 6).

Treatments For Umbilical Hernias With Medications

Rhazes used a paste of frankincense in egg-whites on some cotton-fluff or a pad, and he bound it over the navel. He added galls, pomegranate-peels, acacia, alum, nutmeg, antimony, amber, ceruse, and others used for treating hernias in the groins.

Cumin and laurel berries will dispel gas.

Sulfur, et al. are used when we treat ascites and hernias, both the aqueous, and the fleshy.

Omental and intestinal hernias at the navel were treated with the cautery by Albucasis, Halyabbas, Avicenna, and others. Their technique: Let the patient stand facing the seated surgeon. The patient holds his breath and forces out the hernia as far as he can, and they marked the circumference with ink. Then, with the patient supine, they reduced the contents of the hernia and incised around the ink mark with the cautery. Then they caught the center of the disc of skin with a hook, pulled up and deepened the incision around the bulge down to the abdominal wall. Then they ligated the base of the mass after ascertaining that no intestine was in the protrusion. They repaired the defect by inserting two needles crossing each other at the center, leaving the ends exposed. They wound a tough cord around the defect under the ends of the needles and closed the defect. They tied the knot and left the needle and cord in place until they sloughed on their own. Then they treated the wound as such.

When using that technique, remember to be sure that you have not trapped any intestine. Before you first tie the cord, nick the sac and insert a finger into the belly and withdraw it (ie as you tie the knot).

The operation is challenging, and I do not use it. I leave it to the special skills of the herniotomists. (See Fn 607).

CHAPTER 7

Maladies of the Pelvis (Hips)

T he maladies that can be called surgical are the inguinal hernias (ruptures of the didymus), bladder stones, inflammations of the penis and prepuce, circumcision, castration, hermaphroditism, diseases of the matrix including enlarging the passage or supporting prolapse, extracting a baby, disorders of the anus including stenosis and prolapse. We already have discussed hemorrhoids, thromboses, warts, and fissures.

Ruptures of the Didymus (Hernias)

The ruptures of the didymus that Galen called hernias (*Maladies and Symptoms, Part 2*) are displacements from within the abdomen of intestines or omentum into the didymus and the scrotum. Most often the herniated intestine is the ileum because it is free in the abdomen (Avicenna). Galen said that there are three types (*Unnatural Tumors*): the epiploic (omental), the intestinal, and the combined. The herniation may be small, limited to the groin, where it is called a relaxation. When it is large, it descends and may reach the testicles. Halyabbas called it a rupture (*Sermon 9*, of Part 1),

Causes

The **Immediate Cause** is the dilated opening (Galen, *Maladies*, Part 2 and Avicenna). The enlargement may be a stretch produced by a fall or a blow, or by a sudden lurch, or by strenuous labor, or by shouting, or by abnormal coitus. The enlargement is facilitated by humors that moisten and lubricate the opening (Halyabbas) and it is favored by a congenital weakness of the part, all of which are **Primitive Causes** abetted by excesses of a diet rich in fatty, humid, and gassy foods (Albucasis and Theodoric).

Signs

The patient can feel the rupture and we can see it as it descends. We can detect it by feeling what moves when he coughs, and by reducing what is there when he lies supine. The route is clear: the opening is dilated and what first appears at the groin descends into the scrotum without much effort. At times the hernia does not enter the didymus but passes along with other structures into the scrotum[606], or into the thigh or near the vulva or into the upper abdomen or through the navel as described in the previous chapter.

Another sign is the gurgling sound when you reduce the herniated intestines. When the herniated structure is omentum, there is no gurgle when it is reduced.

Prognoses

A scrotal hernia may threaten life when the herniated intestine is impacted with feces and cannot be reduced. I saw such a patient die and so did Albucasis. You must urgently attend to the treatment and not abandon the patient to a truss.

Avicenna stated that a hernia that does not descend in the didymus (ie not fissural) cannot be cured by desiccation, and Halyabbas added that indeed no other kinds of medications are successful. On the other hand, a dilated opening can be treated by desiccation, especially when it is tender (ie recent) and in children; not so, when it is longstanding and in older people. Theodoric complained about the charlatans who claimed that they could cure patients of all ages with medications alone. I have never seen a complete cure of a large hernia treated by medications, and I am astounded by Lanfranchi's claim to have cured a sexagenarian and a quadragenarian in that way.

Furthermore, although treating hernias with the knife is feasible, the prognosis must be guarded because of the risks for complications: convulsions, severe pain, hemorrhage, damage to the intestines with corrosives, and loss of fertility when a testicle is sacrificed. Lanfranchi said that is why many surgeons who really know how to operate, refuse to do so.

I advise the Reader not to attempt to operate for hernia when the patient is old, has a poor complexion, and coughs. Treat such an one with medicines, and let him survive with his disabilities. Also, I join Bruno and William of Saliceto in advising that the operations be performed only by surgeons who have learned by observing Masters at work, and are themselves skillful, and are equipped with all the necessary armamentarium: razors, spatulas, hooks (large and small), cauteries, needles, stupes, cotton fluff, eggs, cloth, red powder, etc. The best seasons for surgery are the autumn and the spring. Galen advised thorough cleanouts before operation, with enemas and laxatives.

Treatments With Medications

The medications are designed to close the opening or reduce it, and to dry the slippery surface. We employ three measures.

First: When the patient is replete, purge him with phlebotomy if needed, and by laxatives that shrink, such as myrobalans.

Second: Provide for the six non-natural things and the three annexes: fresh air, good foods and beverages, correction of inanition or repletion, exercise and sleep, the emotions,

---

[606]   A fair description of a sliding inguinal hernia (LDR)

the repercussion of foreign matter, baths, time spent out-doors. All those things avoid heat and temper dryness and tend to bring together what has been enlarged and separated. In other words, treat as we do for phlegmatic, aqueous, and gassy aposthems, as I described it in my treatise on ruptures. Here, I shall not repeat a detailed list of the medicines.

However, I will repeat Avicenna's routine treatments of hernias in replete persons.: Avoid things that stuff him more: beans, faseoles, lentils, soup vegetables, and others listed by Rhazes. He avoided raw fruits, turnips, incompletely baked and unleavened breads, pork, fresh cheeses, and milk. The beverages omit undiluted water and new wine, but include rusty water and astringent wine, Fresh-water (ie unmedicated) baths are harmful. Avoid windy places at mid-day. Rain is harmful. Avoid jumping, sobbing, sexual activities, and do not wear a truss. Avoid constipation with use of suppositories, enemas, and laxatives of cassia, tamarinds, and diacatholicon. At bedtime be sober and drink no broths, soups, or hard liquors. Add the herb sage at all meals. After meals, swallow dragées of seeds including nasturtium, coriander, and the like.

Third: Manual reduction of the hernias assisted by enemas, baths, cupping, applications of warm resolutive compresses, and by suspending the patient by his legs and hips. After the reduction, foment with water, wine, and with vinegar heated with galls, cypress-nuts, and alum. Apply a paste spread on a piece of leather shaped as a shield over the site, which you renew daily for nine days; do it early in the morning before the patient arises. When you lift off the adherent pad, hold your fingers over the opening to prevent pulling up the abdominal wall tissues. Hold the pad in place with a binder and truss made of a triple-folded cloth to cover the shield and the groin. Fasten it in back, and both there and in front fasten a strap that passes under the groin. To prevent irritation by the edges of the shield, insert a cloth covered with the white ointment.

When at stool the patient should press a hand against the hernia site and avoid straining. Every morning dose a consolidative potion based on a heavy wine. The patient should be kept abed most the first fifteen days, and gradually return to full activity during the next next fifteen days while wearing a truss.

The paste for the specially fashioned leather shield contains naval pitch, colophony, litharge, ammoniac, opoponax, galbanum, bdellium, mastic, terebinth, bol d'armenie, sangdragon, frankincense, sarcocolla, socotrine aloes, mummy, aristolochia, centaury, consolida, sumach, berberis, cypress-nuts, galls, pomegranate-peels, earthworms, human blood, fish-glue, oak-tree sap, rain water in which leather has been boiled, and vinegar. Boil all the liquids and reduce by half. Dilute the gums with more vinegar, and mix them with the other liquids to make the paste.

A plaster used by Rhazes, Avicenna, Bruno, and Theodoric: Make a fine powder of cypress-nuts, acacia, galls, balaustes, tragacanth, myrrh, sarcocolla, frankincense, gum Arabic, bol d'armenie, and mummy. Add vinegar to make a plaster or a stupe.

The potion noted above contains three consolidas, St. Mary's seal, jacea, plantains, valerian, pimpinelle, cypress-nuts, nutmeg, cannelle, roasted rhubarb, tamarisk-fruit, nasturtium-seeds, cumin in vinegar, coriander, sangdragon, frankincense, mastic, nutmeg, terra sigillata, bol d'armenie, pitch, sarcocolla, and dragacanth. Make a powder and dose one dram with a half cup of sweet wine.

We know of other secret methods for treating hernias.[607] Perhaps one 'discovered' secret was as follows: After reducing the hernia, remand the patient to bed for thirty days. Every morning dose a decoction of iron filings in wine containing wild hepatica. After fifteen days, apply this plaster over the hernia three times every day: chopped amanita, ground in the apostolicon. During the ensuing fifteen days apply a truss over the plaster. God must help, because the basis for the cure is faith; the iron during the initial fifteen days is a well known attractive, and the plaster during the final fifteen days shuts off attractives. The patient sees the smaller mass and thinks he is cured.

Operations

In general there are two methods: In one the surgeon removes the entire didymus and its contents. In the second he plugs the opening between the abdomen and the didymus with scar tissue and prevents the recurrence of herniation.

In the first, as used by Albucasis, Halyabbas, Roger, The Four Masters (ie The Glosses on the Rogerina), Jamier (a follower of Roger), Bruno, Theodoric, and William of Saliceto, the surgeon uses his knife, as follows.

The patient lies supine, securely bound to the table (or bench). The intestine is reduced and the didymus is exposed through an incision at the groin, and is stripped of its surrounding tissues so that the testicle in the sac can be held over the abdomen.

The surgeon transfixes and ligates didymus as tightly as possible. He divides it and discards the testicular portion. To assure hemostasis he cauterizes the ligated stump with a hot iron and returns it within the scrotum. The long ends of the ligature dangle out the wound and will come away during the treatments of the wound with egg-white et al.

A second operation by Albucasis, Roger and his followers, and by Bruno and Theodoric, uses the cautery. The patient is arranged as above, and the testicle is pushed up to lie over the pubic bone, and its circumference is marked with ink, and it is returned into the scrotum. The hot cautery is repeatedly applied at the center making a cruciate wound until it burns through to expose the underlying bone. The burn-wound is then treated with egg-white, etc.

A third operation uses caustics as a potential cautery as described by Theodoric, Master Jean des Creuez of Bologna, Master Andrew of Montpellier, by Master Peter of Orlach at Avignon, and by me. I shall describe it later.

---

[607] Families of nomadic herniotomists and stone-cutters, and cataract-couchers wandered through Italy and France. They thrived as families, passing on 'secrets' which really were not hidden from astute observers. As the centuries passed, the art was taken up by lay surgeons (ie not educated in cathedral schools or universities). By the 16th C they were the premier surgeons, as typified by Paré, Franco, and Wurtz. The story is told in detail by Nicaise (see Henri de Mondeville, and Franco in this translator's bibliography) and by Malgaigne's Introduction to his edition of Parés' works. All of them now are in English editions (LDR).

A fourth method uses a mass-ligature, as described by Roger. After marking the site as above, pass a needle threaded with a stout cord under the entire didymus, lay a small wood rod parallel with the didymus over the site and tie the cord. Tighten it daily until it cuts through the entire didymus from below.

A fifth operation was described by Lanfranchi and Master Peter of Dye. After incising the overlying abdominal wall, grasp the entire didymus and dissect it free and expose the underlying bone. Then divide the entire didymus with a cautery and treat the wound.

A sixth operation uses the gold thread (ie wire)[608] as described by Master Berand Metis. After incising the overlying tissue to expose the didymus, he tied it tightly enough to narrow the passage, using a gold wire, that pleats the didymus.

As I see it, the first four methods are complete and do not fail. The others depend on luck. In general, as Galen wrote (*Therapeutics*) a fail-proof perfect method prevents recurrence, a requirement satisfied by the four. The entire passage is destroyed leaving other tissues to connect the abdomen and the scrotum, which are not a channel that can be dilated, and the connection is sealed by the growth of scar tissues.

I believe that all four measures are successful when treating small hernias. However, recurrences are possible after an initial success when the hernias are large. And when hernias have been treated with Albucasis' method with the three-point cautery, after the eschar is discharged a thick heavy scar remains instead of a simple restraining shield.

Theodoric and others said that an operation was worthless unless it blocked the passage at the groin, and that there is no good reason to save the testicle. I have seen many wives made pregnant by a man with one testicle. However, of the two choices, I prefer the lesser evil.[609] I think that the underlyng essence and nuitritive potential remains in a (deprived) testicle supported by inflow from surrounding tissues. Galen agreed (ibid., Part 5). And those qualities persist even when the testicle shrinks.

---

[608] Joubert's description: This was a beautiful and successful operation which did not remove or impair the testicle. It ligated only the spermatic duct and the sac but spared the vessels, varying the tightness of the ligation acccording to the age of the patient. In any case it is necessary only to narrow the passage into the scrotum to prevent reentry of the intestine.

The technique: Incise at the groin and pass a needle threaded with a gold wire through, over and around the sac, leaving a passage for the spermatic vessels. Twist the ends of the wire and cut them short so not to erode on contact. Then suture and treat the wound. There is no slough and no need to induce pus.

We today (ie 1563) use a gold loop, hollow at one end, into which we can introduce the other end, much like attaching the arms of an extension of a table. We can adjust the size of the passage at will, and it avoids breaking the wire when twisting it, and eliminates the problem with protruding ends. Some rings are made from a hollow tube which weighs less than a solid wire in the scrotum (J).

[609] Not removing the testicle which is deprived of its blood supply (LDR).

Of the four, I prefer the operation that uses a cautery, and I use the knife only when the hernia is very large, when I need to incise the bottom of the scrotum for drainage. Avicenna failed to mention it when he claimed that the cautery was the only instrument for herniorrhapy. However, because the actual cautery excites terror and can cause the patient to weaken during the operation, I use the potential cautery, but I recommend that it be used only by surgeons who have mastered the technique—how often, and how much. I prefer arsenic for its potency as a caustic (as I noted in the chapter on scrofules.) Ignorance in its use can induce fevers and other bad complications. Even a small amount can do a lot of harm, especially when it affects a principal organ. Therefore, always dilute it with vinegar, morel, and other coolants, and prescribe a regimen as if there is fever.

The course of the initial treatment lasts three days, and the applications are made three times a day. If you must back away, apply opium and the juices of morel and cabbage, as we will describe.

The procedure with arsenic begins with a prescription for a healthy diet-regimen and for laxatives. Later, the patient lies supine and the intestines are reduced. The hips and groins are elevated (on a pillow). Lead the testicle into the groin over the pubic bone where its circumference is outlined with ink or lamp-black. After the testicle has slipped back into the scrotum, apply a corrosive in the outlined area, using quick-lime and soft soap tempered with saliva, an amount about as large as a small chestnut. Apply it about a fingerbreadth from the base of the penis, and surround it with a circlet of waxed cloth or a cool adherent gum that will prevent the spread of the corrosive beyond the mark. Cover it with a firm bandage over the hips that passes behind the back as does a truss, and prevents the application from slipping off for an entire day (ie and night).

The following day remove the bandage with the corrosive and expose a dark eschar. Make a small cruciate incision through the eschar, about the size of a rye grain and spread it open and make a tunnel in which you insert some slightly moistened wheat-flour containing powdered arsenic and an equal amount of opium, followed by a pledget of cotton fluff moistened with saline (ie a plug). Cover all with a cloth or some charpie, and surround the opening with populeum ointment. Lay on a pad or cloth moistened with water or vinegar or egg-white, and bandage again to form a shield sutured to a truss made from folded cloth tied front and back. Place the patient supine in bed and bind him to the mattress.

Change the dressings daily and ascertain that there has been no reappearance of herniated bowel. Do not remove the corrosives for two or three days during which you prescribe analgesics for relief of discomforts. Remove the caustics and distend the opening rather than cutting it and causing bleeding. Apply powders, and in time remove the eschar and inspect the cavity to its bottom, and examine the all the surrounding tissues and the didymus. When the testicle and scrotum are inflamed and painful, the opening will drain pale matter. The necrotic didymus will slough and your examining finger can slip through an opening and pass into the scrotum. The intestines cannot descend, and the surfaces of the didymus will fuse within two weeks.

Then, if you so wish, remove the eschar and granulation tissue with applications for one week of pork-lard or butter, or similars. Then you will be able to see the white didymus as a shrunken tube, and that the intestine and the tissues of the abdominal wall are healthy. Apply small amounts of the arsenic on cotton fluff, and apply defensives around the opening, and mitigatives to relieve pain, as oil of peppers, hyoscyamus, and mandragore. You will need them because the cleanly exposed didymus now is more sensitive. The added treatment is brief, avoiding the nearby vessels (ie the spermatic vessels in the didymus). You may apply more caustic within the didymus to destroy it, usually within two weeks. By then the testicle will be inflamed and the pain will spread to the back. When the residual sac is distended, you may make two or three small parallel incisions in the long axis at the bottom of the scrotum, through which to insert, touch, and destroy what remains of the didymus (ie with the arsenic).

If fluid collects in the scrotum, introduce a probe through the opening in the groin and raise the scrotum so it will empty via that opening. Then clean out the remnants of eschar, and use detergents, and await the appearance of pink granulation tissues. Treat the wounds as usual.

The patient should wear a truss for thirty days after the wounds have healed, and before he gradually resumes ambulatory activities. On the other hand, I personally saw Master Peter treat thirty patients without restrictions of bed-rest or activity afoot, as they overlooked the discomforts of the caustics. I disapprove. Although he excised most of the eschar, he could not get at what has to slough on its own. Until then do not plug the opening, except for small dressings over it. You cannot be certain that all the arsenic has escaped until all the eschar has disappeared. The entire process may take eight weeks. On occasion, I have cut short the time by two or three weeks when I was sure that all the eschar was gone.

In spite of all the above, I must admit to the occasional use of the method of Master Louis of Binac of Vienne in the Dauphine, to insure the repairs. After making the initial opening in the didymus, I insert a curved actual cautery, for two or three applications. It has three advantages: it controls bleeding, it allows for a deeper penetration than the arsenic alone, and it does not penetrate the thin eschar of the burn. The maneuver is less painful than that of the caustics. I do not use the cautery as the primary treatment. Why compromise what already is a successful treatment?

The operation is not without risks; do whatever is possible to avoid damaging the patient. The same caveat holds when we remove the eschar with a cautery simply to abbreviate a long course of treatment.

The following complications should be treated during the course:

First: Too much caustic may have been used, or it has burned too deeply. Wash the field and foment with oil of rosat.

Two: The scrotum may be inflamed and be very painful. Mollify with linseeds and fats of pigs, chickens, ducks, et al.

Three: When pus collects. Drain through the bottom of the scrotum and use detergents, as we do elsewhere.

Four: When bleeding interferes, use the red powders, egg-whites, vitriol, and a bit of arsenic. Do not leave the patient's presence until the bleeding has stopped.

Five: If coughing threatens the repair, dose the patient with tragacanth, barley-sugar, and rub the chest with the oil of violets.

Six: Treat constipation with cassia fistula, and use suppositories and enemas.

Seven: Treat diarrhea with constipating troches and astringents.

Stones In The Kidneys And The Bladder

According to Avicenna (Part 3) and others both the kidneys and the bladder participate in forming stones. Galen (*Therapeutics,* Part 3 and *Aliments,* Part 1) said that the body's joints also are involved[610] and the intestines (*Internal Maladies*, Part 6, and the lungs (idem, Part 4). Halyabbas and other authors have added more.

We surgeons will not deal with stones in the kidneys or in other internal organs. There are no surgical cures for them; I can attest to that, in agreement with Bruno and Theodoric. However, we shall discuss them, because the causes are in common with bladder stones.

Halyabbas said that stones form in the body as do bricks in the kiln, and as crusts are deposited on the surfaces of the hot baths. The hard substance begins as layers of adherent viscous matter. The narrow passages retain it and heat it. I will quote from Galen's *Aliments,* Part 1:

"When the passages in the kidneys are abnormally narrow, the raw fluid becomes sticky while it moves slowly, and it thickens and coheres. It solidifies as does 'cake' deposited on the inner surfaces of kettles in which we heat water, and as we can see form on the inner surface of the pipes that drain hot springs. So it is in the kidneys, which by nature are hot and spongy. The liquid part of the matter evaporates and the lees coagulate (*Therapeutics,* Part 4)."

The famous Masters, Avicenna (*Canon,* Part 3,), Alexander (*Practica,* Part 2), and Averroes *(Colliget,* Part 3) all agreed with Galen. However, Serapion (*Breviary,*Part 4) said that even moderate heat can act to thicken the humors and generate the sediment. As I see it, even moderate heat is un-natural, as I explained in *Different Fevers*.

Avicenna said when abnormal heat occurs in the kidneys of young people, stones form rapidly. The same moderate heat engenders stones in the bladders of older people, but over a longer time (*Canon,* Part 1, and in *Aphorisms,* Part 3). The same occurs when old persons are dyspneic. The weaker agents work slowly, and the potent agent is rapid (Galen, *Simple Medications,* Part. 3, and M*aladies and Symptoms,* Part 4). Therefore, the heat works according to its intensity and is **The Efficient Cause** of stones. But Serapion thought that the thick matter is a more important cause. The Masters do not contradict each other, even if it seems so at first glance.

The Cause of the thick matter is strangury, indigestion, and a heavy diet, as part of a patient's bad habits. The faulty elimination of the bad matter and the narrowness of the

---

[610]   Galen had observed stones (urate) in gouty patients (LDR).

passages are causes for retention. The extra work of the kidneys and the bladder, and the use of hot substances account for their increased heat.

Signs Of Stones In The Kidney

Hallyabbas said the patient passes sandy and pink urine in dribbles, and the urine feels hot. Pain in the kidney region and flanks radiates to the scrotum and thighs and legs. The patient tends to sleep on the bad side. If he passes bits of stone, whether or not with the help of medicines, that is a sure sign, and one can be certain that it heralds a cure (ie is in progress).

Signs of Bladder-Stones

The pain is in the bladder. The penis itches, especially at the tip, with intermittent erections. The urine may be cloudy and pale, or watery-thin containing white gravel. Urination may be difficult to initiate. When the diagnosis is in doubt, have the patient lie supine with his hips elevated on a pillow, and have him bounce up and down, and then he may be able to pee.

If that maneuver fails, insert a probe (ie so-called catheter) to reach and push away the impacted stone, and then he will void. All of those things support the diagnosis of bladder-stone, and foretell how we will treat it.

Another maneuver: Have the patient bend forward and press his pubic region inward while you dislodge the stone with your finger in the rectum. That may enable him to urinate. Avicenna said that a bladder stone can cause tenesmus (ie the urge to defecate), and whenever the patient passed bits of stony matter he would have the urge to urinate more.

The pain of a renal stone may resemble intestinal colic and cause difficulties for the physician in the differential diagnosis. It made little difference in the old days; until our epoch the treatments were the same for both: mitigate the pain. Later on they wanted to do more, but the contents of the medicines are about the same (*Internals,* Part 6).

Renal and bladder-stones are different. The renal stones are light, small, and reddish. The bladder-stone is hard, much larger, and lighter in color.

Prognoses

Hippocrates (*Epidemics,* Part 6) wrote that he would not treat patients for renal stones who were older than fifty years. In *Aphorism* #6 he added that treating a painful bladder-stone was not easy in older and feeble folk, and it was better for them to live with the disease than to die with the treatment. Galen agreed (*Commentary).*

Gordon said that when the patient passed thick sandy urine, he no longer had a stone, but if unexpectedly the urine became clear, a stone had formed (ie by condensation).

Avicenna claimed that both renal and bladder-stones were inherited maladies. He added that adults tend to renal stones, whereas the other was true for children and adolescents, beginning to form early in life. And he stated that women tended to bladder stones and other paroxysmal disorders that were spaced at monthly intervals. And he said that a small bladder-stone was more apt to obstruct urination than a very large one that could not impact in the outlet.

Stone disorders may threaten life. The obstruction and retention of urine can lead to hydropsy and death.

There is no way that we can incise and remove a renal stone, and open operations for bladder-stones put the patients at risk for convulsions, hemorrhage and fistulas. That is why I advise most surgeons to delegate those operations to the itinerant stone-cutters. See Fn. 607.

Albucasis said that no surgeon can remove a large bladder-stone without risking permanent damage to the bladder[611]. And when he operates to remove a small stone, he may not be able to find it in the bladder. Therefore, he should operate to remove only medium-size stones.

No one who has not mastered the technique should dare to operate. He must have observed at length Masters at work. Bruno, Theodoric, and William insisted the same. The surgeon must have at hand all the necessary equipment: razors, hooks, long forceps, needles, thread, cotton fluff, linen, eggs, red powder, et al. He must not operate on feeble, depleted, and old persons, or on people who are terrified and piteous. The best risk is a forty-year-old (Lanfranchi). The best seasons are the spring and the autumn.

Treatments: Two Kinds

We opt for one or the other according to where the stones lie. Some can be brought out with medicines. In others, especially for the bladder-stones, you must cut or displace (ie dislodge an impacted stone) (*Techni*, Part3). I always accept Rhazes' advice to give medical treatments a long trial before deciding to operate.

Medical Treatments: Two Kinds

Prophylactic (ie Preservative) Measures aim to eliminate the causes: the thick humors, the local heat, the constricted passages. However, Galen claimed (*Epidemics*, Part 6) that abating moderately excessive heat is not necessary, because the other prophylactic measures will protect the kidneys and the bladder. I quote:

"We arrive at a good state when the humors are thin and the renal outlet (ie the renal pelvis) is soft. Given that, no stones will form. Those conditions are obtained only by attenuative medicines and diets described in my book, *Health*. I plead with nephritic patients to read my book, *Attentuative Diets*. Some have obtained complete relief and others have been improved. The causes are explained in my book *Good Medicine.*"

---

[611]   The need to enlarge the incision in the bladder neck may extend it into the thin bladder wall (LDR).

The 'thinning' treatments work by eliminating the obstruction and by stripping and cleaning the viscous thick humors from the inner surfaces of the passages. But you must anticipate what may happen when those treatments succeed. The calcific matter that is stripped may become stones. Aristotle predicted such in his *Problems,* Part 1, and in *Generation of Animals,* Part 4. In any action, a positive will induce its contrary in like degree. Therefore, follow the rules as to quantity, quality, and timing, as defined by Galen in *Glaucon,* Part 1, and in *Techni,* Part 3.

Prophylactic treatments include dealing with that sort of complication, and with establishing a healthy life-style based on the six non-natural things and the three annexes: air, food and drink, inanition and repletion, exercize and rest, sleep amd wakefulness, emotional states, and avoidance of bad things related to baths and dress, all of which tend to dry and to attenuate. As to diet, I refer to Galen's book, *Attenuative Diets.* I shall not repeat them here.

To summarize: Avoid all things that can engender stones. Rhazes and Avicenna included such heavy foods as partially baked or unleavened breads, beef, sea-birds, fish, unripe and sour large fruits, soft cheeses, all dairy foods, turbid water and wine: in general, all thick and viscous foods that are not easily digested, which tend to burden the abdomen and create gas.

Rhazes and Halyabbas agreed that we should avoid stones by eliminating viscous foods and to favor seeds that clean the kidneys. Abstain from sleeping supine, Do not wear tight belts. Ride horses frequently and work while on horse-back. Use emetics to correct overeating at a meal.

Hermes[612], as confirmed by Arnold and The Conciliator, claimed that if a small image of a lion, engraved or carved, is carried in a sash or a purse, when the sun is in the zodiacal sign of Leo, and when the moon is not in opposition to the planet Saturn, but accompanies it, it will defend against stone. When frankincense is melted or beef-blood is incinerated in the presence of the image during the same astrological hour, and it is mixed with wine and is imbibed, the stone will pass promptly and the patient will pee.

Prophylactic Medicines

These are based on evacuants and irrigants of the passages. When the patient is plethoric, bleed from his basilic or saphenous vens. When he is debilitated, a double evacuation (ie from both ends) is called for—emetic and laxative. The emetics not only

---

[612] Hermes was one of the twelve Olympian Gods; he was variously endowed. Here he is Hermes Trismegistus, the magician, a favorite among the astrologers, among whom were Arnold of Villenova, the 'spiritualist', and Peter of Albano, called 'The Concialator'. See Nicaise's Introduction where he described Guy's authorship of a small treatise on astrology (LDR).

divert (repercuss), they also eliminate some of the antecedent phlegmatic humor that was engendered in the stomach and is the basis for a stone.[613] The bad humor appears in the blood that is carried to the kidneys. Hippocrates prescribed a prophylactic vomit for healthy people once every month, as reported by Galen in *Usam . . .* Part 5, and by Avicenna in several places.

Prophylactic defecations are prescribed during the Spring and Autumn, or oftener if deemed necessary. First digest the phlegmatic matter with simple or compound oxymel (Galen, *Acute Illness*, Part 3), and add squills as a diuretiuc (Heben Mesûe), or prescribe a more potent digestive syrup: roots of the five capillary ferns (adiantum), saxifrage, pimpinelle, filipendula, fraizene (strawberry) centaury (both the marine and the meadow varieties), pansies, calamint, hyssop; seeds of juniper, hedera, fennel, celery, ammi, carrots, the major cold seeds; spica, and acoris; the flowers of camomille, broom and squill; vinegar or verjus, and add honey or sugar to suit one's taste. Dose the syrup with a broth of chick-peas.

After an interval for digestion, prescribe a laxative of acoris pills, herb benedict or the catholicon.

After a successful general laxation, prescribe two kinds of special laxatives and aperitifs to clean the passages. One type is moderately strong: figs, almonds, pistachio, capers, dry raisins, et al. They are in common use and should be dosed early in the course (Galen, *Therapeutics,* Part 9), and *Aliments*, Part 2). They are not charged with forcing nutrition to the liver and the urinary organs. Cresson, acts similarly and it resolves, assists urination, and favors passage of bladder stones (*Medicaments*, Part 8). The Aggregator[614] prefers nettles, as did Avenzoar.

The second and more potent type of special laxatives contain pouliot, fennel, chick-pea broth, etc. They must be correctly used (seldom), dosed during a meal to avoid combusting the blood[615] and overheating the kidneys, and when used judiciously they do not cause undigested food to leave the stomach and go to the liver and the urinary organs (Galen, *Health,* Parts 4 and 6).

Master Arnold made a puree of chick-peas that had soaked in water over night. In the morning he boiled it with parsley, and reduced it by half before adding powdered nard, saffron, and white wine, Many surgeons add mashed dog-tooth violets, or cumin during the winter. Others add the juices of lemons, or oranges, and melon-seeds during the summer. The puree cleans the capillaries of the liver and the passages of the kidneys, and prevents the formation of stones. I approve.

---

[613] The normal humor is combusted in the stomach and becomes the bad phlegmatic matter. It is *not* phlegm (LDR).

[614] The Aggregator: Jacques Dondi of Padua, 1298-1359, a herbalist and contemporary with Guy (LDR).

[615] The reader is reminded that 'blood' means the four humors, When they are combusted, they are converted to the bad matters which engender the diseases (LDR).

I also use a diuretic wine containing betony and saxifrage. Others add very small amounts of spikenard and other diuretics, such as diacalament and diapolytric (*Health,* Part 4, and Avicenna, Part 5). Serapion called it Diacumin (*Breviary* Part 7).

Curative Medical Treatments For Stones

We use all of the prophylactic medications, as Galen suggested (idem, Part 4), but the dosages are different. Bladder stones need more potent medicines (Avicenna).

Begin with a sedative enema, or, if needed, a tub-bath in water that had been boiled with mitigants and softeners such as Arnold' recipe: cynoglossus, berle, cresson, and celery, all partly mashed before cooking them in fresh water. Add a sea-shell and white wine. When it has cooled to tepid, seat the patient in the tub with water up to his navel. It works well to relieve pain and to dilate the passages, and perhaps to allow passage of the stone. Keep the water warm.

Then dose more potent medicines that can cause the stones to fragment. I use gentle laxatives such as cassia fistula and diacatholicon with small amounts of attractives (Avicenna).

After the laxation, prescribe medicines to pass the stone (*Simples,* Part 6). The best time for that is immediately after the patient leaves the tub. They include warm decoctions of chick-peas and dog-tooth. Failing that, go to incisive medicines (idem. Part 5): asparagus, ronce, betony, pouliot, garance, and ashes of glass (highly valued by Master Bertruce of Bologne for use as the last resort). In such cases, Avicenna used the roots of costus, ronce, seeds of guimauve, roots of centaury, minced cardamon, scolopendre, venus-hair fern, rumex, pentaphyla, pouliot, chamaepetys, radish-roots, acoris, souchet, pepper-seeds, incinerated scorpion, molluscs, and sea-urchin shells; beef-blood, droppings of roosters and doves, dried scarabs and cantharides. Others recommend the ashes of cicadas. Be very careful to use only small amounts of the above, lest you ulcerate the bladder.

Avicenna said that the compounds should include five kinds of medications: softeners, penetrants, shrinkers, comforters, and fragmenters, as we find them in his syrup: henna (gremil), venus-hair fern, parsley, figs, and water. Boil down to one-fourth volume. Drink the dose on leaving the tub.

Our noble Serapion prescribed this well-tested recipe for fragmenting[616] the stone: melon-seeds, gremil, carrots, and ground glass. Dose the powder with a decoction of black-peas.

The Cardinal of Naples used a decoction of filipendula, roots of acoris, and saxifrage, all cooked in an alembic.

Avenzoar prescribed an electuary that I think is one of the best: jews'-stone, cherry-gum, sponge-stone, melon-peels, juice of fresh licorice, hazel nuts, carrots, incinerated glass, anise, hyssop, pine-cones, skinned almonds, and a syrup of licorice. Add a very

---

[616] The term 'rompre' was inclusive; fragmenting (rupture) and getting rid of the fragments by urination (LDR).

fine balsam. Keep it in a glass jar and dose six drams every morning with the syrup of violets and hot water.

Master Arnold, prescribed a powder for the Lord of Bellioco: gremil, ammi, anise, fennel, caraway, parsley, anise, celery, lovage, pepper, mountain-siler; seeds of poppy, melon, mauve and juniper; asparagus, radishes; leaves of hedera, pits of peaches and cherries, bitter almonds; roots of saxifrage, galanga, ginger, cannelle, spikenard aromatic reeds, scrapings of licorice, souchet, acoris, red and white sandalwood, jews'-stone, sponge-stone, scrapings of ivory, jaws of pikes(fish), incinerated cicadas, and bull-blood. Dose it with white wine in the morning.

Rhazes used these potent sweet pills: seeds of melon, radish, carrot, and parsley; carpobalsamum, peels of the roots of caper plants, panax, bitter almonds, laurel-berries, acoris, souchet, spikenard, cassia, scolopendre, rue, gentian, aristolochia, cabaret, cardamon, bdellium, ammoniac, serapinum, myrrh, and pepper. Add a light wine to the gums and make pills. Dose with a decoction of chick-peas. Adding a balsam may break the stone.

Avicenna prescribed this very potent recipe: incinerated glass, scorpions, cabbage-roots, molluscs, goat-blood, incinerated shell of a hen's egg that contained an embryo chick, jews'-stone, walnut-gum, acoris, parsley, carrot, pouliot, gum Arabic, seeds of guimauve, and peppers. Make a confit with honey. Dose it with a decoction of centaury and black peas.

William of Saliceto used a diuretic of herbs and roots and added a bit of centaury to make a syrup. Master Odo of Lyon made a decoction of the same plants. It is a medicine in common use as a stone-breaker when taken with warm wine.

After dosing any of the foregone recipes that fragment the stone and induce passage, Avicenna applied an oil of scorpions. He said that scorpions are natural contraries to stone-formation, and it is contrary to snake-venoms. He made plasters, as did Taddeo, with water parsnip, parietory, leaves of wild cucumber, mauve, cabbage, garlic, and pimpinelle.

Theodoric applied this sedative ointment: oil of camomille, rosat-oil, raw egg-yolk. Smear it on a cloth.

Jumping in place may help passage of kidney stones into the bladder. Applying cups on the flank from the kidney toward the bladder, and fomenting with oils of rue and warm castoreum were methods used by Avicenna.

Medicines To Initiate Urination

Urinary retention can be relieved by the foregone medications, especially those that contain cantharides (Galen, *Medications and Temperaments,* Part 3). Rhazes, in his *Almansor,* Part 9, prescribed baths, embrocations, plasters, ointments and humectants applied on the pubis, the penis, and the perineum. He introduced some of them into the penis and the bladder.

Master Jordan injected them directly into the bladder with some balsam. Theodoric added mineral oil. Avicenna added oil of scorpions. Others used dove droppings diluted with a mild lye.

Others anoint the pubis and the flanks with the fat of a rabbit and apply galbanum on the glans of the penis, or use garlic and onions. Another: a suppository of rock salt inserted into the penis. Another: Apply a louse or a house-fly. A commonly used remedy is made from the roots of radishes, parietory, parsley, and nettles, all boiled in wine and the residue is fried in oil and applied on the pubis.

If none of the above is successful, we must resort to operations. I will not carry on with the medications, because the time has come to use instruments.

Two Treatments Of Stone With Manual Procedures

Palliative treatments are called for when the stone is too large to be led into the bladder-neck, and where you can make an incision in flesh that will heal. The bladder sac itself is nervous (ie membranous) and it receives a constant inflow of urine, and it cannot heal; an incision there can be lethal (Hippocrates *Aphorism # 6)* especially in an old man.

Instead, operate in this way: When the patient is in a warm bath, introduce a catheter (a solid rod) or a cannula or a syringe, well lubricated with butter or a non-irritating oil, and displace the stone into the fundus. Or you may be able to displace it from the bladder neck with fingers in the rectum. After all, the stone may have been in the bladder for a long time before it impacted, perhaps as long as forty years (Theodoric).

Passing Urine Through Instruments

The catheter is a long slim introducer like a probe. It may have a blunt tip or a small knob. An argale or syringe is a cannula with a perforation along side and at the tip. The other end flares like a funnel and can be attached to a leather sac or the bladder of a pig or a ram.[617] You may attach a screw (ie tarrier) at the tip to some, or an apparatus as used for instilling an enema.

Halyabbas (Sermon 9, Part 2), Avicenna and Albucais use it this way. After a warm bath or fomentation, they seated the patient and gently introduced the instrument into the penis, which is held up toward the abdomen. When it reaches the perineum near the anus, lower the penis with the cannula until the instrument points at the bladder, where the urethra makes a bend, until you feel it enter an empty space, that indicates you that you have reached the bladder. If the cannula had been fitted with a stylet, or the lumen had been plugged with a wool thread, remove them and the urine will flow. If the instrument had scraped the lining tissue, and blood appears, introduce some white collyrium or some mother's milk.

---

[617]   Nicaise commented that here Guy indicates that he was familiar with a hollow cannula, whereas three centuries later, when Mingelousaulx used one to relieve Cardinal Richelieu's urinary retention (a complication of a perianal abscess), he reported that the Cardinal's medical-surgical staff did no know how to use it. (EN).

Cutting For Bladder Stone

I follow what I learned from the Masters.

In men, first clear the bowel with an enema. The following morning (the patient fasts) he hops in place once or twice hoping to bring the stone into the bladder-neck. He sits on the knees of a sturdy assistant who straddles a bench. The patient's hips are strongly flexed and his knees are held by straps which pass behind back, to spread them and hold them motionless during the entire procedure. Press your right fist into his lower belly above the pubis and insert fingers of your left hand into his rectum to feel and hold the stone in the bladder-neck, between the rectum and the scrotum. Use a razor to incise vertically to the left of the midline, parallel with the skin folds. Avicenna said that an incision in the commissure itself can be fatal. Cut down on the stone, and insert a spoon-end probe under the stone and draw it forth. Clean out the wound and close it with sutures. Apply the red powder and egg-whites and apply a pressure bandage. Carry the patient to bed, and do not disturb the bandage for three days (Roger). Then use the diapalma ointment.

In women the procedure essentially is the same except that you catch the stone with fingers in the vagina.

Roger dressed the wound with egg-yolks during the winter, and with egg-whites in the summer. The Four Masters used a wheat-flour stupe.

The patient should drink very little during the recovery, and limit himself to sour red wine mixed with rusty water. He should eat moderately to reduce the need to defecate.

When a small stone passes into the penis where you can feel it, tie a cord around the penis near the pubis to prevent the stone from retreating into the bladder. Insert a long slim forceps to grasp and remove the stone. Or you may use a screw-tip probe (ie tarriere) to impale and rupture the stone. Then remove the pieces (Albucasis).

Or you may incise in the axis of the penis where you can feel the stone and remove it. Suture the wound and remove the tie around the shaft of the penis.

In all cases avoid abscesses. When the pain is severe, treat with warm baths, and with mitigative oils of camomille, aneth, et al. Apply warm butter over the wound.

If other complications occur, may the Lord Help You.

Diseases of the Penis, Especially impotence and poor performance

These defects involve the genital organs of men, Sterility which befalls women (Serapion, *Breviary,* Part 4) may be due to deprivation of coitus because the man is impotent (ie cold) and cannot perform. Women can be the source of deprivation only when their vulva is obstructed, whereas men are the source of the deprivation because of coldness of complexion (impotence) which prevents erection, or because of physical defects involving the penis and testicles. Whatever is the cause, the result is bad sexual performance (ie malefaction).

Coldness (impotence) is not the same as malefaction. Whereas coldness concerns a person's disposition (ie complexion). Malefaction is physical as well as emotional. In common parlance, impotence reflects the nature of the body, and malefaction a state of the emotions, as if a curse had been applied, or some barrier existed between the man and the woman.

The Signs of impotence due to organic defects are manifest when a man has been castrated, when the penis is underdeveloped or is malformed, or cannot erect, when it is frigid or paralyzed, when the region is hairless, when the penis is wrinkled and discolored as a result of overindulgence in spices and potions, and despite all the inducements to engage in coitus, including massage and warmth, erection does not occur.

When the causes are emotional or spiritual, there are no organic defects, and the patient cannot engage in coitus with his own wife although he may succeed with other women. Or he may be influenced (asceticism) by prayers and avoidance of erotic thoughts and sorcery.

Galen accepted Plato's **Prognosis** (*Commentary on the Aphorisms,* titled 'When a woman cannot conceive'). When a man refuses to impregnate a women, it is a mockery on Nature, and the species will fail. He also said that when the failure is conjoint, it may be a justifiable basis for a divorce. The judge[618] must be circumspect and call for a medical consultation. It is an important matter, and I shall describe it.

The physician must examine the genitalia of the pair. The physician then consults an experienced woman who will attend the couple for three days while they lie in bed together. The matron will provide spices and wines, She will anoint them with warm oils, and engage in suggestive conversation, and encourage them to engage in caresses and embraces. Then the matron will report to the physician what she has witnessed. He will pass on that information to the judge. Abuses are common and all involved must avoid any suspicion of fraud. At risk is the separation of a marriage that the Lord had joined; a just cause is required.

Priapism

This is involuntary and differs from satyrism which is voluntary and desired by the person involved.

The Cause is gas inflating the nervous structures, induced by gassy foods. The gas cannot escape through the skin that has been thickened by its exposure to cold air (ie as an appendage). Galen (*Internals* Part 6) also admitted that the cause may be dilated arteries in the penis.[619]

---

[618] The action is more religious than civil, and the 'court' is entirely ecclesiastical, and the judge probably is a high ranking priest(LDR).

[619] Galen believed that gassy inflation explained the rapidity of erection and detumescence, not possible with blood. If priapism is due to arterial congestion, the blood cannot escape. A return venous circulation was not conceived until the epoch of Harvey (LDR).

The Signs are obvious, and the need for treatment is urgent.

Treatments include the General Regimen. The swelling is relieved by fomentations with rue and agnus castus. Apply cold juices and camphor and Galen's paste, and wrap the organ with a sheet of lead.

Fever and Foul Discharge After Coitus With An Unclean Woman

First wash the penis with oxycrat. Then apply the white ointment with camphor, as we used it in the treatment of ulcerated pustules.

Adherent Foreskin

Use a spatula as we used in elevating a fingernail. Then cover the glans with a cap made from a hollow twig or a lead cannula which allows the patient to void urine and the surgeon to apply butter and oil of almonds on the glans.

Circumcision

Jews, Arabs and other sects circumcise according to their laws. Many individuals undergo it to prevent the accumulation of ordure that can inflame the glans.

The Procedure: Stretch the prepuce with your fingers and cut it off, avoiding the glans. Control bleeding with the red powder or the cautery.

Castration

You may accomplish it by attrition (ie crushing). Seat the patient in a tub of warm water to soften the scotum. Then squash the testicles with your hands until you cannot feel a firm mass.

Or you may use a knife. Incise the skin and expose the testicle. Ligate it and cut it away. Then suture the wound.

A Soft Stretched Scrotum

Cut away some of the skin, sparing the contents. Suture and treat the wound.

Hermaphroditism

Albucasis described two forms in men: A vulva contains two testicles or there is gap between the labia and the testicles on each side.

In Women, where the vulva opens there is a penis and testicles. Treat this by excising them, but take care not to affect the urethral opening.

Disease Of The Matrix

First we shall deal with several kinds of obstructions. Albucasis said that one of them can be treated by surgery. When there is an overgrowth of tissue (ie a [polyp), cut it away or ligate it. When a skin closes the introitus, incise it with a knife or puncture it with a fingernail. Then insert a wood or lead cannula as a drain, and anoint the region with butter or oil. It will allow the woman to urinate; or you may insert a small roll of cloth, about the size of a small penis, replacing it frequently. Or encourage the woman to engage in coitus as a means to keep it open.

Next we deal with a dilated vagina: Avicenna constricted it with a powder of pine-bark, alum, and souchet, all heated with sour wine. Soak a linen cloth and insert it.

Third, we deal with an enlarged clitoris at the intoitus. It may annoy the patient unpleasantly. Avicenna ligated and amputated some of it, short of the base, lest he provoke hemorrhage.

Extraction Of A Fetus

In most normal childbirths, the baby is delivered head-first with its face down. Other presentations are abnormal and may cause difficulties. Sometimes there are twins, or more: five were reported by Avicenna, and Albucasis describes seven or as many as ten. Since those cases usually are managed by the women; there is no need to for us to be involved.

Obstetrics is the province of mid-wives when the delivery is natural.[620] When the delivery is difficult,they apply softening fomentations and ointments. The woman helps by straining, while holding her breath. She may induce sneezing by inhaling pepper,or use euphorbia. The experts say that the Arabic cyclamen tied to the hips will ease a delivery.

When the presentation is not natural, do what you can to bring it to normal after elevating the hips.

When you suspect that the fetus is dead, as is signified by shrinking of the woman's breasts, the fetal mass no longer moves, when the woman's belly is cool, her breath is foul, her eyes are puffy, her lips and her face are pallid, and her belly is inflamed. That condition may have been caused by a fall or a contusion. In such a case, the midwife should insert her well-lubricated hand into the matrix that has been softened with fomentations and pessaries. Helped by sneezing and by prescriptions of abortifacient medicines such as castoreum, myrrh, and rue she delivers the fetus. Or you must insert a speculum with one its two blades movable. Use it to dilate the matrix (ie the cervix) as widely as possible, and pass fingers, forceps, and hooks into the womb to remove an intact or a morcellated fetus. Leave nothing inside.

---

[620]   During the 16th C, obstetrics became a proper part of surgery, and such matters as podalic version, specula, and forceps were discussed in detail by the great surgeons of that era, Paré, Franco, and Wurtz, who were lay surgeons or barber-surgeons. Those matters are discussed in detail in Nicaise's French edition of Pierre Franco, now also in English. The surgeons' interests soon waned, to be renewed in the 18th C. See Bibliography (LDR).

Albucasis said that he had seen a pregnant woman, after a long delay, deliver a dead fetus through her navel; she survived for a long life. You must use your ingenuity when the head, the chest, or the abdomen, or the placenta, or simply edema blocks the passage, and you must open it with your nails or a spatula to expel the fluid and deliver the fetus.

In the case when the mother dies (ie during labor) and you recognize that fact by the signs described in our chapter on death, and you think that the baby is still alive, a Royal Ordinance allows you to open the belly. Use a long incision on the left side, to avoid the liver. Insert your fingers and bring forth the baby, just as the historians tell us Julius Caesar was born.

To Deliver A Retained Placenta

Rhazes and Albucasis urged the patient to sneeze and bear down while she held her breath and pinched her nose. If that failed, they fumigated the matrix with a funnel and bladder,using a decoction of calamint, rue, centaury, camomille, and cassia bark. Again provoke sneezing to help abort the placenta.

Failing that, tell the mid-wife to dip her hand in sesame oil or mucilage of guimalve and to reach into the matrix and remove the placenta. If some of it remains attached, remove what you can and irrigate the remainder with softeners, and inject the basilicon ointment. After a few days, all of the residue will come forth.

Moles Or Fleshy Masses In The Matrix

A mole is a mass of flesh that grows within the womb. It is engendered in two ways (Avicenna). In one, a collection of matter is combusted to become flesh. In the other, the mole is the product of coitus. As if it is a fetus, the mole is nourished by the woman; there is no male contribution, and there is no pregnancy

The Signs

The woman feels no fetal movements, and her legs remain soft past the time of a normal pregnancy.[621]

Treat with softeners, sneezes, abortifacients, pessaries, and instruments to withdraw a fetus.

Prolapse of the Matrix and Rectocoele

Foment the extrusion with sour wine and sprinkle a powder made from the roots of mastic, consolida, sangdragon, bol d'armenie, mummy, myrrh, cypress-nuts, balaustes,

---

[621] Does Guy mean that edematous legs were expected as a normal pregnancy approached term (LDR)?

alum, and ceruse. Mix some of the powder with egg-whites to make a plaster, and apply it on a pad and bandage it in place (ie reducing the prolapse). Keep the patient in bed with her hips elevated, and let her eat sparingly to avoid inflating the prolapse.

Hemorrhoids, figs, scratches, fissures, and ulcers of the matrix and anus have been discussed.

CHAPTER 8

Maladies Of The Thighs, Legs, and The Feet

We have discussed many of these maladies when they occur elsewhere, such as elephantiasis, varices, boils, nails, et al.

We could repeat many fantastic statements about mal-mort (stasis ulcers). Whatever is claimed, the treatments are those for scabies, excepting that scabies appears either as moist or dry. When mal-mort undermines, it resembles gangrene and putrid ulcers. However, for the sake of a complete text, we shall repeat. Treat mal-mort as we do salt phlegm. Prescribe a healthy regimen and diet. Wash with water water decocted with iron filings (from a farrier), vinegar, and a decoction fumitory, parelle, and chelidoine. Then anoint with a liniment of the juices of round aristolochium and bryony that are cooked with rosat oil, vinegar, and soft soap. Then apply an ointment of old pork-lard thinned with vinegar before adding white ointment, sulfur, slum, bol d'armenie and quick-silver. Grind all in a lead bowl.

Painful Heels

When friction by shoes at the heel or frostbite makes sores, (ie blisters) follow Avicenna's treatments: Wash with cool water and with an epitheme of memitte and bol'd'armenie, and follow the rule, and add some of the leather of the offending shoe.

Halyabbas made a powder of the combusted skin of a horse's leg. Galen (*Simple Medicines*, Part 11) incinerated lung of a ram or a hog and made a powder. It relieved the pain. Then he applied galls and acacia in vinegar.

A remedy in common use is the basilicon followed by the white ointment.

HERE WE END TREATISE VI. NOW LETS US REST OUR BODIES
AND SEEK PEACE FOR OUR SOULS!

THE ANTIDOTARY

TREATMENTS

English Translator's Note

G uy's Antidotary is more than a pharmacopoeia. He discussed non-manual treatments in addition to medications. Although he opened Treatise VII with a Chapter on bleedings, and devoted Chapter 3 to Cauteries, most of the Antidotary gives full attention to the medicinals used by a surgeon, always with the approval of the Academic Physicians. He agreed with most of the earlier authors that a manual surgical procedure was called for only when medications fail. That induced the priestly physicians to praise his conservatism during ensuing centuries, while the lay surgeons and the barbers undertook to use their hands and became the Master Operators.

RUBRICS

ENGLISH TRANSLATOR'S NOTE

Guy's Antidotary is more than a pharmacopoeia. He discusses a few manual treatments in addition to medications. Although he opened Treatise VII with a chapter on Bleedings and he devoted Chapter 3 to Cauteries, most of the Antidotary gives full attention to the medicaments used by surgeons, always with the approval or oversight of the Academic Physicians, He agreed with most of the earlier authors that a manual surgical procedure is called for only when the medications fail. That insistence induced the priestly physicians to favor him for his convervatism during ensuing centuries, while the lay surgeons and the barbers undertook to use their hands and instruments, and they became the Master Surgeons!

RUBRICS

Potential

TWO APPENDICES

TREATISE VII
DOCTRINE I GENERAL MATTERS
EIGHT CHAPTERS
CHAPTER 1

Phlebotomy, Cupping, Leeches[622]

I know that I will lengthen my book by dealing in great detail with medications, as well as with the other kinds of treatments applied by surgeons to particular parts of the body, and their techniques, and by citing examples, and by inserting many descriptions of what they considered to be suitable. Should anyone wish for more, I will refer him to *The Continens* (Rhazes), and to *The Great Antidotary* of Azaran, where thousands of the remedies of the Ancients have been collected. I will reject the methods put forth by the empiricists and by the sorcerers, of which a large number can be found in *The Gilbertina* and *The Treasury of The Poor*.[623] Sometimes I will leave the beaten path of the authorities and I will briefly call attention to remedies in common use which I myself have used. In all, the repetitions may be onerous, but useful things can profit by eiteration, and be exposed to emendation

Since phlebotomy is the most common and notable of the (non-medicinal) treatments, and if it is omitted at the outset of treatrments with medications (Galen (*Therapeutics*, Part 9) said all other remedies will be ineffective) I shall begin our chapter with

Phlebotomy

This is an incision in a vein to evacuate blood and its humors. Arnold defined it as such in his *The Special Operation,* where he describes the procedure. Avicenna described it as the most effective General Evacuation (Book 1, Part 4.). In Part 3 he described it as the common evacuant for all humors. Galen (*Aphorism* #6)[624] wrote that it is a useful

---

treatment for all plethoric maladies. Rhazes (*Almansor,* Part 4) said that phlebotomy was a valuable prophylactic for maintaining health as well as a treatment for diseases when properly employed. He added that it was harmful if it depletes a person's vitality and leads to hydropsy and other bad illness, and it can cause premature aging.

Phlebotomy can accomplish great things, as Galen described them in his *Phlebotomy.* He cited a case of a Roman who suffered from bad vision and a contrary case of an Erasistracian, who was called *a Sanguifuge* because he refused phlebotomy. Phlebotomy is a more reliable evacuant than medicines because it can be controlled at will. Whereas after a laxative has been swallowed the outcome cannot be limited.

Galen (idem) established five bases for using phlebotomy 1. Is there a need for an evacuation? 2. Should it be by phlebotomy? 3. Can the patient tolerate it? 4. What veins should be opened? 5. How much blood should be released? Then the surgeon can set the proper time, and proceed after the physician has prescribed a proper regimen (diet, et al).

The First Basis includes the assessment of the two kinds of repletion, and an examination of the veins as to the strength of the pulse, as discussed in *Multitude and Plethora,* and in Part 4 of *Maintenance of Health.* All plethoras called for evacuations (Hippocrates, *Aphorisms*) whether it is for treatments or for prophylaxis, and whether we bleed to eliminate humors or to reduce the volume of blood.

The Second Basis includes the state of repletion in the veins (*Painful Eyes,* Part 1 and *Maintenance of Health*, Part 4); and whether we bleed to eliminate all humors or simply to reduce the volume of blood. We know that the veins contain humors other than the natural ones.

Phlebotomy serves six purposes: Evacuation, Diversion, Attraction, Alteration, Prophylaxis, and Alleviation (is mitigation).

A. Evacuation: The Methodists and others (Galen, *Therapeutics,* Part 2.) said that evacuations served only to relieve plethora. Galen rebutted and claimed that phlebotomy did more than simply reduce the amount of humors. It reduced the severity of a malady, as when a phlegmon (ie edema) appears after a contusion, or when the body suffers pain and sanguinous humors accumulate and the body cannot reject the inflow. Those abnormalities may occur in any one, not just the plethorics, when the impact of a disease is great in any of three ways: 1. the affected part is a principal organ, 2. when the patient's constituion (ie disposition) is weak, and 3. the affection is destructive.

B. Diversion: Galen described this in *Therapeutics,* Part 5, as a treatment for fluxions, or for nose-bleeds. When the epistaxis is from the right nostril, bleed from right hand, etc. Hippocrates sang the same song in *Aphorism # 5.* "When the back of the head suffers pain, bleed from the vein in the center of the forehead." Hippocrates and Galen always considered phlebotomy as a diversive.

C. Attraction: The example is its use to induce delayed menstruation. Bleed from in a vein in the lower part of the body, or from scarifications in the legs, at the time of the expected normal menses.

D. Alteration: The example is described by Galen in his *Commentary on the 23rd Aphorism.* When treating fevers, one bleeds almost to the point of heart-failure, just short of death. At that time, the body will become cooler.

E. For Prophylaxis (ie Preservation of Health) (Idem, #6). Prophylactic phlebotomy has spared the voices of many priests in the choir who seldom suffer quinsy, pleurisy, epilepsy, and apoplexy if they have been bled in the Spring, and they have been spared the bad consequences of contusions and wounds that occur after their bleeds, including phlegmons (ie edema). And they may be spared the ill effects of repletion.

F. Alteration. Galen wrote (*Therapeutics,* Part 11) that phlebotomy is useful not only in periodic fevers, but whenever fever accompanies suppurations, except in old persons and those with weak resistance. Our Nature allows us to reject any grievous invasion that could become an undesirable; that quality reigns supreme. It digests what is digestible and rejects the rest, according to its own assessments.

The Third Basis deals with the suitability of the patient, as I wrote in my small book on the subject. A robust acceptable person has large full veins; his body in neither obese or skinny; his color is not pallid; his flesh is firm. Opposite to him is someone who cannot tolerate bleeding. He lacks blood, and his flesh will melt away. In that group are children younger than fourteen and adults over age seventy. Phlebotomy is permitted only in desperate situations. In such cases, work very carefully, and anticipate problems.

Galen (*Glaucon,* Part 1, and *Therapeutics,* Part 11)) and Rabbi Moses warned that the limitations noted above included persons who were not familiar with phlebotomy (ie frightened), those who have weak stomachs (ie timid), and the victims of diarrhea and flatulence and who suffer from chronic indigestion. Hippocrates also excluded pregnant women before the last months (ie ? 8 or 9) of pregnancy. The most important contraindication (idem, Part 9) is frailty. That can lead to death during a bleed. Therefore, we must bolster a weak person in two ways.

First: Assess the ill effects of other purgations: diarrhea, sweats, et al. Note whether there is colic, spasm, tremors, chronic illness, and the effects of unusual sexual activity, or too frequent use of the baths, or obvious anxiety, sadness, and the effects of insomnia, or of over-work.

Rhazes added (*Almansor,* Part 4) the signs that he risks are small and support the use of phlebotomy: someone who is experienced with the operation, and whose diet includes frequent meat-rich meals and sweet desserts. But he warned against bleeding persons wh0 are abstinent (ie who fast) and have weak organs, who are idle (ie phlegmatic) and suffer from colds and who live in very hot or very cold places.

Do not bleed someone who is drunk or otherwise disturbed until he is well passed that state, because controlling the bleed may be difficult, and the risk of hemorrhage is great.

Master Arnold discussed all of the issues in the decisions for or against bleeding, in his book on *Phlebotomy,* in terms of the natural, non-natural and contra-natural things. He filled an entire chapter with instructions, and provided examples. He did not cast aspersions on the opinions of others, especially on the question of the volume of the bleed. We will discuss that matter in our chapters on Medicines.

All those issues are more Medical than the Surgical, and I will defer them to the Physicians.

The Fourth Basis concerns the selection of the proper vein. All the surgical treatises have dealt with the subject at length. Haly abbas in *Royal Dispositions,* 9 said that there were thirty three veins that can be bled:

Nicaise Inserted This Note

I have attached Figure 5 to supplement the following material, which I also discussed in the Section on Astrology in my Introduction to this Edition (q.v.)

The Ancients had noted that the Sun, the Moon, and the few Planets known to them did not deviate in their motions through the celestial regions, and they remained in a zone which they called the Zodiac, divided into twelve segments. They named the segments according to the constellations that had been identified by their predecessors. (*today, they are different due to the precession of the equinoxes.)*

A Man was considered to be a small world, a microcosm, and all the twelve parts of the celestial world, the macrocosm, corresponded, each of them to a part of the body.

The Ram (Aries) as first in the series, represents the Head; The Bull (Taurus), the Neck; The Twins (Gemini), the shoulders, arms, and hands; The Crab (Cancer), the chest; The Lion (Leo), the stomach (ie the upper abdomen); The Virgin Virgo), the mid-abdomen; The Scales Of Justice (Libra), the pelvis and buttocks; The Scorpion (Scorpio), the genitals; The Bowman (Sagittarius), the thighs; The Goat (Capricorn), the knees; The Water-Bearer (Aquarius), the legs; The Fish (Pisces), the feet.

The 'Nature' of the Zodiacal signs, as described in Champier's *Guidon en Francois,* is: Three are Fiery—Aries, Leo, and Sagittarius
Three are Airy—Gemini, Libra, and Aquarius
Three are Watery—Cancer, Scorpio, and Pisces
Three are Earthy—Taurus, Virgo, and Capricorn

Their Qualities indicate the Zodiacal Time best for Phlebotomy, but always when the moon is in ascension, arising in the East:

Aries is warm and dry, and a good time for bleeding
Taurus is cool and dry, and a bad time for bleeding
Gemini is cool and moist, also a bad time
Leo is warm and dry, also a bad time
Cancer is cool and moist, indifferent for bleeding
Virgo is cool and dry, also indifferent
Libra is warm and moist, a very good time

Scorpio is cool and moist, indifferent
Sagittarius is warm and dry, a very good time
Capricorn is cool and dry, a bad time
Aquarius is warm and moist, indifferent
Pisces is cool and moist, indifferent

Never incise or apply a cautery on an affected part that is governed by its sign on the day of a New Moon, lest hemorrhage ensue. The same is true for bleeding at sunrise, to avoid complications.

In addition to all the above, consider the Zodiacal indications of the proper vein for the phlebotomy: The illustration (Figure 5) depicts Rhazes' practice (*Almansor* Part 7). The surgeon himself will determine the side according to the malady. The numbers that follow here correspond to those in the Figure:

1. The Vein in the Forehead: for headache and ocular disorders
2. The Vein on the glabellum: for ocular discharges
3. The Vein behind (ie below) the Ear: for chronic headache and drowsiness
4. The Vein under the Chin: for ocular pain, pustules on the face and nose, and painful jaws.
5. The Right Cephalic Vein: for headache, and painful right ear, throat and tongue
6. The Right Median (Common) vein: for headache and pain in the ribs and upper abdomen (stomach)
7. The Right Basilic Vein: for headache, pain the right shoulder, liver[625] and to cool the blood in the nostrils (ie probably for treating nosebleeds)
8. The Vein on the dorsum of the right hand between the 3rd (medius) and the 5th (auricularis) fingers[626]: for maladies of the heads and the eyes
9. The Vein in the web between the right thumb (pollex) and index finger: for maladies of the head and eyes.
10. The Vein at the base of the penis; for tympanites (gassy distension)
11. The Veins under the penis: for swelling and pain in the scrotum
12. The Sciatic Vein (ie lateral malleolus): for pain anywhere in the lower extremity and 14. Veins on the lateral dorsum of the foot: for disorders of the kidneys, ophthalmia, amenorrhea, lesions of the testicles, painful hips, flanks, and thighs

---

[625] Nicaise probably followed an erroneous copyist who wrote spleen here instead of liver for a right side vein (LDR).

[626] The labels are vague. The auricularus or 5th finger is defined, but 'medius is not, The middle finger was called impudicus, and the fourth finger was annularis. Perhaps the same prudishness that renames the 3rd finger,caused our auther to name the genitalia as shameful (ie honteuse). I suppose that one of the 2 or 3 veins under the skin of the ulnar side of the hand was suitable (LDR).

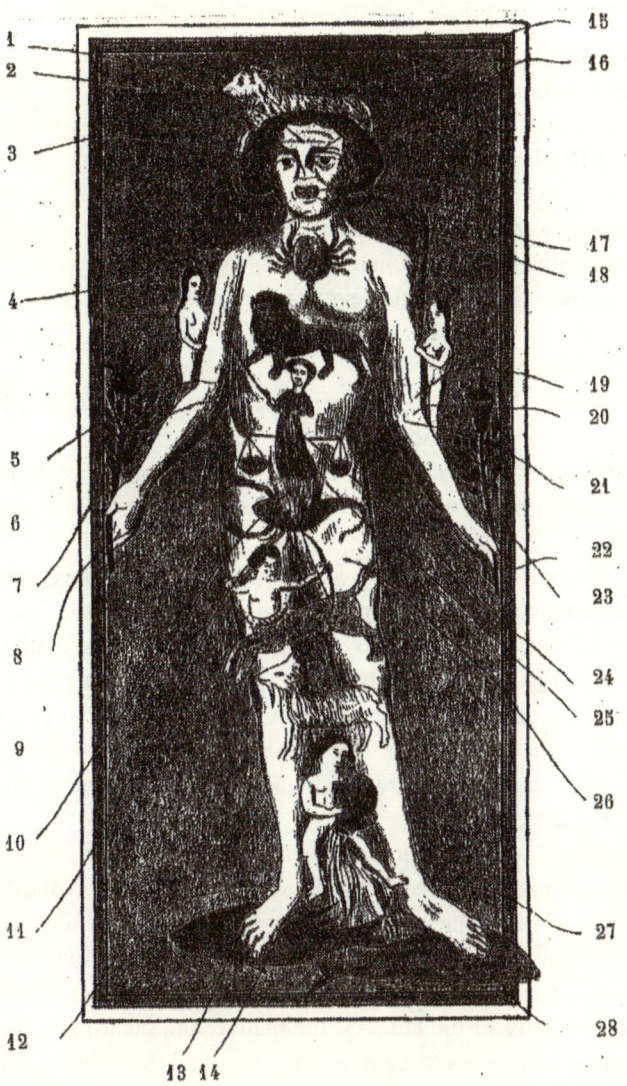

Figure 5. Engraving Showing The Twelve Zodiacal Signs
That Govern The Parts of the Body; and The Veins For Phlebotomy
From The !5th C Latin Folio #6910 In The Bibliotheque Nationale At Paris

15. Veins on the temples: for ocular pain and maladies
16. Veins at the nose at the corner of the eye: for poor vision, headaches, and ailments of the eye
17. Veins on the throat: for pustules on the head and scalp
18. Veins under the tongue: for quinsies, sore throat, and esophagus
19. The left Cepahlic Vein: for the head and the eyes
20. The Left Cardiac (Common) Vein: for the heart and the stomach
21. The Left Basilic Vein: for the head and the spleen
22. The Vein between the left middle and small finger: for the head and eyes
23. The Vein between the left thumb and index finger: for head and the eyes
24. The Vein in the lower abdomen: for the groin
25. The Vein in the axilla (the 'tickle' vein): for pain in the lungs, the chest, and the diaphragm
26. Veins in the thighs: to attract humors from the upper body. They are to be opened after a meal
27. The Saphenous Vein under the medial malleolus: for pain in the hip and kidney, and to induce menstruation
28. The Vein under the lateral malleolus: for sciatica, painful joints, painful testicles (scrotum), morphias, and urinary retention

I will add that much attention was given during the 16th C to the choice of the side to suit the side of the malady, same or opposite, or the side of affected internal organ. See Sprengel, 1793). (EN) two median, two cephalic, two basilic, two axillary, two cubital, and two salvatelle. Thirteen were in the head: two behind (under) the ears, two at the inner corners of the eyes, two organics, two on the dome, the vein at the center of the forehead, the vein at the occiput, the vein at the glabellum, and two under the tongue. There are eight in the lower limbs: two at the knees, two saphenous, two sciatics, two over the arch of the foot.

Albucasis counted twenty-six: Ten at the head, ten at the arms, and six on the legs and feet.

Galen frequently used the three at the elbow: the internal (ie medial) served all parts of the body below the neck; the external (ie lateral) served what was above tbne neck; the in-between vein (ie common) served for all parts. The two veins, at the groins and behind the knees served the lower parts. The veins along the legs and at the ankle were used for treating maladies of the kidneys, the matrix, and the urinary bladder. He used veins at the corners of the eyes to treat ophthalmias, and the veins under the tongue to treat inflamed throats. Other veins were used at times.

I remind the Reader that the General Evacuations should precede the Particular Evacuations whenever possible.

As to **The Arteries:** I hesitate to bleed from them and risk hemorrhage and aneurysms (ie hematomas). Galen used them only for treating pulmonary diseases, when he bled from arteries at the temples and under the ea. Sometimes he bled from the limbs when treating nightmares.

The usual practice is to bleed from the opposite side to attract from phlegmons in the first stages. Later we bleed from the same side when treating full-blown or chronic lesions. In those cases bleed from veins as close to the lesions as possible (*Therapeutics,* Part 5). Hippocrates described the two purposes for all evacuations: to evacuate from nearby lesions, and, as in treating spasms, to divert to the distant and opposite sides. As noted in my *Phlebotomy* the diversion should be made in as straight a line as possible, and not to cross two diameters.[627] For example, when treating nose-bleeds, hemorrhoidal bleeding, and menorrhagia, consider the up-down, right-left, and front-back aspects.

Another example of evacuation and attraction from a nearby vein, use veins on the right hand when treating maladies of the liver, and the left for the spleen. When bleeding to reduce harmful thick blood, the left side is better. But take note that the decision is not based on laterality, but on the fact that there is no vein from the spleen that goes directly to a hand: the anatomy is clear on that point. If you propose to treat spasm, bleed to the opposite side if that is feasible, because that is where the bleeding can be more easily controlled. Avenzoar (*Teysir,* Part 7 of Book 7, on Pleurisy) wrote that the phlebotomy should be made from the basilic vein of the opposite side. However, the young physicians of today argue that it is better to bleed from the same side when we bleed to reduce the volume of blood, basing their doctrine on logic and syllogisms (ie not on common sense). I think their reasoning is faulty, neither rational nor true, and their practice threatens the patient. Whenever you bleed you must be confident that you can control the amount of the loss, as advised by Hippocrates (*Acute Illnesses*, Part 2). And Galen emphasized that the decision was specially important when the pain of the pleurisy reached as high as the shoulder (*Therapeutics,* in several places).

Avicenna, when he bled during the early stage of a disease of repletion, said that phlebotomy was all that was necessary. It was both evacuative and diverting when from the same side, but not in later stages, when suppuration had occurred. When there was spasm and he wished to attract from the lesion, he bled from a distant site on the opposite side. But if he dealt with apoplexy, quinsy, and large threatening and painful abscesses, he bled from the opposite side from the very beginning. He then bled enough almost to make the patient faint; and that is the policy of our school today. However, we must not exceed the patient's tolerance.

Now we return to our topic: Although Avicenna specified which veins to use in particular maladies, we follow Galen's advice to leave that decision to the surgeon.

The Technique of Phlebotomy

Albucasis described three methods: for most cases in the axis of the vein (ie lengthwise); for special cases, incise across the vein; for arteries use ligatures and cauteries. He described three types of phlebotomes: a cultellary lancette, a myrtle-leaf shaped knife (a long lance), and the flammette, as used in bleeding horses.

---

[627] Two Diameters: The body is divided along three axes(diameters): sagittal, transverse, and from side-to-side, dividing front from back (LDR).

The Fifth Basis deals with the amount of blood to take. Galen said (idem, Part 3) that the determination was indefinite. As it is for all medicines, it is a matter for ad hoc decision (ie conjecture). Nevertheless Arnold reminded us that Hippocrates set the rules: determine the amount of the bleed according to the patient's tolerance, by the weather, the geographic region, the age, and the nature of the malady being treated. In respect of the last, consider the intensity of the ailment, and the force of the otflow. You may decide to bleed all at one session, almost to a state of collapse when the patient is robust. But when the patient is feeble, you may bleed small amounts during several episodes. Assess the patient's vitality by feeling his pulse (Galen); when it is irregular and weak, immediately stop the bleed.

Examine the blood as it comes forth. When the color improves, stop the flow (*Regimen for Acute Illnesses*). Galen and Avicenna said that a major bleed should not exceed six pounds; a minor bleed is one-half pound; and average bleed is one pound.

Damascenus said that persons who had regular (ie prophylactic) bleeds since youth, about once a year, can tolerate as many as three times a year when they reach age forty. However, when they reach fifty or sixty, they should not be bled more than once a year.

So much for our discussion of the amounts of bleeds!

The Sixth Basis considers the time when to bleed. Avicenna favored either of two hours, according to the urgency of the case. When it is deemed necessary, do not defer; and when there are good reasons not to bleed, wait for a better hour, or forego the bleed. Sometimes a delay will improve the circumstances and permit the operation (Arnold). He also said that an unnecessary delay may do more harm than good. In such cases you may change to another type of evacuation, such as scarifications in a baby with pleurisy. Avenzoar set an example with that treatment of his own three-year-old child, and saved his life. Averroes reported another similar success (*Colliget,* Book 7). He added that he disapproved of such treatments in general, but would allow it only if there was a raging irregular fever, as affected his son, insisting that such events are rare.

Therefore, no matter the time of day or night, when the disease is overpowering and the patient is robust (except in children) phlebotomy may be performed.

As to the chosen hour, Galen said that it may be set by the location of the malady (*Critical Days,* Part 3).

When the disease is based in the lower parts, and the last meal has been digested, and the residue has been defecated, and the weather is clear and not stormy or rainy, as noted in The Almanac (book of Concordances), and the season is Spring or Autumn, and if winter is approaching, choose a time near mid-day, between the second and third hours.

If the disease is based above the middle of the body, and the moon is clear, choose the seventh, ninth, or eleventh hours when the moon is in ascension. When the moon is waning, choose the seventeenth, nineteenth, or twenty-first hours, except when it is in conjunction or opposition; and it is in a good zodiacal zone (see my book on astronomy).

However, when the malady is based in both halves of the body, (I assume that the Reader has some knowledge of Astrology), and although the causal humor affecting one side is greater than that of the other, the Physicians claim that the humors from above inflame those from below, and have the greater influence. Therefore, the influences of the primary and the secondary stars, of the comets, of the air itself—all of which were called Celestial Signs by Hippocrates—all are important. However, I advise that we look for certain signs, and ignore the uncertain.

The science of Prognosis is full of hypotheses and doubts, as noted in the varying opinions of the Salernitan Physicians, Avicenna, Averroes, et al, as they are in all things in medical practice. In that light let us consider the so-called *Egyptian Days*.[628]

---

[628] The following is what Joubert wrote in 1580. "Guy had good reason to be skeptical about these things (ie Egyptian Days), as was Arnold, who advised the same for everybody. Arnold wrote this to an old authority on kalend days "I have no reason to comsider them to be evil or otherwise supernatural. If such, they would have affected everybody, including kings and pharaohs. If in ancient times they were connected to some constellation, all of that would have changed by now because of the precessions." People who are frightened and avoid doing anything on those fateful days worship the God Aural; they have made me defer phlebotomy even when it was necessary, lest I oncur infamy among the common folk. I never have observed anything bad happen to me or my family. These days we can find a book by an author who teaches mathematics, in which he wrote that in every month there are two days which are Sick, or Evil, or Egyptian. *Sick* because the common folk believe that one cannot escapr except with great difficulty if at all if they become ill on those days. *Evil* because an evil constellation rules if they undertake to do any thing necessary at that time. *Egyptian* because all Egyptians observe them. They believe that there are two such days every month, worse than ten wounds. Some of the adherents sacrifice human blood to Pluto. That practice was condemned because anyone who practiced phlebotomy on those days was sacrificing to Satan. St. Augustine condemned it as a vain superstition, insisting that we not observe Egyptian Days, Kalends, months, or days. Today no one observes them. (Joubert)

There are almanacs that describe good and bad days, as were revealed by Job. Although of it is fantasy, I will continue the matter of Egyptian Days for the sake of the curious Reader, and here I will translate (ie from an Almanac)as follows:

"An Angel revealed the facts of happy and perilous days of the year to Job. Whoever wishes may learn that certain happy days are better for selling, buying, planting, inseminating (ie a wife) to enlarge our race, to begin to build a house, to travel, to make a pilgrimage, to select merchandise, to go to war, or make business trips. He should learn the days. Certainly he should not waste time in travel, or lose time in any gainful enterprise. Remember that Job obeyed the Angel of God, who knew which days were good. He knew what were the good days for a child to be born, or the bad. He knew on what day he should enter school, and when to choose a career that would make him rich if he perservered. Those days occurred in every span of twenty-eight (ie a lunar month). (EN).

Although the matter is not to be taken seriously, it is pert of every-day gossip. Everybody knows this jingle: "The waning moon gathers in the Old Folks; and the New Moon heralds in the Young"[629]

I do not take seriously most of this material. However, Master Arnold noted that phlebotomy was most successful during the third quarter lunation: the blood is not too thick and the rheum does not flow. Arnold paid little attention to the concerns of the Salernitans about what time of the day was better for phlebotomy. He did note that small veins were more easily seen during the early evening hours than during the morning. He also stated that veins on the left side were more apparent during the winter, and those on the right during the summer. The humors that the surgeon wishes to evacuate should be localized accordingly.[630]

The Seventh Basis: Three items.

The Phlebotomist: We agree with Halyabbas (Part 4) that he should be young, adept, clear-sighted, and experienced in bleeding. He should be well supplied with lancettes and pointed (ie myrtle-leaf) blades. He knows how to massage (ie slap) the vein and to tie the band proximal to where he will incise. He knows how to identify the vein by palpation with the ball of his finger. He will grasp the knife with two or three fingers and will gently incise the vein. He will not penetrate or transect it, nor will he injure a nearby artery or nerve. After a sufficient bleed, he will bandage the wound over a pledget of cotton-fluff. He will have at hand a supply of cotton, bandages, and red powder and be able to control hemorrhage, should that ensue.

The Phlebotomy : Before, During, and After.

After deciding that the candidate is suitable for bleeding, and assessing the thickness of his blood[631], the surgeon will warm a cool room with a stove near the patient, and ascertain that the patient has bathed before the phlebotomy, especially if he plans to bleed

---

[629] Arnold also is the source for the verse; he explained it. Young women more often have their menses during the first quarter lunation. The older woman has hers during the last quarter. Those of middle-age, have theirs in between. During the phase when the body is less watery and the joints less rheumatic and swollen, and the blood is thinner, phlebotomy is better suited to reduce the water in the body. The third quarter is when the watery content is moderately reduced and is the best for phlebotomy. Guy summarized Arnold's meanings in two sentences: "When the body is less watery, the rheum is not increased, and the blood thickens. That is when the bleed is more sluggish. (J).

[630] In earlier Latin and French editions of Guy a verse reads,"In Spring and Summer the Right, In Autumn and Winter, The Left". Joubert said that the verse appeared on the margins of the texts, and that it can be found at the conclusion of *The Salernitan Regimen*. Arnold explained that the right side predominated because choleric and sanguine humors were more abundant during the summer, and that the choler was collected in the gall baldder which is on the right side of the body. In the Autumn, melancholy (ie splenic) predominates and remains so during the winter; hence the left side (EN).

[631] Are the veins full and firm? (LDR).

from the hand or foot.[632] When the veins were not easily seen, the *Book of Elhangi*[633] advises that we apply a plaster for a full day, to be removed from the selected site at the time of the operation. The plaster should contain yeast.

You may increase the rate (ie force) of the flow by having the patient drink soup and wine. A robust patient may sit, but a less vigorous patient should recline during the operation with his head slightly elevated.

During the procedure the patient should wear a belt studded with stones, or wear a ring with them, or carry them in a purse (ie hematite). They are useful in preventing hemorrhage. During a bleed (ie from veins at the elbow) the patient should squeeze a rod, or clench a fist, or cough. You may gently slap his back between the shoulders.

When the weather is cold and you suspect that the blood will be thick, make a larger incision. The opposite holds when the weather is warm. If you plan a second bleed (ie after a brief interval), make the initial incision larger. When you treat spasm, or a feeble person, use a smaller incision in an elevated limb.

Do not be amazed when that small incision in the elevated limb brings forth lovely thin blood. Avicenna said that the arrangement allows the clear thin blood to float to the top while the thick troublesome blood is retained.

If the patient faints. Splash cold water and apply cold cloths to resuscitate him.

After The Phlebotomy, when the patient is warm, let him eat a pomegranate and drink cool water (Galen). If he is not warm, give him wine containing sage-leaves, as used by Arnold.

Thepatient should lie supine, turned somewhat to the side of the bleed. Keep shut the windows and doors to dim the bright light (Arnold)

Let him eat a moderate-size meal after one hour. He should not gorge, French-style, lest the bleed resume.[634] The food should be solid and of good quality, that which engenders blood and acts as medicine against his illness, if there be such. The wine should befit the food, not the patient's usual consumption (Jean St. Amand).

Two or three hours later, if he wishes, let him sleep for a short nap. Galen says he should not lie on the side of the bleed (*Therapeutics,* Part 9). Avicenna did not allow sleep immediately after the bleed lest the humors which you have attracted to the bleed subside, and to avoid careless movements during sleep. The restrictions should obtain for three days.

---

[632]   The bath will remove grime or mud that will obscure the vein; the relation between dirt and infection was not a concern in these wounds, although it is implied when 'laudable pus' was preferred as part of the treatment of other wounds (LDR).

[633]   *Elhangi* was Rhazes' *Continens*. The Arabic name was *el Houay,* was spelled in several ways in translations (EN).

[634]   French Style of eating: We are told that the crude table manners and lack of table-ware were condemned by the Italians. The niceties of modern French culinary methods and style of degustation were introduced into France by a daughter of the Medicis when she came to France as the bride of a young king (LDR).

Examining The Shed Blood

Many authors, including Gordon and Henry, have described it, and the many associated fantasies. I shall leave them to the Physicians. Let it suffice that the surgeon should save the blood and measure the amount, and note whether or not the blood is healthy. Healthy blood in the basin means that the body is healthy. If the spilled blood is bad, consider it as good riddance.

Good blood is neither too thin nor too thick. It is foamy, warm, and pure red (ie no flecks), and has a pleasant odor.

Bad blood is opposite the foregone. It tends to be watery, yellowish, and bitter, and it has a bilious odor if it is choleric. It may be thick and dark-yellow and malodorous if it is melancholic. It may be sticky and pale and have the flavor and sharp sweet odor of watery phlegm.

When it is watery and smells of urine, it indicates overindulgence in wine, or weak kidneys. If it is granular or ashen it indicates leprosy.

Examining the clot by washing it is described in the chapter on leprosy

A dark or greenish, or gray, or violet clot signifies corrupt humors and a tendency to abscess, fevers, and pustules. A thick skin over the clot that resists pressure with a blunt instrument-a necessary test-signifies a tendency to obstructions. A fatty clot may indicate coldness or dryness; the difference is the result of the patient's manner of living.

Blood that does not coagulate is not normal, whereas normal coagulation occurs within one-half hour, as it cools (Gordon, *On Black Bile).*

For more information about these matters, especially about prescriptions for diets, regimens, and laxatives, consult the Physicians, and avoid dangerous errors.

Cupping

Cupping is a method of evacuation through the skin. The cup is a small jug with a narrow mouth and a bulging belly. It is made from a horn, or from copper or glass (Avicenna). It may be used with or without scarifications. When one scarifies, he actually withdraws the desired humors from the body; without scarification the evacuation is invisible. Avicenna said that cupping is more effective when used to attract humors that are close to the surface. Halyabbas (Part 2, Sermon 9) compared cupping, phlebotomy, and leeches. Phlebotomy works at a greater depth; cupping works near the surface; leeches work at intermediate levels. Phlebotomy is the most effective evacuant, and leeches are better than cups.

Various Uses For Cups

As a method for prophylactic evacuations and as a treatment, cupping is available in more situations than the others, provided that it is preceded by a general evacuation, and that it fulfills its own special requirements: to use it in the proper cases, and to use the proper techniques.

Cupping With Scarifications

Here we can see what is evacuated, as a substitute for phlebotomy when such is not feasible, as in children younger than age fourteen, in adults older than seventy, in drunkards, et al. (Galen, *Commentary on Treatment of Acute Illness* and Avicenna, *Substitutes For Phlebotomy*).

A proper treatment applies several cups in five or six places.

First: To evacuate from the head and its parts, use the fontanelles (hollows) of the neck, which can supplement phlebotomy at the cephalic vein when treating ocular maladies, inflammations on the face, and halitosis.

Second: Place cups around the shoulders to evacuate from the spiritual organs (ie heart, et al.) to supplement phlebotomy from the medial vein at the elbow when treating disorders of the chest, asthma, pleurisy, and hemoptysis.

Third: Place cups over the kidneys and hips to evacuate matter from the organs of nutrition, to substitute for phlebotomy from the basilic veins, when treating obstructions, abscesses, and pain in the liver and the kidneys, and for scabies anywhere in the body.

Fourth: Place cups on the arms when treating gout and painful joints in the upper limbs.

Fifth: Place cups on the thighs, the legs, and the ankles to subsitute for phlebotomy of the saphenousa veins when inducing menstruation, treating strangury, disorders of thefemale genital organs (ie matrix), the bladder, and for podagra. It helps in treating malignant ulcers.

Dry Cupping—Without Scarification

We aim to attract, by placing cups in eleven places.

First: We use it to treat pain from anywhere below the ribs, and to distract blood from the nose (ie for epistaxis)(*Therapeutics,* Part 5). When the blood comes from the right nostril, apply cups over the liver; when from the left nostril, apply them over the spleen.

Second: Apply cups over the breasts to distract and divert menorrhagia. Galen (*Commentary on The Aphorisms*) in such cases applied large cups below as well as over the breasts.

Third: Galen (*Therapeutics,* Part 13) applied dry cups on the forehead to shrink the uvula and stop the flow of catarrh. The treatment was designed to attract the faulty humors from within toward the surface. Therefore it works better when the cups are applied directly over a swelling (ie aposthem) and near an emunctory. Avicenna insisted that is better than other treatments. Galen (idem) applied cups on the thighs to provoke menstruation and he applied them near swollen joints as well as at distant sites.

Fourth: Apply cups over the origins of the nerves to warm them when treating paralysis (*Canon I* and *Afflicted Places*). That is how Galen disproved Archigenes' claims about the brain being the seat of the vital spirit.[635]

---

[635] In Galen's scheme the Vital Spirit was engendered in the heart, and was carried forth via the arteries (LDR).

Fifth: Place cups on the abdomen when treating colic (*Therapeutics* Part 12). And use a large flame in a cup when the gas is other than intestinal.

Sixth: Avicenna placed cups (*Canon*) over the pubis to treat prolapse of the matrix and intestines (ie cystocoele and rectocoele).

Seventh: Place cups over the ribs to elevate depressed fractured, and over other dislocations.

Eighth: Place cups in the flanks along the course of the ureter to attract renal stones into the bladder. (idem.)

Ninth:Place cups over the ears and over small openings that lead to deeper abscesses (ie fistulas) to extract the dangerous bad matter.

Tenth: Place cups over the neck to enlarge the breathing and swallowing passages, and to treat quinsy.

Elventh: Apply cups over venomous bites, punctures, and pustules, to extract the venom.

The Techniques Of Cupping

There are three elements: The Preliminaries; The Operations; The Followup.

The Preliminaries

Whenever possible—an ancient dictum—use cupping when the moon is full and not yet in decline. Both Galen (*Critical Days,* Part 3) and Albumazar[636] explained the rationale. When the moon is enlarging its attractive power is greater,[637] bringing the humors to the surface of the body. The reverse is true when the moon wanes. Cupping is better during warm seasons, between the second and third hours of the day. Before the application of the cups, the patient should bathe, and the sites should be fomented with warm water for an hour, especially of the blood is thick. I agree with Avicenna that it is not necessary if the blood is thin, when that exercise may cause too much blood to flow and weaken the patient. However, if dry cupping is the plan, these preliminaries can be eschewed, because the aim is to attract the humors to the surface, not to expel them.

The Operations

I use two kinds of cups: the horn-cup with the small opening for suction is for dry cupping; the glass cup for use after scarification. I suck at the horn-cup to pull the soft flesh into it as Nature demands, to fill the vacuum as Aristotle and others have described.

---

[636] Albumazar: Not a person. The title of a book of astrology by Al Kindi (ca 850) LDR).

[637] As observed with the tides (LDR).

The glass cup has no second opening. I insert a bit of dry lint and ignite it with a candle, and immediately set the cup. The flame will consume the air, and Nature will fill the void by attracting the flesh and its humors. Albucasis placed a small ignited candle in the cup before setting it.

Before scarifying, apply a dry cup once or twice to bring humors to the site. Then make several superficial cuts with a lancette or razor to penetrate only the skin. Sponge the area and reapply the cup and empty the air with the flame, and leave it in place for a half-hour, or until the cup is half-filled with blood. Empty it and repeat the application for a longer time, until you are satisfied with total amount of the bleed: usually between a half and a whole pound, to satisfy the treatment for plethora, yet not to enfeeble the patient.

When the initial does not remove enough, rub the tissues around it, and scrape the incisions with your finger-nails, or make new scarifications that will bleed more vigorously.

The cups applied on the breasts or other soft regions may suck in enough flesh to make the removal difficult. In such cases, simply foment the area around the cup with warm water, and loosen the cup for easy removal.

Also be aware that cups placed in the regions over the seats of the Vital Spirit, such as the occiput, may harm the memory; or between the shoulders may harm the heart; or over the hypochondrium may harm the liver. In such cases, splash cold water on the face, and provide a broth of wine and pomegranate.

While the cup is functioning, gently milk the nearby tissues toward it.

After The Cupping

Clean the site and anoint it with rosat-oil or another mitigant, and provide the care we prescribed for patients after phlebotomies.

Leeches

A leech is a worm-like creature shaped like a rat's tail. It has yellowish stripes on its back and some pink on its belly. The best leeches are found in clear (ie running) water, whereas the harmful ones are in stagnant water; they are discolored, have bulging heads, and are venomous.

When To Use Them

Albucasis used leeches where cups cannot be affixed: the lips, the nose, the gums, and places where there is little flesh beneath the skin, as on the fingers or over joints. Avicenna used them to treat serpigo and malignant ulcers. Theodoric and others applied them over abscesses and emunctories when a desired suppuration was delayed or incomplete. Some authors (Halyabbas) applied them on hemorrhoids where they can withdraw blood from deeper levels than do cups.

How To Use Them

Do not use leeches on replete patients until they have completed a general evacuation. Use only freshly gathered leeches that have been isolated in a bowl of clear water during an entire day, when they can expel whatever was in their bellies (ie vomited). Then massage the site of application until it reddens, and smear a bit of blood over it, or make a light scratch to bring forth some blood. Lay on the leech by hand, or with a reed. Apply at least two or three, or more as needed. When they are stuffed with blood, pour some vinegar on their heads, or some salt, or some aloes, or gently saw a horse-hair under them to allow you to detach them.

Then blot the area with vinegar and water. If the bleeding persists, apply a plaster of bol d'armenie, balaustes, and other hemostatics. If you suspect venom, dose a good theriac (Arnold).

CHAPTER 2

Purgative Medicines

G alen said that there are many ways to purge humors in addition to bleeding (*Techni,* Part 3, *Commentary on Aphorism #4,* and elsewhere), as by laxatives, emetics, nasal discharges (ie of rheum), salivation, coughing, urination, vaginal discharges, hemorrhoids, exercises, massages and baths to induce sweating, and by abstinence (fasting) which is purge in reverse. Here we will deal only with evacuations via the belly: laxatives, emetics, and enemas. Those routes and bleeding are concerns for surgeons, and we shall not discuss the treatments that belong to the physicians, except when the responsibilities for treatments involve both professions. We accede to Heben Mesûe who said that the use of laxatives may carry grave risks. Galen (*Medicaments*) said that the most important use for laxatives is to evacuate cacothyme (ie badly combusted humors). I quote from his *Commentary on Aphorism # 1.* "A purge eliminates unpleasant as well as harmful matter." A medicament that empties the belly also is a means of preserving health (ie prophylaxis) when one uses it correctly (*Almansor,* Part 4). Proper use includes the assessment of the need, the dose, the quality, the time for the administration, and how it affects the belly (Halyabbas, *Royal Dispositions,* Part 2). It can destroy a man or seriously deplete him Avicenna wrote (*Section 3, Book 1*) that all laxatives purge and deplete. A purgative is one of the three treatments that require a physician's supervision.[638]

Therefore, the pharmaceutical (the medication) is necessary and useful. Before prescribing it I ask six questions:

First Question : Who needs purgation? Galen (*Purgation*) in opposition to the followers of Asclepiades and Erasistratus, said that we should evacuate all superfluous humors except pure blood by a general purge and we should direct the laxatives against particular humors. We aim to eliminate only the harmful humors, and leave the others (*Commentary on Aphorism # 1*). If there is too much phlegm, get rid of it. If it is choler

---

[638] The Reader will soon be aware that laxatives were not a simple matter. The medications seldom were mild lubricants; more often they were irritating stimulants of peristalsis, and could cause serious complications when administered to a patient with bowel obstructions or intra-abdominal inflammations (appendicitis). And the cramps induced by the medicines could be violent when the patient was severely constipated (LDR).

either red or dark (ie melancholy), eliminate them, and leave the phlegm. The same is true for eliminating an excess of serum and water. However, if whole blood is in excess, a proper general evacuation is called for, and that is by bleeding. I quote Galen's *Use of Medicines.* "Use the choleric laxatives to get rid of choler, the phlegmatics to purge phlegm and hydropsy, and use melancholic laxatives as indicated. If you use them incorrectly, you will void good humors and leave behind the bad ones."

Therefore, in general, use phlebotomy to evacuate a plethora of natural humors engendered by normal digestion, and use medicines to evacuate the non-naturals. Although it may be possible to find somewhere in nature a medicine that will cause a bleed on demand (ie such as hematemesis) as Galen described it (*Medicaments)* in the case of a young man from the suburbs who butchered a piglet. For some reason, he had fed it some herbs and noted that the liver-ducts were full of blood. He thought it meant that the herbs were evacuants, and he gave some of them to friends, and he sent them to Hell. He failed to admit his error, and he was condemned by the courts.[639] So it is, if such herbs do exist, ignore them, as we do poisons.

There are four things to consider when prescribing laxatives:

A.   A. To purge cacothymia. Hippocrates (*Aphorism #2)* sang this well-known song: "Evacuation treats all maladies caused by plethoras, meaning repletion both in quantity and in quality, or both together. Phlebotomy evacuates quantity and laxatives purge specifically, according to what matter is corrupted." In his *Commentary on Aphorism #2 and #6* Galen said, "When the food is unnatural (ie poisoned) both bleeding and laxatives are used."

B.   As to the severity of the malady, Galen (*Therapeutics,* Part 4) wrote that phlebotomy not only reduces the volume of the blood, it lessens the impact of the illness. And that leads us to Hippocrates (*Ulcers)* who wrote that a purge reduces the volume of repletion as well as the force of the illness. He gave examples of the triple force of maladies, as affecting the site (head, wounded abdomen, etc), the severity (as of a wound that requires sutures), and the invasive nature of the malady, such as wounds near joints, or wounds that putrefy.

C.   Purges not only evacuate, they also divert from the affected site by discharging the matter in feces. For example, we purge rheum from on high by eliminating it below. But after the formation of the rheum has been arrested, we should eliminate what already is in the primary site, directly, or from nearby.

D.   Staging the dosage. Give a strong laxative early in the illness (Hippocrates, *Aphorism #2)* and reduce the potency of subsequent doses, yet do not to repeat doses to completely eliminate the bad matter. Nature will digest the residue (Avicenna, *Treatment of Fevers,* Galen in *Commentary, and Therapeutic,* Part 11, and in my Chapter on Phlebotomy.

---

639   Guy's report here is incomplete. The perpetrator confessed only when he was tortured (J).

Second Question: As to suitable candidates. Hippocrates (idem) said that he who is pot-bellied and wide at the hips can better withstand a laxation of the lower abdomen. He who is larger in the upper body can better be evacuated by emetics, excepting if he is asthmatic. Judge the bulk of the patient by the flesh and not by the amount of the spermatic tissues (Albert of Bologna). He also said that persons who are easily provoked to vomit are affected by their own choler.

Other suitable candidates for laxation are persons who habitually over-eat, who exercise little, and who always seem to need to take medicines (Halyabbas, *Conservation of Health,* Fen 3 of Part 1)

Persons who have previously been dosed with laxatives tolerate them better. Galen *Use of Medicines*) advised us to question a patient about his prior experiences with laxation, and how his belly had reacted. You must inform an inexperienced patient who that good health depends on proper defecation. If he is robust and has used medicines, his belly will tolerate the laxatives, he will defecate easily, and he will require less potent laxatives in smaller doses. But if someone is resistant and he needs more potent laxation and objects to it, medicate him prudently.

First: A patient with a weak mid-abdominal region has less success with purgation below.

Second: Do not purge a healthy body. Medicines act by similarity, or in proportion (as taught at Montpellier). A useless laxation that cannot find and eliminate a bad humor may turn against normal tissues and their moist roots, and will irritate the anus.

Third: Do not purge persons who are malnourished lest they defecate normal matter. Master Albert described a leper in whom the laxative cause much misery and worsened his constipatoion. If the bad humors are converted into bad tissues, there is no reason for laxatives, except in emergencies.

Fourth. Do not purge raw and undigested humors (Hippocrates, *Aphorism, #1*). Medicate and evacuate only what has been digested, especially at the onset, and do not try to uproot all of it unless the affection is violent. Galen did not define the furious nature, but our School said that a large quantity of bad matter alone made it furious.

Galen said a continuous fever (ie synoche) with colic, frenzy, quinsy, and anthrax all require urgent evacuations of raw humors.

The bad humors move from one region to another until they reach a principal organ. When that happens, evacuate them at once. But if they settle in one part and cannot be displaced for evacuation until they have been digested, do not attempt to hasten the process (The Commentator—Averroes). However, do not allow them to remain indefinitely and overcome Nature's defenses; help Nature. Avicenna (*Treatment of Putrid Fevers)* cited Hippocrates who claimed that medicines simply assisted Nature, and should not be allowed to interfere. In Part 3 of *Techni*, he added the Nature was the operator and medicines were the administrator.

Fifth: Do not purge what already is empty and feeble. Do not make work when there is no need (Hippocrates). Avicenna added that the surgeon should choose the evacuation.

Sixth: Do not purge when the patient has diarrhea.

Seventh: Do not prescribe laxatives for children or for oldsters. Children are born with the ability to reject superfluidities by natural resolution (Jean de St. Amand), and old people are too feeble for laxation. Galen (*Conservation of Health,* Part 3) avoided the use of aloes and hierapictra in prophylactic laxation. Paul of Aegina said that they could be used as treatments, but Galen said no, except in dire cases. In general, an action that weakens should not be used, including restriction of sweats, baths, coitus, etc.

Eighth: Do not use laxatives that evacuate via the rectum when there are abscesses or eschars there. Galen (*Therapeutics,* Part 13) said, "When inflammation begins there, do not evacuate from below."

Ninth: When those who perform heavy labor are ill, do not use laxatives that compete with their own natural ability to consume the humors. And if they are unusually large and brutish, and barbarian *(Colliget)* they can recover from serious ailments without need for medicines or physicians.

Tenth: Do not purge a pregnant woman lest you contort the matrix and cause an abortion, excepting in cases when the matter is venomous and can spread to principal organs and destroy the patient and her fetus. When you treat a threatened abortion, purge only between the fourth and seventh months, and use gentle medicines. That advice of Averroes contradicts what Hippocrates wrote in his *Aphorisms.*

However, Galen did not forbid the use of all purges in those cases. In *Therapeutics,* Part 8) he wrote about phlebotomy and said that a physician must use his common sense and weigh all the pros and cons before deciding. When there are contradictions, do not favor one side and ignore the others. Remember, that one can mix laxative medicines, or change to different types, or go to evacuations other than by laxatives.

The Third Question deals with the selection of the purgatives, avoiding the harsh and the terrible, and favoring the familiar ones (Galen, *Use of Medicines*). In that regard, both Heben Mesûe and Avicenna described the available types. One type is mostly attractive, such as scammony and turbith. Another is more a constrictive and contractive, as myrobalans. Another is a mild and gentle intestinal stimulant, as are tamarinds and cassia fistula. The fourth type serves mostly as a lubricant, as is the mucilage of psyllium. I will not delve further in the details, and I will refer you to the physicians for more information.

However, I am pleased to cite Heben Mesûe for the laxatives in common use[640] for the evacuation of choler and other humors.

For Choler: The Simples are scammony, rhubarb, aloes, yellow myrobalans, hops, fumitory, violets, milk-whey, rose-water, plums, tamarinds, and cassia fistula. The Compounds contain rose-water diaprunum, diacytoniten, and diadactyl.

For Phlegm: The Simples are turbith, agaric, carthame, colocynth, and chebules. The Compounds contain hierapicra, white wine, herb benedict, coccia pills, diacarthame,

---

[640]   Hereafter, I will list the substances by name and will not provide the amounts. The weight and measures of 1363 as modified through generations of copyists and editors do not correspond with those of 2007. The recipes are repeated in Doctrine II which follows (LDR).

pills containing agaric, iris-roots, marrubium, turbith, hierapicra, colocynth, sarcocolla, myrrh, and a sweet conserve (ie jam).

For Melancholy: The Simples are epithyme, esula, cuscuta, Indian myrobalans, polypode, lapis lazuli, liseron, hops. The Compounds contain Diasenna, imperial cathartic, hierarufinum, hieralogodon, theodorica, reglisse, epithyme, and senna.

For Excess Water (ie hydropsy, edema): Simples are tartar, iris-juice, wild cucumber, milky herbs (barley, et al).

Compounds for Evacuationg All Humors contain conserves of violets, borage, bugloss, lemon-peel, ginger, diatragacanth, diagridon, turbith, senna, and sugar-loaf. Another from Taddeo: catholicon, cassia fistula, tamarinds, seeds of manne, senna, polypode, esula; diagridum, anise, fennel, melon-seeds, cinnamon, syrup of violets and roses.

The Fourth Question deals with the volume of the evacuation. Hippocrates said that complete evacuation is better for the patient and satisfies him, whereas a less complete elimination is not effective. Assess the benefits in the amounts of the discharged bad humors, the intensity of the malady, the affected region, the age, et al. All of them determined what laxative was used. Sleepiness and thirst are signs of a complete evacuation (*Aphorism #14*). Another sign is a change toward normal of the evacuated matter. (*Epidemics*). Rabbi Moses agreed. A sure sign that the disease has been arrested is the appearance of normal feces in several subsequent defecations. Avicenna said that if a small amount of bad matter remains, Nature will resolve it. He estimated that a small evacuation was at least three pounds, that a large one was eight pounds, and that the average was six to eight.

The Fifth Question: When to dose the laxatives. As we described for phlebotomy, there may be an urgent need for prompt action, or the situation may allow you to pick a time after the contents of the intestine have been completely digested. But if the matter is copious or furious, or is at a dangerous place, there will be no opportunity for any delay that may lead to complications.

When you are in position to choose the time, dose the laxative after a meal has been digested, and do not administer it in an early stage of a malady. The Commentator explained that the surgeon should intervene and imitate Nature only when a natural evacuation does not occur early in an illness. A normal complete digestion follows a meal, followed by the separation (ie nutritive chyle from the residue), and the elimination of the residue. Therefore, dosing a laxative should imitate the normal pattern, unless waiting cannot be abided.

When you prescribe prophylactic purges, do so when the moon is waning, and during the Spring (Hippocrates). When it is very hot or very cold, wait until that spell of weather has passes. There are about fifty 'dog-days' during which Sirius, the dog star, rises (Galen *Use of Medicines*, and *Aliments*, Part 2) when the Romans, between the last twenty days of July and the first twenty days of August believed that all medicines are unsafe (*Aphorism #4*).[641] Not

---

[641] Sirius, the dog star, is the brightest of all. The Romans attributed to it great influences on the human body. The so-called 'dog-days' (24 July to 26 August) were when the sun rose with the star. Now, due to the precession of the equinoxes, Sirius rises after the dog-days. The Egyptian New Year began at the end of the dog-days, called the Sothaic or Cynic days in Egyptian-Greek (EN).

all years were considered favorable for the patient, according to the Subtle Doctor (*Colliget*, Part 6).[642] Those who take prophylactic laxatives once or twice every seven years fare better than those that are dosed every year. As we have observed, they seem to become ill at or near those times, a fact for every person to consider for himself.

As to the time of the day: Usually laxative potions are taken in the morning, and pills are taken in the evening; electuaries are taken at any time, as determined by the physicians. A clear day when a breeze comes from the south is preferred, especially during the winter. When it rains, follow instructions from the Table of Concordances.[643] Pick a time while the moon is enlarging up to the time of the full moon (Jean St. Amand) when the humors are more active and ready for evacuation. Also, pick a time when the moon is in a humid zodiacal sign, as are Cancer, Scorpio, Pisces and Libra, and avoid the bad signs and when Jupiter is up, as claimed by Ptolemy (*Centiloquium*). All of this material is known to the astrologers, and I have dealt with it in my book of *Astronomy*.

Question VI: The regimen (diet) before, during, and after the Laxation

Before: Hippocrates provided a diet that stimulated the flow of the digestive juices, and dilated the softened passages (ie the intestines).

Choleric matter is digested by cool Simples: aperitives, as the five capillary herbs (ferns), endives, escaroles, chicory, rostrum porcini, oxalis, the cool seeds, pomegranate juice, vinegar, and water. Compounds contain oxymel, vinegar-syrup, venus hair, adiantum, polytrich, ceterach, scolopendre, ceterach, endive, chicory, escarole, lettuce, rostrum porcini, cold seeds, sandalwood, roses, violets, water lilies, pomegranate juice, and sugar-loaf.

Phlegm is digested by Simples: five-roots (ie dropwort), calamint, pouliot, hyssop, marjolaine, saturia, mint, anise-seeds, fennel, caraway, pepper, ginger, spikenard, honey, and squill-vinegar. Compounds contain oxymel, squill-vinegar, roots of fennel, parsely, celery, cuscuta, asparagus, flowers of dog-tooth soaked in vinegar, hyssop, calamint, oregano, chamedrys, abrotonin, seeds of anise, fennel, and caroway; ammi, ginger, zedoar, honey, chick-pea broth.

Simples for digesting melancholy include borage, bugloss, fumitory, scolopendre, ceterach, adiantum, tamarix, thyme, epithyme, capers, and aromatic wine. Compounds contain syrups of reglisse, fumitory and bugloss; hops-flowers, willow-twigs, bark of fraxinus, tamarix, scolopendre, venus hair, melissa, melon seeds, cuscuta, seeds of nettles, anise, and fennel; reglisse, calamint, aromatic roseau, behen, lemon-peel, spikenard, aloin, broom, raisins, and honey.

Finally: Avicenna advised softening with natural foods before dosing a laxative: loosen the bowels with syrups, and cabbage broth. If the patient is constipated by impacted feces, administer a clyster before giving the laxative.

---

[642] Averroes was called "The Subtle Doctor" and "The Great Commentator" (LDR).

[643] Use the astrologists' almanac when you cannot see the stars (LDR).

During the Dosing Urge the patient not to regurgitate the dose and distract him by massaging his limbs, giving him apples to munch, or the crusts of bread dipped in vinegar (ie sour wine). Do not let him sleep if the laxative is not a solid (ie a pill). If it is solid, allow him to nap until there are signs that the laxative is working; then keep him awake. Avicenna suggested that the patient nap before he takes his medicine, a natural act. When the laxative begins to act, tell the patient to walk. Hippocrates, *(Aphorism # 4)* offered a sip of hellebore to encourage his movement. However, Avicenna advised not to disturb the patient from that moment on, lest you interrupt the laxation. And if you suspect that something is not right, administer an enema rather than repeat the laxative. A double dose on the same day is not advised.

After the Defecation clean the stomach and intestines. Galen (*Therapeutics,* Part 7) used a tea-like beverage. The Parisians use a broth of a beef-bone. At Montpellier they use a chicken-broth. After the cleansing, Jean de St. Amand advised swallowing an astringent before eating solid foods, to warm and comfort the belly.

The diet of a recently purged patient should be mild, with meats like hens and capons. Expand the diet as the patient recovers his strength[644] as stated in *Aphorisms* and in *The Regimen For Acute Illnesses.* We can assess the nature of the defecated product, and have clues as to the cause of the malady, and how to treat it (*Techne,* Part 3).

Emetics

A vomit is a purgation through the mouth induced by an emetic medication. It was long recognized as a prophylactic measure by the ancients (Galen *Usum* . . . . Part 5) who recommended monthly emeses after normal meals. Some advocated it more than once a month.

Emesis also is a therapeutic measure. Avicenna used it to treat chronic illnesses, epilepsy, mania, leprosy, podagra, sciatica, and disorders of the kidneys and bladder. It serves both as an evacuative and as a diversive, although for the most part it purges the stomach and the intestines. The best candidate has a robust upper body and is not a weakling and is not phthisical. Prepare him for an emesis by feeding him a rich meal with greasy and nauseating foods, often with figs, leeks, onions, beans, pork; and with liberal amounts of tea and wine.

Induce emesis in three ways: a mild measure is tickling the palate with a finger or with a greased feather after the patient gulps warm water with oil.

A second way uses a potent emetic: a decoction of the seeds of spinach, radishes, eruca, leeks and onions. Another decoction contains radishes, hellebore, and salt in a jar that is set in the ground for two days. Another is a serving of sliced radishes.

---

[644] Another reference to the demanding and challenging experience of therapeutic laxations (LDR).

A very potent emetic is Nicolas' troche (Myrepsos), made from thapsus, saffron, nux vomica, spurge, juice of cabaret, and honey, taken with warm water.

Dose the emetics after a meal at mid-day, and cover the patient's eyes. After he vomits, rinse his mouth with vinegar and water. He may eat a light meal after a delay of one hour.

Enemas (Clysters)

The word clyster is taken from the name of the stork[645] which treats its own belly-ache by scooping sea-water and infecting it through its cloaca (Galen, *Introduction to Medicines*).

It is a valuable means of rejecting superfluidities from the intestine, and will never reach the mouth or vital organs. It also gets rid of gas and relieves pain in the kidneys and upper abdominal organs. It has three forms.

The Mollitive or Lenitive: a decoction of mauve or its flowers, figs, common oil, and salt.

The Detergent: mercuriale, brancus ursinus, pear-leaves, fat figs, anise, fennel, cabaret, cassia fistula, hierapicra, herb benedict, honey, and oil.

The Restrictive: plantains, roses, balaustes, red powder, bull-fat, and egg-whites.

Enemas may be administered at any time.[646] However, they work better when used before meals. The patient kneels (his belt is loose) and he keeps his mouth open during the injection. After the fluid has been delivered, let him massage his abdomen and recline in the side of the affection, and try to retain the medication for one to two hours.

Suppositories

These are about the size of a small finger, the shape of a candle, and are made from honey heated with salt, and covered with oil. The recipe may include some bile of a mouse to intensify the action. At times a suppository is made from a hard soap and lard, or from ground mercurial, or from a peeled cucumber. They purge and attract bile to the rectum. Avicenna used them to treat colic.

Use them carefully when the anus is ulcerated.

---

[645] Guy's etymology is obscure. The stork is 'cicogna' in Latin and a Greek 'κλψστερ is a syringe (LDR).

[646] The ancient apparatus was a cannula attached to a bladder which was emptied by squeezing. At the end of the 15th C, Gatinaria of Pavia introduced the syringe for enemas. Guy's syringe was not used in that way; rather it was designed for treating urinary dfisorders. In 1668, Regnier de Graaf interposed a long (one or two yards) flexible tube between the bag and the cannula (EN).

CHAPTER 3

Cauteries

A cauterization is a purposeful application of heat by a surgeon to a selected part of the body. The application is of two kinds *The Actual Cautery* applies heat quickly with a metallic instrument, or with hot ashes of an ignited root of aristolochia or asphodel, or with lump of burning sulfur, or with some boiling oil. The application is planned, and is not an accident.

The other method is *The Potential Cautery*. The heat of which is not apparent at the time of the application, yet it becomes obvious as it works. The potential cautery is a caustic substance, a so-called eruptor. Some eruptors, such as quick-lime and soap, or cashew-honey, erode deeply and create eschars. Some work on the surface and produce blisters. They are such as cantharides flammula, and lupins.

Actual cauterizations are more accurate, and are more effective than the potentials, and, as Albucasis claimed, they are simpler and do not harm the surrounding tissues or spread their effects to principal organs. We use the actual cauteries when the patient is not too timid and forbids us to apply heat to divert and destroy. The caustics cause more pain and produce more eschars, and damage more of the region, which in turn attracts more unwanted humors.

Avicenna preferred the gold cautery; Arnold and I use it only when we treat eyes and other delicate regions. For all other uses, I prefer the hot iron rather than the gold or silver, because I can judge their heat by their color, and I do not need the services of a goldsmith.

Albucasis said that cauteries can be used any where, but they work best in persons with cool and moist complexions. However, they work even when they are not contraries, that is, in persons with warm and dry dispositions, because they succeed when complications appear. But when there are no complicatioss, and the person is warm and dry, the cautery is not useful, and may be harmful, as claimed by Albucasis, Bruno, William of Saliceto, and Lanfranchi, and their followers, despite the claims of Avicenna and the beautiful treatises of Albucasis, Hippocrates, and Halyabbas (Part 2, Sermon 9). In our time, the old rules are not followed Henry said that they are abused by idiots who lack knowledge, and who fail to precede the applications with purges, especially in serious cases. The real values of cauteries are hidden by the wrong-doers.who blame it for their failures, and complications. And there are many more abuses, as Albumasar described them in his *Astronomy.*

We consider cauterization to be the last resort of Medical treatments. After the diet, the medications (laxatives), and the bleedings, come cauteries, but not before them, when they can be harmful. Furthermore, a General Evacuation should precede all kinds particular treatments.

Before we use cauteries, we should answer three questions.

I. The Use Actual Cauteries.

There are *Seven General Applications*

A. To comfort the sick parts, especially when they are cool and moist. Galen (*Therapeutics,* Part 4) cited Hippocrates for saying that dryness betokens health and moisture is unhealthy.

B. To limit corruption in any part of the body (Avicenna, Part 4). Both he and Galen cauterized esthiomenes, undermining ulcers, and corrupted bones.

C. To resolve collections of bad matter. Albucasis and Halyabbas applied cauteries over the gouts and other painful joints, for chronic headaches, and for other painful maladies.

D. To control hemorrhage. Avicenna cited Galen (idem. Part 5), and advised long applications to create a thick eschar. Use the hot iron or a hot medicine (ie sulfur, et al.)

E. Master Arnold used them to evacuate and divert chronic lacrimation and other superfluidities. He also used the seton cautery at the nape in the hollow between the neck muscles, and elsewhere in hollows between muscle bellies, two or three fingerbreadths away from the joints in the limbs.

F. Galen used them to incise and drain pus, and to cut veins at the temples for phlebotomy when treating lacrimation, and to seal the openings in the abdominal wall where hernial bulges appear.

G. A cautery can eliminate superfluous matter by draining an abscess, by excising glandules, and by extirpating redundant proud flesh, necrotic or living.

Special Uses

Albucasis named six, Halyabbas named twenty, and Bruno added many more, as did Roger, the Glossators, William of Saliceto, Lanfranchi, and Henry. In our time we classify them according to the eight parts of the body where they are used.

A. *The dome of the head*: To find the spot used by Albucasis, Halyabbas, Bruno, William, Lanfranchi, Roger, and his Glossators, place the palm of your hand over the bridge of the nose. Mark where the tip of your middle finger touches the scalp to indicate where to apply an olivary cautery. Use it to evacuate the brain by diversion of the humors to the lower body. Some authors burn down to the bone, others burn through the outer table as advised by Albucasis for treating mania, epilepsy, headache, lacrimation, and all catarrhs. You may touch the cranial bosses and the occiput to warm and comfort the head in cases of paralysis, tremors, convulsions, and misery (ie depression).

B. *The face* has many places for applications. Use a myrtle-leaf (ie pointed tip) cautery for disorders of the eyelids. Use a needle cautery to touch the hair-pores for permanent depilation, root and all. Use a small cultelary cautery to eliminate redundant proud flesh. Insert a wire-like cautery through a cannula to drain a lachrymal fistula into the nose. Cauterize at the temples to clear other discharges from the eye, Treat nasal polyps with the wire cautery through a cannula. Treat split lips with a small cultelary. Use the same to penetrate a dental cavity to empty it of corrupt matter and relieve pain. Excise some of the uvula with a cutting cautery through a cannula, or coagulate it by lifting it with a spoon-like cautery.

B. *At the neck* use the seton cautery to establish a passage through a fold of skin at the hollow of the nape, when treating by diversion maladies of the eyes (Lanfranchi, alone). Others, especially the Glossators, who were supported by the physicians, treated scotomas and vertigo with applications of round cauteries where Lanfranchi used the seton-cautery. They had to keep open the burn-wounds for a long time. Galen *Therapeutics,* Part 13) applied the cautery at the occiput to treat ocular discharges. What more can I say? I have treated those maladies with both, the setons and the round cauteries,. The four Masters treated rosacea and other inflammations on the face and mouth with a cautery applied under the chin.

C. *At the shoulder* apply at the fontanelles[647] of the arms, three fingerbreadths away from joints, where the underlying tissues are muscles. Use a round claval cautery with flat disc as a stopper (ie to prevent too deep a penetration) to treat maladies of the face and the front of the neck, of the head, and the posterior surface of the neck.

D. *For thoracic maladies* apply a round cautery or a seton in the hollows under the clavicles for asthma and disorders of the esophagus. Apply it in the armpits for maladies of the arms, to deterge the heart, and to palliate the miseries (ie depression, etc.). Use a double-edged cutting cautery to cut between the ribs to drain an empyem: but beware of the risks of the procedure. A fistula may ensue and will weaken the heart, which may fail when it is exposed to air.

E. *To treat maladies of the abdomen* use a round or a seton cautery. Apply it over the stomach or at the sides: the right side for liver ailments, the left for the spleen, or wherever is the site of the pain. Apply it below the navel when treating ascites. Albucasis and Halyabbas used a double or triple claval.

---

[647] Take note that the favored sites for applications were in hollows (ie fontanelles). Cauterizations attracted and diverted the bad humors from the lesion toward the 'burn'. The indentation allowed the surgeon to approach the interior of the region below the burn and be closer to what he wished to attract. Of course, the astrologic indices determined the anatomic regions where the surgeon could find and use a hollow if any existed (LDR).

F.   *For maladies of the pelvis:* Apply cauteries at the groins when treating hernias, and at the pubis for maladies of the bladder. Place it over the back for treating kidneys and gibbosities.

G.   *Apply the seton cautery on the back of the scrotum* when treating aqueous or fleshy hernias.

H.   *Apply cauteries behind the knee*, three fingerbreadths from the joint, where you are certain that you are over true muscle. Use a round claval with an arrestor to purge the entire body.

Additional information about the use of cauteries appears in many of the foregone chapters of this book.

Potential Cauteries are used in many of the same places as the actuals, but they are not comforters, and they deplete the patient. Although they are effective for draining abscesses and for obtaining hemostasis, they are not as good as the hot instruments. They are especially useful in deep fleshy places, but they are hazardous when used near principal organs.

Vesicants are useful when the skin is close to underlying bone, as under the chin, over the nape, at the ankles, and on the hands and feet, where their attraction is effective from a shallow bed.

II. How to Make Cauteries

Actual cauteries usually are metallic instruments. Potential cauteries are caustic medications.

We moderns use only a few types of actual cauteries, as compared with the much more numerous items described by the ancients. William of Saliceto used six or eight; Lanfranchi, ten; Henry, seven.[648] I use six for most purposes, and I own special tools with which I can fashion a cautery for a particular use. All practitioners should have three sizes of all their cauteries, large, medium, and small.

A.   *The Cultelary* is shaped like a knife. The dorsal, has one sharp edge; the ensal is double-edged. I use the cultelary to eliminate redundant granulation tissue, to incise and drain abscesses, and to elongate ulcers.[649]

B.   *The Olivary* is shaped like an olive-tree leaf. William, Lanfranchi, and Henry said that the business-end was shaped like an olive-pit. Halyabbas (Part 2, Sermon 9) used it to cauterize through the scalp—he gave details. I use it at the dome of the cranium and near joints, to relieve pain, and to sweep off

---

[648]   Guy names them as 'Moderns' (LDR).

[649]   The surgeon could 'rectify' a chronic ulcer by incising the margin of the ulcer at both ends of a diameter, to convert its roundness to two apposing sides, which could be brought together for closure (LDR).

corrosion from nerves without cutting them. I use it to burn away diseased bone and expose the healthy.

C. *The Dactylary* is shaped like a date-pit. I use it as I do the olivary when I want a larger size.

D. *The Punctual* is slim and carries a rounded tip which can touch and cauterize only the surface of the skin. One type carries an arrester that prevents a deeper penetration. I use it at the fontanelles of the limbs. Another type is simply a slender rod that can pass through a cannula and cauterize only at the tip, and not endanger the tissues. I use it for such purposes as creating drainage into the nose to treat a lacrimal fistula, or to remove a nasal polyp, or in dental cavities.

E. We use *The Slender Seton Cautery* to perforate (by burning) a fold of tissue grasped by the large perforated forceps. The opening is smaller than what is made by burning through with a node cautery. However, the seton's opening persists longer, and the node cautery may fall out, and is more difficult to dress than the seton.

F. *The Circular Cautery* is a ring with five attachments, each is a node. All five cauterize together, but only to the depth allowed by the ring. I use it to treat pain in the hips or under the arm, and on top of edematous gibbosities.

How To Use Cauteries—The Actuals

Select the proper site, dry it, and mark it. If necessary lay on a plate with a perforation over the site, or insert a cannula. Hold the patient securely, and have ready at least two red-hot cauteries, kept out of view of the patient. Apply them with a gentle twist to prevent adhesion to flesh, but press hard into diseased bone, and very gently on nerves. Maintain contact until the red color fades. Repeat the applications until you have accomplished what you set out to do. Thenn treat the burn wounds.

The Potentials: Some of the caustics, such as quick-lime and lye-soap, create more eschar than others. Use only so much of the caustic as you need; it should be freshly prepared so not to lose its potency. Henry added a small amount of soot. Halyabbas added sel ammoniac. The common folk temper it with saliva. For control, apply it only within a circular opening in an adherent wax cloth or one wet with egg-white (ie allowed to dry in place), or other cool substances. Bandage the whole application securely so not to slip, and leave it for twelve to eighteen hours.

Some caustics are vesicants rather than escharotics. The vesicants include cantharides with yeast and a lard, or with finely minced leaves of flammula, lycopodium, or marsile. The inclusive bandage should remain for six to twelve hours.

III. The Three Stages Of The Process

Before: Albucasis insisted that the body be clean, and not be replete. Galen agreed (*Therapeutics*, Part 3 and 4) not to use resolutives (ie evaporatives) before completing a general evacuation, and he insisted that the patient be fully advised as to the results to be expected in order to insure his cooperation, and endurance. Nevertheless, he suggested that the patient be securely restrained.

After the Cauterization: During the initial three days, apply egg-white and rosat-oil on and around the burn. Then treat to get rid of the eschar, with butter, a small amount of wheat-flour, some unsalted lard, and a gentle suppurative. After the eschar is gone, use detergents and treat the defect as an ordinary ulcer, but keep it open until all the bad matter and fumes are gone, and you have repeated the general purgation. If some of the matter has not been evacuated before the wound closes, it will be dispersed to other parts of the body and engender serious maladies. That warning repeats what we wrote in the prologue, and in notes about *Aphorism #6.,* and about hemorrhoids.

However, should a need arise, you may reapply the cautery at the same site or to the opposite limb. (Arnold).

When a lesion resists treatments and drains for a long time, you may try to direct the drainage to and out through a nearby site.

From the first day through the seventh to ninth after the cauterization, the diet should be cool, and the applications should be comforting. Then return to a diet of contraries.

Usually a cauterization will heal within forty to ninety days (Roger, et al.). That is by the definition the limiting extent of an ulcer (*Aphorism, #6.*) After that time, the site has deteriorated and the patient is chronically ill (ie cacothymic), and whatever remains of the comforters has evaporated (Henry).

You can prevent the ulcer from closing by inserting lumps of wax, or irrigate with decoctions of euphorbia, scammony, colocynth, or hellebore, choosing the medicine that is contrary to the humor you wish to evacuate. Or you may use a sliver of hedera-wood, or a drain made of three-to-six-ply cloth, or a strip of copper or silver bandaged in the ulcer. Change the dressings two or more times every day.

The antecedent general evacuation is more important when we use the potential cauteries, but the patient is not disturbed by the application itself, and need not be restrained. Simply mark the spot clearly. The after-treatments are the same as those for actual cauterization.

The blisters produced by the vesicants require special treatments, although some persons who have had good experiences with them need nothing more than clean baths. In most cases, the blisters should be unroofed with a scissors or opened with a needle. Dress the open sores with cabbage-leaves covered by a cloth, to be replaced as often as needed. The dressings themselves are not vesicants or escharotics. In most cases, the blisters will heal within seven or eight days.

CHAPTER 4

The Formulation Of The Surgeons Medications

P hysicians, especially the surgeons, do well to know how to invent and to
formulate the medicines they need That art is very important when they practice
in remote places where there are no apothecaries. And the apothecary may be ill-equipped,
or he may be impoverished and can ill afford to buy what you need, and he may deal
only with substances in common use. Galen (*Simple Medicines*, Part 1) taught us how
to recognize and identify simples by their consistency, their flavor, and their odor, etc.
Also he taught us how to make compounds in *Composition of Medicines*, Parts 7 and
10. In Part 10 he classified them according to the sites of application (*Miamir*), and in
Catageni he classified them by their types of action. Halyabbas did the same at the end
of his *Techni*.

Galen gave examples in *Miamir,* Part 6: When he was in the battlefield, he needed
some diamoron and could not find any. He filled the need by inventing dianuca as a
gargle to treat sore throats. In *Therapeutics,* Part 11, he used absinth to treat a poor
physician with pneumonia, when, at a late hour during the night, he could not obtain
oil of nard.

I never leave cities without a sack of clysters and commonly used medicines, and I
wander through pleasant meadows and hunt for herbs with which I can make remedies
ad hoc rather than delay treatments while waiting for an apothecary. Those successes
have rewarded me with respect and profit, and with many friends.

Furthermore, I advise you to have several medicines for the same purpose, knowing
that we cannot have all of them. Galen (*Miamir)* in his discussion of infection of the ears,
wrote that a certain medicine may work well at one time, yet may fail at another, and that
it may work for one patient and not for another.

All of this material should be known in the context of various parts of the body,
and of the various complexions of both the patient and the medicine. As Galen insisted
(*Therapeutics,* Part 3) all the particulars cannot be put in books.

Therefore, let us all know that surgical medicaments vary, and how we may recognize them
for what they are.[650] Galen said that there are three principles; Averroes said five (*Complexions).*

---

[650]   As the basis of formulations and recipes (LDR).

First: The complexions and the qualities derive from the basic elements: heat, cold, dryness, and humidity.

Second: We name them according to their primary general function. For example a repercussive may have secondary functions as an attractive, resolvant, softener, maturative, detergent, consolidative, incarnative, and sedative.

Third: We know them by certain particular actions, such as laxatives, or diuretics, or eye-cleaners to remove discharges that obscure vision, or ear-cleaners to restore audition, etc.

Avicenna (*Canon*, Part 2) reminded us that medicines may function as Simples, such as roses, camomilles, or plantains, or as Compounds, such as ointments, oils, plasters, decoctions (waters), epithemes, powders, et al. Whenever feasible, use simples, because a compound may contain substances other than what you want to apply (Galen, idem. Part 3). Master Arnold wrote that you may be successful with simples, and not be able to identify the effective agent when it is hidden in a compound. On the other hand, you may need a compound to do what the simple alone cannot do.

Also, you should take into account the varying effects of medicines in different maladies in different parts of the body, and whether the affected part is noble. You should assess the state of health of the patient as a whole, and where the malady is located, and the kind of tissues are involved. Also, know what caused the malady and/or the complications. And you should know the degree of potency of your medicines: weak, potent, or in between.

Avicenna discussed all of these matters in depth (Part 2) as did Serapion (Part 7), and in the translation of Azaram's *Antidotary,* Part 1, and by St. Amand in his *Aureoles.* I will add that Serapion was wonderfully explicit in his descriptions of special causes (*Servitors*). Now I will proceed with descriptions of how to formulate, purify, boil, and roast our medicines.

The Preparation of Simple Medicaments

First wash and clean them of all foreign matter, to make them clean, or to remove matter that intensifies them and heats them more than desired. We all know about to strip ordure from plants and their roots before we wash them repeatedly, when we make oils and waxes.

Oils and Waxes

For oils we use a jar with openings at both ends. Half-fill it with water and half with oil. Stopper the openings and shake well to emulsify the mixture. Let it stand until the oil floats above the water, which you let run out through the lower opening. Refill with clean water, and repeat the action until the oil is clear.

Another way: Use a spatula to whip the oil and water in a bowl, Set it to rest in sunlight, Spoon off the oil, and repeat with fresh water.

To clean and whiten wax: Melt the wax in a basin, Dip a cold stone or glass ball on which a film of wax will congeal. Peal off the film, and repeat the extraction until all the wax has been removed. Expose the wax in sunlight and it will blanch. Or heat the wax with water and discard the water, repeated ad lib. until the wax is white.

To wash terebinth before using it on nerves: whip it with cold water which is renewed frequently until the medicine is white.

To remove salt from stored butter: do as with terebinthe.

To temper quick-lime: repeat the mix (water) and separation seven to nine times, until the water tastes sweet.

To prepare tuthie: heat it until it crumbles. Then heat it over charcoal embers, and extinguish it by dipping it in cold vinegar, rain-water, rose-water, fennel-water, or marjolaine-water. When it is cool and soft, grind it and wrap the granules in a loose-weave cloth, and collect the fine granules that come through, and discard what remains in the filter. Then wash and refilter, after collecting the earlier filtrate in a basin of cool water. Repeat until nothing remains in the cloth. Allow the sediment in the basin to settle before decanting the water and froth. Repeat the wash until the water is sweet. Let it evaporate and use the dry residue.

Combustion

Use combustion to make a hard mass more brittle and easier to grind, as we did in the first stage for preparing tuthie. Combustion can reduce the corrosive potency and render a caustic useful as a drying agent before consolidation.[651] Galen (idem, Part 3 and *Simple Medicines,* Part 9) described it for converting couperose and verdigris. He heated them in a metal or ceramic bowl by blowing on charcoal embers under the basin until the substances reddened and blistered. When the colors changed and no new blisters appeared, he let them cool, and set them aside for use.

Decoctions (Waters)

This process leaks the active elements from a solid into the water which then is used to make syrups, baths, enemas, gargles, and embrocations. Also, you may use the decoction-process to remove unnecessary elements, as when you prepare cabbage and lentils (idem, Part 3).

Triturition

The grinding prepares a medicine for storage and for use ad hoc, such as ceruse, and litharge. However, the term simply means grinding (Avicenna).

---

[651] 'Drying' meant eliminating pus from the surfaces of a wound, to allow them to cohere, that is to 'consolidate' and heal by scar, to cicatrize (LDR).

Formulating Compound Medicines[652]

The recipes for compounds have varied through the ages (Jean de St. Amand,and Master Stephen Arland of Montpellier), and they have been used in many different ways, as oils, plasters, epithymes, etc.

Oils

An oil is a thick liquid made in three ways (Heben Mesûe, and Azaram): By squeezing, as olive oil which of itself is complete medicine (Galen idem. Part 2), as are oils of walnuts, almonds, myrtle, flax, laurel, or ben, eggs, or wheat. Second: made by decoction and heat or sunlight, as are rosat, camomille, lilies, etc. Third: made by sublimation, as are oils of terebinth, tartar, fraxinus, juniper, etc.

Oils serve two purposes. They help a topical penetrate deeply, and they can mollify the harsher qualities of other elements in a compound. When you use olive oil in cool recipes, use onfacium (from unripe olives, rather than the full oil. And when you use the latter, make sure that it is sweet.

Ointments

An ointment is an oily preparation that does not flow. It can be made by grinding in a mortar, as is the white ointment, and the others that contain minerals. Make the powder and add oil, juices, and vinegar.

Another method uses heat: melt the wax, fats, and oils and add the powders. Use less wax during the winter.

A third method: grind the fats and plants and cook them, and press them though a sieve Ointments will not run, nor will they penetrate as deeply as oils.

Plasters

A plaster is a waxy compound that has been cooked to a semi-solid state. There are three kinds. One contains minerals and oils thickened by heat. You may loosen it with mucilage (diachylon) or keep it as the black plaster.

Another mixes non-mineral substances, as oxycroceum, with gums, resin, pitch, and terebinth. Add vinegar to the powders and let sit overnight. Heat and melt in the morning

---

[652]   Henceforth, as explained in the Preface to Appendix I, I will name the Simples that Guy used as they appear in the text but I will not repeat them when they reappear in recipes for Compounds. They are named again, defined and qualified in the Appendix. Furthermore, the contents of most of the Compounds (ie plasters, etc) were described by Guy earlier in the text, where they have been translated (LDR).

and temper with wine or vinegar. You can judge the proper consistency of a plaster by letting a drop fall into cold water (ie and watching it sink), or onto a marble plate where it does not curdle, or between two fingers where it will remain and not pass through.

The third uses mineral, gums, and powders, similar to our recipe for the Apostle's ointment.

We use plasters for their long-lasting effects as topicals.

Poultices and Cataplasms

These are much alike, excepting the use of flours, decoctions, oil, or honey in recipes for poultices, whereas in cataplasms we also add juices of plants. To make them effective as maturatives they must adhere to the surface and seal the skin-pores to prevent the escape of heat, matter, and spirit. In that way we prevent resolution and promote suppuration which then can be drained. Often they are called plasters (*Medicaments,* Part 4), but they are much easier to prepare, and they are less potent.

Embrocations and Epithymes

These are liquids (Simples or Compounds) with which to foment and to bathe the affected parts. We apply them with soaked sponges or cloths, which we squeeze over the part.

We use them to warm, or cool, and moist: they augment the actions of the resolutives by improving their penetration.

There are other kinds of preparations: distillates, sinapisms (mustard plasters), and liniments. We use them for sake of appearances, to please the patients, and because they are readily available, as noted by Henry. A physician who is fixed in his ways, uses only what has been well-tested. Others may experiment (Arnold). And when a favorable outcome is observed, he will declare that it is due to what he used, as Galen suggested in idem, Part 3. I prefer to leave these matters to the physicians who can better understand the true nature of and the results of the medicines.

CHAPTER 5

The Use of Repercussives In Treating Aposthems (Abscesses)

R epercussives are the most important of all the secondary measures used by surgeons.[653] They resist the enlargement of aposthems by lessening suppuration and the attendant pain and fever, and the ensuing ulcers and fistulas. That defies the popular misconception that a healthy outcome is assured only by suppuration.

Translator's Note

Guy used the following terms to define the functions of Surgical Topical medicines. The potions, laxatives, theriacs, and other internal medicines were the responsibility of the Physicians.

The Actions Of Topical Applications

Repercussives (which include Oppilatives and Constrictives), Attractives, Resolutives (including Evaporatives, et al.) Remollitives (Softeners), Maturatives (Suppuratives), Mondificatives (Detergents), Sedatives and Stupefacients Hemostatics (astringents), Incarnatives Regeneratives, Eruptors (caustics, corrosives) Cicatrizers (Sigillatives), Agglutinatives, Comfortives

All of them are classified as to Warmth, Coolness, Dryness, and Humidity (Moisture), and most of them are ranked by Degrees of Potency (LDR).

---

[653] First are diets (regimens), Second are medicines, Third are manual procedures (LDR).

Repercussive Simples

Repercussive medications act in two ways. They usually repel, cool, thicken. obstruct, and comfort. I disagree with the great doctors, including Halyabbas. Serapion, and Avicenna, who said there is little difference in their actions, I disagree because really they are different. Athough repercussives repel and arrest, they arrest by cooling the matter, that is, by thickening it. The coolants are joubarbes, lettuce, crassula, umbilicus venus, decoctions of lentils, aspen-tree buds, cold water, camphor, and vinegar.

The special actions are oppilative (ie obstructive) by blocking the pores. They are cool and viscous, but are not erosive. The oppilatives are mill-dust, wheat flour, glue, and gums.

Another type are comforters, which bolster the patient's resistance to the inflow. They are rosat oil, myrtle, mastic, myrrh, coriander, sandalwood, hawthorne, aloin, marrubium, centaury, cypress nuts, tamerinds, and saffron.

The constrictives, interceptors, and restrictives are repercussive. They repel the inflow and return it into the body. The cool ones are morel, plantains, grape leaves, verga pastoris, roses, balaustes, sumach, hawthorne, hypocystis, memitte, acacia, myrtle-seeds, pears, quince, nespole, glans, galls, bol d'armenie, argilla, cimolia, terra sigillata, and blood stone.

The warm ones are alum, salt, cypress—tree-nuts, blet, blatta, lupin-flour, and sour wine, all applied under a pressure bandage.

Repercussive Compounds—Four Types[654]

All are combinations of the foregone Simples. They are applied in the early stages of a phlegmon to cool, repel, and trap the repelled matter within the body, and to comfort the affected part. They include oxycrat, liniments, waxes, and ointments.

The Method

Whenever it is possible, first induce a good general evacuation, Then apply the simples and compounds as contraries: cool medicines against warm matter, etc. Mix the medicines ad hoc, and apply them all around the lesion, especially on the side whence came the humors. Renew them frequently until the matter liquefies and the site is altered, and no longer is discolored or indurated. Then go on to the resolutives or maturatives, as the local situation requires.

---

[654] See Foot-note 652 (LDR).

The Attractives

The attractives aim to withdraw the bad matter from the depths and from the noble structures toward the surface where to appear as aposthems or pass through the emunctories, especially when the sources are threatening or venomous, or from sciatica. Also we use them to assist in withdrawing thorns, darts, arrows, and other embedded objects, especially when they are lodged in places where incisions are hazardous, or will not be accepted or tolerated by a patient. They augment the attractive actions of cups or suction (Avicenna). They can be used after the repercussives which have contrary actions. See *Medicaments,* Part 3.[655]

The attractive medicines should have warm complexions and be liquid, so to work in the depths with full power. Galen (idem) said that they work in two ways. In one the power is sui generis, as with dictame, bees-wax, thapsie, leeks, and naveau, serapinum, ammoniacum, euphorbia, garlic, onions (all listed in *Simples,* Part 9, Ch. 2.)

In the others, the attractive action is engendered by the ordure created by yeast and from feces of various kinds. Dove droppings are attractive. Those of the goose are warmer, those of the grouse are cooler. When they fail, use human or pig-feces.

Other kinds of attractives are laxatives, theriacs, et al. Usually those are prescribed by the physicians. Avicenna used incinerated frog meat, aristolochia, and roots of marsh-reeds. Jean de St. Amand culled others from Avicenna's *Canon:* calamint, pouliot, narcissus, centinodia, corrigiola, pyrethrum, costus, peppers, pumice.

Many *Attractive Compounds* may be made from the simples listed above, and others such as amome, xylobalsamum, frankincense, myrrh, resins, lizard-musk, mulberry-tree sap, and the lees of oil of lilies. All of them are used in plasters. Another is the Apostolicon ointment of Nicolas

The Method

First apply a warm gentle oil (ie the patient is kept near a warm hearth). You may find some lower-class person to suck on the spot.[656] Then apply the attractive medicines, and cover them with a cloth smeared with lard. Bandage all with a criss-cross wrap that does not compress the lesion. Change the dressing daily.

---

[655]   Attraction is a basic function in the body. The mesenteric veins attract the chyle from the intestine; the various organs attract what they need from the blood, etc. Here the repercussives have driven the bad matter away from the lesion, and the attractives bring it where it can be evacuated. In that sense, laxatives are attractives, etc. (LDR).

[656]   I am sure that Guy described a treatment to eliminate venom from bites. Joubert's source overlooked the details! (LDR).

Resolutives

Lanfranchi and Henry said that we use resolutives when repercussives have failed or were contra-indicated, or when attractives were successful, as when cups were used with light scarifications.[657]

A resolutive can be a diaphoretic, a rarefactive, or an evaporative—all are the same. They act to separate and to attenuate, and to open the pores to eliminate the bad matter bit-by-bit, until all of it is gone. Therefore, a resolutive must be warm and thin in moderate degree. When too much so, they are bitter, and excite chills, and are desiccative (ie defeat resolution). See *Medicaments,* Part 5.

Simple Resolutives

Camomille, sage (the Egyptians used it in sacrificial rites), guimauve and its oil, wild cucumber, oil of ricinus. When used against cold and fluid matter, they should be potent detergents and desiccatives, such as melilot, aneth, mauves, blette, parietory, fumitory, lime, nettles, hyeble, elder, bran, barley-meal, bean-meal, ers, and soft bread. Lanfranchi and Henry used cumin, hyssop, calamint, oregano, spic, costus, myrrh, mastic, as used in treating contusions (ie hematoma or edema).

Compound Resolutives

Oils, ointments, and plasters use the simples, as in the recipes of Heben Mesûe and Azaram. Other simples were used by William: fennel-seed, aneth, and flours of lupin, fenugreek, and linseeds, and vinegar and oil in plasters and cataplasms. In general, all softeners and maturatives are resolutives when used to treat small amounts of liquid matter, as noted by Lanfranchi and Henry, and can be observed by any surgeon

Method

Foment with decoctions of resolutives. When the region reddens and becomes warm, apply the plasters, etc. twice daily. Be aware that when your resolutives simply extract the liquid elements of the matter and leave behind hard and stony matter, you must resort to softeners.

---

[657] Here the repercussives failed to dispel the bad humors, but the attractives have brought them close to the surface where the resolutives can 'evaporate' them or otherwise eliminate them. Or perhaps the maturatives will cause suppuration that can be drained (LDR).

Softeners (Remollitives)

We use them (ie two types) when the the resolutives fail and when the residue is hard.

First are the ordinary softeners. In *Medicaments,* Part 5, Galen defined three kinds of induration: the congealed, the stretched (taut), and the dry. All may be simples or in composites.

The true softener is effective against the congealed induration. Congelation occurs when matter has been badly treated and it becomes cold and solid. That can be treated successfully only with moderately warm and dry, in the 2nd or 3rd degrees of heat but with 1st degree of dryness.[658] Some authors use slightly moist softeners, that are neither too moist nor too dry. Softeners are better when they are slightly viscous or paste-like, yet not to act as suppuratives and lose whatever evaporative power they may have.

If the induration is cool and dry, use warm and moist softeners, in the proper degrees. If we wish only to evacuate a redundancy, do so. If we wish only to treat dryness, use the proper medicines, as we described them in our chapter on indurated joints, and fractures.

Simple Softeners are listed in *Therapeutics* Part 14. The mildest ones use the lards of goats, kids, and grouse. More potent are the fats of geese and bulls, and the marrow of bones of deer, calves, and pigs, all without salt. You may fortify the recipes with ammoniac, styrax, galbanum, and bdellium. Fresh lards are better. Fresh oils have similar effects, such as the oils of lilies, of roots of guimauve, and wild cucumber (raw or cooked). There are many others.

The Compounds use the simples. Galen added wax, frankincense, bees-wax, and laurel. Avicenna added goat-feces, ripe figs, staphisagre, bean-flour, and bitter almonds.

Rhazes used the diachylon.

Heben Mesûe used the Great Diachylon with additions.

Many other compound softeners are used for treating cold abscesses, glands, and painful joints.

The Method

The treated limb should be warmed and cleaned with decoctions of the simple softeners in water or oil, using wool fleece or wool cloth (the fleece is better). Then apply the medicine on a pad of wool cloth, and bandage it snugly. Renew the dressing daily. If you see no improvement, or if the induration increases, you should proceed with suppuratives and corruptives. Avicenna advised us to make the changes gradually.

Maturatives (Suppuratives)

We can recognize when the evolution of the abscess reaches the stages of suppuration and drainage by the signs described in the chapter on aposthems. That is when we begin to use the suppuratives (maturatives).

---

[658] 'Degrees' are explained in Chapter 8 which follows(LDR).

A maturative topical is not an aperitive[659], as was declared in the Arabic texts. It heats the abscess to exceed that of the limb, and it robs the limb of some of its moisture. That hardly affects the limb, but it causes the matter to suppurate. Suppuration, therefore, represents a change unlike that caused by natural heat when it digests good food to make it chyle. And it is unlike the change caused by external heat when a body putrefies. Here the source of the heat is in the matter itself, the pus., the unhealthy humors.

What is pus; how is it engendered; how do we diagnose and prognosticate? Much of these questions is answered in the chapters on aposthems and ulcers.

Man is a warm and humid beast in whom some matter is always in process of resolution. Therefore, it is apparent that suppurative medicines should be warm and moist with some paste-like and viscous qualities. Natural heat is the resource for maturation and digestion, but not so much as in softeners. Unlike them, their heat is less than 2nd degree, but the degree of humidity exceeds that of the softeners (Galen, idem).

Simple Maturatives

Galen (*Simple Medications* and in *Glaucon* Part 1) gave examples of embrocations and fomentations with warm water and tepid oils, and of cataplasms of wheat-flour and water that are lightly cooked; and of warm partly baked bread. Complete baking dries the bread, and is very desiccative, and useful only in treating phlegmons that are difficult to mature. Use partly baked bread and broth for already warm lesions. A pure maturative oil is more effective than others. Cataplasms of bran and barley-meal are desiccative, and are more resolutive than they are maturative. Well-baked bread is mid-way in degree between flour and barley meal. In that regard, mix a decoction of dry figs with wheat-bran, or mix pork or calf lard with pitch and resin and add oil and wax as a basilicon.

When the matter in the abscess is cool and thick, use a stew of onions, garlic, the bark of guimauve, roots of lilies, and yeast. Or make a paste of fenugreek-flour and various lards. Or use the diachylon.

A frequently used maturative uses the roots of bryony, the white plant itself, parietory, mauve, nigella, branca ursinus, violets, cabbage-leaves, dry raisin-paste, linseeds, honey, butter, and fresh pig lard.

Compound Maturatives

Many have been invented in many forms, using the Simples cited above. For Warm Matter, William of Salieto and Lanfranchi used a paste based on lard and butter.

My favorite is based on wheat flour and saffron.

For Cool, Matter, I use a paste containing egg-yolks and other Simples.

---

[659]   The aperitive brought the abscess to a point. Guy quibbled with the Arabs (LDR).

The Method

First foment the abscess with a decoction of the simples and with lanolin. Then apply warm maturatives plasters covered with stupes or a wool cloth, or with cabbage leaves. Bandage very loosely with a criss-cross technique, taking care not to bear down on where the abscess will point. Change the dressings daily.

Mondificatives (Detergents)

After the 'ripe' abscess has been opened and has been drained with a knife, a hot cautery, or with caustic eruptors, or when you treat an ulcer (ie an old wound) or a contusion, or a complication, or any collection of pus or putrid matter, you must use detergents, abstersives, and irrigations.

A mondificative topical is a detergent and expulsive but not a corrosive. It eliminates pus and necrotic matter and crusts, and it erodes proud flesh.

Abstersive Mondificatives (two kinds) separate waste matter from the normal tissues, and bring it to the surface. Both kinds of abstersives are warm. The so-called lavative type is more gentle, based on honey and flours of wheat, beans, and barley, with many kinds of gums. The *Purgative Abstersives* should be acrid, and contain ers and lupin.

We often find substances that have both kinds of action, as are almonds, ers, and nettle-seeds. Similar to them are squills, niter-scum, stoechas, abrotonin, and many others. Small amounts are used in many forms, chosen according to their potency, weak, moderate, and strong.

Detergent Compounds (Eleven)

The First are used in recently drained abscesses as gentle cleansers, and to reduce the intensity of inflammation. They contain wheat-flour, barley-meal, and spelt mixed with egg-yolks and honey.

The Second is a plaster of the above-listed simples plus rosat oil and washed terebinth. It is suitable for application on nervous tissues.[660]

Third is a plaster of Galen (*Glaucon,* Part 2). It is incarnative as well, and contains frankinscense and myrrh.

The Fourth is called the Bee-Plaster, as used by William, Lanfranchi, and Henry to clean malignant ulcers. It contains carrot juice and wheat-flour. If you add absinthe it may prevent an ulcer from forming a fistula. Temper it with juices of plantain and crassula. You can fortify it with flours of lupin, ers, or fenugreek, as did Theodoric and Bruno. If

---

[660] 'Nervous Tissues': The term defines nerves as cords, tendons, and ligaments alone or accompanying nerves. Nervous tissues were membranous aponeuroses. Nerves themselves were thought to be contractile elements along with the muscles, as well as conduits of sensation (LDR).

you add terebinth, it will be suitable to use on nerves, and added myrrh will help clear out corrupted and malodorous matter.

The Fifth is potent plaster containing resin. It can be used on nerves.The Bolognese use terebinth, resins, honey, myrrh, sarcocolla, and flours of fenugreek and linseeds.

The Sixth is a plaster used at Montpellier. It attracts the pus and expels it: honey, terebinth, yeast, and iris-roots

The Seventh plaster was used by Master Dino. It is based on gums which attack thick pus: galbanum, resin, terebinth, pitch, cow-lard, wax, old oil, and vinegar.

The Eighth is the Apostles' Ointment (because it has twelve ingredients): white wax, resin, ammoniac, opoponax, verdigris, aristolochia, frankincense, myrrh, galbanum, bdellium, litharge, and oil. Heben Mesûe called it the wax ointment. Masters Amerin de la Porta, and Peter of Argenteria at Montpellier called it Gratia Dei because it is very effective in malignant ulcers.

The Ninth is the Egyptian Ointment, used by Galen, Rhazes, Albucasis, and by My Master at Bologna. It cleans granulation tissue: honey, vinegar, verdigris, and alum. It also is called the bicolor because when it is heated it turns red, when it is less effective than when it is raw and green. Although green ointments are not approved by the common folk, we note that after it has done its job as a red ointment, the green color returns.

The Tenth is the Red Greek Plaster, also called the Bicolor (*Glaucon,* Part 2). Master Dino favored it for malignant ulcers. It has oil, vinegar, litharge and verdigris.

The Eleventh is the Ointment of Green Plants. Dino used it to gently deterge wounds. It consumes proud flesh: chelidonium, plantains, scabious, nettles, lovage, centrum galli, goose fat, frankincense, sarcocolla, aloes, and bronze-flower.

There are many more mondificatives. Some of them will be listed with our incarnatives, because they serve both functions.

Analgesics (Sedatives, Mitigants, and Stupefacients)

Pain is a complication that interferes with all of our treatments of abscesses, wounds, and their complications. Here we will deal with pain in wounds and aposthems.

Avicenna said that Pain was the sensation of a contrary. Galen said that although contraries cause pain, other things are painful, too, such as sudden changes in one's Nature brought on by heat or cold, or by a violent blow, or a cut or a stretch (ie distension), or erosion. Nevertheless, Pain is a contrary quality of itself as well as unexpected complication. That is the doctrine of the school at Montpellier.

Averroes (*Colliget,* Part 3) does not support Galen's descriptions in *Maladies and Symptoms,* and in *Unequal Temperaments.* Here we are sailing in deep waters, too deep for us (ie as non-physicians) to navigate.

So, if pain is a sensation of a contrary, then the relief of pain should be with something that engenders an opposite sensation, or by eliminating the original 'contrary' by evacuation or alteration, or by obliterating the sensibility of the affected part. Therefore, we seek relief in two ways:

The first is by the use of true sedatives (Galen, *Medicaments,* and Avicenna. Galen and Averroes said that sedatives should match the Natural heat, or be slightly warmer, They should be liquids that can augment the Natural heat, and can make the faulty humors digestible in a normal fashion.

The best for that purpose are oils and fats: the fats of grouse, duck, goose; the oils of olives (sweet oil), egg-yolks, and most sweet oils.(Azaram, *LargeAntidotary*), aneth, and linseeds which evacuate the faulty humors and give relief when applied directly on the painful place (Avicenna).

Furthermore, Galen (idem, Part 5) agreed that suppuratives, with the same or slightly warmer degree of heat than that of the body, can be sedatives, especially when followed by garden-grown resolutives.[661]

Sedative Compounds are made from *Sedative Simples*

A commonly used sedative plaster uses the pulp of stale white bread, after dipping it in boiling water and squeezing it almost dry, with added egg-yolks, and rosat oil.

Theodoric's, Lanfranchi's, and Henry's sedative plaster used leaves of mauve and powdered bran.

Janvier's plaster was a popular sedative, maturative, and resolutive: mauve-leaves, branca ursina, violets, lizeron, parietory, hyoscyamus, umbilicus venus, lards of pork or duck, flours of wheat, barley, linseeds, and fenugreek, aloin, wine, and honey.

Special mitigants for painful shoulders, abdomens, kidneys, matrices, and nervous places include the Resumptive Ointment: salt-free butter, violet oil, fats of grouse, goose, duck, and donkey, and the bone marrow of a cow. All them must be fresh.

Another is the Martiaton Ointment (Agrippa's): oils of laurel, ben,[662] and nard. It comforts nerves.

Here, I can only state that the total number of claimed sedatives for wounds, abscesses, punctures, painful joints, etc. is overwhelming.

The Method

When the pain is severe, begin with a proper general evacuation with laxatives or phlebotomy. If the cause is a sanguine humor a bleed alone may mitigate the pain, but it will hide the cause. The Doctors say that is because a bleed is not a contrary to sanguine humors. Henry agreed. Then foment and bathe the painful part for one hour, using warm

---

[661] In an earlier chapter, Guy said that a sign of suppuration in an abscess was a rapid decrease of local pain (LDR).

[662] Joubert wrote that the Oil of Ben is muscelin. Later, in Chapter 7, Guy used Behen instead of Ben, whereas the two are different substances. Ben (in the ointment) is from the nuts of the Moringa aptera tree. Behen comes from the roots, in two forms, white or red (J).

water and oil. Gently blot it dry, and apply the sedative plaster, and cover it with stupes or lamb's fleece and bandage it lightly. That is the best treatment.

When that fails and we must do more to relieve such severe pain, as could destroy a patient, we may have to use dangerous medicines to save a life (*Therapeutics,* Part 12). Then we will use the stupefacients.

They do not attack the cause for the pain; they affect only the symptom. As Galen wrote in *Medicaments,* Part 5, a dead man feels no pain. Stupefacients are potent contraries, rather than sedatives. They are cool and contrary-to-nature: opium, roots of mandragore, morel, hyoscyamus, and poppies. All work better when they are dry than when they are green plants. Mix them with saffron, myrrh, styrax, and castoreum, as described for opiates in the Philonium, or as suppositories, or collyria. All are effective.

The amounts and the intervals of the doses must be correct (idem. Part 12), and should be supervised by a physician. The troches prescribed by the Bishop of Rieges, for the Bishop at Marseilles who suffered a painful strangury, led to his collapse and death in his sleep. The medicines did more than simply relieve his pain.

Avicenna (*Continens*) listed his stupefacients: white hysoscyamus, opium, seeds of melon, lettuce, and purslane. He made troches to be taken with a decoction of licorice.

In his *Canon* Part 3, he prescribed this for ulcerated kidneys: seeds of white hyoscyamus, opium, melon, lettuce and purslane.

CHAPTER 6

Topical Medicines For Treating Wounds, Especially Hemostatics

L est it destroy a patient, you must arrest hemorrhage from his wound. His blood and his Vital Spirit[663] are the essence of his Nature, as all physicians know. Blood may be retained in the body in several ways in which medications may play a role We will deal with them here.

Hemostatics

Restrictive Simples—Four Types

These retain cool blood in the limb, and they reduce the acuity (ie erosive quality) of the blood. They constrict the orifices of the tributaries and reduce the flow to the place where it bleeds, and they cause the blood to clot in those tributaries. When it is warm, they dry the surfaces and produce scabs.

The first type is described in *Medicaments,* Part 5. They include cold water, joubarbes, purslane, green tribules, flea-bane, mouse-ear, water-lentils, oxalis, and umbilicus venus.

The second group contains plantains, equisetum, galls, pomegranate-peel, raisin-seeds, rhubarb, bol d'armenie, terra sigillata, and many other astringents and styptics, thick or watery, that do not erode.

Third are sangdragon, frankincense, mastic, resins, glues, wheat-flour, and flour-mill dust.

Fourth are quick-lime, arsenic, couperose, vitriol, et al.

Many Compounds

1. Galen's Powder: egg-white, thapsus, honey, aloes, frankincense.
2. The Red Powder: bol-d'armenie, fried galls, frankincense, aloes, mastic, thapsus.

---

663 A reminder: Blood emanates from the liver, in the veins: the vital spirit comes from the heart via arteries (LDR).

3. A hemostatic powder used by Albucasis, Bruno, and Lanfranchi: frankincense, sangdragon, quick-lime, to be dusted on the suture-line.
4. Another powder from Galen (*Catageni*): wild pomegranate flowers, alum, burnt couperose
5. Roger's and Janvier's powder: clophony, bol d'armenie, mastic, roots of great consolida, and roses.

Incarnatives

There are three elements of incarnation in treating wounds and ulcers[664] First: to bring together and consolidate the separated surfaces, which by definition denote a wound. Second: they engender flesh to replace what has been lost in a wound or ulcer. Third: produce a scar to seal the wound when all that remains to heal are the skin-edges.

In all these actions Nature may be assisted by desiccative medicines, as were described in Treatise III on Wounds. Incarnatives should be dry in the 2nd degree and regenerative in the 1st, and cicatrixative as much as 3rd degree. In all cases, use no more of the incarnatives than is required by the dyscrasia, as when you need an erosive. Your choice should match the complexions of the wound (ie its degrees), of the body, and of the affected part. Suit the age of the patient, the weather, and the needs for contraries.

Simple Incarnatives For Aggregation and Consolidation

Avicenna said that these should be desiccative and thickening agents to treat the wet surfaces of the wound and make them viscous and adherent. They should be styptics, as are sangdragon, aloes sarcocolla, bol d'armenie, terra sigillata, bark of date-palm trees, pomegranate peels, plantains, needles of pine and cypress trees, potentilla, oxalis, leaves of wild-pear trees and sorbs, equisetum, flour-mill dust, burnt barley, terebinth, sour milk, and astringent red wines. All of them can be used in Compounds.

Compounds[665]

1. The Red Powder with terebinth
2. The Black Plaster of Galen
3. The Diapalma Plaster (*Catageni*)—add hog and veal lards
4. The Green Plaster of Master Peter Bonant—add betony, vervaine, pimpinelle, piloselle, hypericon, and cynoglossus.
5. The Centaury Plaster of Peter Argelata—add wax and mothers' milk.
6. Master Dino's Plaster—add betony

---

[664] By definition: An ulcer is a wound that remains unhealed beyond an allotted time, usually forty days (LDR).
[665] see Footnote #652 in Chapter 4, of Doctrine I) (LDR).

7. The King of England's Ointment—add white wax.
8. The Plaster of Count William and Pope Boniface, which the Pope received from Anselm of Genoa, and then gave to the King of France—add pimpinelle, betony, henbane, verbena, vermicularis, white wax and mother's milk.
9. The Glory Ointment of Jean the nephew of Anselm of Genoa—add bedegar and seeds of white roses.
10. The Green Herbal Ointment of Roger, Jamier, and Nicolas (all were Thessalans)—add alleluia, chelidonium, orvale, lovage, and ram-fat.
11. Lanfranchi's Worm Plaster for use in nervous regions—add consolida, cynoglossus, earth-worms, and Greek pitch.

Methods

After the bleeding has stopped and the hematoma no longer enlarges, foment the wound with warm red wine, blot it dry, apply your incarnatives on wine-wet stupes, and bandage 'artfully'.[666]

Incarnatives to Engender Granulation Tissue

Avicenna said that these medicines converted blood clots into tissue with the same complexions, using desiccatives with gentle detergent actions. See *Medicaments,* Part 5.

The Simples are in Three Grades.

The Weak are frankincense, mastic, aloes, colophony, barley-meal, and fenugreek-flour. They can be applied anywhere on the body.

The Potent are aristolochia, iris, the flowers of lupin and ers, climie, tuthy and burnt couperose. Use only small amounts, and apply only on dry surfaces (ie not purulent).

The Extremely Potent are used in deep ulcers: centaury, polium, glue, incinerated sea-shells, lead, burnt antimony, etc. Pitch and resins also engender granulation tissues, and myrrh engenders it on bone.

Many Compounds Have Been Devised[667]

1. The Basilicon Ointment, (Galen's Tetrapharmicon):—add black pitch, resin, wax, cow-fat, oil. Heben Mesûe called it the Major Basilicon when he added frankincense, and Galen called it the Macedonicon.
2. The Dark Ointment (Obfuscum) of Nicolas is made by apothecaries.
3. Heben Mesûe's Gold Ointment, also is a detergent—he added honey.
4. Galen's Green Ointment, used by Avicenna. So-named because he (G) added verdigris.

---

[667] An artful or 'artificial' bandage closed the wound and obviated the need for suture. The 'art' was surgery (LDR).

5. Flax Ointment used by Avicenna as a consolidative I use it as a regenerative. It is applied on a piece of clean pure linen.
6. The Yellow Plaster of Master Peter Bonant is based on seeds of fenugreek.
7. The Precious Ointment from my cartulary is used in all wounds. Add artemisia, scabious, aloin, gallitricum, verbena, fauciole, wild anacarde, and berula. Master Peter Bonant first washed with a decoction of the simples before he applied the ointment.
8. The Gratia Dei Plaster. It adds the herb and chervil and sanabor.
9. The Count's Plaster of Master Aymeri Alesto—add mothers' milk.
10. The Count of Auxerre's Ointment—add ambergris (probably ambrosia).
11. The Diareos Ointment of Master Dino used the usual simples. All the barbers of Montpellier use it.
12. Many Powders based on Rhazes' Powder.
   In addition we use Lanfranchi's powder of frankincense, mastic, and fenugreek.

The Method

Irrigate the wound with warm wine, apply a powder alone or on charpie.Overlay a stupe damp with wine. Wrap a bandage. Change it twice daily.

Cicatrixatives, and Sealants (Sigillatives)

Avicenna used them to dry the edges at the surface and allow a scab to form to protect the wound from contamination while natural skin grows over the scar. The medicines should be astringent simples of two types.

Simples

Galls, pomegranates, Egyptian thorns, ceruse, burnt lead, litharge, cimolea, bol d'armenie, many kinds of clay. All are listed in Galen's *Simples,* Part 9.

The second types are corrosives tempered by combustion and used in small amounts: verdarain, alum, couperose. You may use any astringent that does not corrode, such as centaury, plantains, aristolochia, beef-leather, shoe-soles, barks of oaks and elms. Arnold used iron slag.

Ten Compounds

1. William of Saliceto's, Lanfranchi's, and Henry's Powder—add balaustes, aloes, and silver cadmia.
   Another from the same source—add pine bark, cypress seeds. Avicenna added sandalwood, nenuphar, dry roses, and lanceolette (a sea-worm).
2. The Common White Ointment—add rose-water.

621

3. Rhazes' White Ointment: rosat oil. ceruse, camphor, egg-white. You may add white lead to color the ointment red, as is used by the barbers at the Vatican in Rome.

4. Avicenna's Lime Ointment applied on burns and nerve-wounds. The corrosives are thoroughly washed and tempered.

5. The White Plaster of Ceruse in the *Catageni*: Azaran modified it as did Peter of Avignon, adding ashes of sea-shells.

6. The Bishop of Lyon's Ointment[669]. (He belonged to the House of the Count of Armagnac). It was used in ulcers and fistulas—add pig lard, and tempered mercury (with saliva)

7. The Blue Ointment for facial pustules, scabies, and serpigo. It is based on lard, mercury, alum, live sulfur, bugia, and indigo (the blue) from Baldac.[670]

8. The Diapompholigon Ointment of Theodoric and his followers to treat cancers, achancrimens (? erosive ulcers), erysipelas, and burns.—add red seeds of morel and pompholyx (tuthy).

9. An ointment based on nutritious litharge, used by Rhazes and Avicenna. They added antimony, garance, bran, and dried earth worms.

10. A disc of lead cut to the size of the ulcer and laid in the ulcer after washing it with alum water. Bandage it snugly. The source of this fine treatment is not known, and I cannot take credit for it. But I admit that much credit is due to its simplicity which defends it against the foolish claim by the commoners that debases anything that is not costly.

The Method

Before the redundant proud flesh appears, wash a warm ulcer with warm wine containing balaustes and alum. Blot it dry and apply the cicatrizers. Wrap a bandage

Corrosives, Putrefactives, Caustics, Eruptors

When the maturatives and the detergents cannot eliminate all the foreign substances from an abscess, you must extirpate it with a cautery or with medications. The hot instrument is more reliable when used early; it will leave a smaller defect. The erosive and irritative medications are more painful for a longer time. Nevertheless, many patients dread the knife (or cautery) more than death, and the open operations may not be feasible in some parts of the body, and we are compelled to use extirpative medicaments. Galen called them colliquatives in *Medicaments,* Part 5, but more commonly they are called corrosives or putrefactives. We need not meddle with

---

669 Nicaise said that Lyon may have been Laon or Laudun, in various editions (LDR).

670 Baldac was the home of Acanamusale, often cited by Guy (EN).

the names of the three varieties other than to know that the weaker ones are the corrosives, the more potent are the putrefactives, and the most potent are the caustics (the eruptors). All of them are warm and some are watery; the least warm are the corrosives, the caustics are the most, and the putrefactives are between. The corrosives are suitable for removing superficial accumulations of proud flesh, the putrefactives work well into the subcutaneous deeper and more firm flesh. The caustics work everywhere, surface, deep, soft, and hard.

Henry noted that their actions overlap, as affected by the duration of the application the amounts used, and the complexion of the patient.

Corrosives

Avicenna defined a corrosive as one that resolves small amounts of proud flesh by erosion, when small amounts were applied on stupes of fluffed hemp or on strips of sponge. The commonly used corrosives for that purpose are hermodactyl, alum or the Apostles' ointment. But when the amounts of proud flesh are large, we use couperose, the green ointment, and the Egyptiac. Master Dino made troches of a paste of quick-lime and honey, and dried them in an oven.

Another: Roger's troches of asphodel and orpiment dried in the late summer sun.

Another: the troche of Aldaron described by Andromache and formulated by Avicenna who added dragacanth and vitriol.

Another: The troches of Calidicon, described by Galen: quick-lime, orpiment, salicor, and acacia (ie gum Arabic).

Another: two kinds of capitel (lye). One is commonly use for cleaning the scalp. The other is Dino's, which I have supplemented with quick-lime, sel ammoniac, and a lye from ashes of bean-stalks. I place all of it in a pot with a perforated bottom, set it inside a bowl and bury it in a ditch for seven days. The seepage is a potent caustic. It consumes, cauterizes, and erodes, and rapidly forms eschars.

Putrefactives

Avicenna said that a putrefactive corrupts the tissues and makes a stinking slurry instead of an eschar.[671] The matter is cadaverous, like an esthiomene. It acts on all redundant tissues, of all degrees. (Galen, idem). It resemble the corrosives in that it causes little pain because it acts only on corrupted tissues. The putrefactives are realgar and arsenic. Both are fierce and must be controlled.

Arsenic is controlled ala The Four Masters. They made a paste by adding its powder to cabbage or morelle juice or other cool plants. Let it dry. Repeat the dilution three or

---

[671] Guy suggests that the Corrosives and the Putrefactives work in tandem. The latter melt the matter affected by the former (LDR).

four times before making troches. Realgar is tempered in the same way, and we do the same to temper sublimated arsenic.

The extremely potent are distilled liquids. The best contains sel ammoniac, red and yellow orpiment, couperose, and verdigris. Put all in a glass alembic and distill over a low flame. Replace the distillate and reheat it until the alembic is red. Save the distillate in a sealed glass jar. It is very effective and can erode iron. Use only one drop to necrotize the lining of a fistula or to burn off the roots of excised warts and other excrescences.

I leave other techniques of sublimation and distillation to the alchemists.

Do not use these dangerous and potent medications on feeble bodies, or near principal organs, or on such small soft parts as the penis, lips, eyelids, or fingers. And do not apply a large amount as a single-session treatment. Use repeated small doses, as we advised for removing small glands, or to erupt an abscess.

You may use a powder, or mix them dialthea or the white ointment. After the treatment, wash the region with cool juices and vinegar. When the patient suffers too much, cease the applications and foment with warm oil. After three hours of an application of arsenic (less if the medicine is a decoction) proceed to eliminate the eschar with butter or some maturative ointment.

Caustics, Escharotics, and Eruptors

These erode and cauterize the skin and flesh and change their complexions. The site is mortified and carbonized, without much pain, because the process of cauterization is slow *(Medicaments, Part 5)*.

The weak ones erode only the skin or make blisters or prepare the site for the putrefactives which act better on bared flesh, after the overlying skin has been destroyed by cantharides, hyssop, cashews, garlic, lupin, flammula, marsilia, and anabula. Mix mashed cantharides with yeast and lard. Small amounts of other mild eruptors may be applied alone. Remove them after a half-day and open the blisters. Lay on a cabbage leaf. If the cantharides causes dysuria, the patient should drink milk, or be immersed in a tub up to his navel, the water of which contains mauves, violets, cresson, and parietory, to sit until he is relieved.

There are many potent caustics made from powdered quick-lime tempered with a soft soap or saliva. Use a dose the size of a nut shell. Protect the surrounding tissues against the spread of the caustic by capping it with a walnut shell, or a large acorn, or surround the site with a disc of leather, or of waxed cloth, or a cloth with egg white or another glutinous medicine. After the treatment, cover the area with a cool medication for at least twelve hours. The longer the treatment, the better the eschar. After removing the caustic, separate the eschar with unsalted butter with a small amount of wheat flour, or with other ointments.

CHAPTER 7

Medicines For Treating Fractures And Dislocations.

Prevention of Swelling

T hese medicines may be used as washes (epithemes), plasters, or ointments. Avicenna, (Part 4) explained their purposes; most important is the reduction of swelling (aposthems) and to relieve pain. Also, they may prmote the union by toughening the callus, and they may comfort the limb. On the other hand, some of the medicaments may compress the bulk of the callus; others may soften the induration that persists after the bony lesions heal.

The medicines to reduce the swelling are refrigerants and repercussives, such as egg-white, rosat oil, and myrtle oil applied early in the course.

Agglutinatives

These toughen the callus and reduce its bulk, and support the fracture until the callus has fulfilled its purpose. The Simples are wheat-mill dust, wheaten flour, sangdragon, frankincense, mastic, sarcocolla, and egg-whites.

The Compound is a powder added to an ointment: aloes, myrrh, bol d'armenie, frankincense, acacia, cypress nuts, dragacanth, labdanum, and mill-dust.

Comforters

These are used in the final phase of the treatments, and are applied after fomenting the site with a salted wine heated with roses, aloin, white moss from an oak tree. Then use Lanfranchi's plaster: rosat oil, resin, wax, colophony, mastic, frankincense mastic, cypress nuts, and curcuma. You may use less oil or more of the gums, or add saffron.

Roger laid on a sparadrap of cloth (a sort of waxed cloth) with frankincense, mastic, pitch, mill-dust, bol d'armenie, ram-fat, and wax.

Another is Roger's Apostles' Ointment (as given in an earlier chapter).

Another, taken from Nicolas: the oxycroceum.

Another: Peter Bonant's plaster, used for contusions.

Medicines To Soften Residual Induration

We described these in our chapters on sclerosis and arthritis. However, the formula is complex, and it will be worth our while to repeat some of it.

When used after a fracture has united, the softeners should be humectants more than resolutives (ie evaporatives) (Galen, idem, Part 5) especially when the induration is caused by dryness, when the local nutrition has not been adequate, or when there are complicating nerve-injuries, or when there has been prolonged drainage of pus.

Avicenna said to begin the treatments with embrocations of warm water, followed by lenitive ointments, and plasters, that contain mucilages, gums, and fats and oils. Add vinegar to improve the penetration; use only a small amount that has been well mixed with warm substances; it will do no harm (Galen, *Miamir*, Part 3, and Avicenna (*Canon*, Part 3).

I follow the embrocations of warm water, oil and cow's milk, or I use a decoction of the bark of the roots of guimauve or elm trees, bryony, wild cucumber, meadow-enula, acoris, dates, figs, and flax-seed. Or I use a decoction of the heads, feet, and tripe of sheep. I use a sponge for the embrocation, or I simply rub on lanolin.

After fomenting for an hour, I dry the region and seat the patient near a warm hearth (not too close), and I gently passively flex the limb, and rub on this ointment: lards of pigs, donkeys, mules, bears, marmots, badgers chickens, geese, and ducks; marrow from a deer's femur, fresh milk, oils of nutmeg, sesali, ben, sweet almonds, mucilage of guimauve, fenugreek, linseed, styrax, calamite, bdellium, lanolin, and wax. All the fats must be fresh. If you wish to cool the region, use an ointment of beaver fat with common gums.

Another for use on common folk: dialthea ointment, Agrippa's ointment, and salt-free butter.

After a thorough inunction, apply this plaster: bark of the roots of guimauve, pork-lard, or other softeners. Lanfranchi favored this: fresh pork-lard, fats of geese and chickens; wax, terebinth, common oil, flours of fenugreek and linseeds, bdellium, opoponax, mastic, frankincense, wine, styrax, lanolin, and labdanum.

Ammoniacum is a good softener (*Simples,* Part 11) and it dissolves excessive callus around joints.

These treatments require a long time because passive movements are not as effective as active.

You may use an evaporative (resolutive) of vinegar and marcasite for stiff tendons and sclerosis at the joints.

Instruments and splints may help.

CHAPTER 8

The Degrees of Medicines

Nicaise's Note

I *shall briefly recapitulate what was in my Introduction, concerning the classes and categories of Medicaments as they were in the Middle Ages.*

All things are formed from mixtures of four elements: air, fire, earth, and water, each with its dominant qualities: cool, warm, dry, and moist (ie their Complexions). When the mixture is 'harmonious' the body is 'tempered' and the temperature is that of a tempered body, its temperament.

When the harmony is defective, because one or more, of the qualities predominates, the body is intemperate—its temperature is intemperate.

A medicine is temperate when its effect does not upset the harmony, that is, it does not alter the complexion of the body or the part. In other words, it is not an effective therapeutic agent, and it may serve only as a vehicle. A medicine is intemperate if it changes the body's qualities by exerting its own dominating quality. The medicine can be labeled as warm, cool, dry, or moist.

Therefore, intemperate medicines are labeled and classed according to their elementary qualities, in terms of their own complexions. They warm, cool, dry, or moisten, but not for all of its applications. The intensity of the dominant quality is expressed in four degrees. To be effective they have to exceed the temperature of what they are treating. When the quality is in the first degree, the strength of the medicine is deemed moderate; in the second, it is manifest, in the third it is larger, and in the fourth it may be destructive (ie overwhelming).

Galen, the Prince of Medicine, (*Simple Medicaments,* Part 1) said that a physician cannot formulate a compound medicine for a special purpose unless he knows the qualities (ie degrees) of the simples he uses. To satisfy that need, I shall list the degrees of the simples used in most surgical medicines. The degrees of the compounds derive from those of the simples. Arnold said that the admixture of complexions yielded only an approximation.

The tempered medication (ie 1st degree) matches that of the body and neither warms nor cools.

The intemperate medicine leads to some dominant quality by which the medication is classified.

So it is that we call a medicine warm in the 1st degree, whose effect is not manifest; the same for cool in the 1st deg., and dryness and moisture. And all those whose effects are manifest are said to be in the 2nd deg. The medications with obvious major effects, but not excessive, are of the 3rd deg. And those which chill, or overheat, or burn-dry, are in the 4th deg. (Avicenna, *Canon,* Part1.).

As to dryness, the medications in the 4th deg. obviously burn, and the fourth begins as third reaches its level. As to moistness, there is no 4th deg., which would mean venom or corruption, rather than treatment. However, warmth in the 4th deg., as with garlic and poppies, is not lethal, whereas 4th deg. coolness can be lethal. *The Companion of Concordance,* says that cold is naturally inimical, not heat.

Questions about these topics are inevitable. Can one increase or decrease the degree of a medicine by changing its quantity? Do secondary qualities follow those of the primary? Why can one medication have contrary effects in one but not in another part of the body? One must delve deep to answer such questions.

Therefore to help us to find degrees of medicines, we need an alphabetized catalog. And when the authorities disagree in such matters, the differences may be geographical rather than in the medicines themselves. For the most part, I have accepted Galen's classifications on his *Medicaments,* the Final 6 Parts, as well as the interpretations of Serapion and Avicenna, and items from my own experience.

Translator's Note

The Classification by Degrees in Guy's Chapter 8 repeats most of what Nicaise has placed in his Glossary (my Appendices). I have extended that pharmacopoeia and have inserted all of the Degree Material. I respectfully refer the Reader to that section, Appendix I, which follows Doctrine II, Chapter 8.

Here Ends Doctrine I

DOCTRINE II MEDICINES FOR SPECIAL PARTS: EIGHT CHAPTERS

CHAPTER 1

The Head: Six Remedies For Head-Wounds[672]

1. Theodoric's head potion: canelle, ginger, galanga, cardamon, peppers, cloves, honey and wine. Drink it with this powder: pimpinelle, betony, sanamunda, valerian, gentian-roots, and piloselle.
2. A detergent for the brain and the meninges: rosat-honey, rosat-oil.
3. The Capital Powder of Galen, Dino, and Henry: iris roots, aristolochia, frankincense, myrrh, aloes, sangdragon, and flour of ers.
4. The Betony plaster: wax, resins, terebenth, juice of betony, plantains, and celery.
5. Anselm of Genoa's Head-Plaster: terebinth, wax, resin, vinegar betony, verbena, mothers' milk.
6. Master Peter of Argelata's Ointment to elevate depressed fractures: old oil, bees'-wax. Euphorbia, aristolochia, milk of tithimal.

For tinea: Gordon's Embrocation: litharge, live sulfur, quick-lime, ink vitriol, orpiment, lard, verdigris, hellebore, alum, galls, mercury, pitch, wax, walnut-oil, juice of lapathum, fumitory, scabious, borache.

For alopecia: an ointment of centaury, powdered ashes of incinerated leeches, moles, and mice; honey, ashes of soles of old shoes, pork-lard, verdigris.

---

[672] I believe that it will be clear to the Reader that these brief additions to Guy's extensive pharmacopoeia are 'after-thoughts'. Perhaps they were added to his Antidotary after the the the text had been submitted to copyists, etc. The Eight-Chapter Format is consistent with the rest of his Treatise. The medicinal items are in my Appendix I that follows (LDR).

CHAPTER 2

For The Face

T he preferred ointment everywhere is the yellow ointment containing couperose.

Second is Rhazes' Faro for whitening and cleaning: flours of chick-peas, beans, barley, almond-peels; tragacanth, radish-seeds, and milk. Apply at night, and wash with bran-water in the morning.

Third is the precious French Water: calcined, tartar, mastic, and egg-whites, distilled in an alembic.

Virgins' Milk is used to dry and drain virulent pustules, to blanch dark blemishes and freckles: powdered litharge, good white vinegar, distilled and collected on a triangular piece of felt or on a sachet. Mix the fluid with brine, add rain or spring-water and add milk, Many authors boil the litharge with vinegar. Others add ceruse. Some replace the brine with rock-salt. Others add the scum of niter. Others add alum.

Remedies For Eyes

1. The water of Peter of Spain who used it to clean the eye and improve vision: fennel, rue, chelidonium, verbena, euphrase, claret, and roses.
2. Galen's white collyrium for painful eyes: washed ceruse, sarcocolla, wheat-flour, tragacanth, opium, rain water, mother's milk.
3. The collyrium of tuthy from Montpellier for ophthalmias: tuthy, calaminary stone, cloves, honey, white wine, and camphor.
4. Master Arnold's powder to dry tears and reduce redness, as prescribed for Pope John: tuthy, antimony, pearls, red coral flowers, raw grease from worms. Store it in a bronze flask.
5. Bien-venu's powder which I use for all blemishes: sugar-candy, tuthy, rose-water. Make a paste, put it in a wide-mouth basin, and turn it so to expose the paste to the smoke of burning aloe-wood and frankincense. Evaporate and make a powder to be stored in a bronze box. Apply it wth a silver probe.
6. A collyrium for red-eye and lacrimation. Tuthy, socotrin aloes, camphor, rose-water, and pomegranate wine,

630

For The Nose

Earlier in the text we described medicines for epistaxis, ulcers, and polyps. Master Peter recommended inserting a tent wet with roots of acoris tempered with red juniper oil and a small amount of scammony.

For Earache

I use mothers' milk (as did Galen) with rosat oil, opium and the white collyrium. To mondify ulcers, irrigate with rosat honey and saffron.

Master Peter used this recipe: niter, cardamon, peeled dry figs, and juice of rue. Drip it into the ear to clean out pus and proud flesh.

Toothache

Hold a mouthful of a decoction of pyrethrum or buckthorn in vinegar (Azaram).

Treat dark skin with the Count of Auxerre's wash: sel ammoniac, rock-salt, and alum.

Treat inflammation and ulcers of the gums (ie pyorrhea) with mouth-washes of caprifolium and alum. Or use Dino's recipe pf roses, lentils, sumac, and balaustes, all boiled in vinegar.

CHAPTER 3

The Neck

M ost wounds of the neck receive the routine treatments for wounds. Some maladies receive special attention.

Treat Goiters in two ways. First: Master Dino's powder: scrofularia, ginger, bryony, pyrethrum, serpollet, caprifolium, olives, rock salt, incinerated sponge, cloves, peppers, canelle, and alum. Second: Apply a plaster of diachylon with iris, or a plaster of goat's feces, or an ointment, for phlegmatic aposthems.

CHAPTER 4

Shoulders, Arms and Back

T reat pain in the shoulders and arms with martiaton (Agrippa's) ointment. Treat hump-back (ie gibbosities) with plasters of flame batard, acoris, meadow-enula, juniper, bdellium, castoreum, wine, and wax.

Treat gouty hands as we do other phlegmatic aposthems, with the Montpellier Plaster: red cabbage, a lye made from wine lees, vinegar, and salt.

CHAPTER 5

The Chest

A potion to consume a collection of matter (ie empyema) by sweating was described by Master Aimery in his *Sudatory:* equisetum, roots of osmondia and dragontea, wine and honey. Drink a cupful at bedtime and break into a sweat.

Another is Galen's Potion, useful for all internal disorders: centaury, costus, nepita, caryophyla, pimpinelle, piloselle, cannabis leaves, cabbage-roots, tansy, garance, tormentilla, orvalt, and honey. Dose a cupful at bedtime. It will induce drainage from within. If the patient vomits the dose, a repeat dose will be useless.

CHAPTER 6

The Abdomen

T reat cramps by laying on greasy wool fleece wet with wine, and a decoction of cumin.

Treat contusions with the well-klmnown beverage of Avicenna and Rhazes: mummy, bol d'armenie, and terra sigillata. Mix the powder in a decoction of plantains. The potion described in Chapter 4 also resolves intra-abdominal collections. Fourth: Apply the plaster that we use for contusions.

Ascites

Prescribe diuretics. Use what Galen, and Master Aimery made from cicadas (ie beetles). It was called the Black Cantharides. Cut off their heads and their wings, and incinerate them in an oven and make a powder of the residue. Dose one grain with wine during the evening.

Pain in the Kidneys and Bladder

I prescribe a lye made from the ashes of bean plants. It is a marvelous diuretic, and will clear the urine by ejecting sand and pus.

Rabbi Moses treated ulcers of the kidneys and bladder with a distillate of goat's milk: goat milk, jujubes, sebestes, bol d'armenie, the four cold seeds, seeds of purslane, white peppers, and quince.

Avicenna treated diabetes (ie urinary frequency) with ewe's milk. I prescribed it for the Cardinal of Tulles, and added equisetum, plantains, roses, seeds of guimauve, kekenji, licorice, and shells of acorns,.

Another good treatment is the injection of the penis with milk and various collyria, and I apply a plaster to prevent a rectocoele or a cystocoele.

I refer to our chapter where we cited many good remedies for bladder-stones.

CHAPTER 7

The Genital Structures (Called Shameful)

A paste of egg-yolks and oil of poppies relieves pain in the penis.
Treat penile ulcers with embrocations of alum-water, followed by a plaster of the populeon or white ointments, or oil with egg-whites, or with a powder of burnt lead, ceruse, and aloes.

Treat inflamed testicles with a plaster of mauve, bean-flour, and cumin, all boiled in water.

Treat hernias with three remedies:

First: an electuary of a compote of consolida major, a compote of roses, powders of tragacanth and diacumin, roots of valerian, seeds of nasturtium, bol d'armenie, blood stone, sugar loaf, and rose water.

Second: A plaster of the collagen of sheep-skin boiled in rain-water: naval pitch, colophony, litharge, ammoniac, opoponax, galbanum, bdellium, mastic, serapinum, terebinth, sumac, roots of both consolids, oak-tree sap, blood stone, frankincense, myrrh, aloes, mummy, bol d'armenie, sangdragon, aristolochia, earth-worms, and human blood. Mix them with the collagen.

Third is Bruno's Plaster: cypress nuts, balaustes, tragacanth, myrrh, sarcocolla, frankincense, gum-arabic, sangdragon, mummy, aloes, alum, and vinegar.

Treat Painful Hemorrhoids by fumigating the seated patent with a decoction of tassus barbatus, camomille, and melilot. Then apply the same as an ointment—made in a lead basin.

When the pain is severe, use Alexander's liniment (which I favor): saffron, myrrh, frankincense, lycium, and opium. Add rosat oil and a mucilage of psyllium, and egg-yolks.

For perianal problems use Rhazes' plaster: camomilee, melilot, boiled egg-yolks, fenugreek flour, linseeds, guimauve-roots saffron, myrrh, aloes, and butter.

CHAPTER 8

The Lower Extremity

There are many recipes.

First are those which reduce edema. Foment with sea-water or brine heated with sambucus, elder, thistles, calamint, oregano, aloes, and parietory.

Treat a localized mass with this plaster: bran, bean-flour, dove-droppings, a decoction of asphodels, vinegar, and the juice of red cabbage.

EPILOGUE

Now I have finished this Sermon, that I hope will supplant the navigators who had held the anchor, and had attracted faithful followers, and admitted them to a glorious heaven. Let the same beneficent God who reigns through eternity, do the same for them, and all my Readers who deign to favor me. Amen.

Here ends The Major Surgery, composed in 1363, by Guy de Chauliac
Surgeon, and Master in Medicine of the University At Montpellier

APPENDIX I: THE PHARMACOPEIA

APPENDIX II: SURGICAL INSTRUMENTS

FIVE PLATES

APPENDIX I

THE PHARMACOPOEIA

Translator's Note

N icaise added a Glossary to follow the text of his edition of Joubert's *Guy*. It has three sections: a Glossary of French, Latin, Provencal, and Arabic terms; a long list of medications mentioned by Guy, and a section about surgical instruments. Here I have added Two Appendices. I have eliminated the Glossary because all of Guy's terms have been translated. I have extended Nicaise's list of medicines in the Pharmacopoeia that follows, and revised the section on instruments.

Nicaise counted 750 medical substances mentioned by Guy, of which 260 were specially noted in Chapter 8 of Doctrine I, of Treatise VII, where he listed their Degrees. I have taken most of that chapter and listed the degrees with the entries in the pharmacopoeia, adding a line to each entry; the line identifies the 260 medicines. However, Guy's and Nicaise's explanations about 'degrees' remain, and I suggest that the Reader see them before digging in the Pharmacopoeia. My list provides English, French and Arabic synonyms to identify the medicines, at the risk of making the Appendix too long. The cross-referencing term 'see' may help the Reader without frustrating her/him. Wherever feasible, I provide a proper botanist's name for the plant, at least at one of the synonymic entries. The names that Joubert (or Guy) used has changed many times through the centuries. Nicaise had the good fortune to have an expert consultant, Dr. Saint-Leger, to review and identify all the medieval medications, Yet, during the century since his work, the Latin botanical nomenclature has been revised more than once as the scientists have come to new understandings about the species, the genera and the orders of the world's flora. I have used what seems suitable here, with apologies for my own ignorance of the taxonomy.

The Reader is aware of the size and the complexities of some of the recipes in the text. Most of them came from Arabic source during the several centuries of Arabic ascendancy and were translated into Latin during the 2 or 3 centuries before Guy. His favorites were Rhazes, Halyabbas, Avicenna, Avenzoar, Averroes,and Heben Mesûe, all of them devoted to Galen as their source, reaching them via Aetius, and Paul of Aegina. Guy's clerical medical associates at the University added more recipes. Guy frequently mentioned 'the common' remedies, and we must be aware that the complex recipes with strange and difficult-to-obtain Simples challenged the Apothecaries, and were reserved for the wealthy patients, whereas the common folk had other remedies, which I daresay were equally effective or ineffective.

The medicines indeed were ancient and traditional. In my *Medieval Surgical Pharmacopoeia and Formulary* (qv̇) I printed lists of medicines used in classical Rome and in Byzantine Greece. Most of Guy's medicines can be found in them. More recently, I was able to search two Egyptian surgical papyri from early in the second-millenium BC, the Breasted and the Ebers. We find many of them there. I have marked some of them in our pharmacopoeia here with an **E.**[673]

Guy's Surgery is more a series of instructions for using topical applications than a handbook of manual surgical techniques. That was the Hippocratic Rule until the 16th C when the lay and the barber surgeons wrote the works that began modern European surgery. The treatment of most maladies, other than wounds, began with attention to diets and healthy life-styles. Medicines were used after that delay. The surgeon also intervened with his local applications of medicines. Oral medications were the province of the Physicians, and the surgeon almost always began his treatments with the physician's favorite, the laxatives. They were part of an arcane scheme that insisted certain laxatives were to be used for certain maladies. The Physicians claimed that the selection of the correct laxative and the dosages were matters too complicated for the uneducated therapist; Guy knew better than to bypass his medical consultant. Indeed, his graceful acknowlegment of his subsidiary role was a key to his popularity in the Schools during the ensuing several centuries, and in part explains their lack of attention to Henry de Mondeville, the curmudgeon who wrote that when a physician was not ready at hand, why seek one who would order a laxative, even when none was indicated.

Every Medicine is named, often followed by an **E.** Synonyms follow, in English, et al., followed by *see* and by the Latin botanical name. The degrees are set on another line. Warm (**W**), Cool (**C**), Dry (**D**), Moist (**M**). Example: W-1, H-2. Guy omits numbers when the medicine is a vehicle or when it adapts to another more dominant simple. Example: C and D, without numbers.

The List

Absinth (E*)*	oil of wormwood, aloine. see abrotonin. *Artemisia (sev.species)*
Abrotonin	avrotonin, aurone, southern-wood. see artimesia, et al.
Acacia (E)	Gum Arabic,not Blackthorn plum. sap of *Acacia arabica.* C-3, D-3
Acanthus (E))	bear's breech. see branca ursina. *Acanthus mollis.*
Ache (E)	celery. see apium, *Apium graveolum.* W-1, D-2

---

[673] The Papyrus Ebers. Translated by B. Ebell, Copenhagen, 1937, and the Edwin Smith Surgical Papyrus, Translated by JH Breasted. Univ. Chicago Press 1930. Also see *The Medical Skills of Ancient Egypt,* J. Worth Estes, Canton Ohio, Science History Publications, 1989 (LDR).

Acoris (E)	sweet sedge. see calamus, souchet, squinanthus, sangdragon, flame batard
Adiantum	delicate ferns. see ferns, *Adiantum capillus veneris.*
Aes	bronze, ios. see chalcanthum, see bronze. see vert d'airain.
Agaric	a mushroom. *Polyporus igniarius.*
Agnus castus	also chaste tree. see ammi. *Vitex trifolia.*
Agresta	sour-grape (uva acerba) juice. see verjus, passerile. *Vitis vinifera.* C-2, D-3
Agrimony	hemp. see eupatorium. *Agrimonia eupatorium.* W-2, D-3
Ailes (E)	allium, see garlic et al. W-4, D-4
Airaine	brulé copper oxide, 'flower' of bronze, see vert d'airain et al.
Alcohol	(alcofol) powders made into very dilute collyria for eyes
Aldabat (E)	lizard
Aldaraon	ancorde. see potentilla
Alkanna (E)	henna skin dye *Lawsonia inermis.*
Alkekenji	see kekenji
Alkitron (scent)	cedar resin. see cedar (of Lebanon), see resin
Alleluia	clover et al., see sorrel et al. *Oxalis acetosella*
All Heal	catholicon, panax. several other herbs.
Allium (E)	see garlic et al., see ailes.
Aloes (E)	aloe vera. see pigra. *Aloe socotrina.* W-2 D-1
Aloe	wood also Lignum aloe. wood of *Aquillaria agalocha Not Aloes*
Aloyn	absinth, or a purge made from aloes W-1 D-1
Alum (alun) (E)	from wine lees (potassium tartrate) or from crystals. Alun de roche is sugar-alum (aluminum sulfate), alumen de pluma (feather alum) is halotrichite (iron-aluminum sulfate). W-2 D-3
Amande	almond
Amanita	aimant. a non-deadly mushroom
Amber	see carabe.
Ambergris,	here more likely ambrosia, and not from whales. see artemisia.
Ambrosia	see ambergris
Amethyst	gem stone
Amidum (E)	flour of various grains, usually wheat, amylum, froment. see see farina, see flour, see siligo, see spelt.
Ammi (E)	goat (or gout) weed seeds. see agnus castus. *Ammi visnaga or majus.*

Ammoniacum (E)	hore-hound. see marrubium. *Dorema ammoniacum* or *Bubon gummifer*. W-3 D-1
Amome	cardamon family. see ginger, see zinziber.
Amydon	the 'flowers' of silver and other metals. see litharge
Anabulla	euphorbia et al. see titymal,
Anacardus	cashew nut, elephant-foot tree. *Semicarpus anacordium*. W-4 D-4
Anagallis	see pimpernel, see mourons., *Alsine media*
Ancorde	see potentilla
Andromache	andromeda, a moon-wart
Aneth	sweet anise. see dill, see sifula. *Anethum graveolens*. W-2 D-2
Angelica	see emperor's herb. *Angelica archangelica*.
Anise (E)	leaves and seeds. see pimpinel, see saxigfrage. *Pimpinelle anisum*. W-3 D-2
Anthera	from rose flowers. see roses
Antimony (E)	stibium, as oxides, not as a pure metal C-1 D-2
Aphronitre	saltpeter
Apium (E)	many species. ache, batrachium, berula, patalupi, wild celery, persil, wild parsley, petroselinum, selinum. see all. *Apium graveolens*.
Apricot	chrysomele, Armenian apples
Aqua forte	nitric acid.
Araigne	spider web
Arain	bronze flower, see aes W-3 D-3
Argentine	see potentilla
Argentum Vif (E)	mercury (quick silver). see mercury C-2 M-2
Argilla (E)	clay, see terra sigillata. C-1 D-2
Aristolochium	polyrhizon. see clematis, see crow's foot, see ciclamen, see malum terrae, *Aristolochium rotunda* W-2 D-2
Armoise	see artemisia
Arondelle	droppings from young swallows
Arsenic	sublimate of arsenic sulfate or oxide, see orpiment. W-3 D-2
Artemisia (E)	absinthe. see abrotonin, *Artemisia vulgare*.
Artomel	bread with honey

Arum	see pied de veau (cow's foot), *Arum maculatum,* W-2 D-2
Arundo	marsh-reeds. see reeds, see roseau.
Asafoetida (E)	'stinking' asse. see sagapenum, *Ferula foetida* W-3 D-3
Asarum	wild nard, hazel wort. *Asarum Europaeum*
Asparagus	asperge, *Asparagus officinalis*
Aspen	the tree. see poplar. *Populus tremula*
Asphalt	tar W-2 D-2
Asphodels	see haste royale. *Asphodelus ramosus* W-2 D-2
Athanasia	tansy. *Tanacetum vulgare*
Atrement	atrament. vitriol based. see vitriol et al.
Atriplex	hortensa orache, spinach, *Atriplex hortensis.*
Aumeli	palm-tree seeds
Avellana	hazel-nut. see nux. *Corylis avellana*
Avena	haveron. see oats W-1 D-1
Avrotonin	aurone, see abrotonin, see artemisia W-1 D-2
Axonge	lard. see graisse, et al.
Badger	the fat of
Baguenade	pod of senna
Balaustium	wild pomegranate flowers. see pomegranate et al. C-2 D-2
Balle	Marina moss on palm trees
Balsamodendron (E)	see bdellium, myrrh, and baume *Balsamodendron opobalsamum.*
Balsamita (E)	see costus, see polygonum, and Mary's seal. *Tanacetum balsamita.*
Barecha	cucumber, pumpkin, see melon. *Cucumis astiva,*
Barley (E)	see orges et al.
Basilicon Oint.	contains: bdellium, fats, oil, pix, resin, wax.
Basil(-ium)	herb basil, wall-thyme. *Ocymum basilicum.*
Bats	meat of, chauve souris.
Battitura	see chalcanthum et al.
Baucia	parsnip or carrot, see daucus. W-1 M-1
Baum	crespa see mint
Baume	see balsmodendron
Baurache	see borax
Baye de Laurier (E)	laurel berries. see laurel

Bdellium (E)	balsamic resin. see balsamodendron. *Balsamodendron africanum.* W-1 M-1
Beans (x E)	fava.
Bear	the fat of
Beaver	musk. see castoreum
Bec de Gru	see geranium, Herb Robert C-1 D-1
Bedegar	bendegar, wild roses. eglantiere. see roses. C-1 D-2
Bees	honey bees. apia
Behen	see centaurea. See fn. Tr.VII. Doct I. Ch.5
Belladonna	see henbane et al., see mandragora.
Bellirica	see myrobalans
Belliculus	blata bysantia, belliculli marini. purple and white marine snails, source of royal purple dye.
Belsegensina (E)	see coriander.
Ben (E)	been, nut of tree. see Behen (fn). *Moringa aptera.*
Bendegard	bedegar. gall from stem of eglantine rose. see roses.
Bennett (benedicte)	Herb Bennett. see cicuta.
Benoiste	holy oil
Benzoin	see storax—styrax
Berberis	barberry bush, cortex bugia. see bugia. *Berberis vulgaris.* C-3 D-3
Berula	an herb, a variety of apium. see apium. *Berula angustifolia.*
Beta	beets, bleta, betta, swiss chard, mangel wurzel *Chenopodiacia.*
Betonica	betony, scrophularia. *Betonica officinalis* W-3 D-3
Bezard	see galbanum
Bile (E)	many sources, and human, cranes, falcons, eagles, bulls, et al.
Birds' Tongues	see langue passerini
Bitumen	see tar
Blatta Byzantia	see limace
Blette	chard, see beta
Blood	many animals, humans, goats, et al.
Blood stone	hematite. see sedengi
Bol d'armenie (E))	a clay containing iron oxide, various clays—German, Bohemian, etc.—sometimes fuller's earth, cimoleam. see argilla, see terra sigillata. C-2 D-2
Bone-marrow	midolle, meule. see medulla et al.
Bones (E)	of geese, hens, deer, etc. hooves, ivory, horns, antlers, squid etc., all burned or powdered. see cornu.

Borax	sodium borate, boracis, burud, see nitrum, sal baurachi, sal de nitre.
Borago	borage, bugia. *Borago officianalis.* W-1 M-1
Boüillon	a sedative herb. see tassus barbatus
Boxwood	see buxus
Boyau	denterre detritus from worms.
Bran (E)	son, furfur. see amidum et al.,
Branca Ursina (E)	acanthus, brama, bear's brush. see acanthus. *Acanthus mollis or Heraclium spondylum* W-1 M-1
Bread (E)	crumbs are mica panis. Panata is bread soup. Opirus is whole-wheat bread used in many cataplsasms
Bresillet	*Caesalpina sappan* (later called the Brazil tree)
Bronze	see arain, aes. see vert d'arain.
Broom	see geneste
Bryonia (E)	brionee, viticelle, root of bryony, labrusca, couleuvaé. see clematis et al., see viticella. fesire and sicadis are poisonous. *Bryonia dioecia* White bryony, *Momordica elaterum*, really is a cucumber.
Buchormarien	see ciclamen (cyclamen)
Buckshorn	deerhorn plantain. *Plantago corona*
Bugia	root-bark of barberry bush, cortex bugia. see berberis. C-1 D-1
Bugloss	borrago, blue weed, lingua bovis, lithospermum. *Echium vulgare, Borrago officianalis or Anchusa officianalis* W-1 D-2.
Burud	see borax
Bursa Pastoris	shepherd's purse. sedative *Capsella bursa-pastoris* C and D unrestricted
Butter	from cows, also sweet butter, May butter, bure de mai. W-1 M-3-4
Buxus (Buis)	boxwood, dogwood, ammon's horn. *Cornus amonum and floridum.*
Cabaret	see Asarum
Cabbage	choux, caul, cholet. *Brassica oleracea.* W-1 D-2
Cachimia	climie.
Cadmia (E)	cadmium sulfate, sory, climie. see cathimia et al.
Calament (E)	calamine. thyme. see melissa. *Melissa calaminthus.* W-3 D-3
Calamite	fossil plants. see amber
Calamus	aromaticus, sweet sedge, sweet flag. see acoris,
Calchitis (o)	copper pyrites, chalcathum
Calcide	see colcothar, vitriol

Calcined	tartar see cendres graveles
Calcitrapa	see centaurea
Calf's foot	see arum et al.
Calidicon	see quick-lime
Calomel	mercurous chloride
Calx viva	quick-lime. see lime.
Cambril brûle (E)	red sand (from dry clay)
Camomille	chamomile (several varieties), *Anthemis nobilis and A. pyrethrum.* W-1 D-1
Campanula	rampion. *Campanula indicas.*
Camphor (E)	champhore, caphura. *Laurus camphora.* C-3 D-3
Cancer	crayfish, chancre fluvialis
Canelle	cinnamon
Cannabis (E)	seeds or leaves, see Cheneve. *Cannabis sativa.*
Canne (E)	reeds, grasses. see roseau,
Canne aromatique (E)	reeds. *Arundo donax*
Cantharides (E)	spanish fly. *Eloe vesicatorum.* W-3 D-3
Capers	fruits, shells, bark, oil. *Capparis spinosa.*
Capillus	veneris see venus hair, see adiantum, see ferns et al.
Capitellum	capiteils, potash lye. see lessive, see lixivium et al. W-4
Caprifolium	caprifici, chevrefoile, honeysuckle, fig. see lycium *Lonicera caprifolium.*
Carabe	amber, possibly containing a scarab beetle. see amber.
Caraway	see carvum
Cardamon	see amomie, see balaustium, diazinziber. *Amimum cardamomum.*
Cardo (E)	thistles, cardone benoiste, *Cnicus benedictus*
Carpobalsama (E)	see balsameta.
Carob (E)	St. John's Bread, locust pods. *Ceratonia siliqua*
Carrot	see daucus et al.
Carthame	safflowers, seeds, and diacarthamus. *Carthamus tinctorius*
Carvum	caraway seeds. *Carum carvi.*
Caryophyla	garyophyla, giroflé, sanamunda, herb-benedict, cloves, eugenia, geum, arnaglossa, carnations, chickweed, stellaria, ipia, horse-foot. *Caryophylus aromaticus or Geum urbanum.*
Cashew	see anacardus
Cassia (E)	'bastard' cinnamon and many others., cannelle, see senna. *Cinnamomum cassia.*
Cassia Fistula	the drumstick tree. *Cassia fistula*

Cassonade	moist brown sugar
Castor bean (E)	ricinus, cataputia, cocconidium. Diacastor is a laxative electuary. *Ricinus communis*
Castoreum (E)	beaver's testicles and musk. see musc. W-2 D-2
Cathimia (o E)	lapis calaminaris. see cadmia, see climie, see litharge.
Catholicon	see All Heal
Cauda equina	horse-tail, hippuris. see equisetum.
Caulcide	see centaurea
Cedar (E)	tree of Lebanon and its resin, alkitron. *Cedrus liba*
Cendre de chêne	ashes of Turkish oak (Quercus cerris) lexivium.
Cendres Gravelées	potassium tartrate—lye, calcined tartar W-4 D-4
Centauria	knapweed, jacea, narce, thistle, teazle, behen. al. *Centaurea jacea* and *Erythrea centaurium* W-3 D-3
Centinodium	corrigeola, polygonium. see renoué
Centrum galli	cerebrum galli, chicken brain. see gallitricum
Ceratum	ceroneum, ceroneum. A waxy plaster
Cerdone(E)	cardo, chardun, calendula. see centauria, see thistle, see senecio, see teazle et al. *Centauria centaurium or Cardo benedictus.*
Cerfeuil	chervil, cicely. *Chaerophylum sativum, Myrrhis odorata*
Cerisier	cerice, cherry tree, bark, sap. *Prunus avium.*, *Cerisus vulgarus*
Ceruse (E)	see white lead, see minium, see litharge, galena. C-2 D-2
Cestre	see saxifrage
Ceterach	see bugloss, see scolopendre.
Chalcanthum (E)	copper sulfate and other metallic salts, battitura. see aes, see bronze, see couperose, see vert d'arain, see vert de gris.
Chalmaedrys	kamedrys. see germander, iva, *Teucrium chamaedrys*
Chalmaepitys	another Teucrium
Chanvre (E)	see cannabis
Chardon-bénit	cardonia benedicta. a thistle. *Cnicus benedictus*
Chardon des foullons	see dipsacus. *Dipsacus fullonum*
Chataigne	chestnut. see nux. *Castanea vesca*
Chaussetrape	see centaurea
Chaux	lime. see quick-lime
Chebules	see myrobalans
Chelidonium	celidoine salvage, mentelicum. celandine. see polemon. not memitte *Chelidonium majus or Ficaria ranunculoidis.* W-3 D-3

Chéne	oak, tree etc.
Cheneve	cannabis seeds
	W and D
Cherry	see cerisier.
Chervy	sauvage baucia, *Sium sisarum*
Chevaline	mentastrum a mint *Mentha rotundifoliua*
Chevrefuille	see caprifolium
Chick peas	cicer. see matris Sylvia, and poix ciche *Cicer arietinum.*
Chicory	leaves. scariola, groin du porc, endives. *Cichorium intybus*
Chicotrin	orpin, stonecrop. a *Sedum*
	C and M
Chiendent	dog-tooth, couch-grass. *Cynodon dactylon*
Chou	cabbage
Chrysomele	see apricot
Cicin	castor bean, ricinus
Cicada	incinerated insect, so-called black cantharides. *Cicada plebeia*
Cicer	see chick-pea.
Ciclamen	or cyclamen. pome de terre, earth-nut, panis porcinus, malum terrae. *Conopodium majus, or Cyclamen hederifolium.*
Cicuta	cigue, cowbane, water hemlock, cicuta virosa, benedicta, beneita, hemlock seeds. see herb bennett. *Conium maculatum.*
Cimolea	chimolea, cymolia, terra cimolea. mud containing metallic and stoney bits accumulated under whet-stones. see scoria
	C and D
Cinis	ashes et al. (various woods, bones, crabs, mouse, rabbit, scorpion, sponge, hair, grapevine, oyster shell, sea-shells, seashell, hair, paper, etc.)
Cinnabar (E)	mercuric sulfide. see mercury.
Cinnamon	cannelle, cassia, diazinziber.
	W-3 D-3
Cinq-foil	see tormentilla et al.
Cire wax	
Citrons (citreum)	various citrus fruits and melons. Venarum citrinum is citrus fruit pulp, citron pips were used alone. see oranges, see limes.
Citrullus (E)	gourds, melons. see colocynth, see melons
Claret	an aromatic wine
Clarete	St. Claire's herb, *Valerianella olitoria*
Clavelliere	pansy, violet. *Viola tricolor*
Clay (E)	all types. see argilla, see bol d'armenie, see terra sigillata
Clematis	the flower, ground-ivy, liere, viticella, vitis petit vigne, hedera, flammula. see aristolochium et al. *Clematis vitalis and Hedera helix, etc.*

Climie argenti	metallic oxides, especially silver, gold, zinc and antimony. also cacumia, cadmia, cathimia, spode, tuthy, iron, couperose.litharge, spuma d'argent.
Clover	various, including melilot
Coccia	laxative. see hierorufinum
Coing	quince citonium, diacydomite. *Pirus, cydonia*
	C-2 W-2
Colcothar	see vitriol rosa.
Colle (E)	gelatin, gluten. Vellis vacciniis cowhide as a source
Colocynth (o E)	"bitter apple' or bitter cucumber, cucurbite.citrullus. see melon, *Cucumis colocynthis*
Colophony	pine pitch. pix.
Comfrey	see consolida
Concombre	see cucumber
Connil	rabbit, coney, cuniculus.
Consolida	consoude, greinure, comfrey. *Symphytum officinalis (large) or Brunella vulgare (small)* also *Bellis perennis*
	W and D
Convolvulus	bind-weed. see scammony. *Convolvulus hystrix*
Coquille	sea shells
Copper (E)	flakes, salts, et al. see verdegris
Corail	coral polyps(red and white), sponge stone. see ecume de mer.
	W-2 D-2
Coriander (E)	herb belsegensima, see git, see nigella. *Coriander sativum.*
Cormier	see sorba
Cornu (E)	powdered horns of beef, deer, and goats.
Courge	see cucurbita
Costus (E)	costmary (roots and oils). see balsamita. *Costus arabicus*
	W-3 D-2
Couleuvree	bryony.
	W-2 D-2
Couperose	see chalcanthum et al.
	W-4 D-4
Courges	zucchini, watermelon et al. Juices are elacterina. Watermelon is cocomero. see cucurbite. *Cucurbita maxima.*
Crab	cancer fluvialis, granchia titrata, crab-meat and shells. see crayfish
Crayfish	flesh, juice and ashes.
Crassula	major and minor, sedum, stonecrop, orpine, andrachne, mamilla muris, tegularia, vermicularis. *Sedum purpurascens.*
Cresson	cress, saxifrage
	W-2 D-2

Crocus	see saffron.
Crows' feet	see pied corvin, pes corvus, pes milvi. see aristolochium et al., see clematis. *Ranunculus sceleratus.*
Crystal	crystals, ground glass
Cubebs	a pepper. *Piper cubeba.*
Cucumber (E)	cultivated or wild, momordique, cucumiscelle. see barecha. *Cucumis var., Ecballium elaterium.* W-2 D-2
Cucurbita	melons
Cuivre brulé	see aes, et al.
Cumin (E)	comin, the herb. *Cumin cyminum.* W-3 D-2
Curcuma	turmeric, zedoaria. *Curcuma longa.* W and D
Cuscuta	dodder. see epithyme. *Cuscuta major*
Cuttle-fish bone (E)	squid bone.
Cynoglossus	cinoglossus, hound's-tongue herb, leaves and roots. *Cynoglossus officianale*
Cyperes (E)	succus, papyrus, sedge et al. berries are from *Cyperus longus, C. rotundus.*
Cyprés (E)	the tree, nuts (seeds), bark, leaves, sabina, see vernis. The nuts are from *Cupressus sempervirem.* W-1 D-2
Dactylus (E)	dates and leaves of palm *Phoenix dactilifera.* Diaphoenicon is a laxative electuary of dates. see diamoron.
Damascus plums	see pruneau. plums et al. see sebestes
Darsene	see cinnamon
Dates	see dactylus.
Daucus	carrot, baucia. *Daucius carota*
Deershorn	see buckthorn
Diacalamint	
Diacastoreum	pills
Diacatholicon	
Diachylon	another variety of Mesûe's ointment made of: aloes, aneth, camomille, colle, fenugreek, figwort, lard, linseed oil, litharge, lanolin, resin, squills, terebinth, wax.
Diacumin	
Diacytiton	see coing, quince
Diadragunt	an electuary. see adracanth, tragacanth.
Diagridum	an electuary of scammony.
Diahirundum	from swallow-nests
Dialthea	ointment of Althea

Diamargaritum	see perles. troche based on *Rubis troscicata.*
Diamoron	from mulberries
Dianthus	see caryophylla. also an electuary based on rosemary
Dianucum	from nuts
Diapapaver	from poppies
Diapolytrichum	
Diarhodon	an astringent powder from acacia, anise, blackberry, cicuta, sandalwood (two kinds), terra sigillata.
Diasena	electuary of senna
Diasimpyton	from comfrey (consolida)
Diazinziber	a purgative made from cinnamon, cardamin, cassia, ginger, galanga, nutmeg.
Dictame	diaptamum. see oreganum
Dipsacus (E)	see teazle, see thistle, see cardo, see centaurea. *Cardo fullonum and Dipsacus sylvestris.*
Dog-tooth	a violet
Dogwood	box-wood, stink-weed, punaise. The oil of the berry is cornel. see buxus
Donkey	lard of
Doronic	a thistle, *Doronicum pardalianche*
Dove's foot	see geranium
Dragacanth	see tragacanth, a gum.
Dragantum	iron peroxide. see chalcanthum, see colcothar chalcidis, see vitriol rosa. sometimes confused with dragacanth.
Dragontea (o)	serpentaria. tarragon, dragon weed. *Arum dracunculus*
Duck	fat of
Eagle	fat of, and bile of
Earthworm	see vermis, lumbrici
Eau Aluminium	alum water
	D-4
Eau Ardente	alcohol
Eau des Maréchaux	rust-stained water
Eau Forte	nitric acid
Ecailles d'Arain	see Ecume d'Argenti
Ecailles de Cuivre	battitura aeris, chalcanthum
Ecorce d'Airain	burnt copper. cortex aeris
Ecume d' Argent	spuma argenti, litharge, see climie argenti
Ecume de Cuivre (o)	scoria aeris, see scoria
Ecume d'elgagner	pumice, spuma maris
Écume de Mer	meerschaum, spuma maris, at least 5 varieties, including sponges, algae and corals, halcyon. see corail.
	W-1 D-3

Ecume de Nitre	the scum on the surface of nitric acid
Eels	fat of
Eggs (E)	ova. whole eggs, whites (album), yolks (moel), shells.
Egyptiac Ointment (E)	
Égyptian thorn	see spica. *Mimosa nilotica*
Elder tree	sureau. *Sambucus nigra*
Ellebore	eleborum, ellebre, Christmas rose. see hellebore. *Helleborus album and nigrum, Veratrum album.*
Elm	slippery elm tree bark. *Ulmus fulva*
Embula (o)	see anabula.
Emerald	aluminum silicate gemstones, smaragdus, praze, prassium
Encens	see frankincense
Encre	caustic red ink. encaustrum,. see vitriol rosa, see zez(g)i. *Terminalis vernis.* W-3 D-3
Endive	pig-snout, groin du porc. see chicory, see rostrum du porc.
Enula	inula, elecampani. see spic et al. W-2
Epine noir	sloe, blackthorn. *Prunus spinosum*
Epithyme	see cuscuta. *Cuscuta major and minor.*
Equisetum	asperella, cauda equina, queue equina, horse-tail reed, shave grass. *Equisetum arvense.*
Ers	vetch, see orobe. W-1 D-2
Eruca	charlock. see mustard weed, see senape, see roquette. *Sinapus avensis*
Escarbot	scarab beetle. *Scarabaeus stercorarius*
Escargot	limaicon. see limax
Espic	see spica, fuille de spica
Espinard	spinach
Espurge	spurge, see euphorbia.
Esula	a spurge. see euphorbia et al., see spurge.
Eupatorium	see agrimony, see hemp, see St. John's herb.
Euphorbia	amblete, custos hortis, marsilium, manne, solsequium, esula, titimalle, cataputia. anabulle is the sap. see all named here. *Many varieties of spurges, Euphorbeacea.* W-4 D-4
Euphrasis	eyebright. *Euphrasis officinalis*
Faex cerae	bees-wax. See propolis
Farina de Moulin	far, amylum, farina volatica, mill-dust (fine wheaten flour) also found at bake-ovens, amidum, wheat. C and D

Faseola	horse bean
Fats	lards, see graisse
Fauciole	see plantain. *Plantago lanceolata*
Fueille de nard	Indian spikenard, see spica. *Andropogon schoenanthus*
Fava	feve, fabba, beans or their stalks, especially *Vicia faba*. C and D
Feces (E)	stercus, tordus: sheep, goat, deer, birds, horses, dogs, human, mouse etc.
Feaugére	fougére, bracken, fern. *Pteris aquilina*
Fennel (E)	fenoil,. see nigella, *Foeniculum dolce et vulgare, and Nigella sativa*. W-2 D-2
Fenugreek	Greek fennel, aegoceros, buceros, telis. seeds, ferrugine. seeds, usually powdered. see fennel. *Trigonella foenum-graecum*. W-1 D-1
Fer (E)	iron C-1 D-1
Ferns	filix, beech fern, royal fern etc. see adiantum, see polypode, see polytric, see stag's forn, see venus hair. various *Osmundia*
Fesire (E)	poisonous white bryony. see bryony
Feuchiére	royal fern, water fern *Osmundia regalis* W-2 D-2
Feutre	felt
Ficus (E)	fig, alos, coctana, citonia. caprificus is a wild fig. ashes, see caprifolium, see lycium. *Ficus arboris*. W and D
Fiente (E)	fimus columbinus or equus. pigeon or horse droppings, feces et al. W and D
Filipendula	spirea, dropwort, meadow-sweet. *Spirea filipendula*.
Fisticus	see pistache
Five-finger roots	dropwort
Flambe batard	see acoris.
Flammula	clematis. *Flammula clematis* W-4 D-4
Flax (E)	
Flea bane	pulicaria
Flos (fleur d'arain)	'flower', battitura aeris, aloxan. various films (usually metallic salts) deposited on metals and fluids. see vert d'arain et al.
Flour (xE)	farina, furfur, samich, froment, pigle (coarse-ground). see amidum, wheat
Flower of Salt	scum on sea water. see flos
Frankincense (E)	gum resin.encens, see olibanum, see thus., see manne. *Boswellia cartorii*. W-2 D-4

Fraisne	see fraxinus. *Fraxinus excelsior*
Fraizier	fragara, strawberry. *Fragara vesca*
Fraxinus	bark and flowers of ash tree, fiêne, stone-mint, manna. see tamariscus. *Fraxinus excelsior* C-2 D-2.
Frogs	see grenoüilles
Fumitory	fumiterre, gingidium, perhaps flama, lady's mantle, hen's foot. *Fumaria officinalis or Gingidium.* C-1 D-2
Fungus bedezaris	see bedegar.
Galanga	China root. galingale. *Alpinia galanga. Kampferia galanga*
Galbanum	a ferula resin. *Ferula galbaniflua* W-3 D-2.
Gallie musque	troches containing musk
Galline grasse	lettuce. see lactuca
Galls (E)	oak galls. C-2 D-3
Gallitricum	salvia, centrum galli, crista galli, oculi christis, sclaria. W and D
Garance (E)	madder root, spargula, rubia, valania, woad. *Galium molugo, G. aparime and Rubia tinctorum.* W-3 D-3
Garlic (E)	see ailes, see allium.
Garyophyla	see caryophyla et al. *Geum urbanum*
Geiss	chick pea. see cicer
Geline	grouse. fat of
Geneste	broom plant, see planta geneste,
Genevrier	juniper. see sabine. *Juniperus communis* W and D
Gentian	genista. bitter roots of *Gentiana lutea* W-3 D-3
Geranion	geranium, see gratia dei, Herb Robert.
Germander	chalmedrys, camedrios, yva. see scordium, see polium. *Teucrium polium and montanum.*
Ginger	gingembre, zinziber, genevrier, abel. see amome. *Amomum zingiber.*
Girofle	cloves. see caryophyllus W-3 D-3
Glands	oak acorns C-2 D-2
Glaucium	yellow opium
Glayeul Puant	see spatula foetida

Glu Acamli	alcanach. a gluten. fish glue
Glu Poisson	fish glue, see glu acamli
	W- 1 D-1
Gluten (E)	gelatin from fish or domestic animals, mucilage, animal collagen (not from grains). see colle. Skin
Goat	blood
Gomme albotin	see terebinthe
Goose	fat of
Goutte de Lin	see cuscuta *Cuscuta europea*
Graine	see oak
Graine de Paradis	see cardamon. paprika.
Graisse (E)	animal fat-chicken, geese, pork, turtle,bear, fox, wolf, et al., pinguedo, sepum, ysopus, bovis, gras, sui, adeps. see hyssop, Grassede are the drippings from roasts.
	W and M
Grape	leaves, fruit, skins, raisins, sapa michum (with honey), saramitum. capreoli are the tendrils. see agresta,
Grasses and Reeds (E)	canne, lolium. see roseau.
Gratia dei	grace dieu, see geranium, see et al. *Gratiola officianalis.*
	W-2 D-2
Gravel	wine lees
Gremil	see alkana, hair dye. *Lithospermum officinali*
Grenade	pomegranate. *Punica* granatum
	C-2 DF-2
Grenadier	see balaustium
Grenoüilles	frog meat
Grenouillette	see apium
Groin de Porc	rostrum porcini, *Taraxacum dens leonis*
Groseille	currants or gooseberries. *Rubis grossularia or rubrum.*
Guimauve	mauve, *Althea officianalis*
Gum Arabic (E)	from acacias, white gum (not called Arabic in ancient times)
	C and D
Gumera (	fard, a facial cosmetic
Gum de Chéne	viscous oak sap
Gypsum (E)	gyp, cockel, selenite, alabaster, plaster of Paris, calcium sulfate
	W and D
Hare's Beard	mullein, verbascum. see thapsus
Haste royale	see asphodel
Halhaste	garden Hyssop. see hyssop
Harmel	seeds of rue. *Pegamum harmala*
Hawthorne	*Crategus oxyacantha*
Hedera	ivy. see lievre

Hellebore	see ellebore.
Hematite	rust, emathitis, blood-stone, lapis sanguinis, red ochre, iron Unwashed= W-1 Washed= C-2
Hemlock	see cicuta et al.
Hemp (E)	see agrimony, see cannabis, see St. John's herb, see eupatorium.
Henbane (E)	belladonna, cassilago, chenille, lupini. jusquiame, morel, solathrum, nightshade. see hyoscyamus,
Henna (E)	skin dye. see alkanna
Hepatica	liver-wort, *Marcantis polymorphia* W-2 D-2
Herbe Adhil	see euphrasia
Herbe aux poux	see staphisagre
Herbe aux puces	psyllium. *Plantago psyllium*
Herbe aux vent	wind flower, *Anemone pulsatilla*
Herbe Capillaires	fivevarieties of ferns: Venus Hair, see adiantum, see ferns
Herbe dela paralysie	primula. *Primula veris*
Herb Robert	see geranium
Herbe Saracenique	mentha saracenique. see *Aristolochia clematis*
Herbe Tunix	dianthus caryophyllus
Herisson de mer	Hericium. sea urchin bristles, ericium, hircis, lupis iudaici. see limacons.
Hermodactyl	digitus hermetis. colchicum and related tubers. *Hermodactylus tuberosus*.
Hiere	various laxative compounds
Hiere de ellebore	
Hieralogodon	see coccia
Hierapicra	of Galen, a laxative. see coccia
Hierarufinum	a laxative of colocynth, germander, asafoetida, wild parsley, aristolochia, pepper, cinnamon, saffron, polium, myrrh and honey. see cochia, see pigra.
Hippia	see anagallis
Holy oil	see benoiste
Honey (E)	miel, various kinds. hydromel is a mixture with water. beehive honey contained some wax. see oxymel.
Hops	see houblon
Houblon	hops, lupulus. *Humulus lupulus*
Huiles—Oils (E)	olives, muscade, ben, benedict, fisticum, muscellin, etc.
Hydromel	honey and water
Hyeble	see sambucus. *Sambucus ebukis* W-2 D-2
Hyoscyamus (E)	see henbane et al.

Hypericon	St. John's Wort (herb). mille-fuille, yarrow. *Hypericium perforatum.* W and D
Hypocras	claret wine
Hypocistus	fungus on the roots of cytus plant. *Cytinus hpocistus* C-2 D-2
Hysop humidus	arabic for lanolin. see graisse et al.
Hyssop (x)	herb hasca. *Hyssopus officinalis.*
Indigo de Baldac (E)	indicum. indigo dye. *Indigofera tinctorie*
Ink (E)	see zegi
Ireat	diachylon plus powdered iris
Irin	gladiolus. *Gladiolus communi*
Iris	yreos, powdered orris root, eris ustis is burnt iris flowers, and oil. see gladiolus. Varieties: *Iris germanica, illyrica, florentina, etc.* W-3 D-3
Iron (E)	fer. yellow ochre. see cimolea, see hematite, see irundinum, see limaille, see marcassita, see molo molundini, see scoria.
Isopus des Arabes	lanolion. see hysop
Ive arthetique	iva arthetique. see chamaepetys, teucrium
Jacea	diajaciton is a laxative. see scabious, see centaurea. W and D
Jews stone	herisson, sea urchin. see Pierre judaica
Jonc (E)	juncos, bulrush, sedge. *various Juncus and Scirpus*
Joubarbes (x)	Jove's beard. house leeks, stonecrop, tettesuriz, poireaux. sticado, cepa, leeks, onions, scallions, sempervivum. *Sempervivum tectorum.* C-3 D-1
Jujubes (E)	fruit of the trees. *Zizyphus vulgaris and Z. saturna.*
Juniper	savine, sabine. needles, cones, oil. see vernis. *Juniperus sabina.*
Jusquiaime (E)	caniculata, faba luparia, luparis, chenille. see henbane, see hyoscyamus, see solathrum, see aconite, see marsilium, *Hyoscyamus albus and niger.* C-3 D-3
Kekenji	alkekenji, winter cherry. *Physalis alkekenji.*
Kerua	castor bean, ricinus.
Labdanum (E)	ladanum or laudanum. resin of cistus trees, especially *Cystus creticus.* W-1 M-1
Lac	see milk et al. see whey C and D
Lacque	lacca, lactea. a red resin from litmus. see orobe. *Roccella tinctoria, Coccus lacca*

Lacticinia	see tithymal, and other plants with milky sap
Laictues (o E)	lettuce. lactucca,
	W-3 D-3
Lana succida	hysop, wool-fat. see lanolin, et al.
Langue du chéne	dog's tongue. see cynoglossus
	W-1 M-1
Langue passerine	bird-tongue, swallow tonghue. see polygonum, centinodia, see enouie
Lanolin	suint, lana succida, hysop, ysopus, oesype. see graisse,.
Lapathum	paradella, lappa, lapazio, burdock, rumex et al., sorrel. *Lappatum acutum.*
Lapis lazuli (E)	powdered blue gem, ferrous sulfate.
Lappa	sorrel, see lapathum. *Arctium lappa*
Lard	see graisse
Lauriers	laurel. baccis lauri. berries, leaves of bay trees. Laurin is oil of bay leaves. *Laurus nobilis.*
Laureole	de-barked daphne twig, spurge laurel, medulla milici. *Daphne laureola.*
Lead, white (E)	lead sulfate is galena. see ceruse, see litharge, see plomb.
Leather	shoe-leather, usually soles, semelles
Leeches	incinerated
Leeks	see poireaux
Lentille d'eau	lentigo water-moss, probably sphagnum, moss, lungwort. *Sphagnum cymbifolium*
Lentils	lens, lentes. *Ervum lens*
Lentisc	see pistache
Lessive	a strong soap, a weak lye. see lixivium et al.
Lettuce (E)	laitues, lactucca, lettue. *Lactuca sativa. Lactuca agresta* is wild lettuce (escarolle, endive).
Levain (E)	leaven, bakers' yeast.
	W and M
Levisticus	see lovage.
Licorice	see reglisse
Lie	see tartar
Lierre terrestre	see hedera. *Glechoma hederacea*
Lievre marin	a mollusc or a fish (blenny)
Lily	lys, arcus daemoniacus, crinon. usually oil of bulbs, leaves and roots. *Lilium candidum et al.* 'Lily' often included iris, narcissus and gladiolus.
Limace	limacons, limax, limazun, snails with shells. see herisson. *Escargot*
Limailles (E)	metal filings (bronze, gold, iron, lead, silver).
Lime	see lime stone
Limon	limes, lemons. see citron. *Citrus limonum and medica*

Lime (stone) (E)	calx viva, quick lime, s chaux vive, creta. powdered fresh lime was used in cylotrum.
Lin (E)	flax. linois, semence, semen lini, flax seeds and oil or meal. *Linum usitatissimum.* W and D
Linaire	linaria, penny-wort. see scrophularia. *Linaria vulgaris.*
Liquid amber	see styrax, storax
Litharge (E)	litargerie, yellow lead oxide, scum of melted lead or silver. see ceruse, see climie argenti, amydon adapts all to W-4 and D-4
Litharge nutritum	spumie argentum
Liseron	see convolvulus large and small. *Convolvluus arvensis*us and *sepium*
Livesche	lovage, Alexander's ivy, fabaria, ligusticum, sesali, fabaria, ligusticum, sesali. *Legusticum levisticum.* Black lovage is *Smyrnum dusatrum.*
Lixivium	aqua cineris, very strong soap, lessive, lye, capitellum.
Lizard (E)	lacertus, lucertoli, stellion.
Lotus (E)	see nenuphar
Lovage	see livesche, see siler
Lumbrici (E)	earthworms (or maggots). ges entera, see vermis
Lungwort	poumon, pulmonaria. see borage
Lupin	flowers of faba lupina, kabitegi. *Lupinus album* W-1 D-2.
Lycium	licium. made from caprifolium.
Lycopodium	wolf's foot
Lye (E)	aqua cineris. see capitellum, see lixivium, see savoniere
Macis	mace, myristica, muscade, the shell of nutmeg. *Myristica fragrans.*
Maiden hair	venus hair, capillary herb. see fern, see adiantum.
Malum punicum	pomegranate et al.
Mandragora	mandrake. see henbane et al. Hieromandrea is a potion based on mandragora. *Mandragora officinarium.* C-3 D-3
Manne	ash-tree sap. see fraxinus, see tamarix C-2 D-3
Marcassita	iron pyrites (sulfites). see iron et al. W-2 D-3
Margarita	pearls or daisies.
Marigolds	various. *Calendula officinalis.*
Marjolaine	marjoram, hortensa, amoracus, maiorama, sweet marjoram. see oregano.*Origanum marjorana and.vulgare.*
Marmotaine	the marmot. fats of *Mus alpinus*
Marrow (E)	bone marrow of many sources, medulla.

Marrubium	maruil neir, samsucus, linoscrofon, horehound, see ammoniacum. *Marrubium vulgare.* W-2 D-3
Marsile	marsilium hog-bean or wolf-bean. see hyoscyamus
Massacumia	see vernix of sea-shells. Asperula *odorata*
Mastic	resin and oil. see terebinth. *Pistachia lentiscus* W-2 D-2
Matrisssylvia	see chevrefuille
Mauve	mallow (various), sanaticula, malve, ebiscus malaviscum, cubes, dialthea, diante, mallachee.see guimauve. *Althea officianalis.* W-1 M-1
Meats (E)	many sources: animals, fish, insects
Medulla	see marrow. specific animals
Meerschaum	see ecume de mer,
Mel (E)	miel, honey et al., see oxymel.
Melangiane	melanzane. see henbane
Melilot (E)	sweet clover. *Melilotus indica* W-1 D-1
Melissa	sweet balm. see calamint. *Melissa officinalis.*
Melon (E)	barecha, citrullus, pumpkin, citron, pepo, squash. *Cucumis melo*
Memitte	yellow-horn poppy. *Glaucum flavus*. (Not chelidonium) C-1 D-1
Mente (E)	mint. see calamint. *Mentha sativa, and piperata*
Mente sarracénique	*Achilles ptarmica*
Menthastrum	wild mint, aquatic mint, horse mint, sisimbro. see mente. *Menthastrum sylvestris.*
Mercuriale	linozostis, mercurelle (common weed). *Mercurialis annua.* C-1 M-1
Mercuré doux	see calomel. mercuric chloride
Mercure dulcifié	red oxide of mercury washed with alcohol
Mercuré précipité	red oxide of, or calcined mercuric nitrate, or *Achilla pyrenaica*
Mercury	the metal, quick-silver. see argentum vif, see cinnabar, ochre
Merde d'airain	bronze scoria
Merde de fer	iron scoria C and D
Merde de plomb	lead scoria
Meurier	see mulberry. *Morus noir et alba*
Mezeron	see laureole
Miel	see mel W-2 D-2
Miel anthosat	honey and *Rosemary officinalis*

Milk (E)	human is lactus muliebris. sow's milk is lait de scroppha, albugasse is donkey's milk. includes goat-milk-whey and cheese-water. see whey.
Mill Dust	see farina de Moulin. Also metallic scrapings of granite
Millefuilles	yarrow, hypericon, St. John's herb. *Achillea millefolium, Hypericum perforatum*
Millet	milium, granum. *Panicum miliaceum.*
Minium (E)	lead filings. ceruse, red lead. see merde de plomb. C and D
Mint (E)	see menthe.
Moles	incinerated ashes of the animal
Mollusc	see lievre
Morel	(not the mushroom) see henbane et al., see jusquiaime et al. *Morel officianalis* C-2 D-2
Morgeline	see mourons, chick-weed, *Alsine media*
Mort de diable	devil's bite, snake-root. see scabious. *Scabiosa succisa*
Mourons	scarlet pimpernel. see anagallis, see money wort, see pimpernel.
Moschus	deer musk. see musk et al. *Moschus moschife.*(& muscari plants)
Mouse	feces of
Mouse-ear	see oreille de souris
Mousse	moss, muscus, musceline, muscus aquae, sphagnum, et al. see lentigo, see lungwort, muscus arboris. *Sphagnum cymbifolium* and *Usnea barbata,* C
Mousse des arbes	nuscus arboris, lichens, especially on pines and junipers. *Usnea* Barbata
Mousse de chene	oak moss W and D
Mulberries	mûres, mora, omorusia. see blackberry, see diamoron. see meures *Morus nigra.*
Mummy	flakes and dust from desiccated cadavers. W-2 D-3
Musc	musk: deer, beaver. see castoreum, see moschus, see stellion.
Muscade	oil of *Myristica aromatica.* see Ben W and M
Mustard (E)	eruca, senape. *Synapus alba* W-4 D-4
Mustard Garlic	*Sysimbrium allaria.*
Myrobalans	indian, or yellow. unripe are chebules. emblicus, belliricus. *Myrobalans indica, Terminalia chebula*
Myrrh (E)	mirre, musa, smyrna, resins of commiphora plants. see bdellium, see balsamodendron. *Balsamodendron myrra.* W-2 D-2

Myrtil (E)	myrtle: seeds, leaves, berries, oil, wood, water. *Myrtis communis* C-2 D-2
Myrtil sauvage (E)	wild myrtle, gringon, fragon, ruscus. *Ruscus aculeatus*
Narcissus	daffodils. *Narcissu pseudonarcissus, poeticus, and orientalis*
Nard	spikenard. see spic and oil, see valerian. *Valeriana jatamansi and Nardostachys jatamansi.*
Nasitort	nasturtium, cresson, water-cress, senationes, garden crew. *Nasturtium officinalis, Lepidum sativum*
Naveau Sauvage	water cabbage. *Brassica nava*
Nenufar (E)	lotus, nymphea. farfar. see water-lily et al. C-2 M-2
Nepite cretense	water calamint, *Melissa cretica*
Nespole	medlar, *Mespilus germanicus* C-2 D-3
Nettles	see orties
Nigella	nielle, Roman coriander, git. see fennel et al. *Nigella sativa.*
Nitre (E)	nitrum. see sal de nitre, potassium nitrate, saltpeter
Noisette	hazel nut. see avellana
Noix de inde	nux indica. *Cocos indica* W-1 D-1
Noix muscade	see nutmeg
Noix vomique	nux vomica. *Strychnos, Nux vomica.*
Nombril	see umbilicus venus
Noyer	walnut, nual. see nux, *Juglans regia.*
Nutmeg	nux muscata, noiz muscate, centrum galle, mace, see muscade, *Myristica fragrans or moschtda*
Nux	meats shells and oils of nuts. chestnuts are chatain, hazel nuts are avellanum, nual(noyer) is walnut, brou is walnut shells, amonde is almonds. see acorn, dianucum W-2 D-2
Oats	aegilope. see avena. *Avena sativa*
Ocymum	see basilium
Oeil de Christ	eye of Christ, a speedwell germander. *Veronica chamaedrys, Aster amellus*
Oeufs	eggs, ova C and W
Oignon de rat	a hyacynth, cepa muris. a squill, *Scylla maritimus*
Oils (E)	usually from mature olives, whereas onfacium was made from the thin juice of green olives and was not classed as a real oil. lily, oil of deben, nux, olives, petroleum, roses, violets, et al. see ben. Adapt to the degrees of what they carry

Oing (o)	lard. see axonge
Ointments (x E)	Apostles', Basilicon, Black, Brown, Diachylon, Egyptian Fuscum, Galen's, Green, Martiaton, Mesûes, Moss, Mummy, Nicolas', Palm, Populeum, Rhaze's White, Saracenic, William Somer's, Surgeons', Yellow (citrin). see Text
Olibanum (E)	see thus, see frankincense, see encens. *Thus masculinum* (from Lebanon).
Olives (E)	see onfacium
	leaves are C and D
Omorusia	meure. mulberry et al. *Morus nigra.*
Onfacium	omphacus, infantium, 'oil' of green olives. see oil,
Onions (E)	cepa as distinct from leeks, allium, porrum, scallion. W-3-4 D-2.
Opium (E)	seeds and pods of poppy. see pavot, see philonium. *Papaver somniferum.* C-4 D-4
Opoponax	epoponac, panax. like myrrh a commiphora resin. *Opoponax chironem* W-3 D-3
Or	gold
Orange	fruit or peel, arantium. *Citrus aurantium*
Ordure de ruche	faex cerae. detritus of bee hives, sordes alveari. oxycroceum
Oreille de souris	mouse ear. *various Myosistis*
Orges (E)	barley. hordeum. penidium is barley-sugar cake, ptisan and vitis alba are barley broth. *Hordeum vulgare.* C-1 D-1
Origanum	oregano, wild marjoram. see marjoram et al.*Origanum vulgare*
Orme	elm tree *Ulmus campestris*
Orobe	horobus, bitter vetch, vicia, lacca is the gum. see lacca. *Ervium ervilia and E. lens.*
Orpiment (E)	orphimentum. yellow arsenious oxide,in cylotrum. see arsenic
Orpin	see crassula
Ortie	nettles. *Urtica diecia.* W and D
Orvault	orvale, aurumvalet. see salvia *Salvia sclera*
Os de se seiche (E)	squid bone C and D
Osmonde	Royal fern, osmundia. see fern. *Osmundia regal*
Ostrich	fat of
Oxalis	see ozeille
Oxycrat	water and vinegar

Oxycroceum	see beeswax. see mu, see faex cerae, see apis.
Oxymel	oxysaccharum (a honey-vinegar laxative mixture), ozzizacara, osisatum, oxylaxativum. see mel
Oxyrhodin	oil of roses and vinegar
Ozeille	oxalis. oseille. see rumex
	C-2 d-2
Pain de coucou	cuckoo bread, alleluia, etc. *Oxalis acetosella*
Pain d' herbe a lait	barley-bread on wine, a laxative of a spurge
Pain de porceau	see ciclamen
Palme vert (E)	heart of palm or date palm. see dactylus.
	C-2 D-2
Panax	all-heal. see opopanax
Pansy	see claveliere
Papaver (E)	poppy. See pavot, opium.
Papyrus (E)	see cyperes
	C and D
Parelle	a rumex. see lapathum
	C-2 D-2
Parietory	pellitory, fuchsia. *Parietaria officinalis*
	Various degrees
Paris	fox grape. see solanum. *Paris quadrifolia*
Parsnip	wild. see baucus, see carrot, see daucus *Pastinacia sativa*
Passerille	passula. uva passa. see agresta, see verjus, et al.
Passerines	swallows. (tongues of are herbs. see langue)
Patte de loup	mother-wort. *Leonuris cardiaca*
Pavot	see papaver
	C-2 D-2
Penide	penidium. barley sugar. see orges.
Pentaphyle	pentaphillon. see potentilla
Peppers	piper, serpyllum, Diateron piperion was made of three kinds of peppers. *Polygonum hydropiper*
Peras (E)	see pois.
Perles	pearls or marguerite flowers. leaves **or** oyster-pearls. see margarita.
	C and D
Persil	parsley, selinum is the seed of apium. see petroselinum. *Carum petroselinum.*
Petroleum	mineral-oil
	W-4 D-4
Petroselinium (o)	see persil.
Peuplier	see poplar
	W and D

Phaseole	beans, haricot, *Phaseolus vulgaris*
Pied colombin	dove's foot. Geranium columbium
Pied corvin	pes corvinus, crow's foot. *Ranunculus sceleratus*
Pied de veau	calf's foot. see arum
	W-2 D-2
Pierre assie	pierre de lanterns, pumice
Pierre calaminaire	zinc oxide. calamite
Pierre d'azure	see lapis lazuli
Pierre d' eponge	pebbles found in sponges
Pierre de Lydie	two stones: iron oxide, and black quartz a gem-stone.
Pierre judaique	ossified sea urchins' points
Pierre ponce	see spuma maris
Pierre sanguine	rust bits, scoria, hematite
Pigment	spiced wine. see hypocras
Pignon	pine cones. *Pinus pinea*
Pig-snout	endive, groin du porc. see chicory et al
Piloselle	hawkweed, hieracium, *Hieracium pilosella*.
Pimpernels	scarlet pimpernel, ipia, wood pimpernel is moneywort, a lysimachia. see anagallis, see mourons. *Anagallis arvensis or Stellaria media.*
Pimpinelle	pimprinelle, lesser burnet. see saxifrage, see anise. *Sanguisorbe officinalis.*
Pine (E))	stone pine and others. tree bark, seeds, pine tar, et al. see cortex pini, terebonth. *Pinus pinea.*
Pinguedo	pork lard. see graisse.
Pissenlit	endives. see rostrum porcini
Pistache (E)	see fisticus. *Pistacia vera.*
Pix (E)	pitch, pix alba and nigra, poix, pissa, resin, tar, turpentine, navale, see colophony, see poix grecque.
	W-3 D-3
Planta genesta	broom. papilina bean, bruscus. *Ruscus aculeatus.*
Plantain	plantago, many varieties including water-cress, psyllium, rib-wort, lanceola, lancelette, policaria, quinque nervicium, arnoglossa, waybread, yva. *Plantago psyllium, P. cynops, etc.*
	C-2 D-2
Plomb brûle (E)	alanauch, plumbum ustum, yellow oxide of lead. lead, see litharge
	C-2 D-2
Plum (E)	Damascus, et al. see sebestes, see pruneau
Poire	pear. fruit, flowers, leaves, wine. *Pirus communis.*
Poireux	garlic, see aulx et al. *Allium porrum.*
Pois (E)	pisum, peas, any type, or beans. *Pisum sativum*

Pois ciche	chick pea, *Cicer arietinum*
Poivre	peppers W-4 D-4
Polemonie	see pouliot
Polium	see germander. *Teucrium polium* W-2 D-3
Polycaria	flea-bane. pulicaria. inula. see enula, see spic et al.
Polygonum	knot grass, corrigiola, cesune, geniculata, Solomon's seal. see balsameta, persicaria, langue de passerine. *Polygonum aviculare.*
Polypode	oak fern, beech fern. see fern. *Polypodium vulgare or Gymnocorpium dryopteria.*
Polytriche	beech fern, hair-cap moss, golden maiden hair fern. see adiantum, see fern. *Polytrichium juniperium.*
Pomegranate (E)	malum punicum, pomme gernette, fruit, leaves, flowers or bark (ecorce), wine, water. see balaustium, mellicrate, psidia (the fruit peel). *Punica granatum.*
Pomme	apples, pears, etc. fruit, wood, leaves, bark.
Pomme d'ambre	a perfume, varied contents
Pomme de cedre	cones of cedar tree. *Juniperus oxycedreus*
Pompholyx	see tuthy. white zinc oxide
Poplar	peuplier. Aspen. leaves, buds are oculi populi and bourgeons, bacca. *Populus tremulus.*
Pork	pig meat, lard, skin. see pourceau
Porrum	ptasion, poureaux see garlic, see aulx et al. *Allium porrum* W-3 D-3
Potentilla	see tormentilla.
Pouliot	polial roial, Dragon-tea with mint. mint, pulegium. *Menthe pulegium.*
Pouliot cervin	*Mentha cervina*
Poumon	pulmo. lung-wort, or a conch. *Pulmo marinus*
Pourceaux	pork
Pourpier	portulaca, purslane. *Portulaca oleracea* C-3 M-2
Praevencha	a clematis
Prassium	leeks
Propolis	bee glue and wax, faex cerae, et al., mu. W
Pruneau (x)	plums. prune, plum tree sap (see gummi), seneste, sebeste, Damascus, et al. Diaprunum is an electuary of plums. *Prunus domestica*, also *P. spinosum,* the blackthorn, sloe.

Psidia	pomegranate peel. see pomegranate C-2 D-3
Psilothrum	ashes of grape-wood, a depilatory
Psyllium	glue, see plantain. *Plantago psyllium* C-2 M-1
Pulicaria	fleabane. *Inula dysenterica*
Pumice (E)	lantern stone
Pyrethrum	*Anthea pyretheum* W-3 D-3
Queue de cheval	see equisetum C-1 D-2
Quicklime	lime et al. chalidicon, calx viva. W-4 D-4
Quince	see coing
Quintefeuille	see pentaphylle, *Potentilla reptans* D
Raifort	radish, raiz, horse-radish, rapistrum, raphanus, rave. *Cochlearia armoracia, Raphanus sativus and Nasturtium amphibium* W and D
Raisins	dry or moist. see grape
Ramich	an Arabian compound of aloes, cloves, oak gall, myrrh, nutmeg, onfacium, manna, sandalwood, sorrel.
Rave	turnip. *Brassica rapa* W-2 M-1
Realgar	red arsenic ore. see arsenic, see orpiment W-4+
Reeds (E)	canes, marsh grass. Canne, panicium, darnel, arundo, roseau
Reglisse	licorice. *Glycyrrhiza glabra.* W and D
Renouée	centinodia. see langue passerina. *Polygonum aviculare*
Resin (E)	gumma pini, pine pitch rosin, poix grecque, alkitron. see pix et al. From *Pinus sylvestris* is the source of terebinth.
Rhubarb	*Rhababarum of many species.*
Ribes	gooseberries. various *Grossularia*
Ricinus (E)	castor bean, cicin, oomph. oil, et al.
Ris	rice. *Oryza sativa*
Romarin	rosemary. *Rosmarinus officinalis*
Ronces	blackberry, rubis, mulberry, blackberry, moron, *Ronce nemorosu* C and D
Roquette	rocket-root, see eruca et al. *Brassica eruca*
Rognons	oil from castrated testes,—calves, sheep, et al.

Ronces	blackberry. see rubis et al.
Rosat	rose-water. see sief
Rose	the flower, petals. see roses aromatique
Roseau	marsh reeds. see canne, see arundo, see grasses. *Arundo donax.*
Rose-oil of	oleum rosarum, oil of petals.
Roses aromatique	red roses. powder of petals or whole flowers including the anthers. Eglantiers are wild roses. see bedegar. *Rosa gallica, Rosacea, var. species.*
Rostrum porcinum	see endive, groin du porc, chicory. *Cichorium endivia.*
Rouille de fer	rust. see iron W-2 D-2
Rubis	bramble, framboisier, moron. ronces rouge are unripe, mulberry, blackberry. see Ronces. *Ronces nemorosus et al.*
Rue (x	herb and oil, see moly, see galigan, see oil. Harmel is rue-seed *Ruta angustifolia and R. graveolans,. A. montana.* W-2 D-2
Rumex	patience, sorrel et al., oxalis, lappa, burdoch, dock, acedula, great mullein, shepherd's crook, hare's-beard. *Rumex acetosa et al.*
Rye	secale
Sabin	oils of sambucus
Sabina (E)	cypress, see juniper, see vernis. the resin is sandarac. *Juniperus Sabina*
Safflower	carthamus
Saffron (E)	crocus. *Crocus sativus* W-1 D-2
Sagapenum (E)	serapinum.a ferula plant resin. asefoetida. *Ferula persica, Geum urbanum*
Sage	sauge, salvia. *Salvia officinalis.* W-2 D-2
Saint Mary's Seal	see balsamita
Sal ammoniac	see sel armoniac.
Sal baurachi	see sel de nitre.
Salex	willow tree. see saule
Salicor	alkali, see salsola
Salsola	glass-wort, alkali. samphire, crithimum. *Salsola kali.*
Salt	sel. common or rock, sel gemma, brine, aloxan (brine 'flower'), muriate of soda. W-2 D-2
Sambucus	elder tree, sureau, see ebulus. *Sambucus nigra.*
Samphire	see salsola
Sanabaro	zanbaharia, Spanish carrot
Sanamunda	hare's foot. see caryophylla et al. *Geum urbanum*

Sangdragon	sang de dragunt. sedge. resin of *Calamus draco.* W (var.) D-2
Santal	sandal-wood (two types)
Sapa	a claret wine from late-pick grapes plus honey, sappa michum
Saphir	gem stone
Sarcocolla	argemone. resin of *Pinea mucronata or Astragalus fasciculoformis.* W-2 D-1
Sarieta	satureia. see savory. *Saturia hortense*
Sarment	serment. grape-vine wood. *Vitis vinifera*
Savine (o)	see sabina
Savon	French soap, sapo, soapwort, saponaria, burith, soap, lessive, soap. W-4
Savoniere	white ellebore, or savon
Savory	many varieties. see satureia. *Satureia hortensis.*
Saxifrage	flowers or leaves. see anise, pimpinelle, and cresson
Scabious	devil's bit, knautia arvensis, morsus diabole, snake-root. see jacea. *Centurea scabiosa.* W-2 D-2
Scammony	bindweed, convolvulus, see anabula, see diagridum. *Scammonaciae, several varieties.*
Scarab	beetle. see amber, carabe
Scariola	see chicory et al. *Lactuca scariola*
Scebram	see tithimal, see euphorbia. *Esula minor*
Schoenanthum	schoenanthum, palea camelorum, juncus. see squinanthus et al. *Andropogon schoenanthus*
Scoloprendre	ceterach, hart's-tongue fern, cow-tongue, blue weed. see bugloss et al. *Scolopendrium officinarum*
Scordium	wood sage. see germander, see pilium. *Eucrium scordium.*
Scorsonére	Spanish scorzonera. a chicory. *Scorzonera hispanica*
Scoria	the dross of melted iron. ferrugo, iron filings and rust, merda ferri. cimolea, limailles, iron and other metals. D-4
Scrophularia	pennywort, toad-flax, umbilicus venus, linaria, cymbalaire, see quadrangular. many varieties of *Scrophularaciae nodosum, aquatica, etc. Linaria vulgaris.* W and D
Sea Moss	see lievre marin
Sea Urchin	see herisson
Sebestes (E)	sebesten plums, cordia myxa. see plum et al. *Cordia sebestena*
Sedengi	the 'blood stone', see hematite
Sehan	absinth or stoechas
Seiche	sepia from squids

Sel armoniac	see sal ammoniac, ammonium chlorhydrate (not from ammoniacum). (it is commonly called but is not "arsenic").
Sel décrépite	crystals of sea-salt
Sel de nitre	sal de nitre. saltpeter. see nitre, see borax, potassium nitrate
Sel de verre	sandiver, glass-gall, scum atop melted glass as it cools, sodium carbonate
Sel indien	granular sugar.
Senatio	water hemlock, not senecio. *Sium augustifolium, latifolium, or Cicuta vir.*
Senesson	senecio, also senacio. simissome groundsel, cerdone, centauria. *Senecio vulgaris* C-2
Senisson	nielle. *Nigell asativa*
Senna (E)	sené. cassia varieties. Diasenna is a laxative. *Cassia fistula* and *vulgaris*
Sepia	see seiche
Serapenum	sagapenum. *Ferula persica* W-2 D-2
Serpens	snake meat, skin, venom
Serpollet	serpyllum., wild thyme. *Thymus serpyllum*
Sesali	sisali. mountain lovage, gentian. *Laserpitum siler*
Sesame	seeds of *Sesamum indicum* W-1 D-1
Setaragi	thapsie, fumitory.
Shepherd's Crook	see rumex
Shoe Leather	soles. see leather
Sief	the rosat of Jesus, the ophthalmologist
Sigillum Mariae	St. Mary's seal, see polygonum. *Tamus communis*
Signe celeste	see chelidonium
Siler de montagne	see sesali
Siligo	rye, darnel grass. see flours. *Lolium perenne*
Sisimbre	horse-mint, see menthastrum. *Sisymbrium Sophia*
Skin (E)	see leather, colle, many animals
Sloe	blackthorn.
Solanum	see solathrum
Solathrum	mors canis, egg-plant, nightshade, see henbane, see solanum, see morele. *Solanum nigrum et al.*
Son	flour, bran W and D
Soot (E)	lamp-black, suie, suy. D-4
Sophora	a bean

Sorbier	sorba, cormes. fruits of the mountain ash. see fraxinus *Sorbus domestica, and Sorbus aucuparia.*
Souchet	sedge, acoris, reeds of cypress.
	W-2 D-2
Souci	sun flower, calendula, marigold. *Calendula officinalis*
Spelt (E)	speautre, hard wheat. see amidum, et al. *Triticum speltum*
Spica celtica	spikenard, nard. see nard, see aunee, see enula, see polycaria, see valerian, see Egyptian thorn.*Valeriana officinalis, celtica. Inula conyza et al.*
	W-1 D-2
Spider-web	see araigne
Spinach	see atriplex
	C-1 D-1
Spode	zinc oxide, pompholyx, tuthie, cathimia, climie argenti.
	C-2 D-3
Sponge	W-1 D-2
Spuma d'argent	nutritious litharge, écume d'argent, flower of silver. see litharge, see climie argenti
Spuma maris	magnesium silicate, meerschaum, or an alga. see ecume de mer.
Squid (E)	os sepiae, seiche, sepia, burnt bone.
Squill	hyacinth (wild). *Scilla maritimea.*
	W-2 D-2
Squinanthus	acoris, et al.
	W and D
Staphisagre	cheif d'espurge, pes alauda, jonquarola, larkspur, delphinium, polycaria. *Delphinium staphisagre or inula policaria.*
	W-3 D-3
Stellion	lizard, gaulus, a musk-like excrement of lizards. see musk et al.
Stercus	see feces, see fiente.
Stoechas	lavender, *Stoechas arabica.*
	C-2 D-2
Stonecrop	see crassula. vermicularis
Storax (E)	styrax. balsamic resin. benzoin, *Liquidamber orientale*
	W-1 D or M 1
Strawberry	see fraizere
Sucre candi	see sugar
Sugar	alun de sucre, zuccharum, nabete (powdered) sugar candy, penedium is a barley-sugar, sugar of violets et al.
Suif (x)	tallow
	W and D (ncreasing with age of animal
Sulfur	W-3 D-3

Sumach	a toxicodendron. poisonous or non-poisonous C-2 D-3
Sureau	see elder W-2 D-1
Syco(a)more (E)	sap. leaves, etc. alos called Egyptian mulberry. *Ficus sycomorus*
Taisson	a badger. *Ursa melis*
Tamarind (E)	tamarind-tree fruit. *Tamarindus indica.*
Tamaris (E)	sap of manna. ash tree, fraxinus. *Tamarix africana, Fraxinus ornus, Mysicaria, germanica.*
Tansy	athanasius. *Tanacetum vulgaris* W and D
Tapsie	see thapsus
Tar (E))	navale, pix, poix grecque.
Taraxacum	lion's tooth. *Taraxacum officinalis*
Tartar	potassium bitartrate from wine lees, faex. see lie, cendres, gravelle W-3 D-3
Tassus Barbatus	bouillon, hare's beard. great mullein, rumex et al., verbascum. *Verbascum thapsus.*
Terebinth (E)	olibanum, xylobalsamum, mastic, pistachia, resin, alkitron (a distillate), bito tree, balanites, gum albotim. *Pistacia terebinthus* W and D
Terra sigillata (E)	an astringent trochee of baobab fruit, *Adansonia digitata.* or a reddish clay of Lemnos, fashioned like an Egyptian seal. cimolia, argilla, bol d'armenie, clay. C and D
Terre des meules	mill-stone dust.
Terre selineue	see argilla
Teucrium	germander et al.
Thapsus (E))	tapsie, another scrofularia umbellifer. asa dulcis, verbascum, silphium. *Thapsia villosa and Th. garganilla.* W-4 D-4
Theodoricon	composite laxative (hierapicra, et al)
Thus	encens, frankincense, olibanum. *Boswellia thurifera.*
Thyme (E)	see calamint. *Thymus capitatus, vulgaris*
Tithimalle	see euphorbia et al.
Tongue	see langue
Tormentilla	sarsaparilla, quinquefolum, cranesbill, pie de colomb, pseudoselinon, callipetalon, cinqfoil, doves-foot, potentilla, see geranium maculatum. *Potentilla reptans.* W and D
Tortoise	turtle meat.

Tragacanth (E)	dragacanth, adracanth, gummi, diadragon. *Astragalus gummifer.*
Tremble	tremula. poplar, see(quaking) aspen. *Populus tremula*
Tribul verd	green tribule. *Trapa natans*
Tribule	land and water thistles, water chestnut or Burra Gukaroo. *Tribulus terrestris or T. aqautica.*
Truffe	same as tryphere
Tryphére	an electuary containing truffles and various sweets.
Tunix	betonica
Turbith	turpeth, diatesseron laxative, roots of *Convolvulus or Operculum turpathum*
Turnip	see rave. *Brassica rapa*
Turpentine (E)	terebinth et al.
Turquette	rupture wart. *Herniaria glabra and hirsuta*
Tussilage	colt's foot. *Ungula caballina or farfara*
Tuthie	spode, zinc oxide. see climie argenti C-1 D-3
Tyrie	various snakes, especially venomous
Ulva	marine sedge, sea lettuce
Umbilicus venus	cymbalaria. cotyledon, fern, scrophularia et al. C-3 M-3
Uva	see grapes et al., see agresta.
Uva passa	raisins. sour grapes
Valerian	phu, amatilla, fistra, spikenard. nard, spic, et al. *Valeriana officianalis.*
Venus hair	maidens-hair fern, capillus venus, bed-straw. adiantum, fern. *Adiantum capillus veneris.* W-1 D-1
Verbascum	see thapsus
Verbena	vervain. flowers of *Verbena officinalis* C and D
Verdarain	vert d'arain, ziniar, fleur d'arain, flos aeris, viride. aes, et al., bronze et al., see verd de gris.
Verdegris (E)	verdigris, ziniar, copper 'flower', copper acetate, chloride or sulfate. chalcathum et al., see verd d'arain et al.
Verge de berger	see virga pastoris
Verjus	vinum acerbum, a potion from sour grapes or other sour fruits. agresta, passula, vinum goretum.
Vermicularis	stonecrop, sedum. *Sedum acre*
Vermis (E)	worms, lumbrici. see earthworm W
Vernis	see juniper (tree sap). *Thuia articulata.* or encaustrum. see ink. W-2 D-2

Verre	ground glass
	C-1
Verre de outré mer	lapis lazuli
Vigne (E)	grape vine *Vitis vinifera*
Vinaigre passerille	vinegar.from dry raisins. Acetum passulatum
Vinegar	C-1 D-1
Vinum goretum	styptic wine, agresta, s passula, verjus.
Violet	many varieties. see dogtooth. *Violaria and Hesperis and lunaria*
Virgo pastoris	shepherd's purse, another dipsacus, sanguinary, centinodium, passerinus, proserpinaia, yarrow et al. *Dipsacus. sylvestris.*
Viticella	white bryony, vitis. bryony et al., ivy, clematis et al. *Clematis flammula.*
Vitriol rosa	red iron sulfate, Roman vitriol. couperose, atrament, chalcantum, colcothar, ink.
	W-3 D-3
Voluble	greater bindweed *Convolvulus sepium*
Wasps	vespa
Water	cold. C and M
Water lily (E)	see nenuphar, lotus
Wax (E)	white and red (cera alba and rossa), cere, sire, oxycroceum.
Whey	serum(especially goat's). see milk.
Willow	see salex
Wine	many types. see tartar, gravel, lees
Xylobalsamum (E)	balm of Gilead, terebinth et al.
Xylocaracta	carob
Yeast (E)	see levain
Zedoaria	turmeric. curcuma. *Curcuma longa et al.*
Zegi (E)	zezi. ink, vitriol rosa, encaustrum et al.
Ziniar	see verdegris
Zizyphus	see jujube
Zinziber	see ginger, see amome.
Zurunge	hermodactyl

APPENDIX II: SURGICAL INSTRUMENTS

NICAISE'S NOTE

G uy used many instruments. He listed his armamentarium for removing foreign items from wounds, for trepanation, for operations for hernias and bladder stones, for operations on the eyes and the ears, for fumigation and for dental work. He mentioned many more, but he left no drawings, unless some had been inserted in lost manuscripts. The only illustrated surgical treatise before the invention of printing was that of Albucasis. The first printed edition that contained illustrations of instruments was by Brunschwig. I have reviewed the books that show instruments used by the ancient surgeons, as they were found at Herculaneum and Pompei, and many books printed after the 15th C.[674] Here I will use simple line drawings to illustrate the brief descriptions of what Guy used. My sources are:

> H. Brunschwig, 1497, *Das Buch der Chirurgia*. Strassburg
> A. H. Gersdorf,1542. *Das Feld buch der Wundartzney,* Strassburg, dela Croix,
> > 1573. *Chirurgiae universalis,* Venise, 1573
> Scultetus,1672, *Armamentarium chirurgicum,* Amsterdam
> Arneman,1796, *Instrum de chirurg.* Gottingen
> Rudtorfer,1817, *Instr. De chir.,* Vienna
> P. Savenko,1821, *Note sur la chirurgie dans les premier ages.* In Rev. Med. Hist.
> Vulpes, 1847, *Illustrazione de tutti gli strumenti chirurgica scavati In Ercolane e Pompei,* Naples
> J. Sichel, 1866 *Nouveau recueil de pierres sigillaires*, Paris
> Schoutetten, 1867, *Hist.des instrument. de chir. etc.* Paris

Translator's Note

The instruments are named, followed by brief comments. The numbered items are pictured in the five Plates, as I,#1. Others are described without numbers.

The Instruments

INDEX FOR PLATE I

Needles for Suturing, and Cannulas for guiding the needles *I,#1,2,3,4.*

Lead Needle for use in nosem mouth, anal fistula, etc. *I, #5*

Cataract Needles. *I, #6, 6a.*

Urethral Cannulas, with stents. *I, #7, 7a*

Crossbow for Removing embedded arrows. *V, #1*

Apparatus for Reducing and Splinting Fractures. *V,#2*

Metallic Cap to protect an injured fingertip

Truss to Protect the Lower Abdomen.

Cannulas:

> Cannula to direct a needle (see above)
>
> Fenestrated Cannula to accept a sharp chisel to snip off an excrescence caught in the.window. *I, 8, 8a*
>
> Cannula to introduce a caustic for application only at the uvula. which is trapped in the window.The caustic is applied on a tuft at the end of a piston.
>
> Bronze or silver cannula fro draining ascites.*I, #9*
>
> Cannula for draining ears of children *I, #10*
>
> Cannulas for fumigation. Can be attached to tubes for introduction into the vagina, etc. *I, #11,12*
>
> A solid 'catheter' or sound with a rounded tip for probing wounds

Cauteries: These may be solid hot metal, boiling oilm or a wooden stick with a burning ember at the tip. The potential cauteries are caustics, called eruptors. The Ancients had many varieties, whereas Guy named only six.

> The Cultellary: With one cutting edge, the Dorsal, *I,#14*
>> With two cutting edges, the Ensal *I, #15*
>
> The Olivary, named for the shape of its tip. *I,#17*
>
> The Dactilary. Its tip shaped like a date-pit, is longer and heavier than the Olivary *I,# 17b*
>
> The Punctual. A slim rod with a rounded tip to destroy a node or polyp. Its depth is limited by a lateral pin. It is applied through a perforated plate. I,#17 and 17.
>
> A narrow rod-cautery is applied through a cannula to reach a deep spot and protect the surrounding tissues. *I, #18,18a, 18b*
>
> Guy used some very thin, wire-like silver cauteries for insertion into dental caries, or he used a red-hot silver or bronze probe
>
> A Seton Cautery: This was a long, hot or cold, cutting-edge needle passed through perforations in apposed blades of a forceps that gripped a fold of skin. The needle was armed with a tough ribbon or cord which remained as drain when the apparatus was removed. *I, #19, 20, 21, 22, 23.*

A Circular Cautery: has five buttons or points attached to the undersurface of the ring, to burn a node or other lesion at five places simultaneously. It was applied through a perforated plate *I, #24*

PLATE I

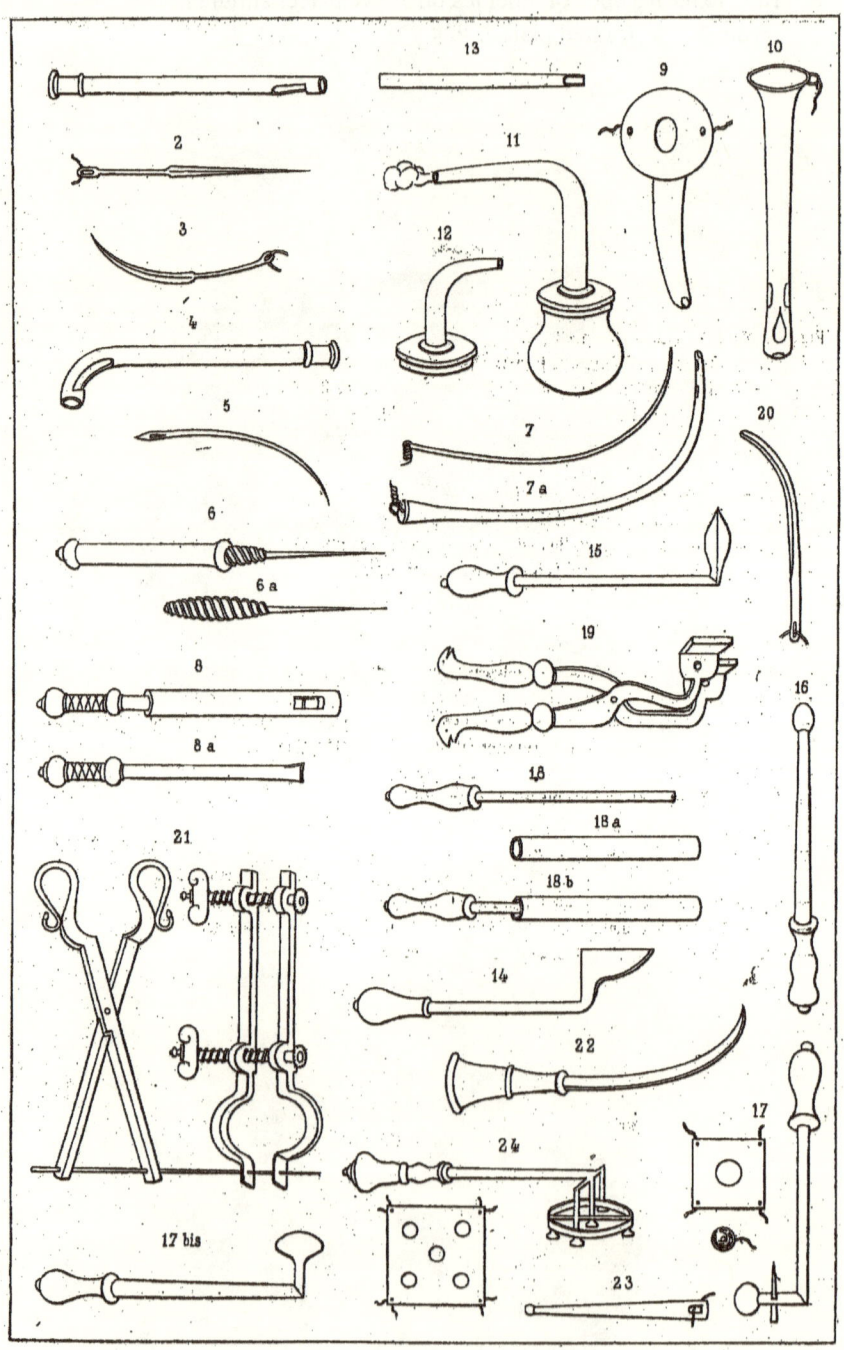

INDEX FOR PLATE II

Scissors and Chisels The drawings here are taken from a 15th C source (Brunschwig).
Surgical scissors were in the collection from Herculaneum, and Albucasis described
two types:

Scissors for excising part of drooped eyelid *II, #25*, and

Scissors for circumcision. *II, #26*

Dilating Scissors for enlarging wounds and ulcers *II, #27*

Separating Chisels, for cutting bone: They were struck with a hammer. *II, 28,29, 30, 31, 32*

Clysters: Guy's consisted of leather sac (a hog's or ram's urinary bladder), attached to a
wood or metal canula *II, #35.*

A syringe as such was not known by Guy. First descrbed by Gatinaria (attributed
to Avicenna). *II, #33.*

Albucasis described a copper syringe for irrigating ears. Syringes for injecting into
the urinary bladder had a narrow perforated end. *II, #34.* Later, an extension
of flexible tubing was added to spew liquid naphtha and ignite an enemy's
ship in battle.

Guy used a kind of injector to apply caustics at the uvula. *II, #36*

Hooks: Guy used many, varying sizes, thickness, shapes, with one or more prongs. Many
of the drawings here are from Albucasis.

Hook for suturing and dissecting. *II, # 37, 38*

Special Hooks for eye-surgery. *II #40, 41, 42*

Special Hook for Eyelid. *II,# 39*

Concave hooks (Scoops) for extracting stones during lithotomy *II, #43,44*

Cups: several sizes. To cover and protect an eye (silver or bronze). To carry a dose of
boiling oil into a bleeding wound (long handle)

Dilators: To enlarge wounds and ulcers without incisions. *II, #45, 46, 47, 48, 49*

Elevators: To realign depressed bone, or to extract teeth. *II, #50, 51, 52*

Funnel: Guy modified a funnel by attaching a cannula to conduct fumigations (vapors)
into ear canals or vaginas. He used it to introduce liquid medications deep within
the nostrils.

Probes: Guy used several instruments for examining: stylets with various tips: a button, an
olive, a flat blade (taste), a rasp. The shafts were straight or flexible; rigid or flexible
(lead). Celsus called them speculums for sounding.

By attaching a twist of cotton or wool-fleece, the probe could deliver a medication.

Guy substituted a pointed stylet (a pen for inscribing a wax plate).

He used a long probe (intromissum) to explore wounds that had entered the thorax.

He used a hollow probe to drain a scrotal hydrocoele, as a trocar.

That device was modified by Paré, Sanctoris, and Canane.

The flat probe (the taste) resembled our tongue-depressor *IV, #86.*

A long lead rod, slightly curved, was used to push food trapped in the esophagus
into the stomach

Fanon: this was a splint, less rigid than wood, made from a stiff reed wrapped in Cloth.
Guy mentioned it for treating *V, 2*

The Glossocome: A casket with pulleys and ropes for reducing fractures. *V, 2* Impulsors:
for pushing through a tube and out the opposite side, there to receive something to
be pulled back. Guy described Paul of Aegina's as 'female', *III, #71.* And a solid rod
to contact and push through a object ahead of it; the 'male', *III,#72*

PLATE II

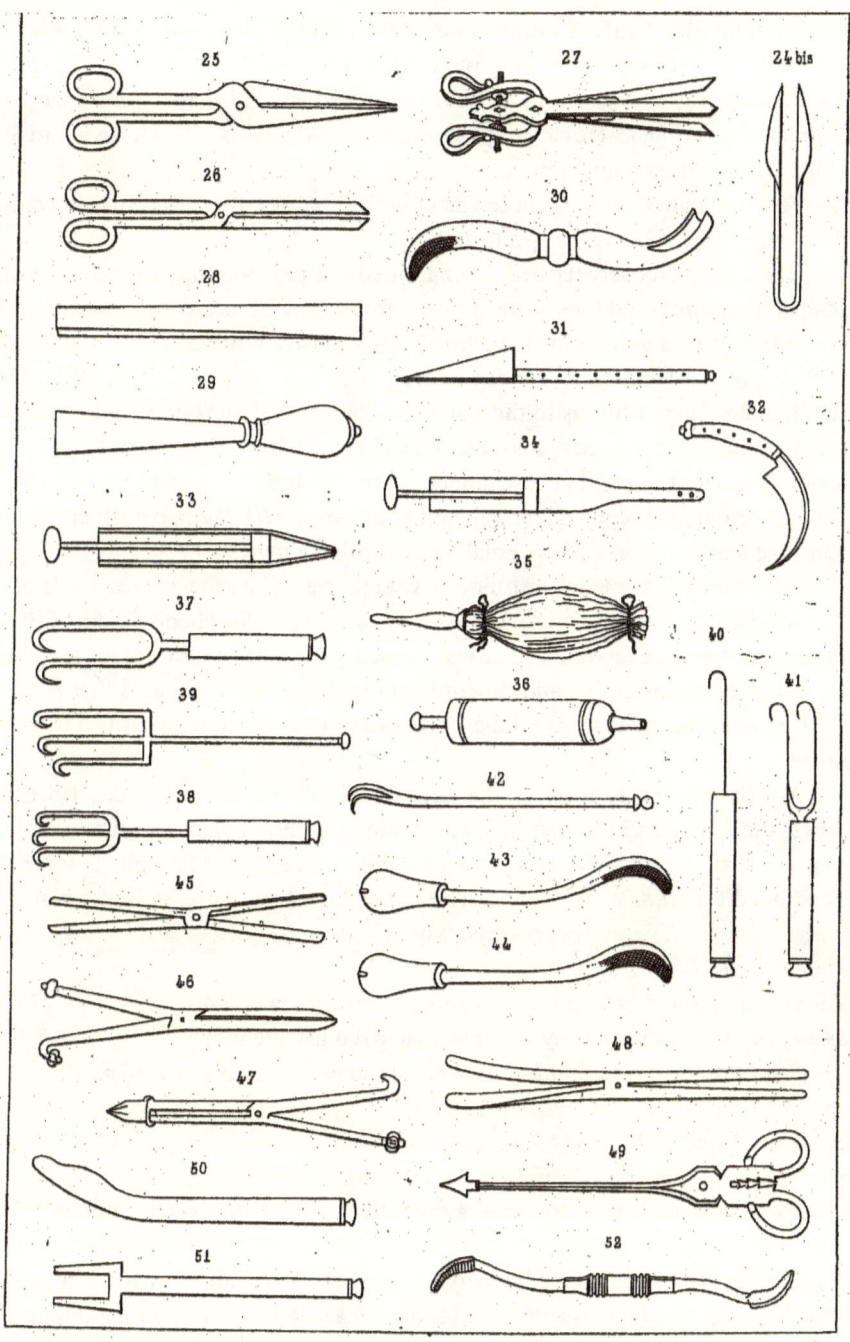

INDEX FOR PLATE III

Knives: A Scythe-like Knife: Cutting edge on the concave surface of a curved blade. It had such special uses as excising part of the uvula and removing tonsils: *III, #53, 54, 55*. Galen used one with a sylet-tip to thread an anal fistula before cutting by withdrawing the blade. He called it a syringotome. Albucasis called it the gourd-knife. Other Arabs called it an embola.

An Amputation knife: a curved razor with a sharp convex edge. In later centuries the scythe-shaped blade was preferred

The Lancet: loosely termed, often limited to a phlebotome. Sometimes called a scalpel (or pilum) used for scarifications or for reopening recent incisions. A sagitella opened an abscess. According to Vedrenes, Celsus used the term scalpel, (later authors included bistouri and knife) for any cutting or pointed instrument.The scalpel had a stiff handle. They differ as to the size of the blade. Vedrenes provided drawings of many scalpels. *III, #56* had a double edged blade (silver) with a rough handle with a slot to attach it to another instrument. Saint-Privat d'Allier found it in 1864. The abscess-draining lancet at Herculaneum and Pompei, *III,#57* had a silver blade and a bronze handle. *III,#58* was a small lancet called a myrtle.

Guy used that name for instruments with points and narrow blades, made of iron.

Friend said that the term, lanceole, appeared after 1220, when William of Brittany described it as a phlebotome that was tapped with a hammer to enter a vein. The terms varied as to origins and functions through the 14th. Guy said that Albucasis advised entering an abscess with an intromssorium, which really was a myrtiform lancet.

We have come to identify the lancet with phlebotomy. Guy described three types: the cultellary; the myrtle-shape blade in common use; and the large lancet (the scissorum) used for bleeding horses. Albucasis described four phlebotomes with fixed handles: *III,#59, 60, 61, 62*. The lenticular chisel-knife incised when struck with a hammer. The tip was covered with a button. *III, #73*.

The Hammer. *III, #74*

The Pennarolum, The carrying case for surgical instruments. *V, #3*

Pessaries: Guy mentioned it only when he discussed uterine moles

Forceps-Pincettes, also called vulsellae. Various sizes and shapes. *III,# 76, 78*

Small forceps to fit in the Pennarol. *III, #77*

Epilatory forceps, *III,#79, 80, and IV, #83*

Pulleys for use in reducing fractures, with lead counterweights.

Pyulque: A suction device. A 'reverse' syringe for draining (by suction) empyemas *II, #36*

Razors: The knives used most by Guy. Some are noted above with Lancets and Phlebotomes. The large razors were used for making incisions and for amputations. They vary as to size, to single or double edges, to convexity, and for straight or curved cutting edges.

The Spathume was a large sword or short dagger. Its blade was shorter than that of the large razor. Guy used several spathumes: single and double-edged, or a flat blade for slapping a lesion rather than incising it. The scythe-knife was used for cutting the uvula. A Spinous knife was used to flick off a wart in the ear-canal. Albucasis recommend it. *III,#63, 64, 66, 67, 69, 70.*

Guy described a Novacular razor for shaving hair.

In Fallopius' epoch (16th C) the knives were razors, phlebotomes, lancets, and gamauts. The last-named was a scythe-like knife with one cutting edge. Later it came to be called the bistoury

Amputation knives in Guy's time were large razors. Albucasis

PLATE III

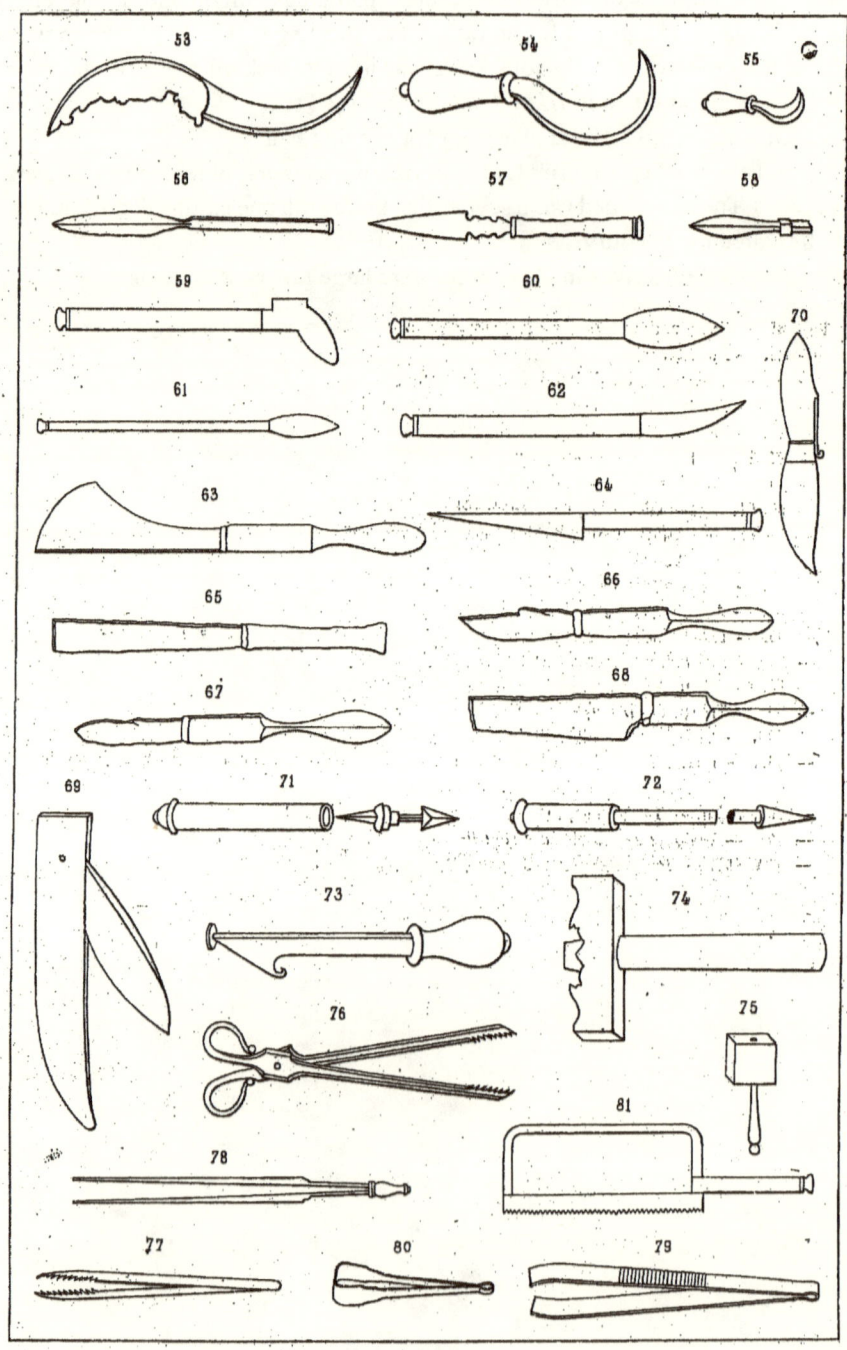

INDEX FOR PLATE IV

Described them but provided no drawings. Celsus had described his own as straight, as are the materials found at Herculaneum and Pompei. *III,#65,66.*

Rasps for scraping: Guy used one for removing teeth. There were many shapes and sizes,. Guy described one that was hollowed, as a gouge.

Saws: Amputations were rare events, and saws were seldom mentioned. *III,#81*

Setons: see clysters above.The term was used in different ways, both before and after Guy's era, with different types of devices, used both in humans and in other animals. Guy described his instruments and how he used them to drain scrotal hydrocoeles, and to reduce edema in the upper body with setons in folds of the skin of the nape and the axillae. His practice was adopted by Pierre Franco and by Gatinaria. *I,#19,20,21, 22,23*

Guy used a grooved probe for treating fistula-in-ano. The trough guided a cultellary cautery. That instrument was improved by Franco and Paré.

Spatulas: blades to mix medicines. Guy's was shaped like a myrtle-leaf. He also used it to apply medicines on the eye. Those from Herculaneum and Pompei had a usable tip at the handle: a button, olive, curette, or elevator.

Speculum: Guy used several kinds. One exposed the inner vagina: *IV,#82,* Another depressed the tongue, *IV #84.*

Traction Apparatus for reducing fractures. *V, #2,* (drawing from Gersdorf). The device was used by Guy.

Taste: see Probe. The taste was a flat wood or metallic blade (resembling a tongue depressor in the 21st C) The surgeon used it to flatten an irregular surface before incising it. *IV,#85.*

Forceps: Guy used many types, especially for removing foreign objects from wounds. Avicenna's had curved blades with teeth. Albucasis used them to crush the head of a dead fetus. *IV,#86*

Albucasis' (another) had 'bird-beak' blades with teethe. He used them to remove arrows. Some were used to extract teeth. *IV, #87,88,89, 90, 91.*

Forceps with gouged blades were used to remove barbed arrows. *IV,#92,93*

Guy described a forceps with long curved blades for removing foreign objects caught in the throat *IV, #94.* He used a long narrow forceps to extract small stones trapped in the penile urethra. He used a small forceps to extract teeth. It resembled the clamps used by coopers to tighten the bindings on barrels. *IV,#95, 96.*

Tents: Guy used them in many ways: to apply detergents, to prevent early closure of a wound, to enlarge an opening, to promote drainage from wounds and abscesses. Sometimes he used perforated cannulas of silver or bronze.

IV, #104,105. He used similar tents of wood or lead to prevent re-adherence of a dissected prepuce, or to dilate a hymen.

Tracheotomy: Guy inserted a cannula between two tracheal rings, thereby distinguishing it from laryngotomy (he called it epiglottis). He intubated the larynx through the throat.

Trepanation: I will not repeat the material presented in Treatise III, Doctrine, II, Chapter1, or what was discussed in the Biographical Section of my (Nicaise) Introduction.

Guy described three trepans: Galen's with a chaperone pin. *IV,#97*

The Parisian with a variable pin, *IV,#98,* and the Bolognese with a tapered lancet-cutter, *IV, #99.*

Guy used a borer-drill (tarrier) to enlarge the opening in a bone in which an arrowhead was embedded. *IV #100,101*

Other Tarriers: A slim one to enter a stone trapped in the penile urethra either to remove it or to break it. *IV,# 102, 102a*

Pipes: short tubes to tighten the splints around a fracture. *V,#2*

Cups: made from ivory, copper or glass, with or without openings for suction *IV #103.*

PLATE IV

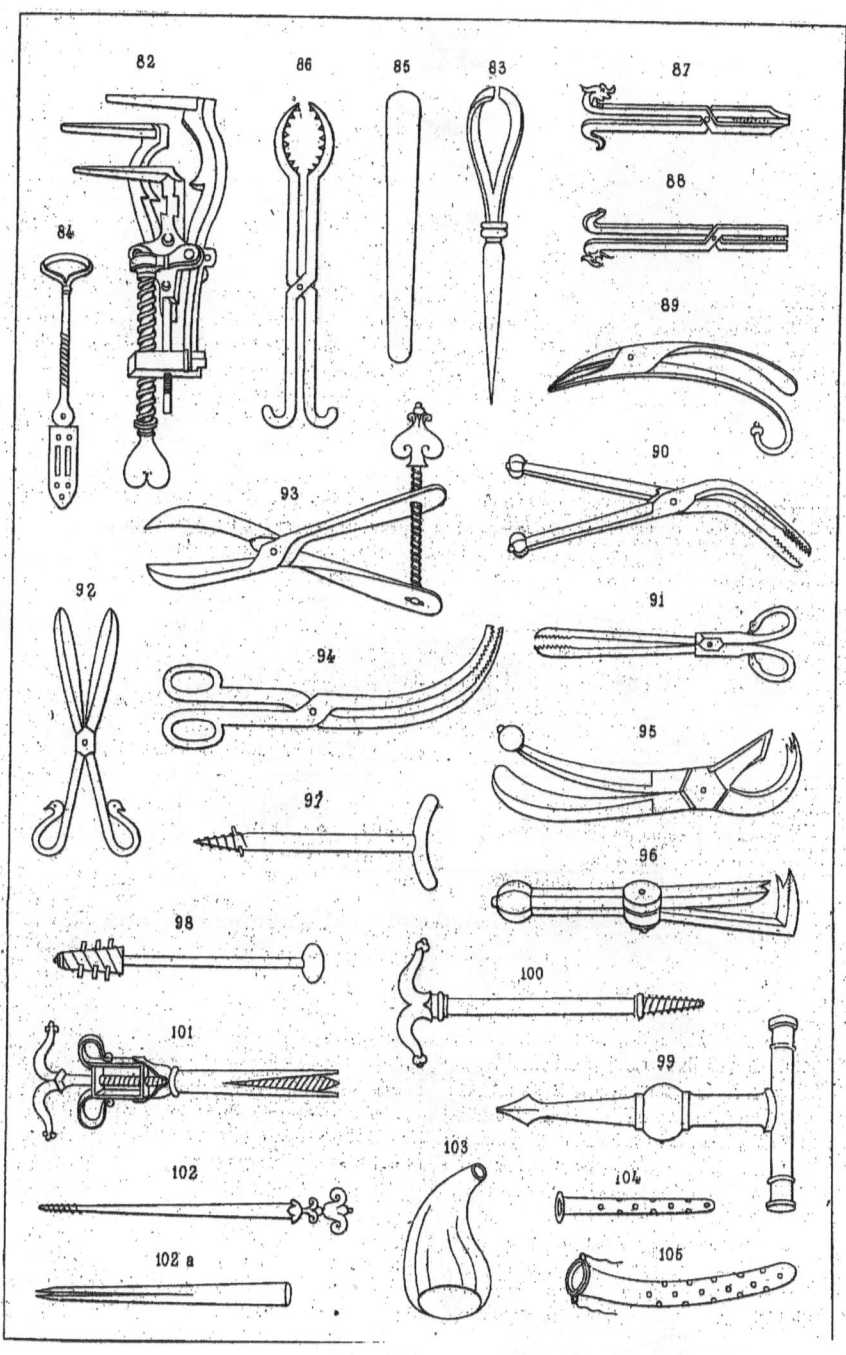

PLATE V

Strong-bow Apparatus For Removing Impacted Arrows

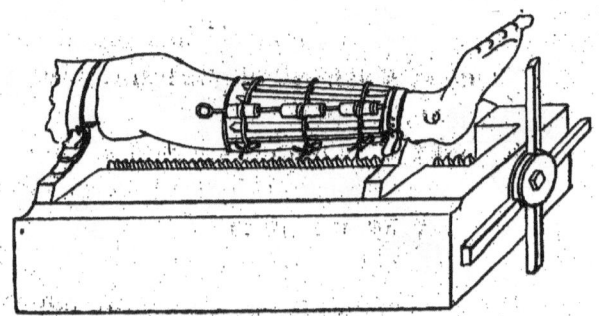

Windlass Box For Reducing and Splinting Fractures

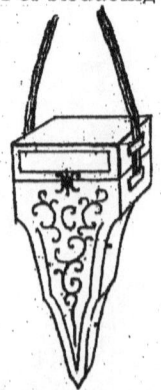

Portable Case (Pennarol) For Surgical Instruments

www.ingramcontent.com/pod-product-compliance
Lightning Source LLC
Chambersburg PA
CBHW031809170526
45157CB00001B/8